Emerging Materials and Technologies

Series Editor: Boris I. Kharissov

The *Emerging Materials and Technologies* series is devoted to highlighting publications centered on emerging advanced materials and novel technologies. Attention is paid to those newly discovered or applied materials with potential to solve pressing societal problems and improve quality of life, corresponding to environmental protection, medicine, communications, energy, transportation, advanced manufacturing, and related areas.

The series takes into account that, under present strong demands for energy, material, and cost savings, as well as heavy contamination problems and worldwide pandemic conditions, the area of emerging materials and related scalable technologies is a highly interdisciplinary field, with the need for researchers, professionals, and academics across the spectrum of engineering and technological disciplines. The main objective of this book series is to attract more attention to these materials and technologies and invite conversation among the international R&D community.

Two-Dimensional Nanomaterials for Fire-Safe Polymers
Yuan Hu and Xin Wang

3D Printing and Bioprinting for Pharmaceutical and Medical Applications
Edited by Jose Luis Pedraz Muñoz, Laura Saenz del Burgo Martínez, Gustavo Puras Ochoa, and Jon Zarate Sesma

Wastewater Treatment with the Fenton Process: Principles and Applications
Dominika Bury, Piotr Marcinowski, Jan Bogacki, Michal Jakubczak, and Agnieszka Maria Jastrzebska

Polymer Processing: Design, Printing and Applications of Multi-Dimensional Techniques
Abhijit Bandyopadhyay and Rahul Chatterjee

Nanomaterials for Energy Applications
Edited by L. Syam Sundar, Shaik Feroz, and Faramarz Djavanroodi

Wastewater Treatment with the Fenton Process: Principles and Applications
Dominika Bury, Piotr Marcinowski, Jan Bogacki, Michal Jakubczak, and Agnieszka Jastrzebska

Mechanical Behavior of Advanced Materials: Modeling and Simulation
Jia Li and Qihong Fang

Shape Memory Polymer Composites: Characterization and Modeling
Nilesh Tiwari and Kanif M. Markad

For more information about this series, please visit: www.routledge.com/Emerging-Materials-and-Technologies/book-series/CRCEMT

Mechanical Behavior of Advanced Materials

Modeling and Simulation

Qihong Fang

Jia Li

CRC Press
Taylor & Francis Group
Boca Raton London New York

CRC Press is an imprint of the
Taylor & Francis Group, an informa business

First edition published 2024
by CRC Press
2385 Executive Center Drive, Suite 320, Boca Raton, FL 33431

and by CRC Press
4 Park Square, Milton Park, Abingdon, Oxon, OX14 4RN

CRC Press is an imprint of Taylor & Francis Group, LLC

ISBN: 978-1-032-12226-7 (hbk)
ISBN: 978-1-032-12681-4 (pbk)
ISBN: 978-1-003-22570-6 (ebk)

DOI: 10.1201/9781003225706

Typeset in Times
by MPS Limited, Dehradun

Contents

Preface....ix
Authors....xi

Chapter 1 Background....1

1.1 Introduction....1
1.2 Theoretical Progress....1
1.2.1 Model Progress....2
1.2.2 Simulation Progress....5
1.3 Outline....7

Chapter 2 Theoretical Methods....13

2.1 Introduction....13
2.2 Microstructure-Based Model....13
2.2.1 Solid Solution Strengthening....14
2.2.2 Dislocation Strengthening....16
2.2.3 Grain Boundary Strengthening....17
2.2.4 Second Phase Strengthening....18
2.3 Continuum-Mechanics-Based Model....20
2.3.1 Diffusion-Induced Stress Model....20
2.3.2 Dislocation-Nanostructure Interaction Model....23
2.4 Multiscale Simulation....24
2.4.1 First-Principles Calculation....24
2.4.2 Molecular Dynamics Simulation....25
2.4.3 Phase Field Simulation....28
2.4.4 Crystal Plasticity Finite Element Simulation....30
2.5 Conclusion....36

Chapter 3 High-Entropy Alloys: Simulation....41

3.1 Introduction....41
3.2 Deformation Mechanism....41
3.2.1 Dislocation Evolution....41
3.2.2 Twinning....45
3.2.3 Phase Transformation....46
3.3 Strengthening Mechanism....49
3.3.1 Grain Boundary Strengthening....49
3.3.2 Solid-Solution Strengthening....51
3.3.3 Second Phase Strengthening....52
3.4 Service Performance....52
3.4.1 Irradiation Resistance....52
3.4.2 Fatigue Performance....61

3.5 Prospects ... 66
3.5.1 Mechanics ... 66
3.5.2 Physics ... 67
3.5.3 Chemistry ... 67
3.5.4 Computing ... 67
3.5.5 Medicine ... 67
3.5.6 Biology ... 67
3.5.7 Environment ... 67
3.5.8 X Applications ... 68

Chapter 4 High-Entropy Alloys: Model ... 72

4.1 Introduction ... 72
4.2 Lattice-distortion Dependent Strength Model ... 72
4.2.1 Modeling Lattice Distortion Strengthening ... 72
4.2.2 Yield Strength ... 74
4.2.3 Model Validation ... 75
4.2.4 Screening Composition for High-strength HEAs ... 79
4.2.5 Remark ... 80
4.3 Irradiation Hardening Model ... 80
4.3.1 Complex Void Hardening ... 80
4.3.2 Dislocation Loop Hardening ... 86
4.4 Hierarchical Multiscale Model ... 91
4.4.1 Multiscale Model Building ... 91
4.4.2 Parameter Transition and Simulation Setup ... 98
4.4.3 Result ... 107
4.4.4 Remark ... 111

Chapter 5 High-Entropy Amorphous Alloys ... 118

5.1 Introduction ... 118
5.2 Indention-Induced Deformation Behavior ... 118
5.2.1 Model ... 118
5.2.2 Results and Discussion ... 120
5.3 Tension-Induced Deformation Behavior ... 126
5.3.1 Model ... 126
5.3.2 Results and Discussion ... 127
5.4 Microstructure-Dependent Deformation Behavior ... 131
5.4.1 Model ... 132
5.4.2 Results and Discussion ... 134
5.5 Conclusions ... 139

Chapter 6 Nickel-Based Superalloys ... 145

6.1 Introduction ... 145
6.2 Mechanical Behavior ... 146

6.2.1 Model and Process Optimization 146
6.2.2 Precipitate Random Distribution 158
6.3 Service Performance 166
6.3.1 Spatial Distribution of Precipitate Size 167
6.3.2 Statistical Distribution of Precipitate Size 167
6.3.3 Creep Rate 168
6.3.4 Quantitative Evaluation 171
6.4 Conclusion 172

Chapter 7 Light Alloys 179

7.1 Introduction 179
7.2 Al Alloys 179
7.2.1 Multilayer Nanotwin Effect 179
7.2.2 Surface Energy 184
7.2.3 9R Phase Stabilization 190
7.2.4 Remarks 194
7.3 Mg Alloys 194
7.3.1 Transverse Propagation of Deformation Twinning 194
7.3.2 Nucleation and Growth Mechanisms of Nanoscale Deformation Twins 205
7.3.3 Transition of Dynamic Recrystallization Mechanism 213
7.3.4 Remarks 225

Chapter 8 Chemomechanical Modeling of Lithiation 231

8.1 Introduction 231
8.2 Basic Modeling 232
8.2.1 Model of Cylindrical Electrolyte 232
8.2.2 Model of Spherical Electrolyte 234
8.2.3 Model of Hollow Spherical Electrolyte 237
8.3 Diffusion-Induced Damage 240
8.3.1 Cylindrical Electrolyte 240
8.3.2 Spherical Electrolyte 245
8.3.3 Hollow Spherical Electrolyte 249
8.4 Conclusions 253

Chapter 9 Nanostructure-Reinforced Composites 259

9.1 Introduction 259
9.2 Inclusions 260
9.2.1 Edge Dislocation Interacting with Nanoscale Inhomogeneity 260
9.2.2 Screw Dislocation Interacting with Nanoscale Inhomogeneity 272

9.3 Nanopores 277
9.3.1 Edge Dislocation Interacting with Nanohole 278
9.3.2 Screw Dislocation Interacting with Nanohole 285
9.3.3 Dislocation Emission from a Crack Tip 287
9.4 Core-Shell Nanowire 291
9.4.1 Screw Dislocation Interacting with Core-Shell Nanowire 292
9.4.2 Screw Dislocations Interacting with Embedded Nanowire 302

Chapter 10 Challenges and Opportunities 312
10.1 Introduction 312
10.2 Modeling and Simulation 312
10.3 Modeling/Simulation-Driven Material Design 314
10.4 Conclusions 318

Index 321

Preface

The emergence of new materials has greatly met the development needs of human society and promote the continuous progress of science and technology. Currently, many theories are proposed, which not only deepen the understanding of the intrinsic laws of materials, but also help to optimize them. Computational solid mechanics can efficiently accomplish material prediction and design, making them a crucial part of modern material research. Essentially, it is an interdisciplinary field that combines material mechanics and computational science to theoretically model and/or simulate the composition, structure, performance, and serviceability of materials. The field encompasses a range of disciplines, including material, physics, computer science, mathematics, chemistry and more.

Computational solid mechanics consists of two main aspects: (i) computational simulation, which establishes mathematical models and conducts numerical calculations based on experimental data; and (ii) computer-aided material design, which predicts and/or designs material structure and performance through theoretical models and calculations. The former enables material research to summarize the specific experimental findings into general and quantitative theories. The latter promotes a more targeted and forward-looking approach to material research and development, facilitating original innovation while significantly enhancing research efficiency.

In recent decades, the study of material has become increasingly complex, with a constant emergence of advanced materials that have promoted the development of computation science. These advanced materials are characterized by either innovative design concepts or exceptional performance, representing significant milestones in modern material development. The relationship between component and performance of advanced materials is highly complicated, requiring the use of advanced computational tools to demonstrate these complex associations. Computational solid mechanics has revolutionized material research theoretically and experimentally. It not only liberates theoretical research from the shackles of analytical derivation, but also fundamentally reforms the experiment, enabling it to be based on a more objective foundation and better suited for revealing objective laws from experimental phenomena. Hence, computational solid mechanics serves as a crucial link between theoretical and experimental research, offering not only novel avenues for theoretical inquiry but also propelling experimental research into new frontiers.

The authors of this book have extensive involvement in the research of modeling and simulation for advanced materials, possessing rich experience in developing and applying computational solid mechanics to predict and design materials. This book is based on some of the authors' work published over the past 20 years and is a distinctive monograph.

This book showcases the authors' significant contributions to explore the macroscopic properties of advanced materials and their correlation with multiscale structures through computational methods. In terms of computational methods, it encompasses the strength model and microstructure-based model, as well as multiscale simulation tools, including first-principles calculation, molecular dynamics simulation, phase field method and crystal plasticity finite element method. In terms of material systems, it covers high-entropy alloys, high-entropy amorphous alloys, nickel-based superalloys, light alloys, Si-based lithium battery, and nanocomposites. By revealing the multiscale microstructure interaction mechanism and developing theoretical models and simulation methods that consider microstructure characteristics, this provides rich scientific basis for assisting advanced alloy design.

This book is divided into ten chapters. The significant progress in microstructure-based model, continuum-mechanics-based model, and multiscale simulation is introduced in Chapter 1. Chapter 2 presents the frameworks of major theoretical methods used in this book. Then, the application of modeling and simulation in different material systems are summarized in chapters. Chapter 3 is focused on the deformation and strengthening mechanism and service performance of high-entropy alloys based on atomic simulations. Chapter 4 covers a lattice-distortion-dependent strength model, an irradiation hardening model, and hierarchical multiscale model to investigate the mechanical behavior of high-entropy alloys. Chapter 5 summarizes the preparation, microstructure evolution and deformation behaviors of high-entropy amorphous alloys using atomic simulations. Chapter 6 treats the effects of precipitate characteristics on the mechanical and service performances of nickel-based superalloys. The intrinsic defect characteristics of Al alloys, as well as the twinning and dynamic recrystallization in Mg alloys are covered in Chapter 7. Chapter 8 is devoted to the model for describing the internal stress field affected by diffusion and dislocation in Si-based lithium battery. Chapter 9 covers the elastic model for the interaction between dislocations and nanostructures in nanocomposites. Future challenges and opportunities in modeling and simulation are summarized in Chapter 10.

At present, the service environment of key components of advanced equipment (such as aircraft engine and nuclear reactor cores) is becoming more and more harsh, which places higher requirements for the comprehensive performance of advanced materials. Computational solid mechanics is constantly developing and improving, and plays an increasingly important role in the development of new materials. The theory and simulation work introduced in this book not only reflects the current state of computational solid mechanics, but also provides reference and basis for developing more advanced computational methods.

Finally, we hope that this book will arouse great interest in scientists, engineers, technical experts and even entrepreneurs, and thus promote the development of modeling and simulation of advanced materials together.

Qihong Fang and Jia Li

Authors

Qihong Fang is a professor in the Department of Mechanics at Hunan University. He received his PhD in 2007 from Hunan University, where he is engaged in the interdisciplinary research of mechanics, mechanical engineering, and materials science. His current research focuses on multiscale mechanics, high-entropy alloys, precision manufacturing, additive manufacturing, high-throughput design and preparation of materials, and their service properties.

Jia Li is a professor in the College of Mechanical and Vehicle Engineering at Hunan University. He received his PhD from Hunan University in 2017. His current research interests focus on the deformation and strengthening behaviours of high-entropy alloys and superalloys using multiscale simulation combined with experiments.

1 Background

1.1 INTRODUCTION

Computational solid mechanics plays a significant role in exploring the mechanical behaviors of materials. Based on the solid mechanics theory, modeling and simulation are conducted to uncover the microstructure evolution under mechanical, thermal, and chemical loading, and construct the relationship between components and properties. The modeling methods represented by elastoplastic constitutive theory and diffusion theory, as well as the simulation methods represented by molecular dynamics (MD) simulation and crystal plastic finite element (CPFE) simulation, have been continuously developed with the progress of solids. Over the past few decades, these modeling and simulation methods have been widely used in the research of advanced metals and alloys. These advanced materials have excellent design concepts, or have outstanding performance advantages over traditional materials, representing the latest progress in materials science.

This book summarizes the work of our research group and mainly concerns the modeling and simulation in advanced solids. The materials covered in this book include high-entropy alloy (HEA), high-entropy amorphous alloy, nickel-based alloy, light alloy, silicon-based electrode, and nanostructure-reinforced composite. We focus on exploring the interaction mechanisms between microstructures, and developing constitutive models to link the microstructure and mechanical properties. Our research achievements not only extend the theory of computational solid mechanics, but also provide a scientific basis for the development of new materials.

1.2 THEORETICAL PROGRESS

To obtain the mechanical behavior of the materials in advance, we propose some model and simulation methods based on the physical laws. How to break the trade-off relationship between strength and toughness is a key issue in developing high performance structural materials. This issue requires general theoretical models that take account of the key microstructures. In terms of modeling, we develop two kinds of mechanical models, including 1) the microstructure-dependent model to construct the relationship between microstructures and macroscopic properties, and 2) the continuum model to depict the elastic/plastic interaction between dislocation and nanostructures. Meanwhile, to uncover the fine-scale mechanical response that is inaccessible to the modeling method, multiscale simulation from quantum scale to macroscale is developed.

DOI: 10.1201/9781003225706-1

1.2.1 Model Progress

1.2.1.1 Microstructure-Based Model

The resistance of the solute atoms, dislocations, grain boundaries, and second phases to dislocation movement significantly contributes to the material strength. Microstructure-based strength models are widely used, and these include solid solution strengthening, dislocation strengthening, grain boundary strengthening, and second phase strengthening.

Solid Solution Strengthening Model Solid solution strengthening is a widely-used method of improving the strength and hardness of alloys by introducing alloying elements into a metallic matrix. The difference of atomic radii between the matrix atom and the solute atom causes lattice distortion, increasing the resistance to dislocation movement. In past years, based on the elastic interaction between dislocations and solute atoms, many solution-strengthening models have been developed, such as the Mott-Nabarro model, Fleischer model, and Labusch model [1–5]. The fundamental principle of these models is determining the critical shear stress required for dislocation motion, which depends on the interaction mode between dislocations and solute atoms. To investigate the solute strengthening effect, an approach coupling the Peierls-Nabarro model and Labusch model is developed. Using volume misfit and chemical misfit as features, a generalized scaling diagram is obtained to quicky evaluate the solid solution strengthening [6]. Furthermore, considering concentration-dependent factors, lattice mismatches, and elastic mismatches, a modified solid solution-strengthening model is proposed. The predicted effect of solution strengthening on yield strength from the model is consistent with experimental results [7]. Compared with traditional alloys, high entropy alloy solid solutions have excellent properties due to their severe lattice distortion. Considering the lattice distortion effect, the analytical model between solution strengthening and yield stress of FCC and BCC HEAs is established. The results further confirm that the atomic radius mismatch dominates the yield stress of high-entropy alloys [8].

Dislocation Strengthening Model Crystal plastic deformation-induced dislocation proliferation and annihilation significantly influence material strength. To evaluate the effects of dislocation density evolution on the strength and plasticity of materials, several dislocation strengthening models have been proposed, such as the Taylor model and the Kocks-Mecking model. Based on the empirical relationship between dislocation density and strength, these dislocation strengthening models are established to estimate the yield strength or flow stress of the material [9]. However, the applications of these strength models still have some limitations. For instance, the influence of dislocation type, structure, and orientation is not considered. Meanwhile, the dislocation behaviors under high temperature or large strain are difficult to describe accurately. Based on the Taylor model, a modified dislocation strengthening model is developed according to the energy gain of different dislocation types. In addition, during crystal plastic deformation, the interaction between new dislocations generated on the slip system and original dislocations leads to the change of work hardening parameter [10]. Meanwhile, the dislocation density evolution is controlled by the competition between dislocation multiplication and annihilation, represented by a thermal activation model. The contribution of dislocation density to strength is determined by considering different types of dislocation effects, which have been verified in experimental studies [10,11]. Furthermore, extensive studies have been conducted on the key parameters of the dislocation strengthening model. These key parameters significantly changed under varying deformation rates, temperatures and material structures [12,13].

Grain Boundary Strengthening Model The essence of grain boundary strengthening is that grain boundary can effectively hinder dislocation movement. The Hall-Petch relationship describes that yield strength decreases with increasing grain size [14–16]. Nevertheless, the inverse Hall-Petch relationship happens when grain size is less than the critical size. In recent years, some new grain boundary strengthening models have been proposed considering the impact of complex microstructures. For example, based on dislocation theory, a revised Hall-Petch relationship is derived, involving the influence of dislocation pile-up and stress concentration at grain boundaries in the eutectic alloys [15]. Considering the interaction between the high-angle grain boundary and the original grain boundary, a strength model related to boundary spacing under large strain is established [17]. Moreover, the influence of chemical ordering degree on strength is studied in the polycrystal multi-principal-element alloy [18], and the modified Hall-Petch relationship involving the abnormal local stacking fault energy originated from the heterogeneous chemical element concentration is elaborately established.

Second Phase Strengthening Model Second phase strengthening is typically classified into two types: dispersion strengthening and precipitation strengthening. Dispersion strengthening refers to the enhancement of material strength by introducing a second phase or refining the size of the dispersed phase. Precipitation strengthening denotes the obstructive effect of the second phase precipitated from the matrix on dislocation movement [19–23]. In order to reveal the interaction mechanism between the second phase and dislocations, two basic second phase strengthening theoretical models are proposed. For the strengthening effect of non-deformable particles, based on the Orowan model, dislocations bend severely when they encounter second phase on slip planes, and eventually form dislocation loops around the second phase. In contrast, for the deformable particle, the Kocks model describes that dislocations cut the second phases on slip planes [24,25]. In these models, the strengthening effect is obtained by increasing the critical shear stress required for dislocation motion. Recently, second phase strengthening models have been further optimized to adapt to the complex microstructure of advanced materials via multiscale simulation methods. The strengthening effect of spherical, rod-like and disc-like precipitates is calculated by dislocation dynamics simulation, and a modified Orowan equation is proposed [26]. Moreover, in order to reveal the impact of space and size distribution of precipitates on yield(1 strength, a probabilistic-dependent strengthening model is proposed. Compared with the prediction from the classical precipitate strengthening models, especially for precipitates with a large size, the predicted yield strength is more accurate [27,28]. Subsequently, a yield-strength predicted model describing the oxide particle strengthening effect is constructed for uncovering the inhibition mechanism of the second phases on dislocation motion. This model not only considers the relative position between the dislocation-slip plane and the center of the oxide particle, but also involves the distribution effect of oxide particle size. Based on the proposed strengthening model, the optimal oxide particle size to maximize the oxide particle strengthening effect is determined [29].

1.2.1.2 Continuum-Mechanics-Based Model

Diffusion-Induced Stress Model The rapid development of new energy demands higher requirements for the safety of capacitors, especially in lithium-ion batteries (LIBs). The diffusion behavior of lithium ions in the solid and liquid phases is an important part of investigating the microstructural evolution of LIBs. This diffusion behavior not only has a significant impact on the performance of LIBs, but also is closely related to their service life. During a long-period service, due to multiple cycles of battery charging and discharging, lithium ions are repeatedly de-embedded in the electrode, inevitably causing changes to the lattice size in the electrode material. The expansion or contraction deformation of the lattice in the electrode material generates a non-uniform stress field. This diffusion-induced stress leads to various failure behaviors such as inelastic deformation of solid particles [30,31], fracture crushing [32], and material interface spalling and detachment [33] of the electrode. Therefore, extensive studies have been conducted to unveil the diffusion stresses.

Currently, the widely used models of diffusion-induced stresses are developed based on the thermal stress model [34]. These models are used to calculate the influence of electrode characteristics [35–37] (particle shape [36,38,39]) and service conditions (charging and discharging mode [36,40], temperature [41]) on diffusion-induced stresses. However, the intrinsic microstructure effects that inevitably exist in the solid electrode are rarely explored, especially the crystal defects represented by dislocations. Prior studies demonstrate that the dislocation-induced stress field releases the diffusion-induced tensile stress, to avoid the formation of cracks during cycling. The results emphasize the importance of dislocation theory for solid electrode fracture. Accordingly, we further develop a diffusion-induced stress model considering the dislocation, size and surface effects, and explore the influence mechanism of microstructure on the stability of electrode materials [42–44].

Dislocation-Nanostructure Interaction Model Nanostructured-reinforced composites are widely applied in the fields of microelectronics and optics. This popularity can be attributed to their excellent mechanical properties and physicochemical properties. The strength theory and deformation mechanism of these nanocomposites are long-standing key scientific problems [45–51]. At the microscopic scale, nanocomposites are non-homogeneous solids containing many different types of inclusions, such as precipitates, oxide dispersion particles, and nanoclusters [52–55]. These inclusions increase the resistance to dislocation movement, thereby affecting the mechanical response of the material. Therefore, the interaction between dislocation and inclusion is one of the fundamental issues in studying the strengthening and toughness mechanism of nanocomposites. To date, extensive models have been established to describe the interaction between dislocations and the micron-scale inclusions. Based on these models, the interaction mechanism between different dislocations (edge or screw dislocations with different Burgers vectors) and varying inclusions (different intrinsic properties and geometrical characteristics) has been investigated [56–58]. However, when the inclusion size reduces to nanoscale, the inclusion shows significant interfacial effect, quantum effect and size effect, in contrast to the micro-inclusion [59–61]. It is critical to consider the surface/interface effects of nano-inclusions in modeling.

However, the mechanical model that integrates the intrinsic properties of nanoscale inclusions and the surface/interface effects is very complex, which results in many challenges to the analytical solution of relevant problems. This is the main reason there are few models to describe the interaction between dislocation and nano-inclusion. We creatively use and develop the complex-function form of elasticity theory, and develop a new series of mechanical models considering the surface-interface effects of nano-inclusion. At the same time, we apply the modern analytic function theory, to obtain analytic solutions of these models, promoting the development of mechanical theory for non-uniform materials. Based on the developed models, the interaction mechanisms between the edge/screw dislocation and nanoinclusion, nanopore, and core-shell nanowire are systematically studied. The basic solutions of the related problems are obtained.

1.2.2 Simulation Progress

Multiscale simulation is an important approach for studying the mechanical behavior and deformation of materials. Using theoretical and empirical models, multiscale simulation can describe the materials' mechanical behavior at different scales, including the microscale (electrons, atoms, molecules) and mesoscale (microstructure) [62–66]. The interaction and coupling between these scales determine the macroscopic properties of materials. Multiscale simulation methods consist of first-principles calculations, MD simulation, dislocation dynamics (DD) simulation, and CPFE simulation [67–74]. Figure 1.1 shows the different time and spatial scales corresponding to different computational simulation methods [75]. Currently, multiscale simulations are widely used to investigate the mechanical properties and elastoplastic behavior of various advanced materials, such as alloys, silicon-based electrode materials, and composite materials.

The material structures at the microscale/mesoscale, such as electronic structure, band structure, electronic orbitals, crystal structure, and phase, greatly affect the mechanical properties of materials. However, the determination of these structures is usually carried out at the nanoscale and on the timescale of nanoseconds, which is difficult to obtain from experimental methods [76,77]. Based on the principles of quantum mechanics, first-principles calculation can calculate the material properties from the perspective of atoms and electrons without relying on experimental data [78]. For example, the effects of element concentration on elastic and plastic properties, generalized stacking-fault energy, surface energy, and electronic properties of high-entropy alloys are investigated by first-principles calculations [79]. However, as it is limited by large-scale simulation, first-principles calculation cannot be used to easily calculate the dynamic mechanical behavior of materials, such as deformation, fracture, and fatigue processes. On the basis of Newtonian

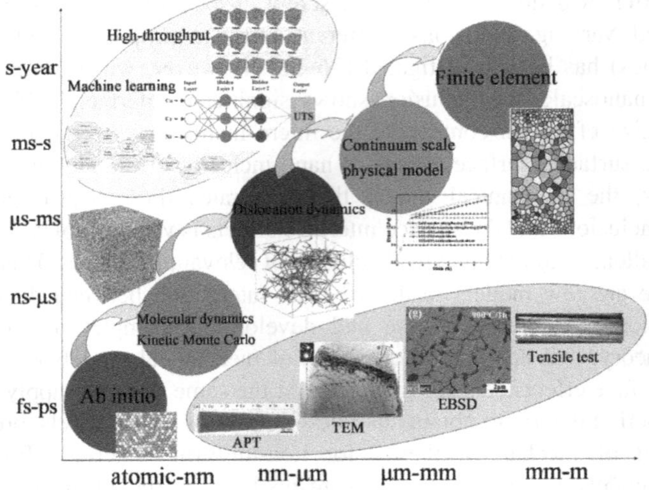

FIGURE 1.1 Different time and spatial scales corresponding to various computational simulation methods [75].

mechanics, MD can simulate the dynamic microstructure evolution under varying temperatures, pressures, and strain rate factors. For example, MD is used to simulate the plastic deformation process and shock response of metallic nanocrystals [80,81]. Meanwhile, the fatigue properties of materials, such as crack propagation, fatigue life and damage evolution, are also investigated by MD [82,83].

At the mesoscale, the dislocation movement and interactions dominate the material deformation mechanism. These interactions are a complex process involving multiple scales and physical fields, from electronic structures at the atomic scale to structures at the mesoscale [84]. However, first-principles calculations and MD methods are limited by computational resources and timescales. DD can provide some information that is difficult to measure from experiments, such as dislocation density, dislocation velocity, and dislocation type, thereby revealing the essential deformation mechanism [85]. For instance, three-dimensional DD is used to investigate heterogeneous lattice strain-induced abnormal dislocation behaviors, providing a significant insight into the strengthening mechanism of strength-plastic synergy [86]. In addition, real materials usually consist of polycrystalline structures. Hence, grain structure, orientation distribution, and grain boundary characteristics should be included during simulation. Based on crystal plasticity constitutive models [87], CPFE can consider the anisotropy of polycrystal materials and simulate the plastic deformation behavior of materials at the mesoscale/macroscopic scale, including stress-strain curves and hardening rate. Moreover, a hierarchical multiscale crystal plasticity framework, including MD simulations, discrete DD, and CPFE, is developed. Based on the framework, effects of nanoscale lattice distortion and microscale dislocation hardening on plastic deformation are studied [88]. At the same time, CPFE can better reveal the microstructure evolution of materials, including analyzing the interactions between grains and the effects of slip systems, and textures on deformation [89].

In summary, multiscale simulation can comprehensively investigate material mechanical behaviors and deformation mechanisms at different scales, including atoms, molecules, dislocations and crystals. Importantly, multiscale simulation can assist researchers to explore the fundamental structures, formation mechanisms, and property of materials in a better way, providing strong support for the design and optimization of advanced materials. However, multiscale simulation also has some limitations, such as high computational costs and difficulties in cross-scale coupling. Therefore, it is necessary to develop new theories, methods, and technologies of multiscale simulation to improve its accuracy, efficiency, and reliability.

1.3 OUTLINE

This book begins with an introduction of modeling and simulation for material research, followed by the summaries of advanced materials in separate chapters. Chapter 1 is the background, which contains introductory information on materials and research methods (Section 1.1), progress on theoretical methods (Section 1.2) and the outline of this book (Section 1.3). Chapter 2 focuses on the main modeling (Section 2.1) and simulation methods (Section 2.2). Chapter 3 and Chapter 4 summarize the work on HEAs using methods of simulation and modeling, respectively. In Chapter 3, atomic simulations are applied to investigate the deformation mechanism (Section 3.2), strengthening mechanism (Section 3.3) and service performance (Section 3.4). In Chapter 4, the lattice-distortion dependent strength model (Section 4.2), irradiation hardening model (Section 4.3), and hierarchical multiscale model (Section 4.4) are included. Chapter 5 focuses on the high-entropy amorphous alloy studied by atomic simulations. We investigate the indention-induced deformation of bulk amorphous (Section 5.2), the tensile-induced deformation of amorphous/crystal composite (Section 5.3), and the microstructure-dependent deformation of polycrystal alloys with different amorphous proportions. Chapter 6 introduces modeling work for the nickel-based superalloy, emphasizing the new strengthening models considering the size and space distribution of precipitates (Section 6.2). Then, a model for predicting high-temperature creep rate is introduced (Section 6.3). Chapter 7 introduces the work on light alloys, including the Al alloy (Section 7.2) and Mg alloy (Section 7.3). Specifically, first-principle calculation is used to study the Al alloy, while the phase field simulation and modeling methods are conducted for Mg alloy. As representative energetic materials, the Si-based lithium battery is introduced in Chapter 8, in which the basic model (Section 8.2) and diffusion-induced damage behavior are presented (Section 8.3). In Chapter 9, the models of nanostructure-reinforced composite are systematically introduced. The interaction between dislocation with inclusion (Section 9.2), nanopore (Section 9.3) and nanowire (Section 9.4) are investigated. Finally, Chapter 10 presents some challenges and opportunities for modeling and simulation methods of advanced materials. The extraordinary potential of artificial intelligence in assisting modeling and simulation methods and modeling/simulation-driven material design is presented in (Section 10.2) and (Section 10.3), respectively.

REFERENCES

[1] Buey D., Hector Jr L. and Ghazisaeidi M. 2018. Core structure and solute strengthening of second-order pyramidal< c+ a> dislocations in Mg-Y alloys. *Acta Materialia*, 147: 1–9.

[2] Leyson G.P.M., Curtin W.A., Hector Jr L.G., et al. 2010. Quantitative prediction of solute strengthening in aluminium alloys. *Nature Materials,* 9: 750–755.

[3] Leyson G.P.M., Hector Jr L. and Curtin W.A. 2012. First-principles prediction of yield stress for basal slip in Mg–Al alloys. *Acta Materialia*, 60: 5197–5203.

[4] Leyson G.P.M., Hector Jr L. and Curtin W.A. 2012. Solute strengthening from first principles and application to aluminum alloys. *Acta Materialia*, 60: 3873–3884.

[5] Wang Z., Fang Q., Li J., et al. 2018. Effect of lattice distortion on solid solution strengthening of BCC high-entropy alloys. *Journal of Materials Science Technology,* 34: 349–354.

[6] Guo, Y., Zhang, S., Wei, B., Legut, D., Germann, T. C., Zhang, H. and Zhang, R. 2019. A generalized solid strengthening rule for biocompatible Zn-based alloys, a comparison with Mg-based alloys. *Physical Chemistry Chemical Physics*, 21: 22629–22638.

[7] Toda-Caraballo, I. and Rivera-Díaz-del-Castillo, P. E. 2015. modeling solid solution hardening in high entropy alloys. *Acta Materialia*, 85: 14–23.

[8] Li, L., Fang, Q., Li, J., Liu, B., Liu, Y. and Liaw, P. K. 2020. Lattice-distortion dependent yield strength in high entropy alloys. *Materials Science and Engineering: A*, 784: 139323.

[9] Lavrentev F. 1980. The type of dislocation interaction as the factor determining work hardening. *Materials Science and Engineering*, 46: 191–208.

[10] Lavrentev F. and Pokhil Y.A. 1975. Relation of dislocation density in different slip systems to work hardening parameters for magnesium crystals. *Materials Science and Engineering*, 18: 261–270.

[11] Mikuriya N., Nishikawa H., Takasaki T., et al. 1977. The Early Stage of Basal Slip in Zinc Single Crystals. *Transactions of the Japan Institute of Metals*, 18: 527–534.

[12] Beyerlein I. and Tomé C. 2008. A dislocation-based constitutive law for pure Zr including temperature effects. *International Journal of Plasticity*, 24: 867–895.

[13] Mecking H. and Kocks U. 1981. Kinetics of flow and strain-hardening. *Acta metallurgica*, 29: 1865–1875.

[14] Hall E. 1951. The deformation and ageing of mild steel: III discussion of results. *Proceedings of the Physical Society. Section B*, 64: 747.

[15] Petch N. 1953. The cleavage strength of polycrystals. *Journal of Iron and Steel Research International*, 174: 25–28.

[16] Thompson A.W. and Baskes M.I. 1973. The influence of grain size on the work hardening of face-center cubic polycrystals. *Philosophical Magazine*, 28: 301–308.

[17] Sevillano J.G., Van Houtte P. and Aernoudt E. 1980. Large strain work hardening and textures. *Progress in Materials Science*, 25: 69–134.

[18] Li, J., Fu, X., Feng, H., Liu, B., Liaw, P. K. and Fang, Q. 2023. Evaluating the solid solution, local chemical ordering, and precipitation strengthening contributions in multi-principal-element alloys. *Journal of Alloys and Compounds*, 938:168521.

[19] He Y., Yang K. and Sha W. 2005. Microstructure and mechanical properties of a 2000 MPa grade co-free maraging steel. *Metallurgical and Materials Transactions A*, 36: 2273–2287.

[20] He Y., Yang K., Liu K., et al. 2006. Age hardening and mechanical properties of a 2400 MPa grade cobalt-free maraging steel. *Metallurgical and Materials Transactions A*, 37: 1107–1116.

[21] Xu W., Rivera-Díaz-del-Castillo P., Yan W., et al. 2010. A new ultrahigh-strength stainless steel strengthened by various coexisting nanoprecipitates. *Acta Materialia*, 58: 4067–4075.

[22] Jiang S., Wang H., Wu Y., et al. 2017. Ultrastrong steel via minimal lattice misfit and high-density nanoprecipitation. *Nature*, 544: 460–464.

[23] He J., Wang H., Huang H., et al. 2016. A precipitation-hardened high-entropy alloy with outstanding tensile properties. *Acta Materialia*, 102: 187–196.

[24] Zhang, Y., and Sills, R. B. 2023. Strengthening via orowan looping of misfitting plate-like precipitates. *Journal of the Mechanics and Physics of Solids*, 173: 105234.

[25] Ghiaasiaan, R., and Shankar, S. 2018. Structure-property models in Al-Zn-Mg-Cu alloys: A critical experimental assessment of shape castings. *Materials Science and Engineering: A*, 733: 235–245.

[26] Zhu A. and Starke Jr E. 1999. Strengthening effect of unshearable particles of finite size: a computer experimental study. *Acta Materialia*, 47: 3263–3269.

[27] Fang Q., Li L., Li J., et al. 2019. A statistical theory of probability-dependent precipitation strengthening in metals and alloys. *Journal of the Mechanics and Physics of Solids*, 122: 177–189.

[28] Li L., Liu F., Tan L., et al. 2022. Uncertainty and statistics of dislocation-precipitate interactions on creep resistance. Cell Reports Physical *Science*, 3: 100704.

[29] Li, F., Li, L., Fang, Q., Li, J., Liu, B. and Liu, Y. 2020. The probability-correlative oxide particle strengthening in solid-solution alloys. *Journal of Materials Science*, 55: 13414–13423.

[30] Huang S., Fan F., Li J., et al. 2013. Stress generation during lithiation of high-capacity electrode particles in lithium ion batteries. *Acta Materialia*, 61: 4354–4364.

[31] Liu P., Sridhar N. and Zhang Y.W. 2012. Lithiation-induced tensile stress and surface cracking in silicon thin film anode for rechargeable lithium battery. *Journal of Applied Physics*, 112: 093507.

[32] Zhao K.J., Pharr M., Vlassak J.J., et al. 2010. Fracture of electrodes in lithium-ion batteries caused by fast charging. *Journal of Applied Physics*, 108: 073517.

[33] Vetter J., Novak P., Wagner M.R., et al. 2005. Ageing mechanisms in lithium-ion batteries. *Journal of Power Sources*, 147: 269–281.

[34] Prussin S. 1961. Generation and distribution of dislocations by solute diffusion. *Journal of Applied Physics*, 32: 1876–1881.

[35] DeLuca C.M., Maute K. and Dunn M.L. 2011. Effects of electrode particle morphology on stress generation in silicon during lithium insertion. *Journal of Power Sources*, 196: 9672–9681.

[36] Wu L., Xiao X., Wen Y., et al. 2016. Three-dimensional finite element study on stress generation in synchrotron X-ray tomography reconstructed nickel-manganese-cobalt based half cell. *Journal of Power Sources*, 336: 8–18.

[37] Zhu M., Park J. and Sastry A.M. 2012. Fracture analysis of the cathode in Li-ion batteries: A simulation study. *Journal of The Electrochemical Society*, 159: A492.

[38] Park J., Lu W. and Sastry A.M. 2010. Numerical simulation of stress evolution in lithium manganese dioxide particles due to coupled phase transition and intercalation. *Journal of The Electrochemical Society*, 158: A201.

[39] Xie Y. and Yuan C. 2015. An integrated anode stress model for commercial LixC6-LiyMn2O4 battery during the cycling operation. *Journal of Power Sources*, 274: 101–113.

[40] Cheng Y.-T. and Verbrugge M.W. 2009. Evolution of stress within a spherical insertion electrode particle under potentiostatic and galvanostatic operation. *Journal of Power Sources*, 190: 453–460.

[41] Ye Y.H., Shi Y.X., Cai N.S., et al. 2012. Electro-thermal modeling and experimental validation for lithium ion battery. *Journal of Power Sources*, 199: 227–238.

[42] Li J., Fang Q., Liu F., et al. 2014. Analytical modeling of dislocation effect on diffusion induced stress in a cylindrical lithium ion battery electrode. *Journal of Power Sources*, 272: 121–127.
[43] Li J., Lu D., Fang Q., et al. 2015. Cooperative surface effect and dislocation effect in lithium ion battery electrode. *Solid State Ionics*, 274: 46–54.
[44] Li J., Fang Q., Wu H., et al. 2015. Investigation into diffusion induced plastic deformation behavior in hollow lithium ion battery electrode revealed by analytical model and atomistic simulation. *Electrochimica Acta*, 178: 597–607.
[45] Dundurs J. and Mura T. 1964. Interaction between an edge dislocation and a circular inclusion. *Journal of the Mechanics and Physics of Solids*, 12: 177–189.
[46] Smith E. 1968. The interaction between dislocations and inhomogeneities—I. *International Journal of Engineering Science*, 6: 129–143.
[47] Deng W. and Meguid S.A. 1999. Analysis of a screw dislocation inside an elliptical inhomogeneity in piezoelectric solids. *International Journal of Solids and Structures*, 36: 1449–1469.
[48] Xiao Z.M. and Chen B.J. 2000. A screw dislocation interacting with a coated fiber. *Mechanics of Materials*, 32: 485–494.
[49] Chi-ping J., You-wen L. and Yao-ling X. 2003. Interaction of a screw dislocation in the interphase layer with the inclusion and matrix. *Applied Mathematics and Mechanics*, 24: 979–988.
[50] Fang Q.H., Liu Y.W. and Wen P.H. 2008. Screw dislocations in a three-phase composite cylinder model with interface stress. *Journal of Applied Mechanics*, 75: 041019.
[51] Li J., Fang Q. and Liu Y. 2013. Crack interaction with a second phase nanoscale circular inclusion in an elastic matrix. *International Journal of Engineering Science*, 72: 89–97.
[52] Gutkin M.Y., Ovid'ko I.A. and Sheinerman A.G. 2003. Misfit dislocations in composites with nanowires. *Journal of Physics: Condensed Matter*, 15: 3539.
[53] Ovid'ko I.A. and Sheinerman A.G. 2006. Nanoparticles as dislocation sources in nanocomposites. *Journal of Physics: Condensed Matter*, 18: L225.
[54] Vollath D. and Szabó D.V. 2004. Synthesis and properties of nanocomposites. *Advanced Engineering Materials*, 6: 117–127.
[55] Fang Q.H., Liu Y.W. and Chen J.H. 2008. Misfit dislocation dipoles and critical parameters of buried strained nanoscale inhomogeneity. *Applied Physics Letters*, 92: 121923.
[56] Ting T.C.T. 2000. Recent developments in anisotropic elasticity. *International Journal of Solids and Structures*, 37: 401–409.
[57] Ting T.C.-t. and Ting T.C.-t. 1996. *Anisotropic elasticity: theory and applications*. Oxford University Press on Demand.
[58] Fang Q.H. and Liu Y.W. 2006. Size-dependent interaction between an edge dislocation and a nanoscale inhomogeneity with interface effects. *Acta Materialia*, 54: 4213–4220.
[59] Miller R.E. and Shenoy V.B. 2000. Size-dependent elastic properties of nanosized structural elements. *Nanotechnology*, 11: 139–147.
[60] Fang Q.H., Liu Y.W., Jin B., et al. 2009. Interaction between a dislocation and a core–shell nanowire with interface effects. *International Journal of Solids and Structures*, 46: 1539–1546.
[61] Fang Q.H. and Liu Y.W. 2006. Size-dependent elastic interaction of a screw dislocation with a circular nano-inhomogeneity incorporating interface stress. *Scripta Materialia*, 55: 99–102.
[62] Fish J., Wagner G.J. and Keten S. 2021. Mesoscopic and multiscale modeling in materials. *Nature Materials*, 20: 774–786.

[63] Horstemeyer M.F. 2010. Multiscale modeling: a review. *Practical aspects of computational chemistry: methods, concepts and applications*, 87–135.

[64] Steinhauser M.O. 2017. *Computational multiscale modeling of fluids and solids*, Springer.

[65] Weinan E. and Engquist B. 2003. Multiscale modeling and computation. *Notices of the AMS*, 50: 1062–1070.

[66] Peng J., Li L., Li F., et al. 2021. The predicted rate-dependent deformation behaviour and multistage strain hardening in a model heterostructured body-centered cubic high entropy alloy. *International Journal of Plasticity*, 145: 103073.

[67] Neugebauer J. and Hickel T. 2013. Density functional theory in materials science. *Wiley Interdisciplinary Reviews: Computational Molecular Science*, 3: 438–448.

[68] Perez D., Uberuaga B.P., Shim Y., et al. 2009. Accelerated molecular dynamics methods: introduction and recent developments. *Annual Reports in Computational Chemistry*, 5: 79–98.

[69] Roters F., Eisenlohr P., Bieler T.R., et al. 2011. *Crystal plasticity finite element methods: in materials science and engineering*, John Wiley & Sons.

[70] Xu S. 2022. Recent progress in the phase-field dislocation dynamics method. *Computational Materials Science*, 210: 111419.

[71] Fang Q., Chen Y., Li J., et al. 2019. Probing the phase transformation and dislocation evolution in dual-phase high-entropy alloys. *International Journal of Plasticity*, 114: 161–173.

[72] Chen Y., Fang Q., Luo S., et al. 2022. Unraveling a novel precipitate enrichment dependent strengthening behaviour in nickel-based superalloy. *International Journal of Plasticity*, 155: 103333.

[73] Fang Q., Lu W., Chen Y., et al. 2022. Hierarchical multiscale crystal plasticity framework for plasticity and strain hardening of multi-principal element alloys. *Journal of the Mechanics and Physics of Solids*, 169: 105067.

[74] Zhao Q., Li J., Fang Q., et al. 2019. Effect of Al solute concentration on mechanical properties of AlxFeCuCrNi high-entropy alloys: A first-principles study. *Physica B: Condensed Matter*, 566: 30–37.

[75] Li J., Fang Q. and Liaw P.K. 2021. Microstructures and properties of high-entropy materials: modeling, simulation, and experiments. *Advanced Engineering Materials*, 23: 2001044.

[76] Liu X.Y., Andersson D.A. and Uberuaga B.P. 2012. First-principles DFT modeling of nuclear fuel materials. *Journal of Materials Science*, 47: 7367–7384.

[77] Schleder G.R., Padilha A.C.M., Acosta C.M., et al. 2019. From DFT to machine learning: recent approaches to materials science–a review. *Journal of Physics: Materials*, 2: 032001.

[78] Zhao, S., Osetsky, Y. and Zhang, Y. 2017. Preferential diffusion in concentrated solid solution alloys: NiFe, NiCo and NiCoCr. *Acta Materialia*, 128: 391–399.

[79] Zhao, Q., Li, J., Fang, Q. and Feng, H. 2019. Effect of Al solute concentration on mechanical properties of AlxFeCuCrNi high-entropy alloys: A first-principles study. *Physica B: Condensed Matter*, 566: 30–37.

[80] Yamakov V., Wolf D., Phillpot S.R., et al. 2004. Deformation-mechanism map for nanocrystalline metals by molecular-dynamics simulation. *Nature Materials*, 3: 43–47.

[81] Peng, J., Li, J. and Mohammadzadeh, R. 2022. Role of lattice resistance in the shock dynamics of fcc-structured high entropy alloy. Materials Today *Communications*, 33: 104884.

[82] Tang, T., Kim, S. and Horstemeyer, M. F. 2010. Fatigue crack growth in magnesium single crystals under cyclic loading: Molecular dynamics simulation. *Computational Materials Science*, 48: 426–439.

[83] Peng, J., Li, F., Liu, B., Liu, Y., Fang, Q., Li, J. and Liaw, P. K. 2020. Mechanical properties and deformation behavior of a refractory multiprincipal element alloy under cycle loading. *Journal of Micromechanics and Molecular Physics*, 5: 2050014.
[84] Yamakov V., Wolf D., Phillpot S.R., et al. 2003. Dislocation–dislocation and dislocation–twin reactions in nanocrystalline Al by molecular dynamics simulation. *Acta Materialia*, 51: 4135–4147.
[85] Zhang, X., Lu, S., Zhang, B., Tian, X., Kan, Q. and Kang, G. 2021. Dislocation–grain boundary interaction-based discrete dislocation dynamics modeling and its application to bicrystals with different misorientations. *Acta Materialia*, 202: 88–98.
[86] Li, J., Chen, Y., He, Q., et al. 2022. Heterogeneous lattice strain strengthening in severely distorted crystalline solids. *Proceedings of the National Academy of Sciences*, 119: e2200607119.
[87] Xie C., He J.M., Zhu B.W., et al. 2018. Transition of dynamic recrystallization mechanisms of as-cast AZ31 Mg alloys during hot compression. *International Journal of Plasticity*, 111: 211–233.
[88] Fang, Q., Lu, W., Chen, Y., Feng, H., Liaw, P. K. and Li, J. 2022. Hierarchical multiscale crystal plasticity framework for plasticity and strain hardening of multi-principal element alloys. *Journal of the Mechanics and Physics of Solids*, 169: 105067.
[89] Khan A.S. and Liu J. 2016. A deformation mechanism based crystal plasticity model of ultrafine-grained/nanocrystalline FCC polycrystals. *International Journal of Plasticity*, 86: 56–69.

2 Theoretical Methods

2.1 INTRODUCTION

Physics-mechanism-dependent modeling and simulation have been developed urgently for understanding and predicting the mechanical behavior of advanced materials. In Chapter 1, the progress of theoretical methods has been summarized. This chapter introduces the framework of the modeling and simulation used in the following chapters, including the microstructure-based model, continuum mechanics model, and multiscale simulation.

2.2 MICROSTRUCTURE-BASED MODEL

To predict the mechanical properties of materials, the constitutive model based on microstructure is used to describe the stress-strain response [1–7]. The deformation is divided into the elastic stage and plastic stage. Then the total strain rate is expressed as:

$$\dot{\boldsymbol{\varepsilon}} = \dot{\boldsymbol{\varepsilon}}_e + \dot{\boldsymbol{\varepsilon}}_p \tag{2.1}$$

The elastic strain rate $\dot{\boldsymbol{\varepsilon}}_e$ is linearly related to the applied stress, and described by the classical Hooke's law $\dot{\boldsymbol{\varepsilon}}_e = \mathbf{M}: \dot{\boldsymbol{\sigma}}$, where $\mathbf{M}$ is the elastic tensor. The plastic strain rate $\dot{\boldsymbol{\varepsilon}}_p$ is described using the following equation:

$$\dot{\boldsymbol{\varepsilon}}_p = \frac{3}{2}\frac{\dot{\varepsilon}_p}{\sigma}\boldsymbol{\sigma}' \tag{2.2}$$

where $\dot{\varepsilon}_p = \sqrt{2\dot{\varepsilon}_{p_{ij}}\dot{\varepsilon}_{p_{ij}}/3}$ is the Von Mises equivalent plastic strain rate, $\sigma = \sqrt{3\sigma_{ij}^{'}\sigma_{ij}^{'}/2}$ is the Von Mises equivalent stress. $\boldsymbol{\sigma}' = \sigma_{ij} - \sigma_{kk}\delta_{ij}/3$ denotes the deviatoric stress tensor. The relationship between the equivalent plastic strain rate and equivalent stress is:

$$\dot{\varepsilon}_p = \dot{\varepsilon}\left(\frac{\sigma}{\sigma_{\text{flow}}}\right)^m \tag{2.3}$$

where $\dot{\varepsilon} = \sqrt{2\dot{\varepsilon}_{ij}^{'}\dot{\varepsilon}_{ij}^{'}/3}$ represents the equivalent strain rate, $\dot{\varepsilon}'_{ij} = \dot{\varepsilon}_{ij} - \dot{\varepsilon}_{kk}\delta_{ij}/3$ is the deviatoric strain rate, m denotes the strain-rate sensitivity coefficient. σ_{flow} is determined by the resistance to dislocation movement that comes from

DOI: 10.1201/9781003225706-2

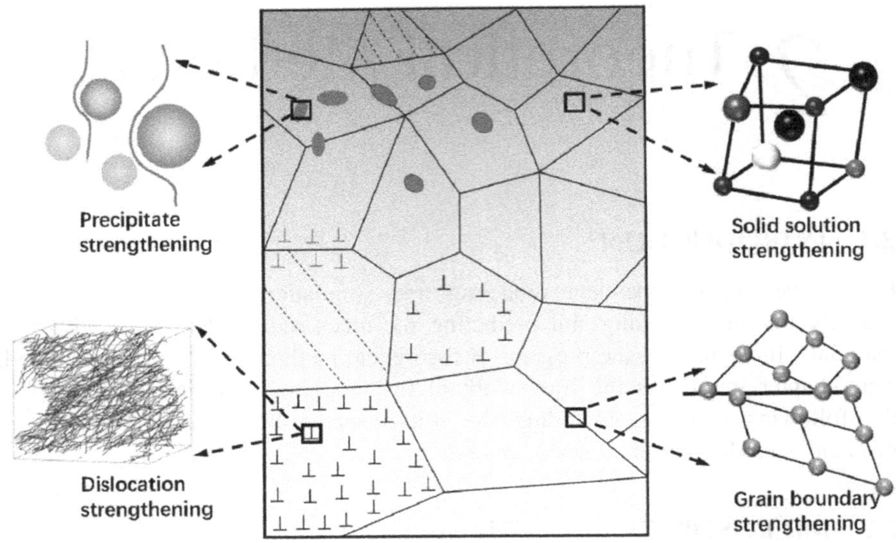

FIGURE 2.1 Schematic diagram of the key strengthening mechanism.

microstructures [8,9]. In alloys, the dislocation movement is always inhibited by the solute atom, dislocation, grain boundary (GB) and second phase, thereby contributing to the strength. Thus, the flow stress σ_{flow} is generally given by

$$\sigma_{flow} = \sigma_{ss} + \sigma_{dis} + \sigma_{gb} + \sigma_{sp} \tag{2.4}$$

where σ_{ss}, σ_{dis}, σ_{gb}, and σ_{sp} denote the contribution of the solute atom, dislocation, GB, and second phase particle, respectively (Figure 2.1). In Chapters 4 and 6, the microstructure-based model is used to analyze the relationship between the strength and microstructures of high-entropy alloys (HEAs) and nickel-based alloys.

2.2.1 Solid Solution Strengthening

The solute atom in the matrix induces the lattice distortion (Figure 2.2), thereby producing extra resistance to the dislocation motion. This strengthening mechanism derived from the solute atoms is called solid solution strengthening [10,11].

1. When the concentration of solid solution element is low, the solid solution strengthening derives from the elastic interaction between the dislocation stress field and the local stress field induced by the solute atom. The interaction force is given by [12,13]:

$$F = Gb^2 f \tag{2.5}$$

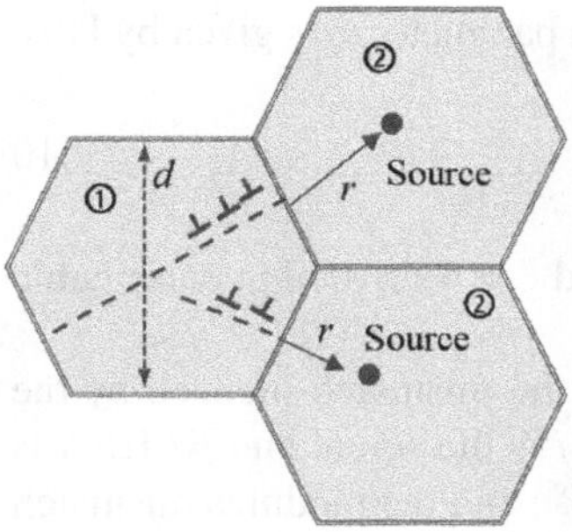

FIGURE 2.2 Diagrams of lattice distortion [16].

where G is the shear modulus of alloys, b is the absolute value of the Burgers vector, and f is the mismatch constant.

2. If the concentration of solid solution element is high, the contribution of solid solution strengthening mechanism to the strength is calculated by [14]:

$$\Delta\sigma_{ss} = AF^{4/3}c^{2/3}E_L^{-1/3}/b^2 \tag{2.6}$$

where A (≈ 0.04) is a constant determined by the intrinsic characteristics of material [15], c is the solute concentration, $E_L = Gb^2/2$ is the dislocation line tensor. Combining Eqs. 2.5 and 2.6, the contribution of the i-element on the solid solution strengthening is given by:

$$\Delta\sigma_i = AGf_i^{4/3}c_i^{2/3} \tag{2.7}$$

where $f_i = \sqrt{\delta_{G_i}^2 + \beta^2\delta_{r_i}^2}$, $\delta_{G_i} = (1/G_i)dG_i/dc_i$, $\delta_{r_i} = (1/r_i)dr_i/dc_i$. G_i and r_i are shear modulus and atomic size of the i-element, respectively. The value of β depends on the dislocation type: for edge dislocation, $\beta \geq 16$; for screw dislocation, $2 < \beta < 4$.

3. For a HEA that contains multiple principal elements, the solution strengthening is hard to predict due to the difficulty in distinguishing between solvents and solutes. In recent years, models based on Vegard's law have been widely accepted [16–18]. For the HEA composed with n elements, the contribution of solid solution strengthening is estimated by

$$\sigma_{ss} = \sum_{i=1}^{n} c_i\sigma_{ss}^i \tag{2.8}$$

where c_i is the atomic fraction of the i-element. σ_{ss}^i is the strength contribution derived from the mismatch induced by i-element. σ_{ss}^i is expressed by:

$$\sigma_{ss}^i = AGc_i^{2/3}\delta_i^{4/3} \tag{2.9}$$

where $G = \sum_i^n c_i G_i$ is the shear modulus. The mismatch parameter δ_i is given by [19]:

$$\delta_i = \xi\,(\delta G_i^2 + \beta^2 \delta r_i^2)^{1/2} \qquad 2.10$$

where $\xi = 1$ for face-centered cubic (FCC) matrix and $\xi = 4$ for body-center cubic (BCC) matrix [19].

For a HEA composed with elements i, j, k, and l, the mismatch induced by the i-element can be estimated by assuming that element i is the solute and *jkl* HEA is the solvent. The expressions of atomic size mismatch δr_i and the modulus mismatch δG_i are given by:

$$\delta r_i = \frac{\delta r_{ijkl}^{ave} - \delta r_{jkl}^{ave}}{c_i} \qquad (2.11)$$

$$\delta G_i = \frac{\delta G_{ijkl}^{ave} - \delta G_{jkl}^{ave}}{c_i} \qquad (2.12)$$

where δr_{ijkl}^{ave} and δr_{jkl}^{ave} are the average atomic size mismatch of *ijkl* HEA and *jkl* HEA, respectively, while δG_{ijkl}^{ave} and δG_{jkl}^{ave} are the average modulus mismatch of *ijkl* HEA and *jkl* HEA, respectively. The δr^{ave} and δG^{ave} are calculated by:

$$\delta r^{ave} = \sum_i^n \sum_j^n c_i c_j \delta r_{ij} = (c_1, c_2, \ldots, c_n)\begin{pmatrix} \delta r_{11} & \delta r_{12} & \ldots & \delta r_{1n} \\ \delta r_{21} & \delta r_{22} & \ldots & \delta r_{2n} \\ \vdots & \ldots & \ddots & \vdots \\ \delta r_{n1} & \delta r_{n2} & \ldots & \delta r_{nn} \end{pmatrix}\begin{pmatrix} c_1 \\ c_2 \\ \vdots \\ c_n \end{pmatrix} \qquad (2.13)$$

$$\delta G^{ave} = \sum_i^n \sum_j^n c_i c_j \delta G_{ij} = (c_1, c_2, \ldots, c_n)\begin{pmatrix} \delta G_{11} & \delta G_{12} & \ldots & \delta G_{1n} \\ \delta G_{21} & \delta G_{22} & \ldots & \delta G_{2n} \\ \vdots & \ldots & \ddots & \vdots \\ \delta G_{n1} & \delta G_{n2} & \ldots & \delta G_{nn} \end{pmatrix}\begin{pmatrix} c_1 \\ c_2 \\ \vdots \\ c_n \end{pmatrix} \qquad (2.14)$$

where $\delta G_{ij} = 2(G_i - G_j)/(G_i + G_j)$ and $\delta r_{ij} = 2(r_i - r_j)/(r_i + r_j)$ are the modulus mismatch and the atomic size mismatch among the atom i and j, respectively. G_i and G_j represent the shear modulus of pure metals composed with elements i and j, respectively. r_i and r_j are the atomic size of atom i and j, respectively.

2.2.2 Dislocation Strengthening

With the increase of plastic deformation, the density of dislocations inside the grain increases. This trend improves the probability of the interaction between dislocations, thus causing the obstacles to the dislocation movement and resulting in the working hardening. This phenomenon is called dislocation strengthening.

According to the Taylor relation [20,21], the contribution of dislocation slip resistance to the critical shear stress (CSS) is written as:

$$\tau_n = b\mu\sqrt{h_n \rho_n} \tag{2.15}$$

where μ is shear modulus, and h_n is the hardening coefficient [22]. ρ_n is the dislocation density determined by the proliferation and annihilation of dislocations, and given by:

$$\dot{\rho}_n = \dot{\gamma}\,[k_1\sqrt{\rho_n} - k_2(\dot{\varepsilon}, T)\rho_n] \tag{2.16}$$

where $\dot{\varepsilon}$ is the applied strain rate. k_1 and k_2 are the proliferation coefficient and the annihilation coefficient, respectively. The relationship between the two coefficients is:

$$\frac{k_2(\dot{\varepsilon}, T)}{k_1} = \frac{\chi b}{g}\left[1 - \frac{kT}{Db^3}\ln\left(\frac{\dot{\varepsilon}}{\dot{\varepsilon}_0}\right)\right] \tag{2.17}$$

where $\dot{\varepsilon}_0$, D, g, k, and χ are the reference strain rate, drag stress, normalized activation energy, Boltzmann constant, and interaction parameter, respectively. Eq. 2.17 claims the nonlinear correlation between temperature and applied strain rate-dependent proliferation coefficient and annihilation coefficient.

2.2.3 Grain Boundary Strengthening

GBs are prevalent microstructures in metallic materials, which prevent the dislocation transferring from one grain to adjacent grains. Therefore, GBs can serve as effective barriers to dislocation movement, thereby enhancing the strength. During deformation, dislocations on the same slip plane always pile up in front of GBs, forming the dislocation pile-up groups (Figure 2.3). After sufficient dislocation accumulation, the stress concentration will activate the dislocation in adjacent grains, resulting in macroscopic yielding [23].

Assume that $\tau*$ is the stress required to activate the dislocations in the adjacent grains. When the applied stress τ_{app} meets $(\tau_{app} - \tau_0)(d/4r)^{1/2} = \tau*$, the dislocation in the adjacent grain will be activated, where τ_0 is the intrinsic stress resisting the movement of dislocation in the deformed grain. The effect of the stress concentration induced by the dislocation pile-up is represented as $(d/4r)^{1/2}$, which increases with the increasing number of dislocations. Thus, the applied shear stress is expressed as:

$$\tau_{app} = \tau_0 + kd^{-1/2} \tag{2.18}$$

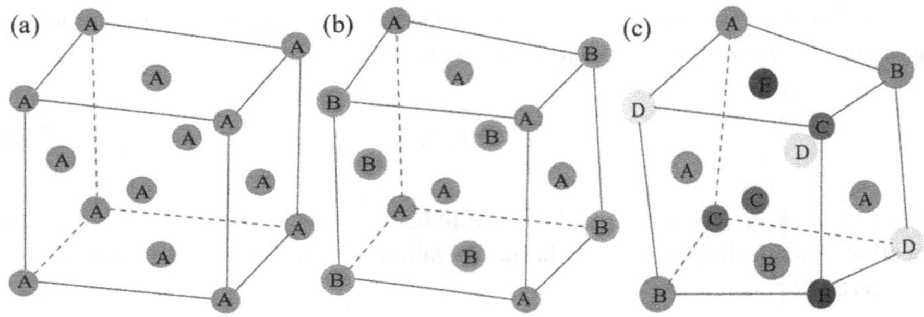

FIGURE 2.3 Schematic diagram of grain boundary strengthening mechanism.

where $k = 2\tau_* r^{1/2} d^{-1/2}$. Hence, the contribution of GB strengthening is written as:

$$\sigma_{app} = \sigma_0 + kd^{-1/2} \tag{2.19}$$

This law demonstrates that the yield strength of polycrystalline alloys increases linearly with $d^{-1/2}$, which has been confirmed in many materials [24,25].

2.2.4 Second Phase Strengthening

The strengthening effect derived from the interaction between dislocation and second phase particle relies on the type of particles. Specifically, it includes the dispersion strengthening and the precipitation strengthening [26]. The disperse particle is non-deformable and always introduced by powder sintering or internal oxidation. The precipitate is deformable and generally induced by the aging treatment [27,28]. According to the physical properties and distribution laws of the second phase particles, two mechanisms are proposed to describe the interaction between dislocations and second phase particles, i.e., the Orowan mechanism and the cutting mechanism. When the elastic modulus of the second phase particle is smaller than that of the matrix, the cutting mechanism dominates the dislocation-particle interaction. Otherwise, the Orowan mechanism controls the dislocation-particle interaction. When the particle size is small, the cutting mechanism dominates the strengthening behavior, and the strengthening effect increases with the particle size. When the particle size is large, the Orowan mechanism dominates it, and this effect increases with decreasing particle size. There is a critical size between the two mechanisms, and the mechanism changes when the critical size is exceeded (Figure 2.4).

Orowan Mechanism. In terms of the Orowan mechanism, the dislocation will bend due to the dislocation-particle interaction. The bending degree of the dislocation line is related to the applied stress. When there are second phase particles interspersed on the slip plane of the dislocation, an additional stress is required when the dislocation passes through the particles [29]. When the applied stress is high enough, the dislocation lines between particles bend into an approximate semicircle. As the dislocation continues to move on the slip plane, the dislocation extrudes between the particles. As the applied shear stress further increases, the

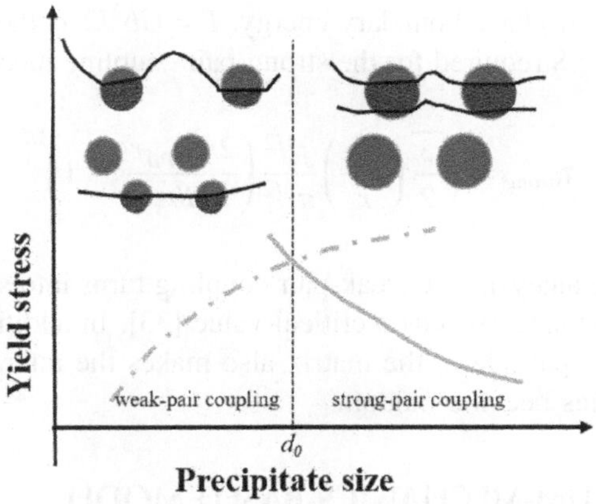

FIGURE 2.4 The schematic diagram of the weak-pair coupling cutting mechanism and the strong-pair coupling cutting mechanism [31].

dislocation line continues to bend between the particles until the dislocations with reverse Burgers vectors cancel each other. Finally, dislocation rings are left surrounding the particles. The shear strength associated with dislocation bending is written as [30]:

$$\tau_{orowan} = \frac{3Gb}{2L} \tag{2.20}$$

where $L = \left(\frac{2\pi}{3f}\right)^{1/2} r$ is the averaged spacing between particles. r and f are the size and volume fraction of the particle, respectively.

Cutting Mechanism. When the size of the second phase particle is smaller than the critical size, the shear stress needed for the dislocation to bypass the precipitate is greater than the stress needed for the cutting mechanism, so the dislocation will shear through the precipitate. The cutting mechanism is distinguished into the weak-pair and strong-pair coupling cutting mechanisms, depending on the dislocation spacing and the precipitate size (Figure 2.4) [31]. When the precipitate diameter is smaller than the dislocation spacing, the leading and trailing dislocations are located in different precipitates. In this case, the dislocation-particle interaction is controlled by the weak-pair coupling mechanism. However, when the precipitate diameter is larger than the dislocation spacing, the leading and the trailing dislocations move in the same precipitate at the same time. In this case, the strong coupling mechanism controls the dislocation-particle interaction [32].

The CSS required for the weak-pair coupling mechanism is written as:

$$\tau_{weak} = \frac{\gamma_{APB}}{2b}\left[\left(\frac{6\gamma_{APB} f r}{2\pi T}\right)^{1/2} - f\right] \tag{2.21}$$

where γ_{APB} is the anti-phase boundary energy. $T = Gb^2/2$ is the tension of the dislocation line. The CSS required for the strong-pair coupling mechanism is given by:

$$\tau_{strong} = \sqrt{\frac{3}{2}}\left(\frac{Gb}{r}\right)\frac{f^{1/2}}{\pi^{3/2}}\left(\frac{2\pi\gamma_{APB}r}{Gb^2} - 1\right)^{1/2} \tag{2.22}$$

Based on the above analysis, the weak-pair coupling turns into strong-pair coupling when the precipitate size exceeds a critical value [33]. In addition, the distribution of the second phase particle in the matrix also makes the interaction between dislocation and particles become random.

2.3 CONTINUUM-MECHANICS-BASED MODEL

2.3.1 Diffusion-Induced Stress Model

The distribution of diffusion-induced stress is one of the reasons leading to the microstructure evolution of electrode materials. In lithium-ion batteries (LIBs), the stress field controlled by the lithium-ion diffusion and dislocations determines the lifespan of electrode. It is crucial to establish a theoretical model to analyze the coupling effects of lithium diffusion-induced stress and dislocation-induced stress.

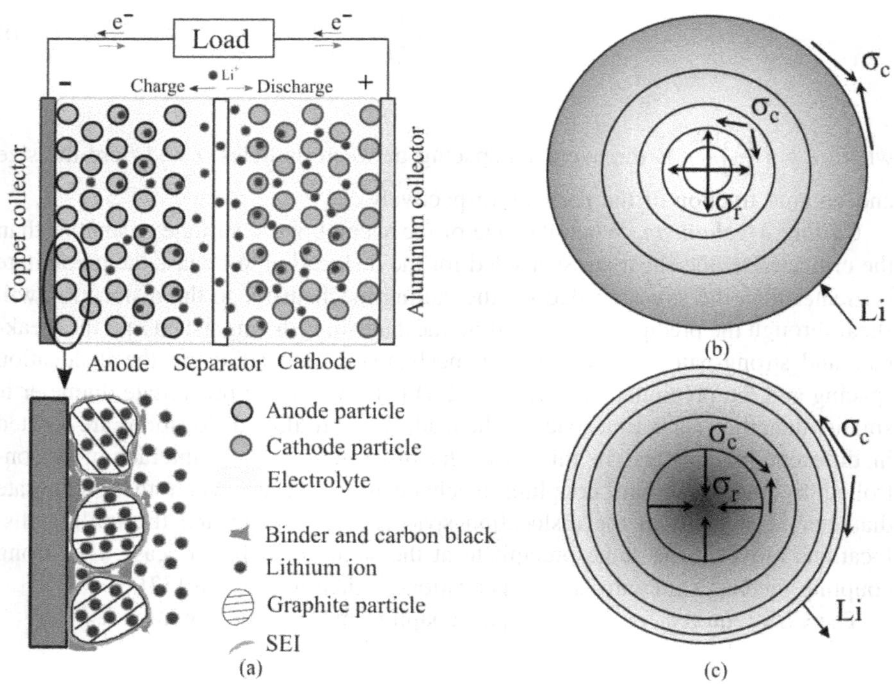

FIGURE 2.5 (a) A schematic of the composition of the lithium-ion battery. Stress and deformation generated by lithium-ion (b) insertion and (c) extraction [40].

The model of LIB is shown in Figure 2.5. Generally, the total strain ε_{ij} of a electrode consists of three parts: the elastic strain ε_{ij}^{e}, the diffusion-induced strain ε_{ij}^{di}, and the thermal strain ε_{ij}^{T}:

$$\varepsilon_{ij} = \varepsilon_{ij}^{e} + \varepsilon_{ij}^{di} + \varepsilon_{ij}^{T} \tag{2.23}$$

Assuming that the electrode is an isotropic linear elastic solid, the elastic strain ε_{ij}^{e} is described by Hooke's law:

$$\varepsilon_{ij}^{e} = \frac{1}{E}[(1 + \upsilon)\sigma_{ij} - \upsilon\sigma_{kk}\delta_{ij}] \tag{2.24}$$

where E and υ are the elastic modulus and Poisson's ratio, respectively. σ_{ij} is the component of stress, and δ_{ij} is Dirac-delat function.

$$\delta_{ij} = \begin{cases} 0 & i \neq j \\ 1 & i = j \end{cases} \tag{2.25}$$

In the positive and negative active layers, the lithium concentration gradient of the active material can cause diffusion-induced deformation. The diffusion-induced strain of the active material is determined by the change state of the concentration at the corresponding location [34]:

$$\varepsilon_{ij}^{di} = \frac{1}{3}\Omega C(x_i, t)\delta_{ij} \tag{2.26}$$

where Ω is the partial molar volume of solute, and $C(x_i, t)$ is the molar concentration of lithium-ion.

In addition, an increase in temperature will cause thermal deformation of the material. The resulting thermal strain is calculated using the following formula:

$$\varepsilon_{ij}^{T} = \alpha_{ij}\Delta T\delta_{ij} \tag{2.27}$$

where α_{ij} is the coefficient of thermal expansion, and ΔT is the temperature increment.

In the Cartesian coordinate system, the total strain ε_{ij} is decomposed into three components:

$$\begin{cases} \varepsilon_x = \frac{1}{E}[\sigma_x - \upsilon(\sigma_y + \sigma_z)] + \frac{1}{3}\Omega C(x, t) + \alpha_x\Delta T \\ \varepsilon_y = \frac{1}{E}[\sigma_y - \upsilon(\sigma_x + \sigma_z)] + \frac{1}{3}\Omega C(y, t) + \alpha_y\Delta T \\ \varepsilon_z = \frac{1}{E}[\sigma_z - \upsilon(\sigma_x + \sigma_y)] + \frac{1}{3}\Omega C(z, t) + \alpha_z\Delta T \end{cases} \tag{2.28}$$

where σ_x, σ_y, and σ_z are the components of stress in three directions of the coordinate system. In addition, the relationship between strain and displacement is given by $\varepsilon_x = \frac{\partial u}{\partial x}$, $\varepsilon_y = \frac{\partial v}{\partial y}$, $\varepsilon_z = \frac{\partial w}{\partial z}$, where u, v, and w are the displacements in the x, y, and z directions, respectively.

In solids, atomic diffusion is very slow compared to the elastic deformation response. In this case, the mechanical equilibrium always occurs before diffusion equilibrium. Therefore, for static equilibrium problems, the stress component satisfies the following equilibrium conditions:

$$\frac{\partial \sigma_x}{\partial x} = \frac{\partial \sigma_y}{\partial y} = \frac{\partial \sigma_z}{\partial z} = 0 \tag{2.29}$$

Combining the above Eqs. 2.28–2.29, one can obtain the stress components σ_x, σ_y, and σ_z without the dislocation effect. The Von Mises stress is given by:

$$\sigma_v = \sqrt{\frac{(\sigma_x - \sigma_y)^2 + (\sigma_y - \sigma_z)^2 + (\sigma_z - \sigma_x)^2}{2}} \tag{2.30}$$

The above model has been developed to solve the problem of diffusion-induced stress in the cylindrical electrode [35] and spherical electrode [36,37].

The diffusion behavior will induce the change of lithium-ion concentration, thereby affecting the dislocation density at the corresponding position in the electrode [38]. The lithium-ion concentration at x_i direction is $C(x_i, t)$, and the concentration at point $x_i + \Delta x_i$ is $C(x_i, t) + (\partial C/\partial x_i)dx_i$. The value of x_i ($i = 1, 2, 3$) denotes the coordinate direction of Cartesian coordinate system. The dislocation density $\rho(x_i, t)$ is calculated by:

$$\rho(x_i, t) = \frac{\beta}{b} C(x_i, t) \tag{2.31}$$

where β is the solute lattice contraction coefficient, and b is the Burgers vector.

When the diffusion-induced stress is greater than the stress required for dislocation nucleation, the dislocation will be activated [38]. Assuming that the interaction between dislocations and lithium-ion is determined by the long-range effect of dislocations, the relationship between stress and dislocation density is obtained [39].

$$\sigma_\tau(x_i, t) = M\alpha\mu b\sqrt{\rho(x_i, t)} \tag{2.32}$$

where M is the Taylor orientation factor, α is the empirical constant, and μ is the shear modulus. The dislocation-induced stress hinders the intercalation and expulsion of lithium ions. Considering the effect of dislocation nucleation, the stress component is calculated by the following equation:

$$\sigma_{x_i} = \sigma_{x_0} - \sigma_\tau(x_i, t) \tag{2.33}$$

where the first term is the stress component under static equilibrium condition, while the last term represents the stress component affected by the dislocation.

Finally, the stress at any location and any time, induced by the dislocation and lithium-ion diffusion, can be deduced based on the above equations. In Chapter 8 of this book, the cylindrical electrolyte model, the spherical electrolyte model, and the hollow spherical electrolyte model are discussed in detail.

2.3.2 Dislocation-Nanostructure Interaction Model

The mechanical property of the nanostructure-strengthening composite material is controlled by the interaction between dislocations and various nanostructures, such as the nanoinclusion, nanopore, and nanowire. Several models have been developed to describe the dislocation-nanostructure interaction, by assuming that the inclusion has an ideal interface without defects. However, for practical composites, interface defects are inevitably introduced during manufacturing and long-range service. Therefore, understanding the interaction between the dislocation and nanostructure with interfacial defects is crucial for developing new nanostructure-strengthening composite materials.

Extensive models based on the elastic theories have been developed to describe the dislocation-nanostructure interaction. This elastic interaction is an anti-plane elastic problem, in which the effective displacement and stress are related to x and y, and independent of the variable z. For the anti-plane problem, the elastic equilibrium equation for the anti-plane shear is:

$$\frac{\partial^2 w}{\partial x^2} + \frac{\partial^2 w}{\partial y^2} = 0 \tag{2.34}$$

where w represents the anti-plane displacement. Its general solution is represented by an analytical function $f(z)$ defined in the complex potential plane $z = x + iy$:

$$w = \mathrm{Re}[f(z)]/G \tag{2.35}$$

where "Re" refers to the real part of the function, and G denotes the shear modulus. The rectangular coordinate component of shear stress is expressed as:

$$\tau_{xz} = \frac{\partial w}{\partial x},\ \tau_{yz} = \frac{\partial w}{\partial y} \tag{2.36}$$

$$\tau_{xz} - i\tau_{yz} = Gf'(z) \tag{2.37}$$

After coordinate transformation, Eq. 2.37 is represented in the polar coordinates as:

$$\tau_{rz} - i\tau_{\theta z} = G\exp[i\theta]f'(z) \tag{2.38}$$

where τ_{rz} and $\tau_{\theta z}$ represent the anti-plane shear stress in the directions r and θ in polar coordinates, respectively. In the elastic interaction problem between the dislocation and nanostructure, the equilibrium and constitutive equations in nanostructures and matrix are consistent with classical elasticity. Meanwhile, due to the presence of surface/interface stresses, it is necessary to establish new local boundary conditions. The key for solving the above problem is to find the analytic function $f(z)$ under new local boundary conditions. In Chapter 9 of this book, the elastic interaction between dislocations and nanostructures is studied, including the nano-inhomogeneity, nanowire and nanopore.

2.4 MULTISCALE SIMULATION

In order to reveal the underlying physical mechanisms, to facilitate the development of modeling, multiscale simulations are necessary for establishing the relationship between the components and properties of materials from atomic scale to macroscopic scale. These simulation methods include the first-principles calculation, molecular dynamics (MD) simulation, phase field method, and crystal plasticity finite element method (CPFEM). In this section, the details of various simulation methods are introduced.

2.4.1 First-Principles Calculation

The first-principles calculation is conducted using the quantum mechanics theory. Currently, the first-principles calculation based on density functional theory (DFT) is the most precise simulation method. By calculating the electron interaction in the atomic model, the ground-state properties of materials are calculated [41]. The electronic structure of the material is obtained by solving the quantum-mechanical Schrödinger equation, which describes the law of microscopic particles [42]. However, solving the Schrödinger equation is difficult, because the interaction terms between electrons and nucleus are hard to decouple. To solve this problem, an adiabatic approximation is proposed, whereby the mass of the nucleus is three orders of magnitude larger than the electron [43]. In this case, the motion of nucleus is negligible, and the state of atomic model is determined by the motion of the electrons. Therefore, the electronic structure of the material is captured just by solving the Schrödinger equation for electrons.

For a system containing multiple electrons, solving the Schrödinger equation is more difficult. Therefore, the single-electron approximation is proposed, which assumes that there is only one electron in the system, and the electron moves under the average potential field of all other electrons. Finally, the complex electron problem is simplified as a single electron problem and a mean-field approximation problem.

Based on the adiabatic approximation and the single-electron approximation, the first-principles calculation is conducted by solving the Schrödinger equation using self-consistent calculation:

$$-(\hbar^2/2m)\Delta^2\psi + V(r)\psi = E\psi \tag{2.39}$$

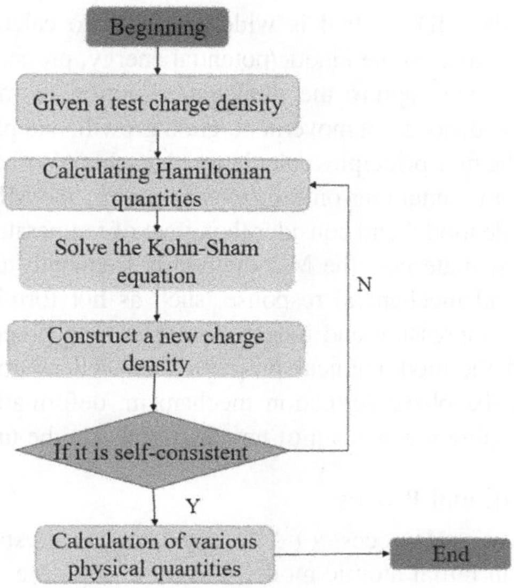

FIGURE 2.6 Schematic diagram of the self-consistent iterative process based on DFT.

where $\hbar = h/2\pi$, h is Planck's constant, m is the mass of the particle, ψ is the intrinsic wave function, $V(r)$ is the potential energy of the particle in the force field, E is the intrinsic energy, and Δ is the splitting operator.

The Hartree-Fork is a classical mean-field approximation, which treats the electrons as moving in the ion potential field and the mean potential field of other electrons [44]. However, this approximation neglects the change and correlation effects between electrons, which limits the accuracy of the calculation.

To solve the above problem, the DFT is introduced [45], which is based on the theorem that all ground-state properties are functionally related to the charge density [46,47]. Subsequent studies reveal that the multi-electron problem is replaced by a fully equivalent set of self-consistent single-electron equations, that is the Kohn-Sham equation [48]. Figure 2.6 shows the self-consistent iterative process based on DFT. In Chapters 3 and 7 of this book, the DFT method is used to study the mechanical properties of HEAs and light alloys. The key material parameters that determine the mechanical properties are analyzed, such as the lattice constant, elastic constant and the defect energy.

2.4.2 Molecular Dynamics Simulation

To overcome the drawbacks of first-principles calculation, that is being unable to simulate atomic motion and the scale limitations, a large-scale atomic simulation method, also known as MD simulation, is developed. In MD simulation, the law of atom motion is described by the classical Newtonian mechanics theory. By solving the equations of motion of the system through specific integration algorithms, the trajectories and states of the atoms are analyzed. The thermodynamic quantities of the system are calculated using statistical physics. By constructing an atomic model containing millions of atoms, the MD method is widely utilized to calculate the microscale thermodynamic state, such as the kinetic/potential energy, pressure, and temperature. More importantly, it can capture the dynamic evolution of microstructure at the nanoscale, such as the dislocation movement, grain growth, and phase transformation.

Compared with the first-principles calculation that can only obtain the ground-state properties of a system containing only dozens of atoms, the MD method is able to simulate a larger-scale model and consider the effect of temperature. When exploring the properties of new materials, the MD method is frequently used to simulate the formation process and mechanical response, such as hot forming and mechanical processing, tensile/compression and indentation/scratching processes. By analyzing the mechanical and thermodynamical properties as well as corresponding microstructure evolution, the phase formation mechanism, deformation mechanism and strengthening/toughening mechanism of new materials can be uncovered [49,50].

2.4.2.1 Computational Process

The typical computational process of MD simulation is illustrated in Figure 2.7. After constructing an initial atomic model, the static structure would be optimized to an energy-minimal state [51]. Generally, the energy minimization of the system is performed by iteratively adjusting atom coordinates. The iterations would be

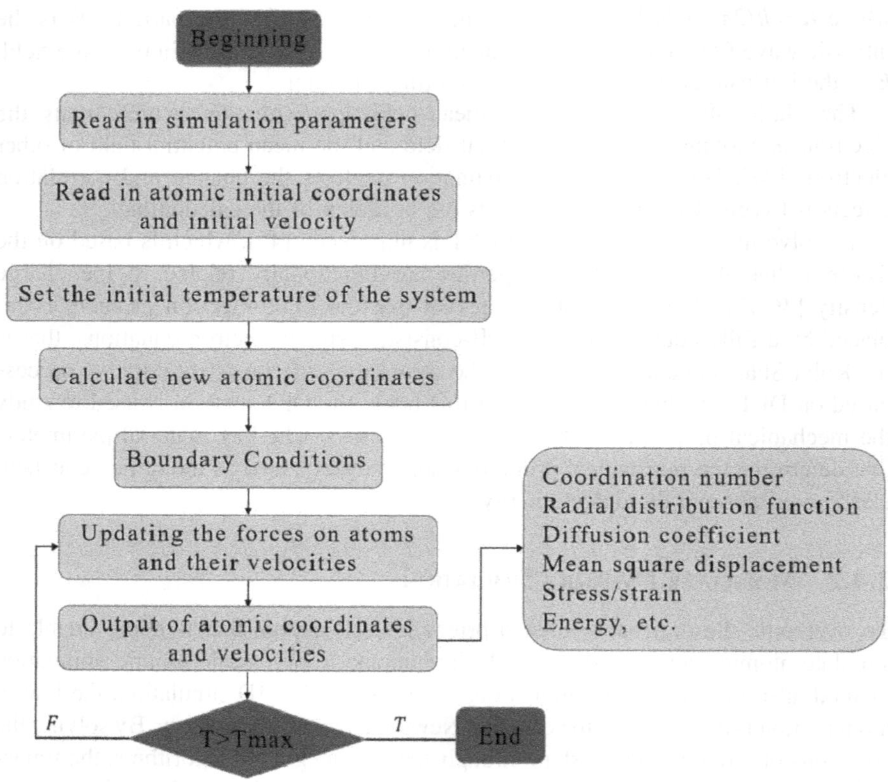

FIGURE 2.7 Flow chart of molecular dynamics simulation.

terminated when the preset stopping criteria is satisfied. At this time, the configuration will hopefully be in a local minimum of potential energy. More precisely, the configuration should approximate a critical point for the objective function:

$$E(r_1, r_2, \ldots, r_N) = \Sigma_{i,j} E_{\text{pair}}(r_i, r_j) + \Sigma_{ij} E_{\text{bond}}(r_i, r_j) + \Sigma_{ijk} E_{\text{angle}}(r_i, r_j, r_k) + \Sigma_{ijkl} E_{\text{dihedral}}(r_i, r_j, r_k, r_l) + \Sigma_{ijkl} E_{\text{improper}}(r_i, r_j, r_k, r_l) + \Sigma_i E_{\text{fix}}(r_i)n \tag{2.40}$$

where $E_{\text{pair}}(r_i, r_j)$ is the non-bonded pairwise interactions. The following fifth terms represent the bond, angle, dihedral, and improper interactions, respectively. is the potential energy due to the constraints or applied force to atoms.

After energy minimization, all atomic initial velocity is set according to the target temperature. Given the boundary and initial condition, the system enters a dynamic equilibrium state after sufficient simulation time, that is the relaxation process. The relaxed model is used to perform MD simulation under the designated mechanical and thermodynamical loading, in which the desired data is produced. The relaxation and production are conducted under the constraint of ensembles, which perform a time integration to update the position and velocity for atoms. Typical ensembles include the microcanonical ensemble (NVE), canonical ensemble (NVT), and constant-pressure, constant-temperature ensemble (NPT). The NVE ensemble restrains the atomic number N, total volume V and total energy E of the specific atomic group. In NVT ensemble, the atomic number N, total volume V and temperature T are constrained. For NPT ensemble, the atomic number N, pressure P and temperature T are constrained.

The accuracy of MD simulation depends on the function that is used to describe the interaction potential between atoms [52]. The commonly used potential functions include the Lennard-Jones (LJ) potential, Morse potential, Tersoff potential, embedded atom method (EAM) potential, and modified EAM (MEAM) potential. For metallic materials, EAM potential is most widely used [53,54], because of its balance between computational efficiency and accuracy. In EAM potential function, the basic idea is that the electron density of any point in space is the sum of electron density contributed by nearby atoms at this point, thus forming a background of electron density. The interaction between atoms and these electrons is equivalent to embedding it into the electronic background. The EAM potential function calculates the potential energy of each atom as the sum of two-body potential and multi-body potential. The total energy E_i of an atom i is given by:

$$E_i = F_\alpha\left(\sum_{i \neq j} \rho_\beta(r_{ij})\right) + \frac{1}{2}\sum_{i \neq j} \phi_{\alpha\beta}(r_{ij}) \tag{2.41}$$

where α and β are the element type of atom i and j, respectively. ϕ is the pair potential, which is the function of atomic type and atomic distance r_{ij}. F_α is the multi-body potential depending on the atomic type and electron density at atom i. ρ_β is the electron density at atom i contributed from the atom j, which relies on the element type of atom j and atomic distance r_{ij}.

When the simulation system involves multiple elements, all interatomic interaction is hard to be described just by one single potential function. In this case, it is necessary to combine multiple potential functions [55]. For example, the plastic deformation of the AlCrCuFeNi HEA is studied by combining the Morse potential function [55] and the EAM potentials of FeNiCr [56], Cu [57], and Al [58]. The simulation results are well consistent with experiments, indicating the feasibility of using joint potential functions. In Chapters 3 and 5 of this book, MD simulations are applied to investigate the deformation and strengthening mechanisms of HEAs and high-entropy amorphous.

2.4.3 Phase Field Simulation

The phase field method is a semi-phenomenological method based on classical thermodynamic and kinetic theories, which quantitatively studies the evolution of microstructures and the essential origin of the relationship between microstructures and their properties. In the most widely used phase field model, the conservative field variables (concentration field variable) and non-conservative field variables (structure order parameters) are used as the governing variables [59,60]. Currently, the Cahn-Hilliard [61] diffusion equation is used to depict the evolution of the concentration field, and the Ginzburg-Landau equation [62] are employed to derive the evolution of the structure field. So far, the phase field method has been used to simulate various physical phenomena [63–67], such as crystal formation [68–70], grain growth [71], dislocation motion [72], dislocation-solute interactions [73], crack propagation [74], solid-state phase transition [75], and multielement interdiffusion [76].

2.4.3.1 Kinetic equation

The phase field method is a technique employed to investigate diffusion interfaces. It is assumed that the phase interface is characterized by a certain thickness. The concepts of diffuse and sharp interfaces are illustrated in Figure 2.8. When the system undergoes a continuous phase transition, each state variable presents a state of the system [75]. A phase field variable η is defined as a spatially continuous function, which has a certain value in each phase, such as $\eta = 0$ represents phase A and $\eta = 1$ represents phase B. The variable η changes smoothly crossing the

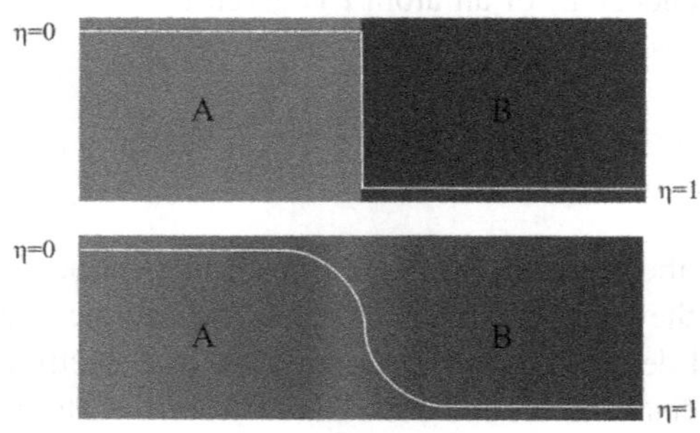

FIGURE 2.8 Conceptual illustration of a diffuse interface and a sharp interface, with η representing the phase field variable.

interface between two phases. Phase field variables are divided into two types: the conserved variables related to the local concentration, and the non-conserved variables related to the local microstructure.

In the phase field theory, the thermodynamics of the phase transition governs the characteristics of the diffusion interface. The microstructure evolution is governed by the minimum value of free energy of the system. For a phase interface model, the free energy usually consists of three parts: bulk free energy, interfacial or gradient energy, and elastic strain energy. The volume fraction of the equilibrium phase is determined by the bulk free energy, whereas the competition between interfacial energy and elastic energy governs the shape and volume fraction of the equilibrium phase [77]. In contrast to classical thermodynamics, where material properties remain uniform across the system, the Gibbs free energy in the phase field model relies on the phase field variables and their spatial gradients. Phase field variables or state variables specify material properties and characteristics, such as its crystal structure (η) or concentration (c). The mathematical expression of the Gibbs free energy involves the state variables η and c. The variable η is a non-conserved quantity, while c is a conserved quantity. In general, the phase field variable typically ranges between 0 and 1. For example, for the single-phase field model used to study solidification, the liquid and solid phase are usually represented by a constant phase field variable. When multiple phases coexist in a system, N_v phase field variables are required to represent N_v phases. For instance, for the phase transition from austenite to martensite, one austenite can transit to three different martensite variants. Therefore, three-phase field variables are required.

Considering a system involving two phases A and B, the phase field variable η is defined to represent the two phases, and it serves as an analogy to the crystal structure of the material. As mentioned above, the variable $\eta = 0$ for phase A and $\eta = 1$ for phase B. The value of η varies smoothly across the interface between the two phases. The Gibbs free energy of the system is calculated by:

$$G = \int_V \left[f_v(\eta) + \frac{1}{2}\beta(\nabla\eta)^2 \right] \mathrm{d}V \tag{2.42}$$

where $\frac{1}{2}\beta(\nabla\eta)^2$ represents the phase interface. $f_v(\eta)$ represents the sum of the bulk chemical energy density f^v and the elastic energy density f^{el}. The two minimum values in the f^v represent two stable phases, and f^{el} represents the effect of elastic stress. β denotes the thickness of the interface. The change in Gibbs free energy is calculated by:

$$\delta G = \int_V [f_v'(\eta)\delta\eta + \beta\nabla\eta\delta(\nabla\eta)]\mathrm{d}V \tag{2.43}$$

Assuming the integral on the boundary is equal to 0, the following equation is derived:

$$\frac{\delta G}{\delta\eta} = \int_V (f_v'(\eta) + \beta\nabla^2\eta)\mathrm{d}V \tag{2.44}$$

If N_v phase field variables are used to represent N_v different phases, the time evolution of each variable in the phase field is determined by solving the following equation:

$$\frac{\partial \eta_p}{\partial t} = -L_{pq} \frac{\delta G}{\delta \eta_p} \tag{2.45}$$

Likewise, the temporal variation of the phase field variable for the conserved variable c is determined by solving the following equation:

$$\frac{\partial c_i}{\partial t} = \nabla^2 M_{ij} \nabla \frac{\delta G}{\delta c_i} \tag{2.46}$$

where L_{pq} and M_{ij} are associated with the mobility of the phase interface. Eq. 2.45 is the Ginzburg-Landau equation used to describe the evolution of non-conserved quantities, while Eq. 2.46 is the Cahn-Hilliard equation used to describe the evolution of conserved quantities.

2.4.3.2 Solution

There are three main methods for solving phase field equations, namely, finite difference method, finite element method, and finite volume method.

The key to the finite difference method is to divide the solution domain into finite grids, and then use Taylor series expansion to discretize the derivatives of the governing equations and boundary condition functions. This method is suitable for the simple solution domain, but it is difficult to solve a problem with complex boundary conditions.

The fundamental concept of the finite volume method is to divide the solution domain into finite volume elements, and integrate the solution equation for each volume element to obtain a series of discrete equations. Then define the interpolation function between volume elements to solve the discrete equations. This method is generally applicable to fluid motion with complex boundary conditions [78].

For the finite element method, the continuous solution domain is discretized into a finite number of elements. An approximate function for each element is used to replace the unknown function to be solved on the solution domain. The approximation function is usually expressed by the interpolation function of the unknown function and its derivative at each node of the element. In this way, a continuous problem with infinite degree of freedom is turned into a discrete problem with finite degree of freedom. In this book, phase field simulation is used in Chapter 7, to investigate the transverse propagation of deformation twinning for hexagonal-closed packed crystals.

2.4.4 Crystal Plasticity Finite Element Simulation

The CPFEM is an important tool for bridging the mesoscale dislocation dynamic and macroscopic mechanical response [79–81]. Traditional finite element simulation generally uses the isotropic material model, which does not consider the fact that the plastic deformation of crystalline material is driven by the dislocation slip on discrete slip systems. Accordingly, the CPFEM is developed by considering the slip systems in a crystalline material [82]. It can describe the microscale shear behavior, and simulate the anisotropic plastic deformation under complex boundary conditions. Current CPFEM is able to take into account all slip systems in crystalline materials [82–84]. Based on CPFEM, the deformation mechanism of various materials with complex structure are revealed [85–90].

The essence of CPFE theory is to relate the deformation and microscopic shear using the constitutive equations based on the continuous media mechanics. The constitutive equation is phenomenological or physics-based. 1) The phenomenological constitutive equations assume that the plastic deformation is only dominated by dislocation slip. Therefore, the CPFE model based on phenomenological constitutive equation is unavailable when other deformation mechanisms, such as the twinning and phase transformation, need to be considered. Besides, for complex problems such as the small-scale deformation and the model with interface effect, the phenomenological constitutive model will no longer be applicable. 2) The CPFE model based on physics-based constitutive equation overcomes these limitations. The physics-based model relies on internal variables [91,92]. Because the plastic deformation is mainly carried by the dislocation, the dislocation density is the most important internal variable. However, the volume fraction and nucleation rate of the twin and phase are also used to describe the plasticity contributed from twining and phase transformation. In Chapter 4 of this book, a CPFE model is developed to study the mechanical behavior under uniaxial tensile loading for single crystal HEA.

2.4.4.1 Flow Kinematics

In this section, the different deformation paths that form the CPFE geometrical framework are discussed. Firstly, the finite deformation measures are considered for the small deformation problems. The relative position of two adjacent material points in the reference configuration is represented by d$\mathbf{x}$. For the deformed structure, the initial vector d$\mathbf{x}$ is mapped to the reference configuration, d$\mathbf{y}$ = d$\mathbf{x}$ + d$\mathbf{u}$, where d$\mathbf{u}$ represents the differential of total displacement vector $\mathbf{u}$. The vector before and after deformation is determined by the total deformation gradient $\mathbf{F}$:

$$\mathrm{dy} = \left(\frac{\partial \mathrm{y}}{\partial \mathrm{x}}\right)\mathrm{dx} = \left(\mathbf{I} + \frac{\partial \mathrm{u}}{\partial \mathrm{x}}\right)\mathrm{dx} = \mathbf{F}\mathrm{dx} \tag{2.47}$$

In the above equation, the tensor $\mathbf{I}$ represents the second-order identical tensor. The second-order tensor formed by the partial derivatives of $\mathbf{u}$ with respect to x is the total deformation tensor. The deformation gradient is decomposed by introducing a deformed intermediate state, which is given by:

$$\mathbf{F} = \mathbf{F}_e\mathbf{F}_p \tag{2.48}$$

As mentioned above, the total deformation is divided into two parts: elastic deformation and plastic deformation. $\mathbf{F_e}$ and $\mathbf{F_p}$ are the elastic and plastic

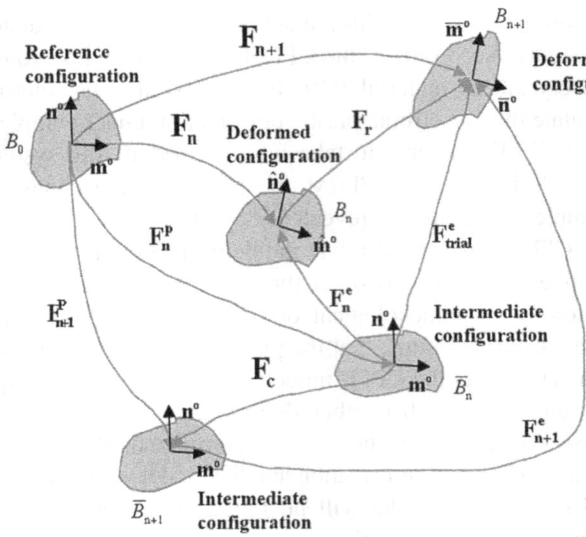

FIGURE 2.9 Various configurations of the body during deformation based on crystal plasticity theory [93].

deformation gradients, respectively. The decomposition process of deformation is shown in Figure 2.9 [93]. A velocity gradient tensor **L** is introduced to describe the change rate of the deformation gradient with respect to time:

$$\mathbf{L} = \nabla \otimes v = \dot{\mathbf{F}}\mathbf{F}^{-1} \tag{2.49}$$

The velocity gradient tensor **L** is further divided into the elastic part $\mathbf{L_e}$ and plastic part $\mathbf{L_p}$:

$$\mathbf{L} = \mathbf{L}_e + \mathbf{L}_p \tag{2.50}$$

where $\mathbf{L_e} = \dot{\mathbf{F}}_e\mathbf{F}_e^{-1}$ is the elastic velocity gradient, $\mathbf{L_p} = \dot{\mathbf{F}}_p\mathbf{F}_p^{-1}$ is the plastic velocity gradient. The plastic velocity gradient tensor $\mathbf{L_p}$ is determined by the shear strain contributed by different plastic deformation mechanisms.Five polygons are connected to each other. The slip system of polygon B0 in the upper left corner is the Reference configuration. Starting from B0, the arrows connect the other four polygons (Bn/Bn+1 and), which denote deformed configuration and intermediate configuration, respectively.

2.4.4.2 Constitutive models

2.4.4.2.1 Phenomenological Constitutive Model

Consider the case where the plastic deformation is controlled by the dislocation activity, and the contributions of other deformation mechanisms are neglected. The relation of the macroscopic velocity gradient $\mathbf{L_p}$ and microscale deformation γ^α is expressed as:

$$\mathbf{L_p} = \dot{\mathbf{F}}_p\mathbf{F}_p^{-1} = \sum_{\alpha=1}^{N_{slip}} \dot{\gamma}^\alpha \mathbf{P}^\alpha \tag{2.51}$$

where $\dot{\gamma}^\alpha$ is the shear plastic deformation rate of slip system α, N_{slip} is the quantity of slip systems, and $\mathbf{P}^\alpha$ is the Schmid tensor on the corresponding slip system, which is represented as:

$$\mathbf{P}^\alpha = \mathbf{m}_0^\alpha \otimes \mathbf{n}_0^\alpha \tag{2.52}$$

where $\mathbf{m}_0^\alpha$ and $\mathbf{n}_0^\alpha$ represent the unit vectors of slip direction and normal direction of the slip plane, respectively. In crystal plasticity theory, it is always assumed that the plastic deformation of metals is controlled by the dislocation slip in a specified slip system. For the rate-independent model, the yield surface f^α is defined for each slip system:

$$f^\alpha = |\tau^\alpha| - s^\alpha \tag{2.53}$$

where s^α is the slip resistance on the slip system α. Subsequently, based on the yield plane and hardening model, the shear plastic deformation rate $\dot{\gamma}^\alpha$ is obtained, which describes the evolution of slip resistance.

The resolved shear stress τ on slip system α is written as:

$$\tau^\alpha = (\mathbf{F}_e^T\mathbf{F}_e\mathbf{T})\cdot\mathbf{P}^\alpha \tag{2.54}$$

According to the power-law form of plastic deformation criterion, the shear strain rate $\dot{\gamma}^\alpha$ on the slip system α is defined as follows:

$$\dot{\gamma}^\alpha = \dot{\gamma}_0 \left| \frac{\tau^\alpha - \tau_b}{\tau_{sr}^\alpha} \right|^{1/m} sign(\tau^\alpha - \tau_b) \tag{2.55}$$

where $\dot{\gamma}_0$ is the reference shear rate, τ_b is back stress, τ^α is the shear stress applied on slip, τ_{sr}^α (> 0) is the critical resolved shear stress (CRSS) representing the state of the slip plane, and m is the strain rate sensitivity coefficient. Based on commonly used hardening models, $\dot{\tau}_{sr}^\alpha$ is expressed as:

$$\dot{\tau}_{sr}^\alpha = \sum_\beta |\dot{\gamma}^\beta| h^{\alpha\beta} \tag{2.56}$$

where $h^{\alpha\beta} = [q_0 + (1 - q_0)\delta^{\alpha\beta}]h^\beta$ is the hardening matrix influenced by potential hardening and self-hardening in the slip system. $\delta^{\alpha\beta}$ is the Kronecker delta. q_0 represents the potential hardening parameter. h^β is the self-hardening parameter, which is given by:

$$h^{\beta} = h_0 \left| 1 - \frac{\tau_{sr}^{\beta}}{s_s^{\beta}} \right|^{r^{\beta}} \tag{2.57}$$

where h_0 is the initial hardening parameter, r^{β} is the hardening index for the slip system β, and s_s^{β} is the slip resistance at the hardening saturation on the slip system β. In the phenomenological constitutive model, the state of material deformation is only determined by the CRSS τ_{sr}^{α} rather than the practice lattice defects [94,95]. The issue pushes the development of physics-based constitutive model with the consideration of microscale mechanisms.

2.4.4.2.2 Physics-Based Constitutive Model

Compared with the phenomenological constitutive model, the physics-based constitutive model is more dependent on the internal variable. According to the method of Kalidindi [96], the effect of dislocation density, deformation twinning, and phase transformation are jointly considered as the dominant factors of plastic velocity gradient [97]:

$$\begin{aligned}\mathbf{L_p} &= \left(1 - \sum_{\beta=1}^{N_{tw}} f^{\beta} - \sum_{\chi=1}^{N_{tr}} f^{\chi}\right) \sum_{\alpha=1}^{N_s} \dot{\gamma}^{\alpha} \mathbf{m}^{\alpha} \otimes \mathbf{n}^{\alpha} \\ &+ \sum_{\beta=1}^{N_{tw}} \dot{\gamma}^{\beta} \mathbf{m}_{tw}^{\beta} \otimes \mathbf{n}_{tw}^{\beta} + \sum_{\chi=1}^{N_{tr}} \dot{\gamma}^{\chi} \mathbf{m}_{tr}^{\chi} \otimes \mathbf{n}_{tr}^{\chi}\end{aligned} \tag{2.58}$$

where f^{β} is the volume fractions of twins in twin system β, and f^{χ} is the volume fractions of the new phases in phase transformation system χ. According to the dislocation-based model, the shear rate on the slip system α is described by the Orowan equation:

$$\dot{\gamma}^{\alpha} = \rho_e b_s v_0 \exp\left[-\frac{Q_s}{k_B T}\left\{1 - \left(\frac{|\tau_{eff}^{\alpha}|}{\tau_{sol}}\right)^{p}\right\}^{q}\right] sign(\tau^{\alpha}) \tag{2.59}$$

where τ_{eff}^{α} is the effective RSS on the slip system α, b_s is the magnitude of the Burgers vector, τ_{sol} denotes the solid solution strengthening, v_0 is the dislocation slip velocity, k_B is the Boltzmann constant, T is the temperature, Q_s is the activation energy of dislocation slip, and p and q are the fitting parameters. According to the relationship between the RSS and the passing stress τ_{pass}^{α}, the value of effective RSS can be determined. The effective shear stress τ_{eff}^{α} is computed using the following equations:

$$\tau_{\text{eff}} = \begin{cases} |\tau| - \tau_{\text{pass}} & \text{for} \quad |\tau| > \tau_{\text{pass}} \\ 0 & \text{for} \quad |\tau| \leqslant \tau_{\text{pass}} \end{cases} \tag{2.60}$$

The passing stress τ_{pass}^{α} is given by:

$$\tau_{pass}^{\alpha} = Gb\left(\sum_{\alpha'=1}^{N_s} \xi_{\alpha\alpha'}\left(\rho_e^{\alpha'} + \rho_d^{\alpha'}\right)\right)^{1/2} \tag{2.61}$$

where $\xi_{\alpha\alpha'}$ is interaction matrix between slip systems α and α', G is the shear modulus. $\rho_e^{\alpha'}$ and $\rho_d^{\alpha'}$ represent the density of edge dislocation and dislocation dipole, respectively. Their values are related to the average free path and the climbing speed of dislocations.

The twinning nucleation rate of twin system β is calculated as follows:

$$\dot{N}_{tw}^{\beta} = \dot{N}_0\left(1 - \exp\left[-\frac{V_{cs}}{k_B T}(\tau_r - \tau^{\beta})\right]\right)\exp\left[-\left(\frac{\hat{\tau}_{tw}}{\tau^{\beta}}\right)^{A}\right] \tag{2.62}$$

where $\dot{N}_0$ is the quantity density of potential twin nuclei per unit time, V_{cs} is the cross-slip activation volume, A is the fitting parameter that reflects the sharpness of the transition between the non-twinning to the twinning state. τ_r is the stress required to bring the two parts to within the critical distance to form the twin nucleus, τ^{β} is the resolved shear stress of twin system β, $\hat{\tau}_{tw}$ is the critical shear stress for twinning. The evolution of twin volume fraction is given by:

$$\dot{f}^{\beta} = \left(1 - \sum_{\beta=1}^{N_{tw}} f^{\beta} - \sum_{\chi=1}^{N_{tr}} f^{\chi}\right) V^{\beta}\dot{N}_{tw}^{\beta} \tag{2.63}$$

where $V^{\beta} = \frac{\pi}{4}\Lambda_{tw}^2 t_{tw}$ is the volume of the new twin lath. t_{tw} is the average thickness of twins, and Λ_{tw} is the mean free paths affected by deformation twins.

In addition, the effect of phase transformation on plastic deformation is related to the nucleation rate of new phases. Here, taking the ε-martensite transformation as an example, the nucleation rate is calculated by:

$$\dot{N}_{tr}^{\chi} = \dot{N}_0\left(1 - \exp\left[-\frac{V_{cs}}{k_B T}(\tau_r - \tau^{\chi})\right]\right)\exp\left[-\left(\frac{\widehat{\tau_{tr}}}{\tau^{\chi}}\right)^{B}\right] \tag{2.64}$$

Assuming that the probability of twin nucleus and martensite nucleus produced by the dislocation reaction is the same, the initial nucleation rate $\dot{N}_0$ is given by Eq. 2.64. $\hat{\tau}_{tr}$ and τ^{χ} are the CSS and RSS, respectively.

The evolution of ε-martensite volume fraction is calculated by:

$$\dot{f}^{\chi} = \left(1 - \sum_{\beta=1}^{N_{tw}} f^{\beta} - \sum_{\chi=1}^{N_{tr}} f^{\chi}\right) V^{\chi}\dot{N}_{tr}^{\chi} \tag{2.65}$$

where $V^{\chi} = \frac{\pi}{4}\Lambda_{tr}^2 t_{tr}$ is the volume of the new ε-martensite lath. t_{tr} is the average thickness of the new phase, and Λ_{tr} is the mean free paths affected by phase transformation.

2.5 CONCLUSION

The above modeling and simulation methods are applicable to the investigation of macroscopic mechanical properties and microscopic deformation mechanism. The models for the prediction of material properties are established based on the characteristics of microstructures. The multi-scale simulation methods are capable of uncovering the underlying physical mechanism of mechanical responses. After years of development, these calculation methods have been successfully applied to solve various mechanical problems in traditional solid materials. However, since advanced materials have new microstructures and exhibit unique performance, the computational methods need to be further developed. In the following chapters, we provide a detailed introduction to some newly developed models, and demonstrate some application cases of computational methods in different advanced materials.

REFERENCES

[1] Estrin Y. and Mecking H. 1984. A unified phenomenological description of work hardening and creep based on one-parameter models. *Acta Metallurgica*, 32(1): 57–70.
[2] Estrin Y., Toth L.S., Molinari A., et al. 1998. A dislocation-based model for all hardening stages in large strain deformation. *Acta Materialia*, 46(15): 5509–5522.
[3] Kocks U. and Mecking H. 2003. Physics and phenomenology of strain hardening: the FCC case. *Progress in Materials Science*, 48(3): 171–273.
[4] Li J., Lu W., Chen S., et al. 2020. Revealing extra strengthening and strain hardening in heterogeneous two-phase nanostructures. *International Journal of Plasticity*, 126: 102626.
[5] Li J. and Soh A. 2012. Modeling of the plastic deformation of nanostructured materials with grain size gradient. *International Journal of Plasticity*, 39: 88–102.
[6] Estrin Y., Arndt S., Heilmaier M., et al. 1999. Deformation behaviour of particle-strengthened alloys: a Voronoi mesh approach. *Acta Materialia*, 47(2): 595–606.
[7] Estrin Y. and Kubin L.P. 1989. Collective dislocation behaviour in dilute alloys and the Portevin—Le Chatelier effect. *Journal of the Mechanical Behavior of Materials*, 2(3-4): 255–292.
[8] Lee C., Maresca F., Feng R., et al. 2021. Strength can be controlled by edge dislocations in refractory high-entropy alloys. *Nature Communications*, 12(1): 5474.
[9] Ma E. and Wu X. 2019. Tailoring heterogeneities in high-entropy alloys to promote strength–ductility synergy. *Nature Communications*, 10(1): 5623.
[10] Labusch R. 1970. A statistical theory of solid solution hardening. *Physica Status Solidi (b)*, 41(2): 659–669.
[11] Maresca F. and Curtin W.A. 2020. Mechanistic origin of high strength in refractory BCC high entropy alloys up to 1900K. *Acta Materialia*, 182: 235–249.
[12] Fleischer R.L. 1963. Substitutional solution hardening. *Acta Metallurgica*, 11(3): 203–209.
[13] Senkov O.N., Scott J.M., Senkova S.V., et al. 2011. Microstructure and room temperature properties of a high-entropy TaNbHfZrTi alloy. *Journal of Alloys and Compounds*, 509(20): 6043–6048.

[14] Yao H.W., Qiao J.W., Hawk J.A., et al. 2017. Mechanical properties of refractory high-entropy alloys: experiments and modeling. *Journal of Alloys and Compounds*, 696: 1139–1150.

[15] Lee C., Song G., Gao M.C., et al. 2018. Lattice distortion in a strong and ductile refractory high-entropy alloy. *Acta Materialia*, 160: 158–172.

[16] Li L., Fang Q., Li J., et al. 2020. Lattice-distortion dependent yield strength in high entropy alloys. *Materials Science and Engineering: A*, 784: 139323.

[17] Toda-Caraballo I. and Rivera-Díaz-del-Castillo P.E. 2015. Modelling solid solution hardening in high entropy alloys. *Acta Materialia*, 85: 14–23.

[18] Vegard L. 1916. VI. Results of crystal analysis. *The London, Edinburgh, and Dublin Philosophical Magazine and Journal of Science*, 32(187): 65–96.

[19] Toda-Caraballo I. and Rivera-Díaz-del-Castillo P.E.J. 2015. Modelling solid solution hardening in high entropy alloys. *Acta Materialia*, 85: 14–23.

[20] Franciosi P., Berveiller M. and Zaoui A. 1980. Latent hardening in copper and aluminium single crystals. *Acta Metallurgica*, 28(3): 273–283.

[21] Taylor G.I. 1934. The mechanism of plastic deformation of crystals. Part I—Theoretical. Proceedings of the Royal Society of London. *Series A, Containing Papers of a Mathematical and Physical Character*, 145(855): 362–387.

[22] Lavrentev F.F. 1980. The type of dislocation interaction as the factor determining work hardening. *Materials Science and Engineering*, 46(2): 191–208.

[23] Courtney T.H. 2005. *Mechanical behavior of materials*, Waveland Press.

[24] Hall E.O. 1951. The deformation and ageing of mild steel: III discussion of results. *Proceedings of the Physical Society. Section B*, 64(9): 747.

[25] Petch N.J. 1953. The cleavage strength of polycrystals. *Journal of Iron & Steel Inst*itute, 174: 25–28.

[26] Gladman T. 1999. Precipitation hardening in metals. *Materials Science and Technology*, 15(1): 30–36.

[27] Liu G., Zhang G.J., Wang R.H., et al. 2007. Heat treatment-modulated coupling effect of multi-scale second-phase particles on the ductile fracture of aged aluminum alloys. *Acta Materialia*, 55(1): 273–284.

[28] Zhao Y., Guan K., Yang Z., et al. 2020. The effect of subsequent heat treatment on the evolution behavior of second phase particles and mechanical properties of the Inconel 718 superalloy manufactured by selective laser melting. *Materials Science and Engineering: A*, 794: 139931.

[29] Liu J., Gong Q., Chen H., et al. 2022. Strengthening caused by precipitates with different morphologies in Cu–Cr–Ti alloys: The role of dislocation bending angle. *Materials Science and Engineering: A*, 840: 142927.

[30] Reppich B. 1993. Particle Strengthening, Chapter 7. *Materials Science and Technology*, 311–357.

[31] Hüther W. and Reppich B. 1978. Interaction of dislocations with coherent, stress-free, ordered particles. *International Journal of Materials Research*, 69(10): 628–634.

[32] Sun K., Huang P. and Wang F. 2022. The bimodal nanocoherent precipitates leads to superior strength-ductility synergy in a novel CoCrNi-based medium entropy alloy. *Journal of alloys and compounds*, 909: 164809.

[33] Collins D.M. and Stone H.J. 2014. A modelling approach to yield strength optimisation in a nickel-base superalloy. *International Journal of Plasticity*, 54: 96–112.

[34] Li Q., Wang Y., Li H., et al. 2021. Stress and its influencing factors in positive particles of lithium-ion battery during charging. *International Journal of Energy Research*, 45(3): 3913–3928.

[35] Bhandakkar T.K. and Gao H. 2010. Cohesive modeling of crack nucleation under diffusion induced stresses in a thin strip: implications on the critical size for flaw tolerant battery electrodes. *International Journal of Solids and Structures*, 47(10): 1424–1434.

[36] DeLuca C.M., Maute K. and Dunn M.L. 2011. Effects of electrode particle morphology on stress generation in silicon during lithium insertion. *Journal of Power Sources*, 196(22): 9672–9681.

[37] Deshpande R., Cheng Y.-T. and Verbrugge M.W. 2010. Modeling diffusion-induced stress in nanowire electrode structures. *Journal of Power Sources*, 195(15): 5081–5088.

[38] Prussin S. 1961. Generation and distribution of dislocations by solute diffusion. *Journal of Applied Physics*, 32(10): 1876–1881.

[39] Estrin Y. 1998. Dislocation theory based constitutive modelling: foundations and applications. *Journal of Materials Processing Technology*, 80: 33–39.

[40] Clerici D., Mocera F. and Somà A. 2020. Analytical solution for coupled diffusion induced stress model for lithium-ion battery. *Energies*, 13(7): 1717.

[41] Liu Y., Ren H., Hu W.-C., et al. 2016. First-principles calculations of strengthening compounds in magnesium alloy: a general review. *Journal of Materials Science & Technology*, 32(12): 1222–1231.

[42] Filippov A.E. 1996. Nonlinear nonlocal Schrödinger equation in the context of quantum mechanics. *Physics Letters A*, 215(1-2): 32–39.

[43] Kuznetsov A.M. and Medvedev I.G. 2006. Does really Born–Oppenheimer approximation break down in charge transfer processes? An exactly solvable model. *Chemical physics*, 324(1): 148–159.

[44] Chaikin P.M., Lubensky T.C. and Witten T.A. 1995. *Principles of condensed matter physics* Vol. 10, Cambridge University Press, Cambridge.

[45] Hohenberg P. and Kohn W. 1964. Density functional theory (DFT). *Physics Reviews*, 136(1964): B864.

[46] Dreizler R.M. and Gross E.K.U. 2012. *Density functional theory: an approach to the quantum many-body problem*, Springer Science & Business Media.

[47] Robert G. and Parr G. 1989. *Density-functional theory of atoms and molecules*, Oxford University Press, 1.

[48] Kohn W. and Sham L.J. 1965. Self-consistent equations including exchange and correlation effects. *Physical review*, 140(4A): A1133.

[49] Fang Q., Chen Y., Li J., et al. 2019. Probing the phase transformation and dislocation evolution in dual-phase high-entropy alloys. *International Journal of Plasticity*, 114: 161–173.

[50] Li J., Chen H., Li S., et al. 2019. Tuning the mechanical behavior of high-entropy alloys via controlling cooling rates. *Materials Science and Engineering: A*, 760: 359–365.

[51] Smith W.R. and Qi W. 2018. Molecular simulation of chemical reaction equilibrium by computationally efficient free energy minimization. *ACS Central Science*, 4(9): 1185–1193.

[52] Daw M.S. and Baskes M.I. 1984. Embedded-atom method: derivation and application to impurities, surfaces, and other defects in metals. *Physical Review B*, 29(12): 6443.

[53] Zhou X.W., Wadley H.N.G., Johnson R.A., et al. 2001. Atomic scale structure of sputtered metal multilayers. *Acta Materialia*, 49(19): 4005–4015.

[54] Farkas D. and Caro A. 2018. Model interatomic potentials and lattice strain in a high-entropy alloy. *Journal of Materials Research*, 33(19): 3218–3225.

[55] Wang Z., Li J., Fang Q., et al. 2017. Investigation into nanoscratching mechanical response of AlCrCuFeNi high-entropy alloys using atomic simulations. *Applied Surface Science*, 416: 470–481.

[56] Bonny G., Castin N. and Terentyev D. 2013. Interatomic potential for studying ageing under irradiation in stainless steels: the FeNiCr model alloy. *Modelling and Simulation in Materials Science and Engineering*, 21(8): 085004.

[57] Mishin Y., Mehl M.J., Papaconstantopoulos D.A., et al. 2001. Structural stability and lattice defects in copper: Ab initio, tight-binding, and embedded-atom calculations. *Physical Review B*, 63(22): 224106.
[58] Winey J.M., Kubota A. and Gupta Y.M. 2009. A thermodynamic approach to determine accurate potentials for molecular dynamics simulations: thermoelastic response of aluminum. *Modelling and Simulation in Materials Science and Engineering*, 17(5): 055004
[59] Landau L.D., Abrikosov A. and Halatnikov L. 1956. On the quantum theory of fields. *Il Nuovo Cimento*, 3(1 Suppl.): 80–104.
[60] Cahn J.W. and Hilliard J.E. 1958. Free energy of a nonuniform system. I. Interfacial free energy. *The Journal of Chemical Physics*, 28(2): 258–267.
[61] Cahn J.W. 1962. On spinodal decomposition in cubic crystals. *Acta metallurgica*, 10(3): 179–183.
[62] Shen J. and Yang X. 2010. Numerical approximations of Allen-Cahn and Cahn-Hilliard equations. *Discrete Continuous Dynamic System*, 28(4): 1669–1691.
[63] Boettinger W.J., Warren J.A., Beckermann C., et al. 2002. Phase-field simulation of solidification. *Annual Review of Materials Research*, 32(1): 163–194.
[64] Mamivand M., Zaeem M.A. and El Kadiri H. 2013. A review on phase field modeling of martensitic phase transformation. *Computational Materials Science*, 77: 304–311.
[65] Qin R. and Bhadeshia H. 2010. Phase field method. *Materials Science and Technology*, 26(7): 803–811.
[66] Artemev A., Wang Y. and Khachaturyan A.G. 2000. Three-dimensional phase field model and simulation of martensitic transformation in multilayer systems under applied stresses. *Acta Materialia*, 48(10): 2503–2518.
[67] Levitas V.I. and Preston D.L. 2002. Three-dimensional Landau theory for multivariant stress-induced martensitic phase transformations. I. Austenite↔ martensite. *Physical Review B*, 66(13): 134206.
[68] Diepers H.J., Beckermann C. and Steinbach I. 1999. Simulation of convection and ripening in a binary alloy mush using the phase-field method. *Acta Materialia*, 47(13): 3663–3678.
[69] Loginova I., Amberg G. and Ågren J. 2001. Phase-field simulations of non-isothermal binary alloy solidification. *Acta Materialia*, 49(4): 573–581.
[70] Tong X., Beckermann C., Karma A., et al. 2001. Phase-field simulations of dendritic crystal growth in a forced flow. *Physical Review E*, 63(6): 061601.
[71] Fan D., Chen L.Q. and Chen S.P.P. 1998. Numerical simulation of Zener pinning with growing second-phase particles. *Journal of the American Ceramic Society*, 81(3): 526–532.
[72] Pi Z.P., Fang Q.H., Jiang C., et al. 2017. Stress dependence of the dislocation core structure and loop nucleation for face-centered-cubic metals. *Acta Materialia*, 131: 380–390.
[73] Takahashi A., Suzuki T., Nomoto A., et al. 2018. Influence of spinodal decomposition structures on the strength of Fe-Cr alloys: A dislocation dynamics study. *Acta Materialia*, 146: 160–170.
[74] Spatschek R., Brener E. and Karma A. 2011. Phase field modeling of crack propagation. *Philosophical Magazine*, 91(1): 75–95.
[75] Chen L.-Q. and Khachaturyan A.G. 1991. Computer simulation of structural transformations during precipitation of an ordered intermetallic phase. *Acta Metallurgica et Materialia*, 39(11): 2533–2551.
[76] Chen L.-Q. 2002. Phase-field models for microstructure evolution. *Annual Review of Materials Research*, 32(1): 113–140.
[77] Moelans N., Blanpain B. and Wollants P. 2008. Quantitative phase-field approach for simulating grain growth in anisotropic systems with arbitrary inclination and misorientation dependence. *Physical Review Letters*, 101(2): 025502.

[78] Eymard R., Gallouët T. and Herbin R. 2000. Finite volume methods. *Handbook of numerical analysis*, 7: 713–1018.
[79] Courant R. 1943. Variational methods for the solution of problems of equilibrium and vibrations.
[80] Zienkiewicz O.C. 1967. The finite element method in structural and continum mechanics: numerical solution of problems in structural and continuum mechanics. (No Title).
[81] Zienkiewicz O.C. and Taylor R.L. 2005. *The finite element method for solid and structural mechanics*, Elsevier.
[82] Peirce D., Asaro R.J. and Needleman A. 1982. An analysis of nonuniform and localized deformation in ductile single crystals. *Acta Metallurgica*, 30(6): 1087–1119.
[83] Becker R. 1991. Analysis of texture evolution in channel die compression—I. Effects of grain interaction. *Acta Metallurgica et Materialia*, 39(6): 1211–1230.
[84] Becker R., Butler J.F., Hu H., et al. 1991. Analysis of an aluminum single crystal with unstable initial orientation (001)[110] in channel die compression. *Metallurgical Transactions A*, 22: 45–58.
[85] Harren S.V. and Asaro R.J. 1989. Nonuniform deformations in polycrystals and aspects of the validity of the Taylor model. *Journal of the Mechanics and Physics of Solids*, 37(2): 191–232.
[86] Harren S.V., Deve H.E. and Asaro R.J. 1988. Shear band formation in plane strain compression. *Acta Metallurgica*, 36(9): 2435–2480.
[87] Eisenlohr P. and Roters F. 2008. Selecting a set of discrete orientations for accurate texture reconstruction. *Computational Materials Science*, 42(4): 670–678.
[88] Melchior M.A. and Delannay L. 2006. A texture discretization technique adapted to polycrystalline aggregates with non-uniform grain size. *Computational Materials Science*, 37(4): 557–564.
[89] Raabe D. and Roters F. 2004. Using texture components in crystal plasticity finite element simulations. *International Journal of Plasticity*, 20(3): 339–361.
[90] Zisu Z., Roters F., Raabe D., et al. 2001. Introduction of a texture component crystal plasticity finite element method for anisotropy simulations. *Advanced Engineering Materials*, 3: 984–989
[91] Bieler T.R., Eisenlohr P., Roters F., et al. 2009. The role of heterogeneous deformation on damage nucleation at grain boundaries in single phase metals. *International Journal of Plasticity*, 25(9): 1655–1683.
[92] Ma A., Roters F. and Raabe D. 2006. On the consideration of interactions between dislocations and grain boundaries in crystal plasticity finite element modeling–theory, experiments, and simulations. *Acta materialia*, 54(8): 2181–2194.
[93] Yaghoobi M., Ganesan S., Sundar S., et al. 2019. PRISMS-Plasticity: An open-source crystal plasticity finite element software. *Computational Materials Science*, 169: 109078.
[94] Ganesan S., Yaghoobi M., Githens A., et al. 2021. The effects of heat treatment on the response of WE43 Mg alloy: crystal plasticity finite element simulation and SEM-DIC experiment. *International Journal of Plasticity*, 137: 102917.
[95] Mecking H. and Kocks U.F. 1981. Kinetics of flow and strain-hardening. *Acta Metallurgica*, 29(11): 1865–1875.
[96] Kalidindi S.R. 1998. Incorporation of deformation twinning in crystal plasticity models. *Journal of the Mechanics and Physics of Solids*, 46(2): 267–290.
[97] Wong S.L., Madivala M., Prahl U., et al. 2016. A crystal plasticity model for twinning-and transformation-induced plasticity. *Acta Materialia*, 118: 140–151.

3 High-Entropy Alloys

Simulation

3.1 INTRODUCTION

High-entropy alloys (HEAs) with multiprincipal elements exhibit exceptional properties in sharp contrast to conventional alloys [1,2], such as high hardness, high strength, and high irradiation resistance [3–8]. As a thriving advanced material, the understanding of deformation behavior is limited through experiments [3–8]. Driven by this issue, some simulation methods are widely adopted to further derive our cognition in HEAs. In this chapter, the deformation and strengthening mechanism and service performance of HEAs are reviewed, and the future research in other fields is evaluated.

3.2 DEFORMATION MECHANISM

The deformation behaviors of HEAs are dominated by the evolution of the main microscopic defects, such as dislocation, twin, and phase. The atomic simulations are usually used to investigate their deformation processes based on the microstructure evolution.

3.2.1 Dislocation Evolution

Generally speaking, the dislocation nucleation and slip are the intrinsic deformation behaviors of HEAs [9–11]. However, only a few studies on the deformation of HEAs from a dislocation perspective are focused on in this book. Dislocation acts an effective obstacle for enhancing the strength of materials. For example, the dislocation evolution of $AlCrFeCuNi_{1.4}$ HEA under the tensile deformation process studied by means of molecular dynamic (MD) simulation [12]. Here, the preparation process of AlCrFeCuNi1.4 HEA is simulated firstly based on melting and casting, which are the commonly used synthesizing methods. Figure 3.1a gives a detailed simulation flow chart.

Initially, the AlCrFeCuNi1.4 HEA is FCC crystal structure at 300 K. All atoms of the HEA are heated to 1500 K at 0.04 K/fs, and then they are relaxed to an equilibrium position at 1500 K. After this, a quenching process is performed by decreasing the temperature to 300 K at 0.04 K/fs, and then the HEA is relaxed to an equilibrium state at room temperature. As shown in Figure 3.1b, the model of $AlCrFeCuNi_{1.4}$ HEA includes 8×10^4 atoms, and its size is 20a × 20a × 50a, where a is the lattice constant. Periodic boundary conditions are used in all directions. The system temperature is about 300 K at

DOI: 10.1201/9781003225706-3

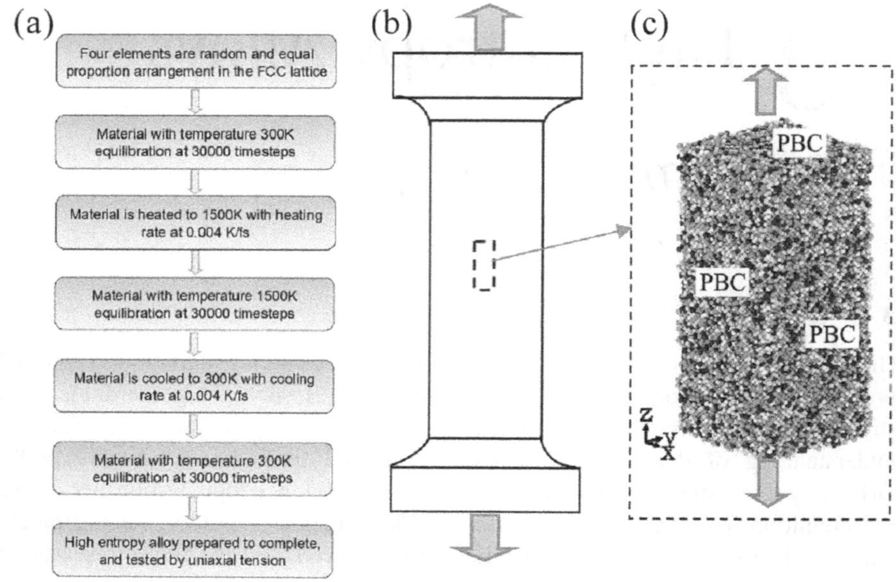

FIGURE 3.1 (a) MD simulation flow chart of preparing $AlCrFeCuNi_{1.4}$ HEA. (b) Schematic diagram of tensile specimen, and (c) the MD simulated $AlCrFeCuNi_{1.4}$ HEA tensile sample, in which the loading directions are shown in arrows. ● Cr, ● Fe, ○ Ni, ○ Cu, ○ Al.

an isobaric-isothermal ensemble to eliminate thermal effect. The motion equation is integrated by the velocity-Verlet algorithm at the time step of 1 fs. The model is relaxed firstly for 100 ps at 300 K to achieve an equilibrium state. Then the uniaxial tensile loading is applied along the z direction with the strain rate of $1 \times 10^9\ s^{-1}$.

Based on the MD simulation, the dislocation gliding and pinning are main plastic deformation mechanisms of AlCrFeCuNi1.4 HEA. The strong dislocation pinning occurs due to severe lattice distortion and solid solution effect. A larger number of stair-rod dislocations and Hirth dislocations in the HEA are generated than those of the Al and Cr metals, and then lead to the Lomer–Cottrell lock, thereby enhancing the strength of AlCrFeCuNi1.4 HEA (Figure 3.2).

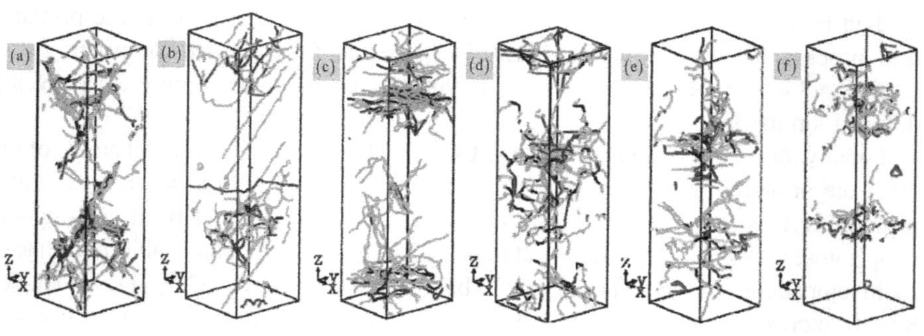

FIGURE 3.2 The evolution of dislocations in the $AlCrFeCuNi_{1.4}$ HEA (a) and the Al (b), Cu (c), Fe (d), Ni (e), Cr (f) metals when the strain is 25% [12].

Through the regulation of cooling rate gradient, a series of $Cu_{0.5}CoNiCrAl$ HEA are prepared, and the corresponding plastic deformation behavior is studied [13]. The results show that the gradient cooling rate produces abundant vacancies and causes the dislocation nucleation point to surge. At the same time, the disordered structure, high potential energy and chemical segregation in the HEA reduce the nucleation barrier for the dislocation motion. These phenomena make the HEA prone to trigger a multi-plane dislocation slip and significantly improve the plastic deformation ability.

In addition, the alloying effect on the physical properties is explored to reveal the characteristics of dislocation motion. The generalized stacking-fault energies of $Al_xFeCuCrNi$ HEAs with different Al concentrations are studied by first-principles calculations [14]. The cell model of HEA is required to contain many atoms due to random multi-component feature, which makes first-principles calculation difficult and limits its feasibility. Here, since the atomic ratio between Al and other elements is low in the $Al_{0.1}FeCuCrNi$ HEA, an $8 \times 8 \times 8$ supercell including 2048 atoms is built (Figure 3.3a). Virtual crystal approximation (VCA) method is used to improve computational efficiency, which has been shown to be reliable for the study of alloys composed of refractory elements. Figure 3.3b presents the FCC supercell model of $Al_{0.1}FeCuCrNi$ HEA based on VCA method.

A model of supercell with 12 closed-packed (111) atomic layers is presented by means of the first-principles calculation (Figure 3.4a). The stacking fault can be

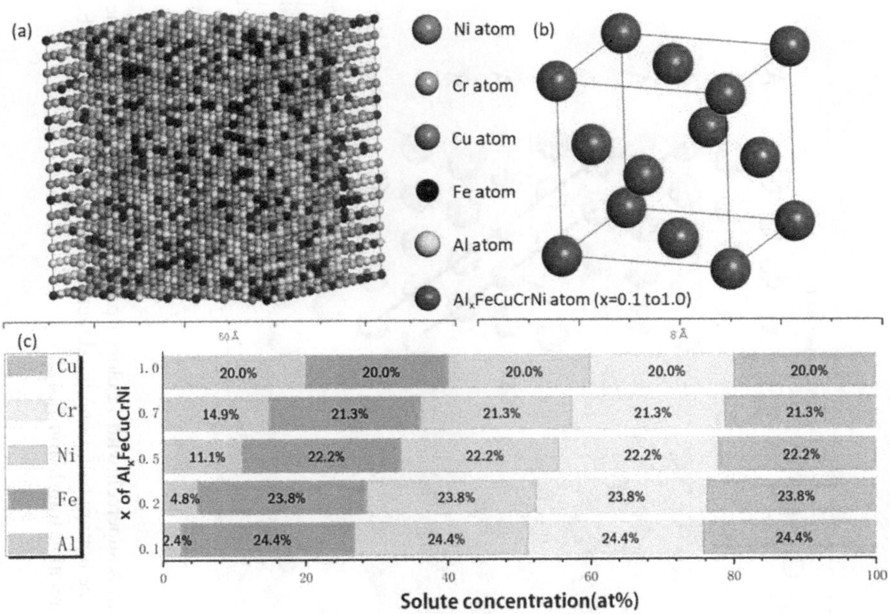

FIGURE 3.3 (a) FCC supercell model of $Al_{0.1}FeCuCrNi$ HEA, containing 2048 atoms distributed by the required atomic ratio. (b) VCA model of $Al_xFeCuCrNi$ HEAs with a variation of x from 0.1 to 1.0. (c) Element concentrations of $Al_{0.1}FeCuCrNi$, $Al_{0.2}FeCuCrNi$, $Al_{0.5}FeCuCrNi$, $Al_{0.7}FeCuCrNi$, and AlFeCuCrNi HEAs.

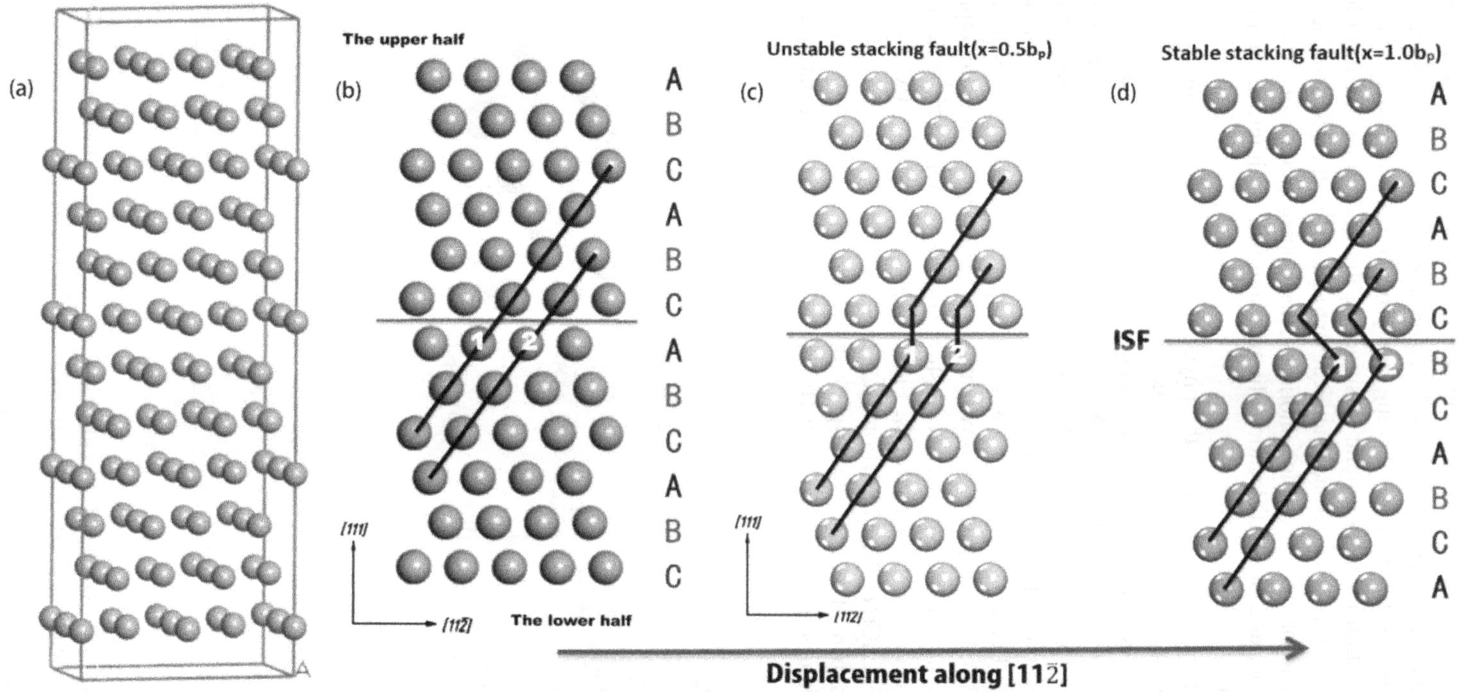

FIGURE 3.4 (a) A supercell with 12 closed-packed (111) atomic layers. The mixed atoms of VCA models are colored. The relative movement between the top half and the bottom half along the direction of [11$\bar{2}$] by 0 b_p (b), 0.5 b_p (c), and 1 b_p (d), where the movement of atoms "1" and "2" is given as an example of the relative translation [14].

formed by changing the (111) stacking plane from 'ABCABCABCABC' to 'ABCABC] [BCABC', where '] [' means the intrinsic stacking fault position (Figure 3.4b). The unstable stacking fault and stable stacking fault are presented in the $Al_xFeCuCrNi$ HEA (Figure 3.4c, d). The shearing deformation is conducted by the relative movement between the top half and the bottom half along the direction of $[11\bar{2}]$. The unstable stacking fault energy and intrinsic stacking fault energy are calculated, showing that the unstable stacking fault energy in $Al_xFeCuCrNi$ HEAs increases dramatically from 133 to 489 mJ/m^2 with increasing Al concentration, promoting dislocation nucleation but inhibiting nanocrack formation. The addition of Al results in the increase of intrinsic stacking fault energy and it reaches a maximum value of 50.5 mJ/m^2 at X=0.7. The ratio of intrinsic stacking fault energy and unstable stacking fault energy is calculated to analyze the partial dislocation nucleation and full dislocation dissociation. The lower value of AlFeCuCrNi than that of $Al_{0.1}FeCuCrNi$ HEA indicates a trend to the dissociation of full dislocation into partial dislocation, which demonstrates that the partial dislocation dominates the deformation mechanism of $Al_{0.1}FeCuCrNi$ HEA.

3.2.2 Twinning

The deformation twinning shows a great contribution to the outstanding damage resistance of CrMnFeCoNi HEA, owing to the formation of stable structures accompanied by significant energy consumption [15]. It is observed that the deformation twinning occurs at the crack under the tensile loading. The deformation twins show the representative {111} structures with highly coherent twin boundaries. Thus, the deformation twinning hinders crack propagation and further promotes the outstanding damage resistance of HEA. In addition, dislocation motion and dislocation pile-up are suppressed by the twin boundary in the $CoNiFeCrAl_{0.6}Ti_{0.4}$ HEA, thereby improving strength significantly [16].

Besides contributing to superior mechanical properties, the twinning plays an important role in the wear properties of HEAs. The nanoscratching response of AlCrCuFeNi HEA is investigated using MD simulation. Figure 3.5 shows the nanoscratching model of AlCrCuFeNi HEA. The model comprises a workpiece of FCC AlCrCuFeNi HEA, and an indenter of rigid diamond. The size of HEA is set as $20 \times 20 \times 40$ nm^3 with the three orientations in x-[1 0 0], y-[0 1 0] and z-[0 0 1], and the radius of spherical indenter is 6 nm. Three regions are divided for the workpiece, including the boundary atom region to fix the workpiece, the thermostat atom region to imitate heat dissipation, and the Newtonian atom region. The workpiece temperature is set at 293 K initially.

The results reveal that the outstanding wear properties of AlCrCuFeNi HEAs are attributed to the multiple strengthening mechanisms, including deformation twinning strengthening, dislocation strengthening, and solute strengthening. During the nanoscratching of AlCrCuFeNi HEA, the plastic deformation mechanism is mainly dominated by the twinning, stacking fault nucleation, and dislocation slip due to the low stacking fault energy [17], as shown in Figure 3.6. In addition, the plastic deformation mechanism of NiCuFeCoAl HEA under compression loading also relies on the deformation twinning [18].

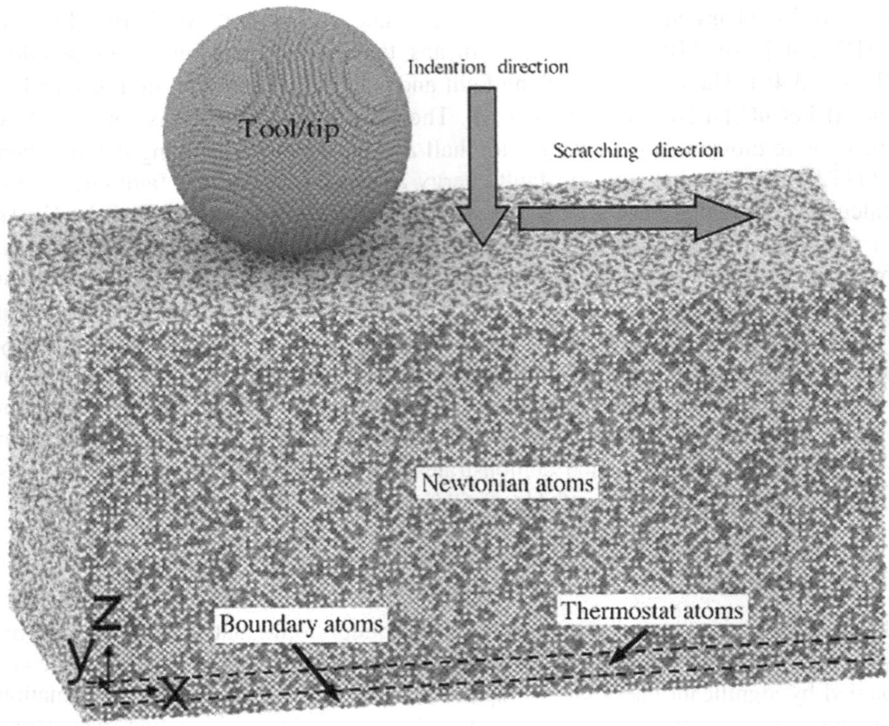

FIGURE 3.5 Schematic diagram of MD simulation nanoscratching model of AlCrCuFeNi HEA. Al, Cr, Cu, Fe, Ni [17].

FIGURE 3.6 The deformation twinning in the AlCrCuFeNi HEA subjected to scratching [17].

3.2.3 Phase Transformation

Phase transformation is frequently observed in HEAs under tensile deformation [17], and causes their good plasticity and high strength. The strain induced fcc→ bcc phase transformation in the $Co_{25}Ni_{25}Fe_{25}Al_{7.5}Cu_{17.5}$ HEA is investigated by atomic simulation. All simulations are conducted by open-source LAMMPS code. The MD-simulated $Co_{25}Ni_{25}Fe_{25}Al_{7.5}Cu_{17.5}$ HEA sample with FCC structure

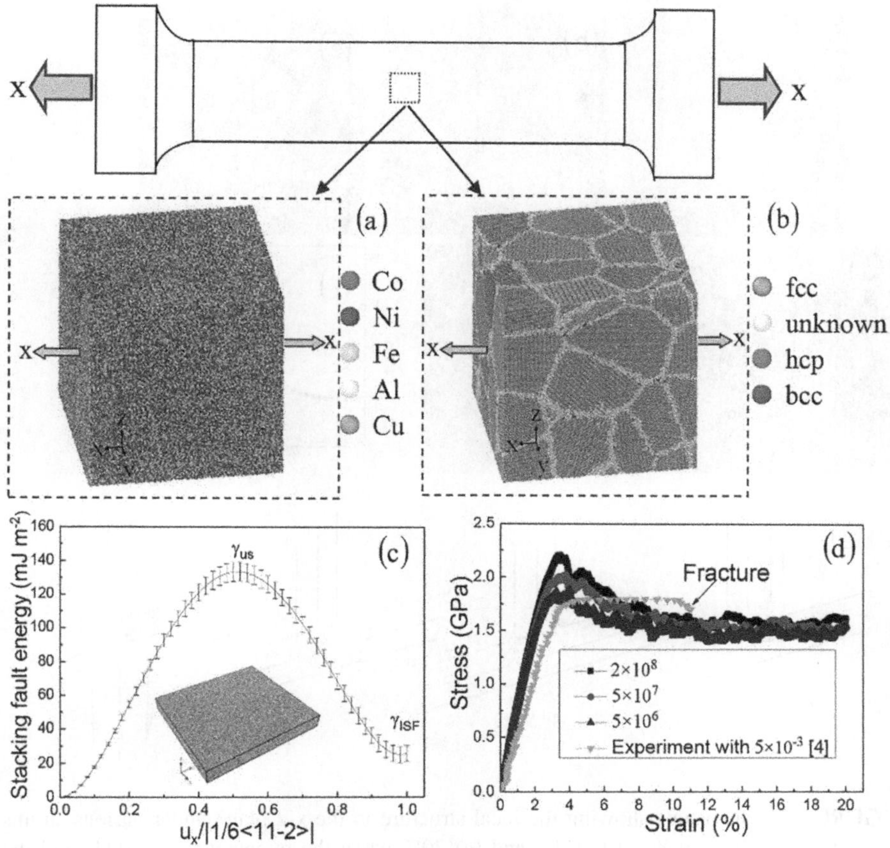

FIGURE 3.7 MD model of $Co_{25}Ni_{25}Fe_{25}Al_{7.5}Cu_{17.5}$ HEA, in which the atoms are colored by the atom type (a) and CNA (b). Stacking fault energy of the HEA along the <112(−)> direction (c). The compressive stress-strain curves of the HEA obtained by MD simulation (d).

contains 27 randomly orientated Voronoi grains with an average size of 10 nm, and the total atomic number is 2,290,000 (Figure 3.7a,b). The dimension of sample is set as $30 \times 30 \times 30\,nm^3$. To achieve a random distribution of elements, Ni atoms are replaced with Co (Fe, Al, Cu) atoms randomly until the desired composition is obtained. The simulation is conducted at 300 K with time steps of 1 fs under the Nose–Hoover thermostat. Periodic boundary conditions are used in all directions. The $Co_{25}Ni_{25}Fe_{25}Al_{7.5}Cu_{17.5}$ HEA undergoes energy minimum initially, and then it is heated to 300 K under the isothermal-isobaric ensemble. The uniaxial tensile deformation is loaded under the constant strain rates of 5×10^6, 5×10^7, and $2 \times 10^8\ s^{-1}$ (Figure 3.7a). The open-source software OVITO is used for data visualization. The common-neighbor analysis (CNA) is applied for local atomic structure analyses, and four types of structures are colored: FCC structure for green, BCC structure for blue, stacking faults for red, and grain boundaries or dislocation cores for white.

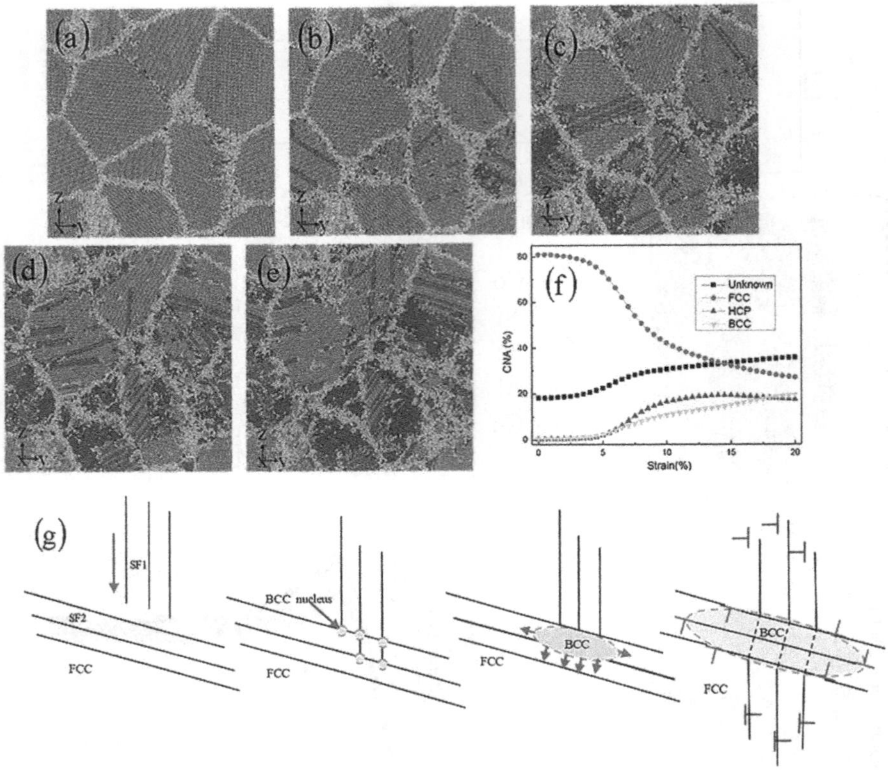

FIGURE 3.8 Snapshots showing the local structure in the x-z plane under various strains of (a) 0%, (b) 3%, (c) 8%, (d) 17%, and (e) 20% when the temperature is 300 K and the strain rate is 5×10^6 s^{-1}. The green, red, blue, and white atoms mean FCC, HCP, BCC, and other atoms. (f) Evolution in the content of FCC, HCP, BCC, and other atoms with the strain increasing. (g) The formation and growth of BCC phase, where BCC crystal is produced from FCC crystal by the interaction between two groups of dislocations [19].

Figure 3.8 shows the CoNiFeAlCu HEA undergoing FCC-to-BCC phase transformation during the compressive deformation, and leads to the softening owing the volumetric expansion [19]. The BCC structure increases and achieves maximum at the strain of 20%, indicating that the FCC-to-BCC phase transformation process is more likely to occur at high strain rates. In addition, the crystal orientation in the FCC-to-BCC phase transformation is investigated: (i) in [100] directions, the FCC-to-BCC phase transformation occurs; (ii) in [110] directions, deformation twining takes place; (iii) in [111] directions, the FCC-to-HCP phase transformation appears. The finding gives a path to develop high-performance HEAs.

Figure 3.9 depicts the effect of phase volume fraction on deformation behavior of nanocrystalline dual phase CoCrFeMnNi HEA based on MD simulation [20]. During tensile deformation of the dual phase CoCrFeMnNi HEA, the HCP-to-FCC phase transformation occurs firstly, followed by the FCC-to-HCP phase transformation. The

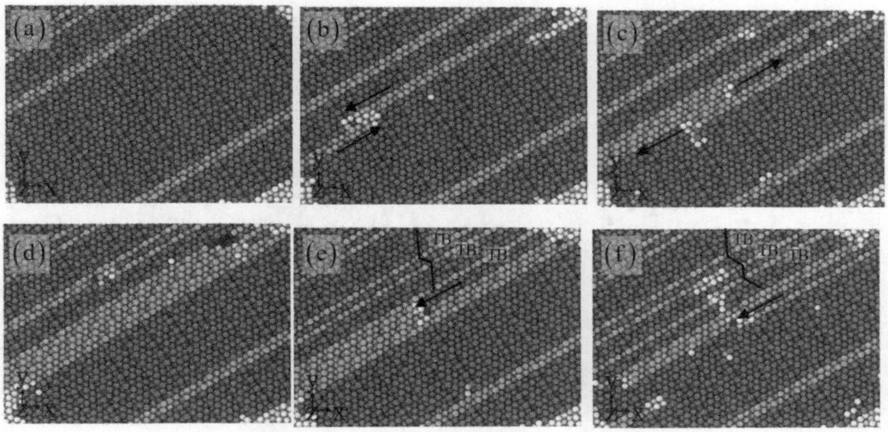

FIGURE 3.9 The HCP-FCC (a-c) and FCC-HCP phase transformations (d-f) in the CoCrFeMnNi HEA, where the arrows mean the moving direction of partial dislocations and the twin boundary is represented by "TB" [20].

dislocations are emitted into the grain from the boundary, and the new dislocation continuously slips towards opposite grain boundary, thereby yielding to the FCC phase (Figures 3.9b,c). With the increasing strain, a stacking fault is created by the partial dislocations propagating in the FCC phase (Figures 3.9d-f), forming the FCC phase. This process improves the plasticity induced by the phase transformation and the higher number of slip systems activated in the FCC phase.

3.3 STRENGTHENING MECHANISM

3.3.1 Grain Boundary Strengthening

Grain boundary would hinder dislocation movement strongly and result in a strengthening effect. The relationship of strength and grain size follows the Hall-Petch law, *i.e.*, strength increases with decreasing grain size. The Hall-Petch coefficient of conventional alloy is less than 600 MPa/$\mu m^{0.5}$, and that of FeCrNiCoMn HEA is as high as 677 MPa/$\mu m^{0.5}$ [21]. It suggests that the grain boundary strengthening is stronger in HEA.

Considering a grain size range from 5 to 20 nm, the uniaxial tension deformation mechanisms of the FeCrNi multi-principal-element alloys (MPEAs) with different ordered phase degrees are studied via MD simulation. As shown in Figure 3.10b-g, a sample with the size of 89.9 × 89.9 × 2.5 nm^3 and atomic number of 1,720,000 is built by Voronoi method, in which the average grain size is set as 5, 7, 10, 15, and 20 nm. The z axis corresponds to the <111> direction, and the periodic boundary condition is used at x, y, and z directions. The Cr atoms are randomly replaced by the Fe and Ni atoms to construct a FeCrNi MPEA sample with desired random elements (Figure 3.10b). The square regions of Cr element segregation are created to achieve the FeCrNi MPEA sample with local chemical ordering (Figure 3.10c-g). To regulate the short-range order degree, the size of Cr rich region in the alloys is set as 0, 1, 2, and 5 nm, and the corresponding volume fractions (VFs) are 0%,

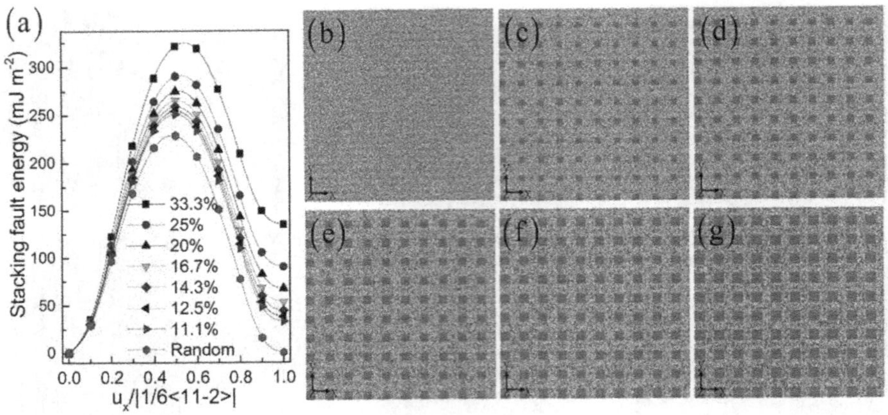

FIGURE 3.10 Stacking fault energy of MPEAs with different VFs of ordered structure (a). The FeCrNi MPEA samples with desired random elements (b) and different VFs of ordered structure: 11.1% (c), 16.7% (d), 20% (e), 25% (f), and 33.3% (g).

11.1%, 16.7%, 20%, 25%, and 33.3%. Figure 3.10a gives the stacking fault energy of the MPEAs with different VFs of ordered structure. According to the size of local ordered structure, the samples are divided into three types: the random solid solution with the size less than 0.5 nm, the local chemical ordering with size between 0.5 and 3 nm, and the precipitation with size between 3 and 100 nm.

To investigate the effect of local chemical short-range order, the average stress-strain curves of polycrystal FeCrNi MPEA are presented in Figure 3.11 [22]. As the degree of local chemical ordering increases, the yield strength improves and the flow stress decreases. The ordered phase strongly influences the strain partitioning due to local lattice distortion. The large deformation gradient occurs in a sample with small grain size owing to the activation of more plastic behavior, and a low deformation gradient is induced in the sample with a large grain attributed to the weak grain boundary linkage characteristics.

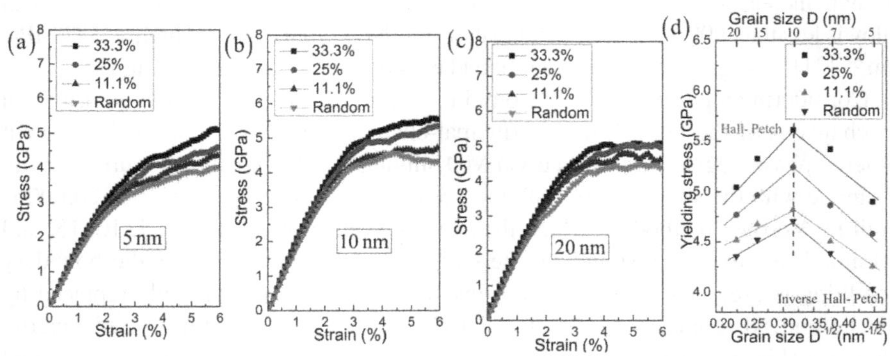

FIGURE 3.11 (a-c) Average stress-strain curves for various grain sizes. (d) The evolution of yielding strength with average grain size [22].

3.3.2 Solid-Solution Strengthening

By introducing a variety of metallic elements or non-metallic elements into a matrix material, a solid solution is formed, resulting in lattice distortion to effectively hinder the dislocation slip, which improves the material strength. For the HEA, the lattice distortion degree is more serious than that of the traditional alloy, and thus makes the dislocation slip inside the alloy overcome a large resistance. Generally, the solution strengthening is divided into two categories: displacement solution strengthening and gap solution strengthening. Since there are many principal elements in HEAs, there is no significant difference between the traditional "solvent" and "solute" among the principal elements. Hence, the concept of displacement solution strengthening fails naturally in HEAs [23–27].

It is reported that by the introduction of H, C, B, N, O, the lattice distortion of HEAs is effectively enhanced owning to the strong solution strengthening. For example, the solution strengthening of FeCoNiCrMn HEA is improved by the addition of 0.1% C element through the limited migration of twin boundaries, thus increasing the dislocation cross-slip. This trend would simultaneously enhance strength and plasticity of HEAs [28].

In addition, our recent work evaluates the contribution of solid solution strengthening, local chemical ordering strengthening, and precipitation strengthening in the FeCrNi MPEA via MD simulation. The model and method have been described in Section 3.3.1. Figure 3.12 shows the solid solution strengthening effect of FeCrNi MPEA [22]. As the size of the ordered structure increases, the Cr content reduces significantly in the matrix (Figure 3.12a). In addition, the low-strain regions increase obviously (Figure 3.12c), causing a low solid solution strengthening. As the size of ordered structure increases, the solution strengthening effect is impaired, but the ordered strengthening effect is improved.

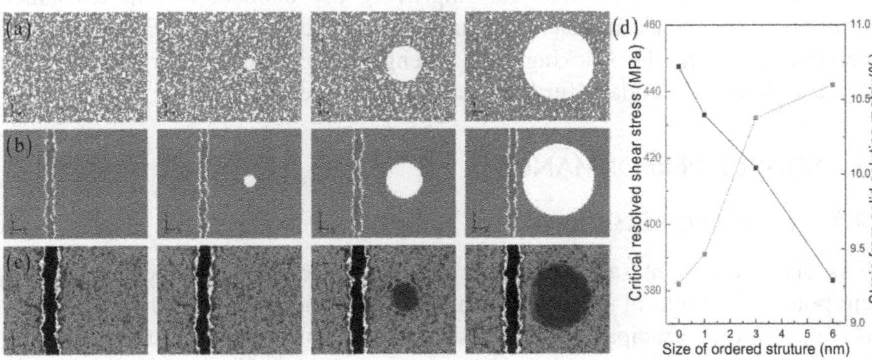

FIGURE 3.12 The solid solution phase-dislocation interaction of FeCrNi MPEA. The element distribution (a), microstructure evolution (b), and strain distribution (c) for the alloy. The change of critical resolved shear stress and strain from solution matrix with the size of ordered structure (d) [22].

3.3.3 Second Phase Strengthening

In addition to grain boundary strengthening and solid-solution strengthening, the second phase is introduced into metallic materials to hinder dislocation from moving and improve the mechanical performance. Based on the generation pathway of the second phase, it is divided into precipitation strengthening and dispersion strengthening [29–33]. The (FeCoNiCr)94Ti2Al4 HEA with a large amount of L12 structure Ni3(Ti,Al) phase is successfully prepared along with appropriate deformation and heat treatment process. The yield strength is increased to 650 MPa at room temperature, the tensile strength is over 1 GPa, and the fracture elongation is close to 40%. Here, the contribution from second phase strengthening to the yield strength exceeds 300 MPa [31].

In addition, the interaction between the precipitate and dislocation during the shear deformation of $Co_{25}Ni_{25}Fe_{25}Al_{7.5}Cu_{17.5}$ HEA has been reported via MD simulation. Figure 3.13a,b presents a schematic diagram and MD model consisting of the $Co_{25}Ni_{25}Fe_{25}Al_{7.5}Cu_{17.5}$ HEA matrix with a nanoscale coherent precipitate, respectively. The atomic number of the HEA matrix is 1,270,000, and the sample size is $40.8 \times 19.4 \times 18.7$ nm^3. The spherical coherent precipitate with a radius of 3 nm is located on the plane of dislocation slip. The edge dislocation is created about 5 nm away from the precipitate on the <110> slip plane. The *x*, *y*, and *z* axes are consistent with the $[11\bar{0}]$, $[11\bar{2}]$, and [1 1 1] directions, respectively.

Based the simulations, the synergistic enhancement mechanisms from the coherent precipitated phase and atomic lattice distortion are revealed in $Co_{25}Ni_{25}$-$Fe_{25}Al_{7.5}Cu_{17.5}$ HEA. The critical stress required for dislocations to overcome precipitates under different temperatures, chemical disorders, precipitate spacing, precipitate sizes, elemental segregation, and dislocation-cutting number is investigated. Compared with traditional metals and alloys, more severe lattice distortions are generated within the HEA matrix (Figure 3.13), thereby causing high local tensile/compressive stress for improving the dislocation slip resistance. The obvious difference in stacking fault energies between the HEA matrix and precipitate leads to the stacking fault strengthening, and the formation of the antiphase domain boundary further increases the strength [34].

3.4 SERVICE PERFORMANCE

3.4.1 Irradiation Resistance

HEAs show excellent radiation resistance and mechanical properties, making them potential structural materials in the field of nuclear energy. The HEAs composed of multiple principal components have an atomic-scale residual strain when these different elements are alloyed into a same lattice. The residual strain significantly affects the deformation mechanism, and dominates the mechanical properties [35]. Severe lattice distortion in HEAs results in high lattice distortion energy, and thus reduces the defects, such as vacancies, dislocations, stacking faults, and twins compared to traditional alloys. The influence of lattice distortion on the radiation-induced microstructure determines the radiation resistance of

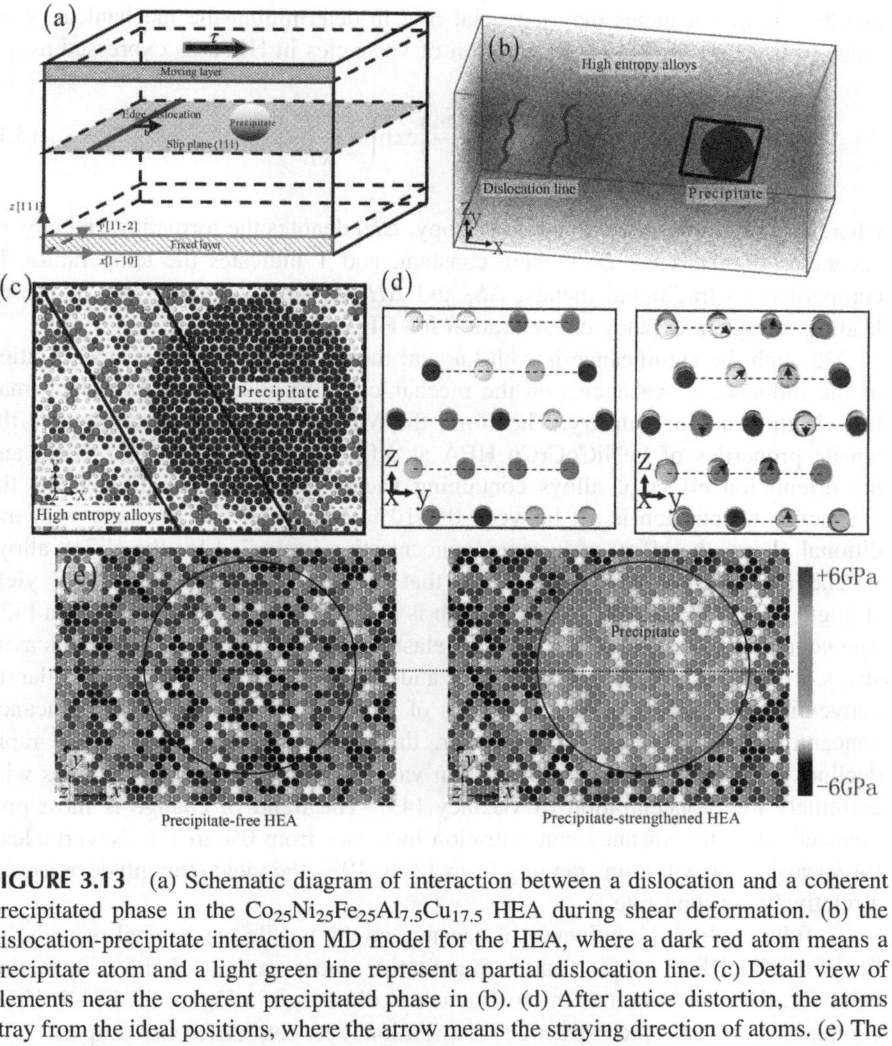

FIGURE 3.13 (a) Schematic diagram of interaction between a dislocation and a coherent precipitated phase in the $Co_{25}Ni_{25}Fe_{25}Al_{7.5}Cu_{17.5}$ HEA during shear deformation. (b) the dislocation-precipitate interaction MD model for the HEA, where a dark red atom means a precipitate atom and a light green line represent a partial dislocation line. (c) Detail view of elements near the coherent precipitated phase in (b). (d) After lattice distortion, the atoms stray from the ideal positions, where the arrow means the straying direction of atoms. (e) The stress of τ_{xz} in the precipitate-free and precipitate-strengthened HEA. Atoms in areas of high tensile stress, high compressive stress, and no stress or low stress are colored by red, blue, and green, respectively [34].

HEAs. Peierls-potential and dislocation-line energies change continuously in HEAs due to severe lattice distortion, presenting different potential energy barriers to hinder dislocation movement [36]. Considering the high cost of experiments and the difficulty of in-situ observation of microstructure evolution with time and load, simulation and modeling are important methods for studying the mechanical behavior of irradiated alloys [37–39], especially HEAs.

During the service of HEAs under irradiation condition, a significant number of vacancies are created due to collisions of high-energy particles. These vacancies affect the properties throughout proliferation and annihilation, ultimately affecting the microstructure and mechanical properties [40,41]. The generation, density, and distribution of vacancies play a crucial role in determining the mechanical properties [40]. Moreover, the concentration of vacancies in HEAs is expressed as

$$C_v = \exp\left(\frac{\Delta S_f}{k}\right)\exp\left(-\frac{\Delta H_f}{kT}\right) \tag{3.1}$$

where ΔS_f represents the formation entropy, ΔH_f denotes the formation enthalpy of vacancies, k means the Boltzmann constant, and T indicates the temperature. In comparison to traditional metals, ΔS_f and ΔH_f are comparatively low in HEAs, leading to higher vacancy concentration for HEAs.

Although the significance is well-known, there is no atomic-level investigation of the influence of vacancies on the mechanical behaviour of HEAs, which may impede their use in industry. Therefore, the MD method is used to simulate the tensile properties of FeNiCoCrCu HEA at different vacancy concentrations, and the orientation effect of alloys containing vacancy defects is studied. Here, the vacancy concentration is set between 0%-10%. Meanwhile, to compare with traditional alloys, the effect of vacancy concentration is studied for the Ni718 alloy.

The previous findings demonstrate that the FeNiCoCrCu HEA has a yield strength of 16 GPa approximately, which is similar to our result of 17.2 GPa [42]. The equation (b) is used to compute the elastic modulus, where (c) represents axial stress, E denotes the elastic modulus, and (a) signifies axial strain. Similar to conventional alloys, the elastic modulus of HEA slightly decreases with vacancy concentration increases [43]. Moreover, the yield strength experiences a rapid decline, causing the corresponding strain value to decrease, even in samples with extremely low concentration of vacancy [43]. The trend of change is most pronounced when the vacancy concentration increases from 0% to 1%. Nevertheless, for vacancy concentrations ranging from 1% to 10%, the yield strength decreases at a relatively constant rate.

To fully explore the influence of vacancy on the tensile mechanical response of FeNiCoCrCu HEA samples and a quantitative understanding of the deformation behavior, the distribution of microstructure is illustrated in Figure 3.14 [44]. Upon comparison of the microstructure distribution between different samples, it is observed that a considerable amount of stacking faults consisting of HCP structure are formed when the vacancy content is 0% (Figure 3.14a,b). Deformation twinning with an extremely small thickness is present in all samples, leading to a mild strengthening effect [44]. In Figure 3.14b, the quantity of stacking faults and deformation twins decreases as the vacancy content increases. Moreover, the coherence and width of stacking faults are impeded by the clustering of vacancies, leading to the creation of shortened and dispersed dislocation. The threshold stress required for dislocation sliding, which is contributed by stacking faults, relies on the thickness of stacking faults [45]. An expression for the strengthening effect contributed by parallel-spaced stacking faults is (b) [46], where d represents the spacing between stacking faults, and k is a constant specific to the material. The reason for maximum yield strength occurring at the vacancy content of 0% is the inverse relationship between stacking fault strengthening and stacking fault spacing.

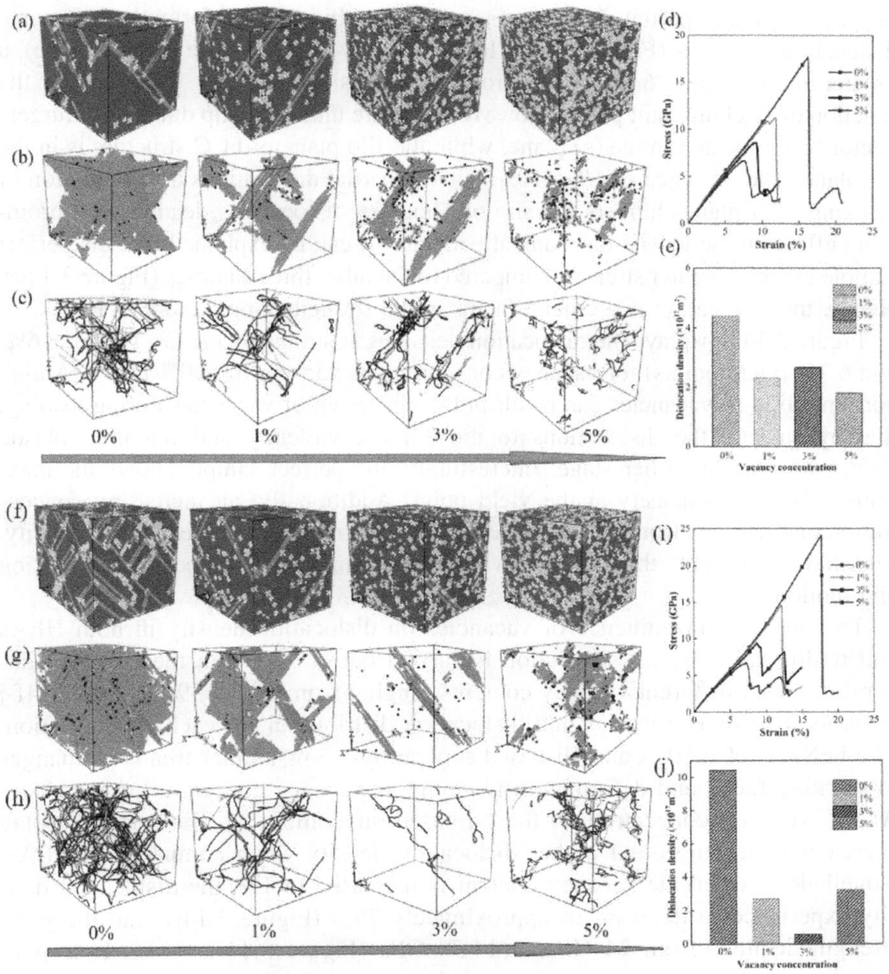

FIGURE 3.14 For FeNiCoCrCu HEA, (a) the microstructure and (b) the defect at various vacancy concentrations from 0% to 5% at the yield point. The different colors correspond to different crystal structures, such as green representing the FCC, red represents the HCP, blue is the BCC, and white means other structures. (c) The dislocation evolution with different types of dislocations represented by different colored lines. For example, the 1/6<112> Shockley dislocation is green, the 1/6<110> stair-rod dislocation is pink, the 1/6<110> perfect dislocation is blue, the 1/3<111> frank dislocation is sky-blue, and other dislocations is red. Stress-strain curves (d) and the dislocation density (e) are also shown for different vacancy concentrations. For Ni-based superalloys, the microstructure (f), defect structures (g) in Ni at the yield point for different vacancy concentrations, including 0%, 1%, 3%, and 5%. Panel (h) displays the dislocation evolution. (i) and (j) show the stress-strain curve and dislocation density, respectively [47].

In the sample with vacancy content of 0%, the dense distribution of stacking faults means that a large number of slip systems are activated. Numerous long Shockley partial dislocations are formed, which constitute a dense dislocation network. The existence of vacancies disrupts the continuance of the slip plane, thus, the formation of short Shockley partial dislocations occurs, resulting in the formation of sparse dislocation networks (Figure 3.14c). In the vacancy-free structure (Figure 3.14c), a greater number of 1/6<110> stair-rod partial dislocations are generated at the junction of stacking fault planes, however, they are unable to slip due to the Burgers vector being located in the (c) plane, while the slip plane of FCC structure is in the 1st plane. The presence of 1/6<110> stair-rod partial dislocations at the junction of stacking fault planes hinders the slip of Shockley dislocations, leading to a prominent effect on the tensile mechanical behavior. It can be explained that the perfect sample exhibits greater strength compared to the other three samples (Figure 3.14d). Hence, the dislocation is a crucial factor in the strengthening factors of HEAs.

Figure 3.14e displays the dislocation densities at strains of 17.7%, 11.2%, 8.6%, and 6.7% (yield points) for vacancy concentrations varying from 0% to 5%. A high concentration of vacancies can result in the clustering of vacancies during loading, thereby causing the dislocations to move more violently and initiating plastic deformation at an earlier stage. Interestingly, the perfect sample shows the maximum dislocation density at the yield point. Additionally, an increasing concentration of vacancies from 0% to 1% leads to a 50% reduction in dislocation density. This trend highlights the ability of vacancies to hinder the dislocation nucleation and motion.

To compare the influence of vacancies on dislocation density in both HEAs and traditional alloys, a simulation is carried out at 300 K in the Ni-based superalloys with different vacancy contents ranging from 0% to 5%. Figure 3.14f-j presents the microstructures and dislocation distribution for each concentration. The FeNiCrCoCu HEA and Ni-based superalloys show similar trends in changes to stacking faults and deformation twinning as vacancy concentration increases. At 0% vacancy concentration, the Ni-based superalloys are known to display a greater yield stress and higher dislocation density when compared to HEAs. Nonetheless, when the vacancy content is increased to 1%, the dislocation density experiences a decrease of approximately 75% (Figure 3.14j), and the yield strength declines from 24 GPa to 14.58 GPa (Figure 3.14i), which is a more significant reduction compared to that observed in the FeNiCrCoCu HEA. The results suggest that HEAs may have the ability to mitigate the detrimental effects of vacancies more effectively than conventional alloys.

To study the influence of vacancy content on different grain orientations of radiation-resistant HEAs containing vacancy defects, uniaxial tensile tests are conducted along three different loading directions. The relationship between stress and strain for varying crystal orientations at 0% and 1% vacancy concentration are depicted in Figure 3.15. The phenomenon highlights a remarkable anisotropy existing across the studied orientations. For the perfect sample, the highest yield strength and strain are exhibited along the [100] direction. In line with previous investigations, the yield strain remains comparable between the [111] and [110] directions, but the yield stress in the former is twice that of the latter. Additionally, with the introduction of vacancy, the yield stress of samples with a [100] loading direction decreases significantly from 17.9 GPa to 11.2 GPa when the concentration of vacancies rises from 0% to 1% [43]. However, the yield stresses for HEAs with [110] and [111] loading directions are relatively unaffected by vacancy. These

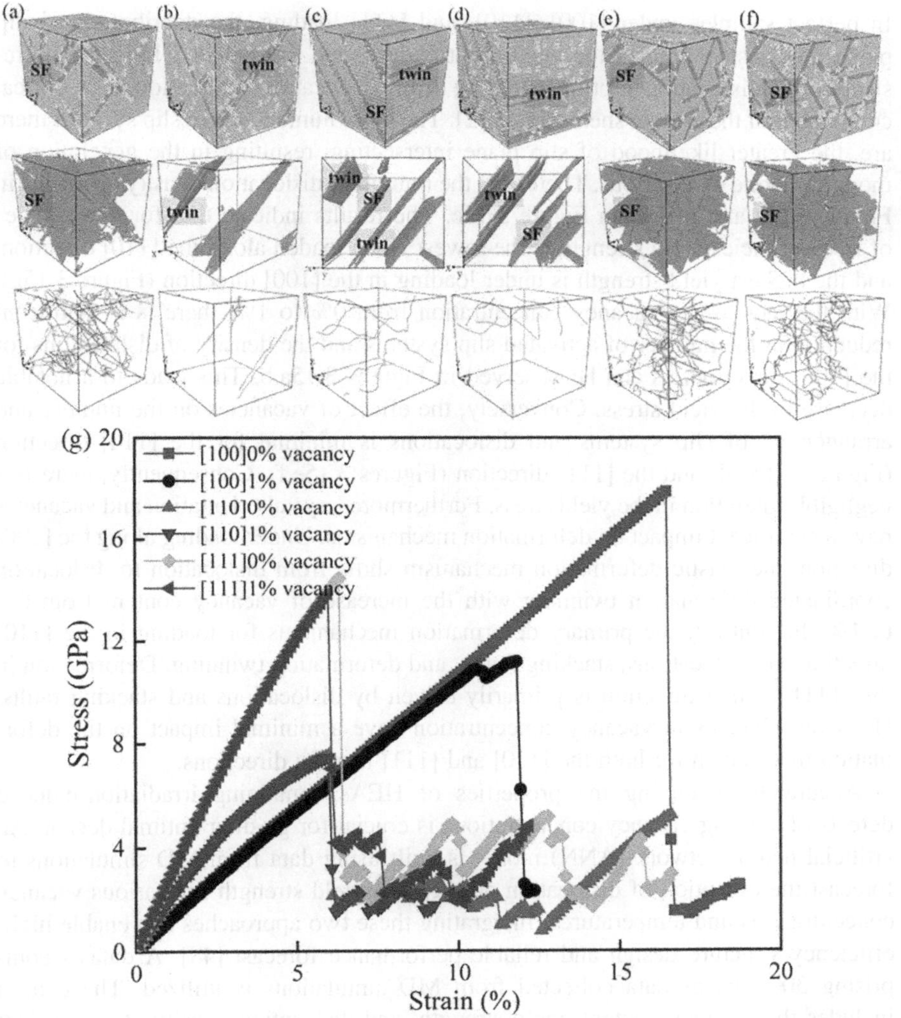

FIGURE 3.15 Snapshots are taken to represent the evolution of each sample, including the perfect sample with crystal orientations of x-[100] and z-[001] (a), the sample with a vacancy content of 1% and crystal orientations of x-[100] and z-[001] (b), the free vacancy sample with crystal orientations of x-[110] and z-[001] (c), the sample with 1% vacancy content and crystal orientations of [110]x and z-[001] (d), the free vacancy sample with crystal orientations of x-[111] and z-[11-2] (e), and the sample with 1% vacancy content and crystal orientations of x-[111] and z-[11-2] (f). (g) illustrates the stress-strain relationships at 0% and 1% vacancy content for uniaxial loading in the [100], [110], and [111] directions [47].

results demonstrate significant anisotropy of vacancies on tensile properties, which is notably distinct from that observed in γ-TiAl alloys.

The significant anisotropy outlined previously is strongly linked to the microstructural characteristics. When exposed to loading along the [100], [110], and [111] crystallographic directions, plastic deformation involves 8, 6, and 4 slip systems.

In perfect samples under [100], [110], and [111] loading, the distribution of slip planes and dislocations at the yield point is described in Figures 3.15a,c,e. The resistance of dislocation motion resulting from dislocation interaction is a critical component of the critical shear stress [12]. The more number active slip systems there are, the greater likelihood of slip plane intersecting, resulting in the generation of more immobile dislocations. Therefore, the immobile-dislocation density is highest in Figure 3.14a and lowest in Figure 3.15c. The results indicate that for the samples of free vacancies yield strength is the lowest when loaded along the [110] direction, and the highest yield strength is under loading in the [100] direction (Figure 3.15g). With the increase in vacancy concentration from 0% to 1%, there is a significant reduction in the number of activated slip systems and the density of dislocations for the [100] direction, as can be observed in Figures 3.15a,b. This leads to a notable decrease in the yield stress. Conversely, the effect of vacancies on the number and arrangement of slip systems and dislocations is minimal for the [110] direction (Figures 3.15c,d) and the [111] direction (Figures 3.15e,f). Consequently, there is a negligible alteration in the yield stress. Furthermore, crystal orientation and vacancies have a significant impact on deformation mechanisms. Under loading along the [100] direction, the plastic deformation mechanism shifts from dislocation to dislocation coordinated deformation twinning with the increase of vacancy content from 0% to 1%. In contrast, the primary deformation mechanisms for loading in the [110] direction are dislocations, stacking faults, and deformation twinning. Deformation in the [111] loading direction is primarily driven by dislocations and stacking faults. However, changes in vacancy concentration have a minimal impact on the deformation mechanism for both the [110] and [111] loading directions.

Accurately predicting the properties of HEAs containing irradiation-induced defects of varying vacancy concentrations is crucial for guiding optimal design. An artificial neural network (ANN) model is built using data from MD simulations to forecast the evolution of dislocation density and yield strength for various vacancy concentrations and temperatures. Integrating these two approaches can enable high-efficiency structure design and reliable performance forecast [48]. A dataset comprising 300 sets of data collected from MD simulations is utilized. The dataset includes the vacancy content, yield strength, and dislocation density at six distinct temperatures. The input variables for the model include the vacancy content in conjunction with temperature, while the yield strength and dislocation information serve as output variables. The data is partitioned into training, validation, and testing sets. The ANN model exhibits a correlation coefficient exceeding 0.9 between the predicted and target data for diverse datasets. This suggests that the ANN model employed in this research is highly precise.

Using the ANN model, the 3D distributions of yield strength and dislocation density in yield state under various vacancy contents and temperature effects are acquired (Figure 3.16). These distributions are displayed at the top of the graphs. For vacancy content below 5%, the dislocation density initially decreases before experiencing a slight increase. However, when the vacancy content surpasses 5%, the dislocation density undergoes a rapid increase at 800 K and attains its maximum value. With the increasing vacancy concentration, the same trend is observed at other temperatures, where the dislocation density increases at higher vacancy

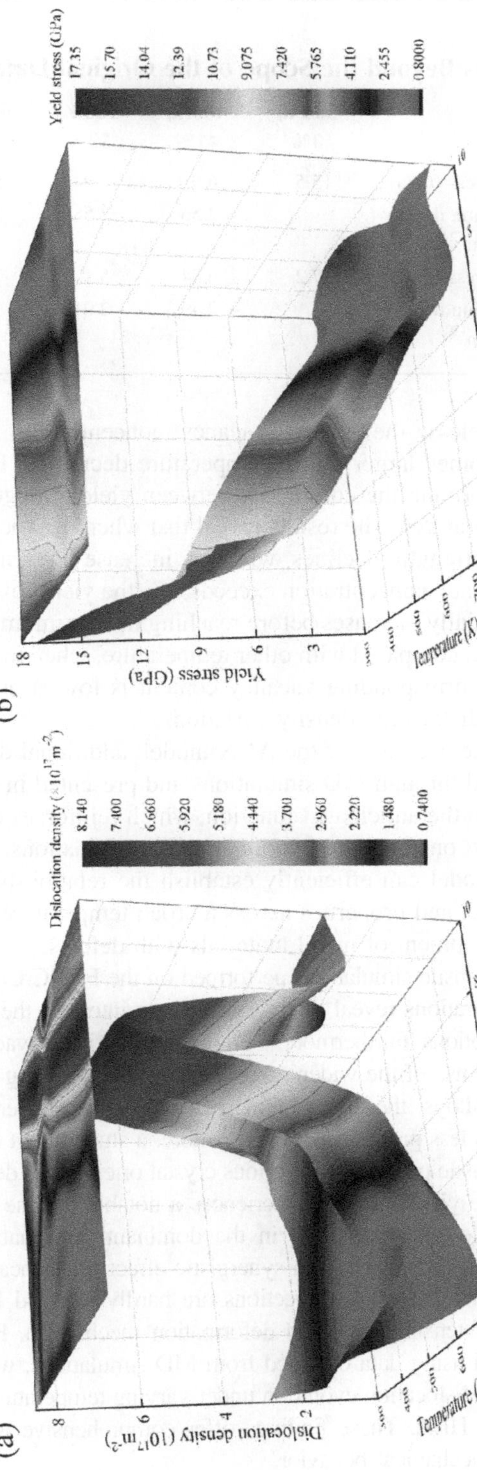

FIGURE 3.16 ANN prediction results of (a) dislocation density and (b) yield strength with changes in vacancy concentration and temperature [47].

TABLE 3.1
Validation Data That is Beyond the Scope of the Original Dataset [47]

Sample		300 K, 0%	400 K, 41%	500 K, 53%	600 K, 27%	700 K, 13%
Validating results	Yield stress (GPa)	15.5	6.9	4.94	6.96	4.39
	Dislocation density ($\times 10^{17} m^{-2}$)		2.56	2.58	1.27	2.6
Predicted results	Yield stress (GPa)	17.2	6.84	5.1	7.15	4.19
	Dislocation density ($\times 10^{17} m^{-2}$)		2.49	2.01	1.12	2.4

concentrations. Nonetheless, the critical vacancy concentration for dislocation density to increase becomes larger as the temperature decreases. From the ANN model, we can also determine the correlation between yield strength and vacancy content at various temperatures. The results reveal that when the vacancy content is below 6%, the yield strength declines with an increase in temperature [49]. However, when the vacancy concentration exceeds 6%, the yield strength exhibits a different behavior: it initially increases before reaching its maximum value and then decreases again. Notably, compared with other temperature, when the yield strength begins to increase, the corresponding vacancy content is lowest at 800 K, which agrees with the rule of dislocation density variation.

To further validate the accuracy of the ANN model, additional data beyond the preset dataset is obtained through MD simulations and presented in Table 3.1. The obtained data aligns with the anticipated function, which reinforces the accuracy of the model. Hence, based on the above findings and confirmations, it can be concluded that the ANN model can efficiently establish the relationship between the irradiation-induced defect and properties across a broad temperature range. In turn, this can guide the development of novel materials with defects.

The outcomes of the tensile simulations performed on the FeNiCrCoCu HEAs with different vacancy concentrations reveal that vacancies can interrupt the continuance of stacking faults and dislocations. Furthermore, a high concentration of vacancies can lead to decreased strength because of the widening of stacking fault spacing [47]. However, compared to traditional alloys, the effect of increased vacancy concentrations on the yield strength of HEAs is less pronounced. Furthermore, a study about the influence of vacancy defects on the tensile properties of various crystal orientations demonstrates that in the [100] direction, the yield strength experiences a notable decline as the vacancy concentration increases, leading to a shift in the dominant deformation mechanism which is transformed from dislocation to a synergistic effect of dislocations and twinning. Conversely, the [110] and [111] directions are hardly affected by the vacancy concentration in terms of tensile stress and deformation mechanism. Furthermore, an ANN model is developed using data obtained from MD simulations, which accurately forecasts the strength and dislocation evolution under varying temperatures and vacancy contents in FeNiCrCoCu HEA. These findings offer comprehensive insights into the impact of vacancies on mechanical behavior.

3.4.2 Fatigue Performance

Due to their distinctive structure and exceptional performance, HEAs have gained substantial potential for diverse applications in different conditions, resulting in improved serviceability [50]. However, long-term material failure, which is mostly attributed to fatigue, poses a serious threat to human safety. The characteristics of the microstructure, including grain size, grain boundary characteristics, type of defects, and precipitation typically determine the fatigue properties of materials [51]. Designing material microstructure to enhance fatigue performance remains one of the most arduous tasks for both industrial applications and scientific research, ranging from nano to engineering scales.

To examine the particular impact of microstructure characteristics on the fatigue behaviors under cyclic deformation, diverse tests and simulations are conducted. An atomic-level analysis of the microstructure during cyclic loading is essential in understanding the fatigue performance. The cyclic deformation mechanism of Ni-based superalloys under different environmental temperatures and loading rates has been investigated using the MD method [52], providing an effective explanation for the atomic-level fatigue micro-mechanism of these superalloys. However, limited research has been conducted on the correlation between microstructure and fatigue properties of refractory MPEAs, with only a handful of experiments conducted. Due to this scarcity of experimental and simulation data, it is challenging to perform a comprehensive analysis of MPEA fatigue performance from the microstructural perspective. Therefore, to ensure the long-term use of these materials, expanding relevant simulation and theoretical research on MPEA fatigue performance is necessary.

We studied the effects of different grain sizes on work hardening and fatigue damage behavior of nanocrystalline refractory MPEA (TaNbZr) under low cycle loading by MD simulation [53]. According to prior research, when the grain size decreases to 10-15 nm, the grain boundary effect transitions from the Hall-Petch positive relation to the inverse relation [52]. Thus, four different samples with an average grain size of 8-20 nm are selected. The models can contain approximately 1,339,000 atoms and have dimensions of $30 \times 30 \times 30$ nm3, consisting of 6, 12, 30, and 90 grains, respectively. For low-cycle fatigue simulations, a fully reversed triangular waveform of strain is applied to the nanocrystalline refractory sample. The cycle is represented by stretching and compressing 100 times with strain rate of 1×10^9 s^{-1} along the x-axis direction. The strain amplitude under loading is ±10%. Meanwhile, the microstructural evolution is observed using the Ovito software [54].

The size of grains can significantly impact various properties of HEAs, such as strain-hardening, phase transformation, and twinning behavior [55]. By analyzing the strain range of 7% to 15%, the average flow stress can be determined for different grain sizes. For an 8 nm grain size, the average flow stress is 3.23 GPa, for a 12 nm grain size it is 3.41 GPa, for a 16 nm grain size it is 3.21 GPa, and for a 20 nm grain size it is 3.19 GPa. As a result, among all four-grain size samples, the sample with the average grain size of 12 nm demonstrates the highest average flow stress. The results indicate a clear manifestation of the positive Hall-Petch relationship when grain sizes are 12-20 nm. The simulation results once again prove that the Hall-Petch relationship can be transformed from inverse into positive as the grain size increases [50,56]. Upon reaching a strain of 15%, a comparison of the model before loading reveals that the migration and slip of grain boundaries are a crucial factor leading to plastic deformation of the sample with an 8 nm average grain size [50,57]. The primary mechanism that governs plastic deformation changes with variations in grain size. Deformation twinning is the dominant mechanism when the average grain sizes of samples are 16 and 20 nm. However, when the grain size is 12 nm, there is a simultaneous occurrence of both deformation twin and grain growth jointly dominating plastic deformation [58,59]. The identification results of dislocation configurations show that screw dislocations exhibit a greater tendency to nucleate, and slip compared to edge dislocations among all partial dislocations. With decreasing grain size, the dislocation density notably reduces at 15% strain, because of the reduced average dislocation motion path for very small grain sizes. Moreover, a high dislocation density is an important reason for the enhancement of strain-hardening capability [60].

The stress-strain hysteresis loops display a linear increase in stress during the elastic stage, followed by a gradual increase or decrease at the yielding and plastic stages until the strain reaches 10%, which is consistent with the behavior observed in single tensile loading (Figure 3.17). Moreover, the occurrence of strain-softening or strain-hardening after the yielding stage is influenced by both the cycle number and grain size [58]. If the sample conforms to the inverse Hall-Petch relationship, the strain-hardening degree intensifies with the progress of cyclic loading. The phenomenon of strain-softening is noticeable at the yielding stage when the grain size is 12 nm. When the grain size of the sample is within the range that satisfies the positive Hall-Petch law, there is a strain softening phenomenon at fewer cycles, which transforms into strain hardening as the number of cycles increases. The strain-hardening or softening (that is evident in stress-strain curves) is mainly ascribed to alterations in microstructure that occur during cyclic loading, such as grain boundary transfer, dislocation nucleation/motion, deformation twinning, and grain growth [61,62]. The present cyclic stress response accords with earlier findings on nanocrystalline metals that are established by means of experiments and MD simulations [61].

Figure 3.18 shows the variations of dislocation characteristics and distribution in free strain sample with increasing cycle number. The impact of cycle number on grain boundary configurations is conspicuous. The grain boundary coarsening arises during the first cycle of loading due to the occurrence of grain boundary slip, which notably promotes the grain boundary diffusion process. In comparison to the 8 nm and 12 nm samples, the grain boundary coarsening and grain growth are less pronounced in the 20 nm sample in the cyclic process. The coarsening process of grain boundary is accompanied by twinning nucleation, resulting in dislocation pinning, which enhances the strength and strain-hardening degree. Deformation twins are often formed during the loading process of BCC HEAs [63], as one of the dominant factors for plastic deformation [64]. In this study, the generation of a coherent twin boundary is depicted at zero strain after the end of the initial cycle (as depicted in Figure 3.19a). Furthermore, the coherent twin boundary remains exceptionally stable under cyclic deformation as a result of precise atom distribution,

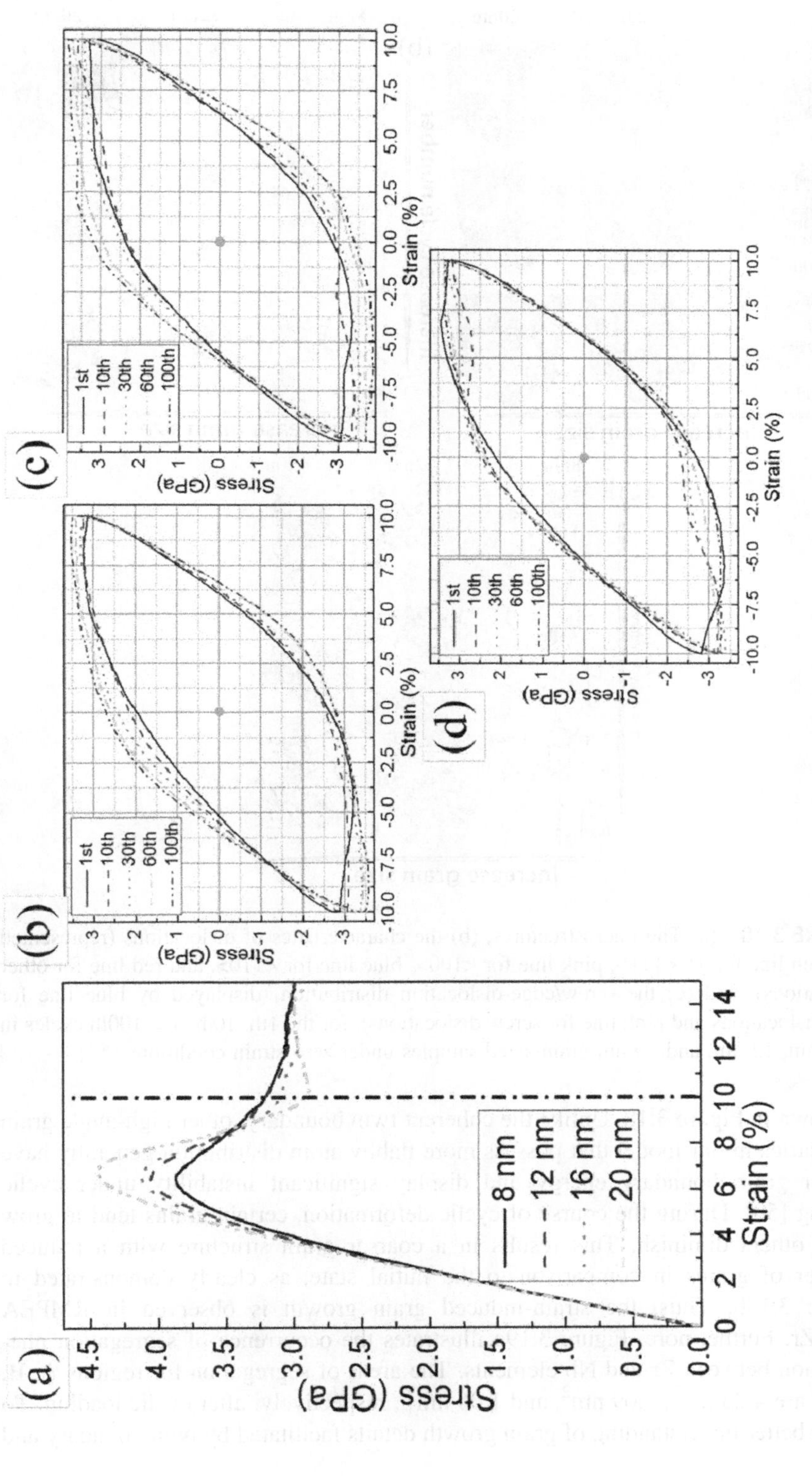

FIGURE 3.17 Uniaxial tension stress-strain relationship (a). Cyclic loading stress-strain relationship for grain size 8 (b), 12 (c), and 20 nm (d) [53].

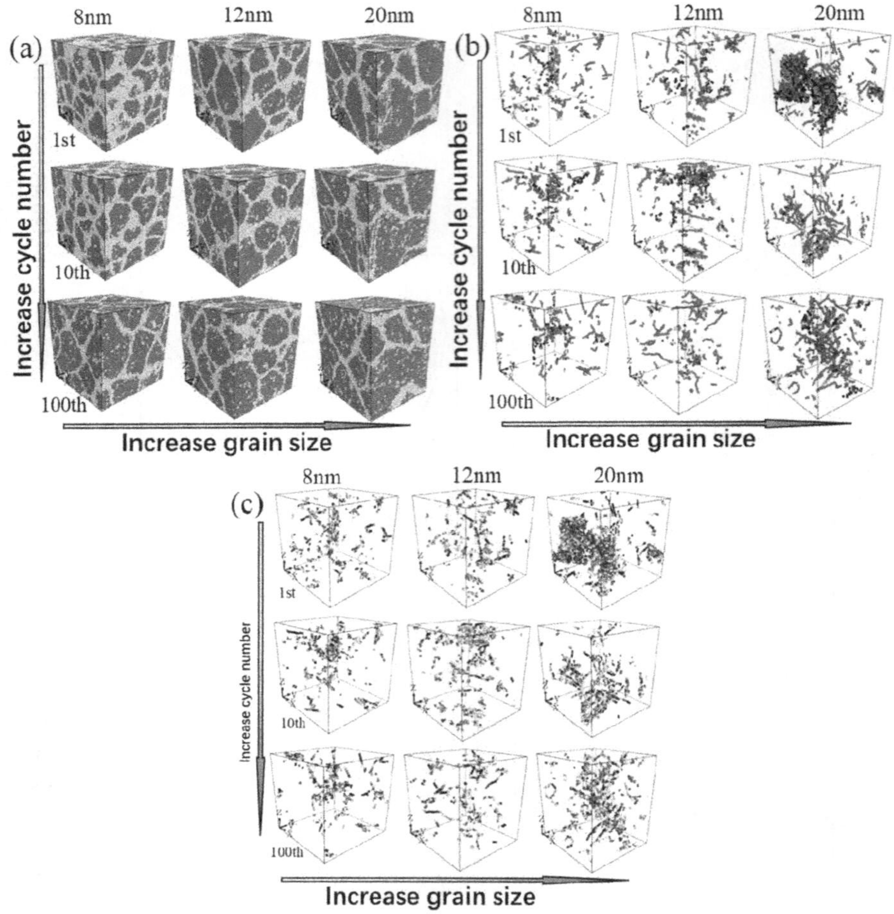

FIGURE 3.18 (a) The microstructures, (b) the characteristics of dislocations (represented by green line for 1/2<111>, pink line for <100>, blue line for <110>, and red line for other orientations), and (c) the screw/edge-dislocation distribution (displayed by blue line for edge dislocations and pink line for screw dislocations), for the 1th, 10th, and 100th cycles in the 8 nm, 12 nm, and 20 nm grain-sized samples under zero strain conditions [53].

as shown in Figure 3.18. Unlike the coherent twin boundary, other high-angle grain boundaries in our model that possess more flabby atom distribution generally have greater grain boundary energy and display significant instability under cyclic loading [50]. During the course of cyclic deformation, certain grains tend to grow while others diminish. This results in a coarser grain structure with a reduced number of grains in comparison to the initial state, as clearly demonstrated in Figure 3.19b. Thus, the strain-induced grain growth is observed in RMPEA TaNbZr. Furthermore, Figure 3.19c illustrates the occurrence of segregation phenomenon between Zr and Nb elements. The areas of segregation for regions A, B, and C are 4.25 nm^2, 1.09 nm^2, and 1.06 nm^2 , respectively, after cyclic loading. To gain a better understanding of grain growth details facilitated by twin boundary and

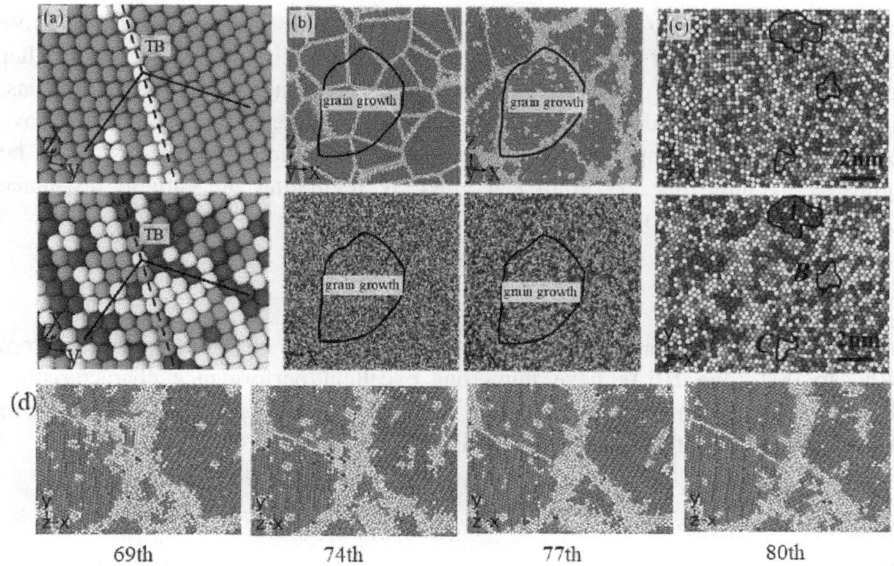

FIGURE 3.19 (a) A magnified view of deformation twinning during the 10th cycle, and (b) comparison of grain morphology to the initial structure at strain of 0% in a sample with a grain size of 8 nm after the hundredth cycle. (c) Variation of element distribution before and after cyclic loading. (d) Microstructures observed at the 69th, 74th, 77th, and 80th cycles in a 12 nm grain-sized sample under zero strain conditions [53].

grain boundary deformation, an explicit evolution of microstructure is presented in the sample containing grains with a size of 12 nm. When the cycle reaches 69 times, a coherent twin boundary is exhibited, and ultimately stabilizes after 10 cycles, as shown in Figure 3.19d.

In the case of high cycle numbers, the dislocation nucleation and annihilation constantly occur owing to the constraints of the grain boundaries [61]. The distribution and features of dislocations are investigated for TaNbZr refractory MPEAs at varying grain sizes and cycle numbers, as depicted in Figure 3.18b,c. It is observed that when the grain size is less than 12 nm, the degree of damage does not significantly increase in samples after 100 cycles, because of the suppression of the formation of numerous dislocation nucleation sites at high grain boundary density. Conversely, when the average grain size of sample is 20 nm, significant damage accumulation is observed as a result of the reduced resistance of dislocation nucleation caused by grain boundaries [50]. Apart from deformation twinning and grain growth, plastic deformation in all the samples shown in Figure 3.18b is attributed to the movement of 1/2<111> full dislocations. To gain deeper insights into how grain size and cycle number affect the distribution of different dislocation types, their respective structures are illustrated in Figure 3.18c. It can be observed that after cyclic loading, screw dislocations dominate in the sample of 8 nm grain size, whereas with an increase in grain size, the fraction of edge dislocations increases and gradually dominates the cyclic deformation behavior. Furthermore, when the grain size is 20 nm, the fraction of edge dislocations progressively declines with an increase in the number of cycles due to the consumption of massive dislocations during the grain growth process. Nevertheless, the tiny grains smaller than 12 nm acting as the sources of dislocation constantly generate dislocations, resulting in a sustained rise in dislocation density. Therefore, based on the above analysis, it can be concluded that the fatigue deformation mechanism can be adjusted by adjusting the grain size, thereby regulating the fatigue resistance behavior of refractory medium entropy alloys.

3.5 PROSPECTS

As opposed to traditional alloys with a single principal component ranging from 40% to 90%, the HEAs have numerous excellent performance characteristics. Based on the performance classification (Figure 3.20), the potential applications of HEAs can be classified into '7+X' fields [65].

3.5.1 Mechanics

Contrary to conventional alloys containing a single main component, the HEAs are synthesized by combining conventional and unconventional physical metallurgical techniques in a complex manner. The mechanical properties of HEAs, such as strength, ductility, wear resistance, creep, and fatigue, are significantly influenced by factors, including constituent elements, phase structure, grain size, and so on. Current research is focused on investigating novel approaches to enhance the mechanical properties of HEAs, such as the development of eutectic and gradient structures [66,67].

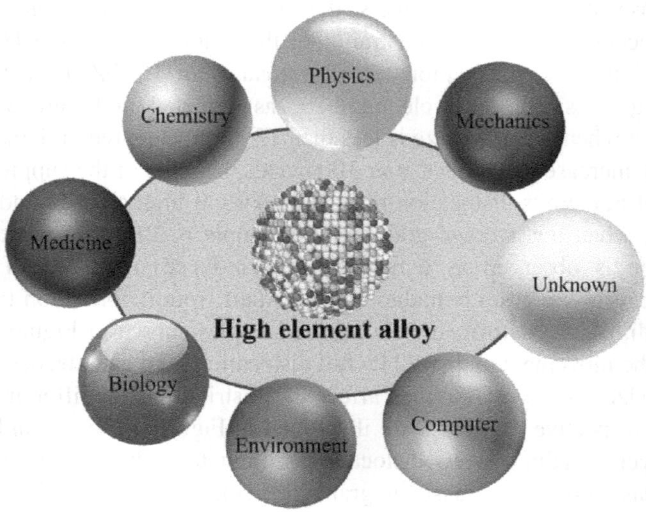

FIGURE 3.20 Potential engineering applications of HEAs [65].

3.5.2 Physics

The HEAs possess distinctive optical and magnetic characteristics that are contingent upon their element composition and phase structure. In HEAs, thermal and electrical conduction are affected by the lattice distortion and configuration entropy, which are critical factors. Future research needs to focus on understanding how the magnetic moments, spin, hall coefficient, and lattice vibration impact their physical properties [68–71].

3.5.3 Chemistry

The HEAs show great promise as corrosion-resistant and catalytic materials. To explore novel superior-performance HEAs, future work should focus on understanding their fundamental corrosion resistance from an atomic scale, realizing collaborative designs for mechanical properties and corrosion resistance, expanding the types of supporting substances to improve catalytic performance, and bridging theory and experiment gaps in catalytic applications of HEAs.

3.5.4 Computing

The integration of machine learning and multi-scale modeling has the potential to significantly promote the development of high-performance HEAs, reduce resource costs, and expand their capabilities in mechanics, physics, and chemistry. Nevertheless, current multi-scale computation techniques for HEAs do not encompass meso-scale microstructures, such as short-range order, dislocation walls, and dislocation cells, which pose significant challenges for future research. As a result, machine learning offers a solution to overcome such limitations.

3.5.5 Medicine

The HEAs are increasingly used in medical applications due to their exceptional mechanical properties. Future investigations should optimize processing techniques to avoid elemental segregation and improve coating approaches for the extensive composition range of HEAs.

3.5.6 Biology

The Cu/Ag-bearing HEAs show excellent antibacterial and antivirus performance, making them promising materials for bio-applications. Additional research is required to refine the development concepts and processing means for creating new bio-HEAs.

3.5.7 Environment

The exceptional hydrogen storage capability and decolorization performance of HEAs can be derived from their distinct crystal structure, phase stability, and four core effects.

However, challenges exist that restrict large-scale commercial use of HEAs, such as the presence of high-density elements like Zr or Hf decreasing hydrogen-storage capability. Low-density elements are introduced to increase hydrogen-storage capacity in the future. Additionally, it is critical to manage the reaction speed between azo dyes and powder HEAs to minimize usage and enhance the recyclability of HEAs.

3.5.8 X Applications

The HEAs will impact multiple disciplines, including geoscience, anthropology, economics, psychology, art, linguistics, and cognitive science. Their infinite combination space requires new vocabulary to describe, promoting the development of phonology and human understanding.

In summary, the HEAs represent a new frontier in material science, formed by combining conventional and unconventional physical metallurgy in a sophisticated manner. Their unique properties create vast opportunities while also presenting significant challenges, calling for ongoing and future research.

REFERENCES

[1] Yeh J.W., Chen S.K., Lin S.J., et al. 2004. Nanostructured high-entropy alloys with multiple principal elements: novel alloy design concepts and outcomes. *Advanced Engineering Materials*, 6(5): 299–303.

[2] Cantor B., Chang I., Knight P., et al. 2004. Microstructural development in equiatomic multicomponent alloys. *Materials Science and Engineering: A*, 375: 213–218.

[3] Thurston K.V., Gludovatz B., Yu Q., et al. 2019. Temperature and load-ratio dependent fatigue-crack growth in the CrMnFeCoNi high-entropy alloy. *Journal of Alloys and Compounds*, 794: 525–533.

[4] Li Z., Pradeep K.G., Deng Y., et al. 2016. Metastable high-entropy dual-phase alloys overcome the strength–ductility trade-off. *Nature*, 534(7606): 227–230.

[5] Zhang Z., Mao M., Wang J., et al. 2015. Nanoscale origins of the damage tolerance of the high-entropy alloy CrMnFeCoNi. *Nature Communications*, 6(1): 10143.

[6] Shi P., Li R., Li Y., et al. 2021. Hierarchical crack buffering triples ductility in eutectic herringbone high-entropy alloys. *Science*, 373(6557): 912–918.

[7] Zhang M., Zhu L., Zhou G., et al. 2021. Single-layered organic photovoltaics with double cascading charge transport pathways: 18% efficiencies. *Nature Communications*, 12(1): 309.

[8] Liang Y.-J., Wang L., Wen Y., et al. 2018. High-content ductile coherent nanoprecipitates achieve ultrastrong high-entropy alloys. *Nature Communications*, 9(1): 4063.

[9] Ma E. 2020. Unusual dislocation behavior in high-entropy alloys. *Scripta Materialia*, 181: 127–133.

[10] Hu Y., Shu L., Yang Q., et al. 2018. Dislocation avalanche mechanism in slowly compressed high entropy alloy nanopillars. *Communications Physics*, 1(1): 61.

[11] Li J., Chen Y., He Q., et al. 2022. Heterogeneous lattice strain strengthening in severely distorted crystalline solids. *Proceedings of the National Academy of Sciences*, 119(25): e2200607119.

[12] Li J., Fang Q., Liu B., et al. 2016. Mechanical behaviors of AlCrFeCuNi high-entropy alloys under uniaxial tension via molecular dynamics simulation. *RSC Advances*, 6(80): 76409–76419.

[13] Feng S., Li L., Chan K.C., et al. 2019. Tuning deformation behavior of Cu0. 5CoNiCrAl high-entropy alloy via cooling rate gradient: An atomistic study. *Intermetallics*, 112: 106553.
[14] Zhao Q., Li J., Fang Q., et al. 2019. Effect of Al solute concentration on mechanical properties of AlxFeCuCrNi high-entropy alloys: A first-principles study. *Physica B: Condensed Matter*, 566: 30–37.
[15] Zhang Z., Mao M.M., Wang J., et al. 2015. Nanoscale origins of the damage tolerance of the high-entropy alloy CrMnFeCoNi. Nature *Communications*, 6(1): 1–6.
[16] Fu Z., Chen W., Fang S., et al. 2013. Alloying behavior and deformation twinning in a CoNiFeCrAl0.6Ti0.4 high entropy alloy processed by spark plasma sintering. *Journal of Alloys and Compounds*, 553: 316–323.
[17] Wang Z., Li J., Fang Q., et al. 2017. Investigation into nanoscratching mechanical response of AlCrCuFeNi high-entropy alloys using atomic simulations. *Applied Surface Science*, 416: 470–481.
[18] Li W., Fan H., Tang J., et al. 2019. Effects of alloying on deformation twinning in high entropy alloys. *Materials Science and Engineering: A*, 763: 138143.
[19] Li J., Fang Q., Liu B., et al. 2018. Transformation induced softening and plasticity in high entropy alloys. *Acta Materialia*, 147: 35–41.
[20] Fang Q., Chen Y., Li J., et al. 2019. Probing the phase transformation and dislocation evolution in dual-phase high-entropy alloys. *International Journal of Plasticity*, 114: 161–173.
[21] Liu W.H., Wu Y., He J.Y., et al. 2013. Grain growth and the Hall–Petch relationship in a high-entropy FeCrNiCoMn alloy. *Scripta Materialia*, 68(7): 526–529.
[22] Li J., Fu X., Feng H., et al. 2023. Evaluating the solid solution, local chemical ordering, and precipitation strengthening contributions in multi-principal-element alloys. *Journal of Alloys and Compounds*, 938: 168521.
[23] Wang Z., Fang Q., Li J., et al. 2018. Effect of lattice distortion on solid solution strengthening of BCC high-entropy alloys. *Journal of Materials Science & Technology*, 34(2): 349–354.
[24] Wen C., Wang C., Zhang Y., et al. 2021. Modeling solid solution strengthening in high entropy alloys using machine learning. *Acta Materialia*, 212: 116917.
[25] Toda-Caraballo I. and Rivera-Díaz-del-Castillo P.E.J. 2015. Modelling solid solution hardening in high entropy alloys. *Acta Materialia*, 85: 14–23.
[26] Coury F.G., Kaufman M. and Clarke A.J. 2019. Solid-solution strengthening in refractory high entropy alloys. *Acta Materialia*, 175: 66–81.
[27] LaRosa C.R., Shih M., Varvenne C., et al. 2019. Solid solution strengthening theories of high-entropy alloys. *Materials Characterization*, 151: 310–317.
[28] Stepanov N.D., Shaysultanov D.G., Chernichenko R.S., et al. 2017. Effect of thermomechanical processing on microstructure and mechanical properties of the carbon-containing CoCrFeNiMn high entropy alloy. *Journal of Alloys and Compounds*, 693: 394–405.
[29] Liu W.H., Yang T. and Liu C.T. 2018. Precipitation hardening in CoCrFeNi-based high entropy alloys. *Materials Chemistry and Physics*, 210: 2–11.
[30] Tong Y., Chen D., Han B., et al. 2019. Outstanding tensile properties of a precipitation-strengthened FeCoNiCrTi0. 2 high-entropy alloy at room and cryogenic temperatures. *Acta Materialia*, 165: 228–240.
[31] He J.Y., Wang H., Huang H.L., et al. 2016. A precipitation-hardened high-entropy alloy with outstanding tensile properties. *Acta Materialia*, 102: 187–196.
[32] Zhou K., Wang Z., He F., et al. 2020. A precipitation-strengthened high-entropy alloy for additive manufacturing. *Additive Manufacturing*, 35: 101410.

[33] Liu L., Zhang Y., Han J., et al. 2021. Nanoprecipitate-strengthened high-entropy alloys. *Advanced Science*, 8(23): 2100870.
[34] Li J., Chen H., Fang Q., et al. 2020. Unraveling the dislocation–precipitate interactions in high-entropy alloys. *International Journal of Plasticity*, 133: 102819.
[35] Senkov O.N., Isheim D., Seidman D.N., et al. 2016. Development of a refractory high entropy superalloy. *Entropy*, 18(3): 102.
[36] Ritchie R.O. 2011. The conflicts between strength and toughness. *Nature Materials*, 10(11): 817–822.
[37] Chen Y., Li J., Liu B., et al. 2022. Unraveling hot deformation behavior and microstructure evolution of nanolamellar TiAl/Ti_3Al composites. *Intermetallics*, 150: 107685.
[38] Fang Q., Chen Y., Li J., et al. 2018. Microstructure and mechanical properties of FeCoCrNiNbX high-entropy alloy coatings. *Physica B: Condensed Matter*, 550: 112–116.
[39] Chen Y., Fang Q., Luo S., et al. 2022. Unraveling a novel precipitate enrichment dependent strengthening behaviour in nickel-based superalloy. *International Journal of Plasticity*, 155: 103333.
[40] Xu Q., Guan H.Q., Zhong Z.H., et al. 2021. Irradiation resistance mechanism of the CoCrFeMnNi equiatomic high-entropy alloy. *Scientific Reports*, 11(1): 1–8.
[41] Lin Y., Yang T., Lang L., et al. 2020. Enhanced radiation tolerance of the Ni-Co-Cr-Fe high-entropy alloy as revealed from primary damage. *Acta Materialia*, 196: 133–143.
[42] Held L.I. 2021. Designing FeNiCr (CoCu) High Entropy Alloys Using Molecular Dynamics-a Study of the Enhanced Mechanical Properties of a Novel Group of Composites, Rochester Institute of Technology.
[43] Ruicheng F., Hui C., Haiyan L., et al. 2018. Effects of vacancy concentration and temperature on mechanical properties of single-crystal γ-TiAl based on molecular dynamics simulation. *High Temperature Materials and Processes*, 37(2): 113–120.
[44] Peng J., Li L., Li F., et al. 2021. The predicted rate-dependent deformation behaviour and multistage strain hardening in a model heterostructured body-centered cubic high entropy alloy. *International Journal of Plasticity*, 145: 103073.
[45] Li J., Chen H., Fang Q., et al. 2020. Unraveling the dislocation–precipitate interactions in high-entropy alloys. *International Journal of Plasticity*, 133: 102819.
[46] Jian W.W., Cheng G.M., Xu W.Z., et al. 2013. Physics and model of strengthening by parallel stacking faults. *Applied Physics Letters*, 103(13): 133108.
[47] Peng J., Xie B., Zeng X., et al. 2022. Vacancy dependent mechanical behaviors of high-entropy alloy. *International Journal of Mechanical Sciences*, 218: 107065.
[48] Li L., Xie B., Fang Q., et al. 2021. Machine learning approach to design high entropy alloys with heterogeneous grain structures. *Metallurgical and Materials Transactions A*, 52(2): 439–448.
[49] Li L., Chen H., Fang Q., et al. 2020. Effects of temperature and strain rate on plastic deformation mechanisms of nanocrystalline high-entropy alloys. *Intermetallics*, 120: 106741.
[50] Li J., Li L., Jiang C., et al. 2020. Probing deformation mechanisms of gradient nanostructured CrCoNi medium entropy alloy. *Journal of Materials Science & Technology*, 57: 85–91.
[51] Vayssette B., Saintier N., Brugger C., et al. 2019. Numerical modelling of surface roughness effect on the fatigue behavior of Ti-6Al-4V obtained by additive manufacturing. *International Journal of Fatigue*, 123: 180–195.
[52] Chen B., Wu W.-P., Chen M.-X., et al. 2020. Molecular dynamics study of fatigue mechanical properties and microstructural evolution of Ni-based single crystal superalloys under cyclic loading. *Computational Materials Science*, 185: 109954.

[53] Peng J., Li F., Liu B., et al. 2020. Mechanical properties and deformation behavior of a refractory multiprincipal element alloy under cycle loading. *Journal of Micromechanics and Molecular Physics*, 5(04): 2050014.
[54] Stukowski A. 2009. Visualization and analysis of atomistic simulation data with OVITO–the Open Visualization Tool. *Modelling and Simulation in Materials Science and Engineering*, 18(1): 015012.
[55] Wu S.W., Wang G., Yi J., et al. 2017. Strong grain-size effect on deformation twinning of an Al0. 1CoCrFeNi high-entropy alloy. *Materials Research Letters*, 5(4): 276–283.
[56] Chen S., Aitken Z.H., Wu Z., et al. 2020. Hall-Petch and inverse Hall-Petch relations in high-entropy CoNiFeAlxCu1-x alloys. *Materials Science and Engineering: A*, 773: 138873.
[57] Zhang L. and Shibuta Y. 2020. Inverse Hall-Petch relationship of high-entropy alloy by atomistic simulation. *Materials Letters*, 274: 128024.
[58] Juan C.-C., Tsai M.-H., Tsai C.-W., et al. 2016. Simultaneously increasing the strength and ductility of a refractory high-entropy alloy via grain refining. *Materials Letters*, 184: 200–203.
[59] Chen S., Tseng K.-K., Tong Y., et al. 2019. Grain growth and Hall-Petch relationship in a refractory HfNbTaZrTi high-entropy alloy. *Journal of Alloys and Compounds*, 795: 19–26.
[60] Shi R., Nie Z., Fan Q., et al. 2018. Correlation between dislocation-density-based strain hardening and microstructural evolution in dual phase TC6 titanium alloy. *Materials Science and Engineering: A*, 715: 101–107.
[61] Panzarino J.F., Ramos J.J. and Rupert T.J. 2015. Quantitative tracking of grain structure evolution in a nanocrystalline metal during cyclic loading. *Modelling and Simulation in Materials Science and Engineering*, 23(2): 025005.
[62] Bahadur F., Bisis K. and Gurao N.P. 2020. Micro-mechanisms of microstructural damage due to low cycle fatigue in CoCuFeMnNi high entropy alloy. *International Journal of Fatigue*, 130: 105258.
[63] Rogal L., Wdowik U.D., Szczerba M., et al. 2021. Deformation induced twinning in hcp/bcc Al10Hf25Nb5Sc10Ti25Zr25 high entropy alloy–microstructure and mechanical properties. *Materials Science and Engineering: A*, 802: 140449.
[64] Gao Y., Zhang Y. and Wang Y. 2020. Determination of twinning path from broken symmetry: A revisit to deformation twinning in bcc metals. *Acta Materialia*, 196: 280–294.
[65] Chen Y., Xie B., Liu B., et al. 2022. A focused review on engineering application of multi-principal element alloy. *Frontiers in Materials*, 8: 625.
[66] Chen G., Qiao J.W., Jiao Z.M., et al. 2019. Strength-ductility synergy of Al0. 1CoCrFeNi high-entropy alloys with gradient hierarchical structures. *Scripta Materialia*, 167: 95–100.
[67] Jin X., Zhou Y., Zhang L., et al. 2018. A novel Fe20Co20Ni41Al19 eutectic high entropy alloy with excellent tensile properties. *Materials Letters*, 216: 144–146.
[68] Tsai M.-H. 2013. Physical properties of high entropy alloys. *Entropy*, 15(12): 5338–5345.
[69] Niu C., Zaddach A.J., Oni A.A., et al. 2015. Spin-driven ordering of Cr in the equiatomic high entropy alloy NiFeCrCo. *Applied Physics Letters*, 106(16): 161906.
[70] Kao Y.-F., Chen S.-K., Chen T.-J., et al. 2011. Electrical, magnetic, and Hall properties of AlxCoCrFeNi high-entropy alloys. *Journal of Alloys and Compounds*, 509(5): 1607–1614.
[71] Körmann F., Ikeda Y., Grabowski B., et al. 2017. Phonon broadening in high entropy alloys. *Npj Computational Materials*, 3(1): 1–9.

4 High-Entropy Alloys

Model

4.1 INTRODUCTION

The mechanical behavior of high-entropy alloys (HEAs) relies on their composition and microstructure. However, the complexity of the composition and the uncertainty of the microstructure cause great challenges to the modeling for HEAs [1]. For example, severe lattice distortion and the lattice distortion-induced inherent non-uniform strain fields in HEAs lead to complex dislocations and microstructural characteristics, resulting in unique mechanical behaviors [2,3]. This phenomenon is difficult to describe using traditional strengthening and toughening mechanisms. Therefore, it is necessary to use multiscale modeling to reveal the interaction mechanism between dislocation and microstructure considering severe lattice distortion in HEAs. Meanwhile, HEAs usually have good mechanical properties in complex environments, such as irradiated condition [4]. Nevertheless, the related micro-mechanisms are still unclear. Hence, using modeling to investigate the relevant deformation mechanisms of HEAs in complex environments is very crucial. In this chapter, lattice-distortion dependent strength model, irradiation hardening model, and hierarchical multiscale simulation are introduced in detail.

4.2 LATTICE-DISTORTION DEPENDENT STRENGTH MODEL

4.2.1 Modeling Lattice Distortion Strengthening

The lattice distortion-induced stress in dilute alloys is very small. For dilute solid solutions, the simple expressions for solid solution strengthening have been established [5,6]. Although the solid solution strengthening models have a good applicability in traditional metals, they are not good enough for the strength of HEAs. This is due to the lattice distortion effect being particularly remarkable in HEAs compared with traditional alloys [7,8] (Figure 4.1).

Lattice distortion arises from the modulus mismatch and atomic-size mismatch in the alloys. The mismatch for atomic size and modulus between *i* and *j* atoms is expressed as [9]:

$$\delta r_{ij} = 2(r_i - r_j)/(r_i + r_j) \tag{4.1}$$

$$\delta G_{ij} = 2(G_i - G_j)/(G_i + G_j) \tag{4.2}$$

DOI: 10.1201/9781003225706-4

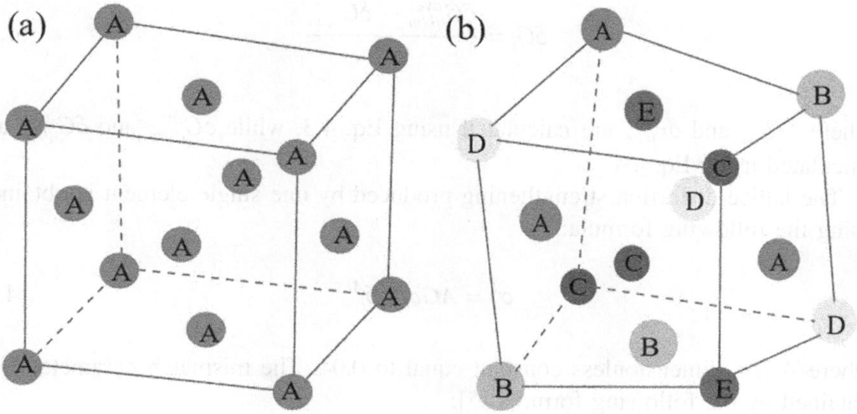

FIGURE 4.1 Sketch map of face-centered cubic (FCC) alloy: (a) Alloy with single element; (b) five-principal-element HEAs [8].

Assuming that HEAs are randomly and uniformly distributed, the average atomic radius and modulus mismatch are calculated by the following equation:

$$\delta r^{ave} = \sum_{i}^{n}\sum_{j}^{n} c_i c_j \delta r_{ij} = (c_1, c_2, \cdots, c_n)\begin{pmatrix} \delta r_{11} & \delta r_{12} & \cdots & \delta r_{1n} \\ \delta r_{21} & \delta r_{22} & \cdots & \delta r_{2n} \\ \vdots & \vdots & \ddots & \vdots \\ \delta r_{n1} & \delta r_{n2} & \cdots & \delta r_{nn} \end{pmatrix}\begin{pmatrix} c_1 \\ c_2 \\ \vdots \\ c_n \end{pmatrix} \tag{4.3}$$

$$\delta G^{ave} = \sum_{i}^{n}\sum_{j}^{n} c_i c_j \delta G_{ij} = (c_1, c_2, \cdots, c_n)\begin{pmatrix} \delta G_{11} & \delta G_{12} & \cdots & \delta G_{1n} \\ \delta G_{21} & \delta G_{22} & \cdots & \delta G_{2n} \\ \vdots & \vdots & \ddots & \vdots \\ \delta G_{n1} & \delta G_{n2} & \cdots & \delta G_{nn} \end{pmatrix}\begin{pmatrix} c_1 \\ c_2 \\ \vdots \\ c_n \end{pmatrix} \tag{4.4}$$

where δG_{ij} and δr_{ij} denote mismatch parameters between elements i and j. Here, δr_{nn} and δG_{nn} are equal to zero. The mismatch probability for i and i is $c_i c_i$, and the probability of a mismatch between j and i is $c_i c_j$, which is the same as the probability of mismatch parameters between i and j.

Based on the average mismatch parameters calculated above, we then explore the mismatches caused by individual elements in HEAs. Assuming *ijklm* HEAs consist of alloy *jklm* and element *i*. The mismatch parameters in *ijklm* HEAs are obtained by:

$$\delta r_i = \frac{\delta r_{ijklm}^{ave} - \delta r_{jklm}^{ave}}{\delta c_i} \tag{4.5}$$

$$\delta G_i = \frac{\delta G_{ijklm}^{ave} - \delta G_{jklm}^{ave}}{\delta c_i} \tag{4.6}$$

where δr_{ijklm}^{ave} and δr_{jklm}^{ave} are calculated using Eq. 4.3, while δG_{ijklm}^{ave} and δG_{ijklm}^{ave} are calculated using Eq. 4.4.

The lattice distortion strengthening produced by one single element is obtained using the following formula:

$$\sigma_f^i = AGc_i^{2/3}\delta_i^{4/3} \tag{4.7}$$

where A is a dimensionless constant equal to 0.04. The mismatch parameter δ_i is obtained by the following formula [5]:

$$\delta_i = \xi\,(\delta G_i^2 + \beta^2\delta r_i^2)^{1/2} \tag{4.8}$$

where the modulus mismatch parameter δG_i and size mismatch parameter δr are obtained using Eqs. 4.5–4.6. ξ depends on the crystal structure, while β depends on the type of dislocations. Based on Vegard's law, the lattice distortion strengthening in HEAs is obtained by the following formula:

$$\sigma_f = \sum c_i \sigma_f^i \tag{4.9}$$

4.2.2 Yield Strength

In order to calculate the yield strength of HEAs, this section should consider the hindering effect from grain boundaries and precipitation on dislocation motion in the alloy. The grain boundary strengthening is obtained by the following formula:

$$\sigma_{GB} = k/\sqrt{d} \tag{4.10}$$

The precipitation strengthening is mainly controlled by the looping and cutting mechanisms. The expression for precipitation strengthening is given by [10]:

$$\sigma_{pre} = \frac{0.4MGb}{\pi\sqrt{1-v}}\frac{\ln(2\bar{r}/b)}{L_p} \tag{4.11}$$

where L_p represents average precipitation spacing:

$$L_p = \left(\sqrt{\frac{\pi}{4f}} - 1\right)\cdot 2\bar{r} \tag{4.12}$$

Combining lattice distortion strengthening, grain boundary strengthening, and precipitation strengthening, the strength is obtained by the following formula:

$$\sigma_y = \sigma_f + \sigma_{GB} + \sigma_{pre} \tag{4.13}$$

4.2.3 Model Validation

Figure 4.2(a) compares the calculation with experimental data for Al0.3CrCoFeNi HEA [11]. The Hall-Patch coefficient k is reported to be 824 MPa·μm$^{0.5}$ according to previous work [12]. The theoretical model of lattice distortion strengthening, incorporates atomic radii and shear moduli of each element (Table 4.1). The results confirm that the yield strength predicted by the proposed model agrees well with experimental data. Here, the yield stress is made up of grain boundary strengthening, lattice distortion strengthening, and precipitate strengthening. The yield stress value obtained experimentally is 1147 MPa, while the current model predicts a value of 1128 MPa, a deviation of only 1.7%.

To further confirm the plausibility of the current model, the predicted results from lattice distortion strengthening are compared with experimental data. The contribution of lattice distortion strengthening is 165 MPa in the experimental result [12], and the model calculation result is 177 MPa, a difference of only 7.2% (Figure 4.2b). This result shows that the prediction value agrees well with experimental result. Figure 4.2c presents the correlation of experiments and calculation for different FCC HEAs [13]. Here, the pink area indicates that the deviation value between the experiment and the prediction is less than 10%, and the black line indicates that the lattice distortion strengthening from experiments and theoretical model is entirely consistent. The theoretical model can capture most previously published results, where the error of CoCrFeNiMn is 7.2%, CoCrNiMn is 15%, CrCoFeNi is 11.7%, FeNiCoMn is 6%, FeNiMn is 17.8%, and CoNiMn is 0.7%. These results suggest that lattice distortion strengthening is dependent on composition, in accordance with the published work [2]. Therefore, the proposed theoretical model can capture the primary features of lattice distortion strengthening as a function of alloy composition in FCC HEAs.

To assess the plausibility of the theory for BCC HEAs, theoretical results are compared with published work for BCC HEAs. Figure 4.2d shows that for BCC TaNbHfZrTi HEA, the model predicts a yield stress of 960 MPa, whereas the experimental yield stress is 1,073 MPa or 929 MPa [9,14]. The two different experimental results originate from two distinct works in the literature [9,14], and the variance in the outcomes is mainly due to differences in the preparation and processing technology of alloys. Additionally, the yield strength in TaNbHfZrTi HEAs comes from grain boundary strengthening (22 MPa) and lattice distortion strengthening (938 MPa), indicating that lattice distortion strengthening dominates the yield stress. Figure 4.2e illustrates a comparison of published work and the theoretical model in individual BCC HEAs. The pink area represents the error between published work and the calculation which is less than 10%. The results indicate that the model can match most of the published work.

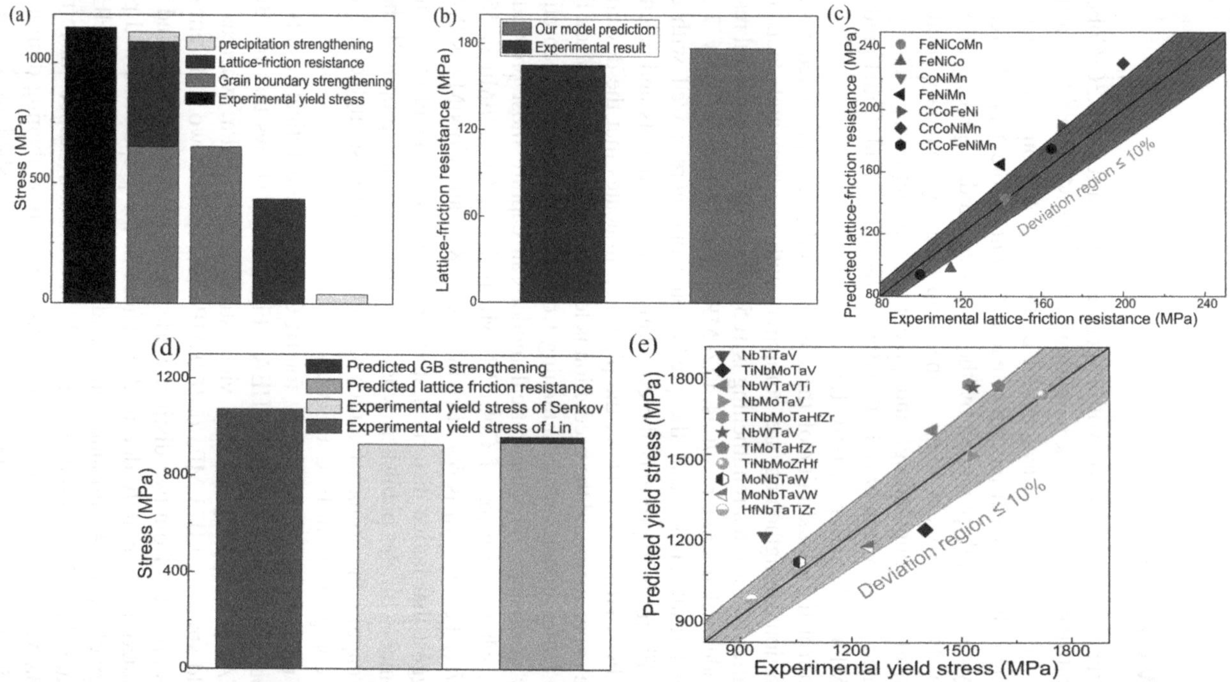

FIGURE 4.2 Model validation [8]. (a) Experimental results vs. predicted results; (b) Comparison of lattice distortion strengthening of CrCoFeNiMn HEA: theoretical results vs. experimental results; (c) Deviation relationship between experimental and predicted results of lattice distortion strengthening for various FCC HEAs [13]. (d) Comparison of experimental yield stress and theoretical results for body-centered cubic (BCC) TaNbHfZrTi HEAs [9,14]. (e) Deviation relationship of the yield stress between experimental and theoretical calculation in BCC HEAs [15].

TABLE 4.1
Parameters of the Constituent Elements

Elements	Al	Cr	Fe	Ni	Co
Atomic fraction, c (%)	6.98	23.26	23.26	23.26	23.26
Atomic radius, r (pm)	143	128	126	124	125
Shear modulus, G (GPa)	26	115	82	76	75

Published experimental results reveal that the crystal structure of $Al_xCrCoFeNiMn$ HEAs changes from original FCC to the dual-phase structures [16]. Figure 4.3a presents the comparison of the predicted lattice distortion stress and experiment. The phenomenon demonstrates the predictions capture the published work. Therefore, the proposed model accurately predicts the trend of yield stress and FCC to BCC phase transition as Al concentration increases. Figure 4.3b-d illustrates the trend of atomic size mismatch, shear modulus mismatch, and strengthening by various elements with increasing Al concentration. The mismatch parameters by Al elements are obviously larger than others, the strengthening contribution produced from Al elements increased radically to dominate the lattice distortion stress. In Figure 4.3d, when the value of $x < 7$, the stress produced by each element remains invariable.

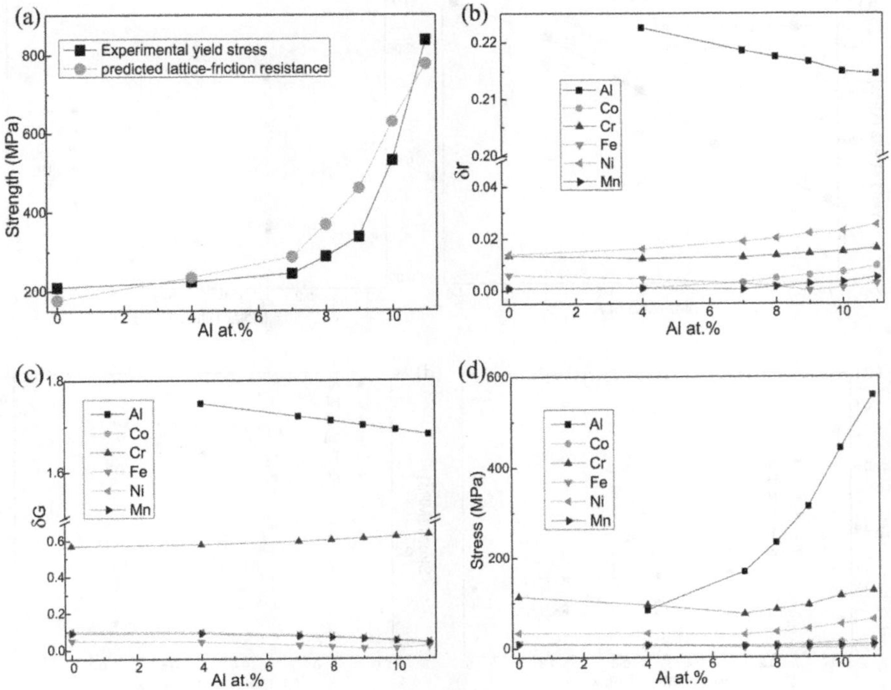

FIGURE 4.3 Lattice distortion strengthening in the $Al_xCrCoFeNiMn$ [8].

As the Al content increases, the stress contributed from each element increases significantly, originating from the occurrence of BCC phase. Therefore, the current model can capture the trend of strengthening as the formation of BCC structure. It identifies the effect of element on lattice distortion stress.

Prior studies show that $Al_xHfNbTaTiZr$ HEAs are BCC alloys when the values of x range from 0 to 1 [14]. In Figure 4.4a, as Al content increases, yield stress of the current model matches the experimental results well. In addition, calculation results exhibit an enhancing tendency with increasing Al content, which is consistent with previous experiments [14]. All calculation results are lower than the published work because of the underestimation of elastic constant. The elastic constant exhibits an enhancing tendency in the published work with the increase of Al content, but the shear modulus used in the current model exhibits a decreasing tendency based on mixing law. However, the proposed model still captures the main feature of strengthening in this alloy system. Figure 4.4b and c illustrates the evolution of mismatch parameters that is heavily influenced by Al content. Figure 4.4d depicts the trend of lattice distortion stress with various elements. Compared to the lattice distortion stress produced by other elements, the Zr element provides the greatest contribution to the lattice distortion stress, whereas the Ti element provides the minimal contribution. The contribution of Al elements to lattice distortion strengthening increases at high Al concentration.

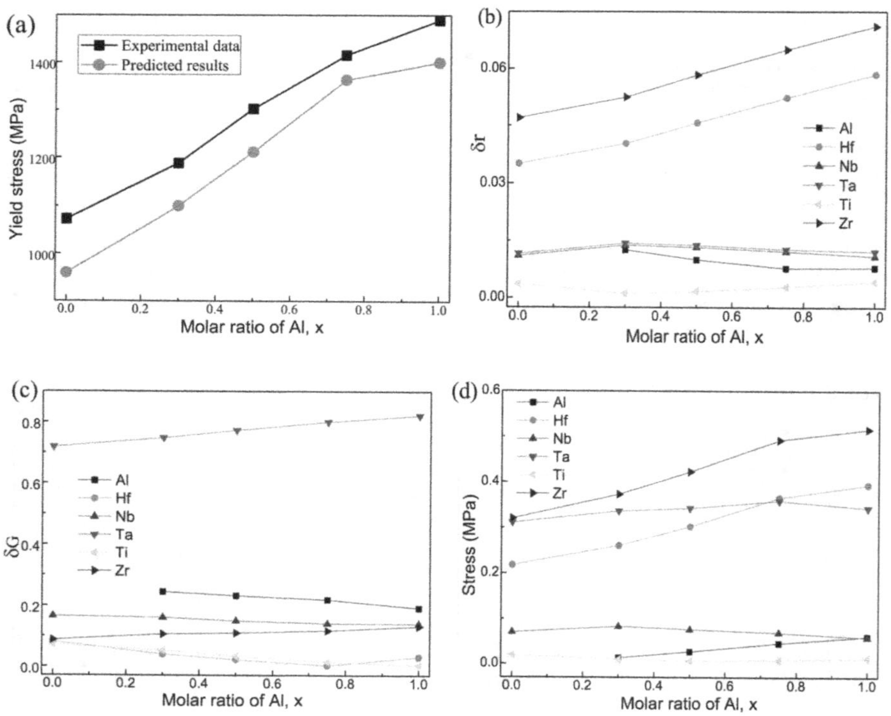

FIGURE 4.4 Lattice distortion strengthening in $Al_xHfNbTaTiZr$ HEAs [8].

4.2.4 Screening Composition for High-strength HEAs

Previous studies have suggested that size mismatch is enhanced as the number of elements increases [17], but it is not always reasonable from recent research [18]. Figure 4.5a depicts individual BCC HEAs calculated using Eq. 4.3. Surprisingly, increasing the element number may increase or decrease the size mismatch, contradicting with published work [17]. Notably, Nb decreases the size mismatch of TiMoTaHfZr because of its close atomic radius value to other elements [9,15]. Therefore, the difference in atomic size determines the degree of atomic radius mismatch and further dominates lattice distortion stress, which aligns with the conclusions of earlier studies [19]. Figure 4.5b illustrates the atomic radius mismatch calculated based on Eq. 4.3 for individual BCC HEAs and binary alloys. The results reveal that the size mismatch of traditional alloy is similar with that of HEAs, agreeing well with published research [18].

The proposed model can evaluate the influence of additional elements on the property of alloys. Figure 4.5c-d demonstrates that the shear modulus and atomic radius are dependent on lattice distortion strengthening. The contour on Figure 4.5c shows lattice distortion strengthening at 1440 MPa for NbMoTaV. The letter "X" represents various types of elements. For lattice distortion stress smaller than 1290 MPa in NbMoTaVX, the element features of X must be located at the red area in Figure 4.5c. For instance, Ti element falls into the red area, and the lattice distortion strengthening of TiNbMoTaV is 1137 MPa, smaller than that of

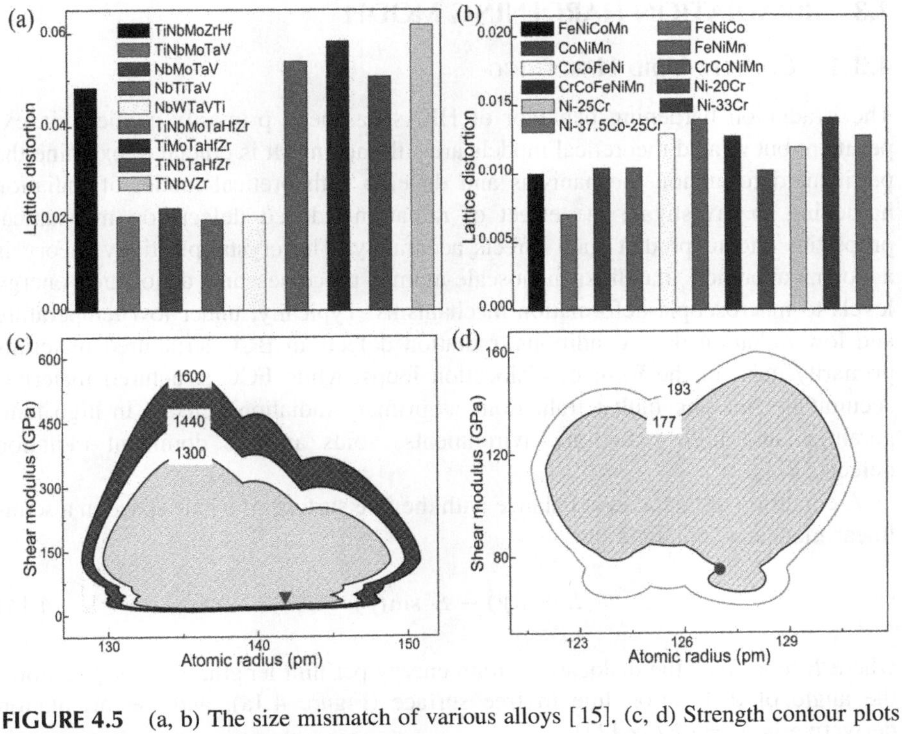

FIGURE 4.5 (a, b) The size mismatch of various alloys [15]. (c, d) Strength contour plots in NbMoTaVX and CoCrFeNiX [8].

NbMoTaV [15]. If one selects the NbMoTaVX HEA with lattice distortion strengthening greater than 1440 MPa but less than 1600 MPa, X should be located in the green area in Figure 4.5c. CoCrFeNiMn HEA falls at the border of the pink area, providing a lattice distortion strengthening of 177 MPa [20]. For CoCrFeNiX HEA with a lattice distortion stress smaller than this value, element X must be inside the pink area. Therefore, the proposed theory could offer theoretical assistance for the discovery of advanced HEAs.

4.2.5 Remark

In this section, a lattice distortion strengthening model is constructed to describe the lattice distortion of HEAs, where the elements' features (atomic radius and shear modulus) in HEAs are used as functional elements. The model reveals the effect of size mismatch and elastic mismatch of different compositions on yield strength in HEAs. The yield strength calculated based on the proposed theory is consistent with published work for both FCC and BCC HEAs. Further, it is elucidated that the size mismatch is significantly higher than the elastic mismatch to the lattice distortion strength, thus dominating the mechanical performance of alloys. By constructing the strength contour, the contribution of different element types to the yield strength is evaluated, offering a scientific basis for screening and discovering high-strength HEAs.

4.3 IRRADIATION HARDENING MODEL

4.3.1 Complex Void Hardening

The irradiation hardening behavior of HEAs has been primarily studied via experiment but related theoretical models are still lacking. It is crucial to examine the pertinent deformation mechanisms and develop a theoretical model of radiation hardening to investigate the effect of radiation-induced defects on mechanical properties and to predict such effects accurately. The crystal-plasticity theory is a potent approach that links nanoscale-atomic processes and dislocation-energy levels to macroscopic-deformation mechanisms. Typically, under low temperature and low radiation dose conditions, radiation defects in BCC structured materials primarily exist in the form of dislocation loops, while FCC structured materials accumulate stacking fault tetrahedrons as primary radiation defects. In high temperatures and high radiation environments, voids are the dominant radiation defects [21].

A condition of dislocation balance with the free surface of a half-space in a semi-linear manner is obtained by:

$$E\cos(\gamma) - E'\sin(\gamma) = 0 \tag{4.14}$$

where E represents the dislocation strain energy per unit length, $\gamma = \alpha + \beta$ denotes the angle of dislocation line to free surface (Figure 4.1a), and the orientation derivative is $E' = \partial E / \partial \gamma$ [22].

The strain energy associated with a mixed character dislocation is denoted by

$$E = \frac{\mu b^2}{4\pi}\left(\cos^2\theta + \frac{\sin^2\theta}{1-\nu}\right)\ln\frac{R}{r_c} \tag{4.15}$$

For edge dislocation, the relationship between the angles α and β is:

$$(1 - \nu\cos^2\alpha)\cos(\alpha+\beta) - \nu\sin 2\alpha\,\sin(\alpha+\beta) = 0 \tag{4.16}$$

Figure 4.6a shows an illustration of edge dislocation bypassing the rhombic void [23]. The dislocation segment IA, moves a distance S in the Burgers vector direction to form the UI' segment. The UI' (QA') segment length is given by $\Delta m = S\sin\beta/\sin(\alpha+\beta)$. The UI' (QA') segment strain energy is denoted by $\Delta w = \frac{\Delta m\mu b^2}{4\pi}\left(\cos^2\alpha + \frac{\sin^2\alpha}{1-\nu}\right)\ln\left(\frac{R}{r_c}\right)$. The segment IA evolves into UI'QA' configuration. The work done by the shear stress is approximated as $\tau_c^{\mathrm{rhom}} bLS$, which corresponds to the increase in dislocation energy resulting from the formation of the additional segments, UI' and QA' ($\tau_c^{\mathrm{rhom}} bLS = 2\Delta w$) [22]. The critical resolved shear stress (CRSS) for dislocation bypassing rhombic voids is estimated by [23]:

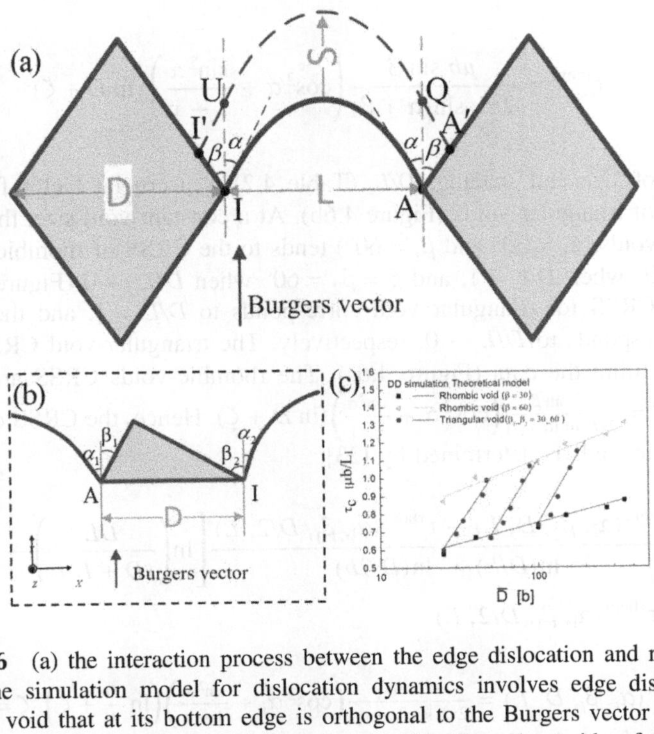

FIGURE 4.6 (a) the interaction process between the edge dislocation and rhombic void [23]. (b) the simulation model for dislocation dynamics involves edge dislocation and a triangular void that at its bottom edge is orthogonal to the Burgers vector [23]. (c) the CRSS is the function of harmonic mean for rhombic/triangular voids of $\beta_1 = 30°$ and $\beta_1 = 60°$ [23,24].

TABLE 4.2
Void Spacing and Size Utilized in This Study Are Expressed in Terms of the Burgers-Vector Magnitude [23]

D (b)	L (b)	D/L
50	50	1
50	75	0.67
50	100	0.5
50	200	0.25
50	400	0.125
100	100	1
100	150	0.67
100	200	0.5
100	400	0.25
100	800	0.125
200	200	1
200	300	0.67
200	400	0.5
200	800	0.25
200	1,600	0.125

$$\tau_c^{\text{rhom}} = \frac{\mu b \sin\beta}{2\pi L \sin(\alpha+\beta)}\left(\cos^2\alpha + \frac{\sin^2\alpha}{1-v}\right)(\ln \bar{D} + \zeta) \tag{4.17}$$

The ratio of size and spacing, *D/L*, (Table 4.2) is a crucial factor for affecting the CRSS of triangular voids (Figure 4.6b). At a constant void size, the CRSS for triangular voids ($\beta_1 = 30°$ and $\beta_1 = 60°$) tends to the CRSS of rhombic voids with $\beta = \beta_1 = 30°$ when $D/L = 1$, and $\beta = \beta_2 = 60°$ when $D/L \to 0$ (Figure 4.6c). The minimum CRSS for triangular void corresponds to $D/L = 1$, and the maximum CRSS corresponds to $D/L \to 0$, respectively. The triangular-void CRSS is determined by fitting the data (Figure 4.6c). The rhombic-voids CRSS are computed using $\tau_c^{\text{rhom}} = \frac{\mu b \sin\beta}{2\pi L \sin(\alpha+\beta)}\left(\cos^2\alpha + \frac{\sin^2\alpha}{1-v}\right)(\ln \bar{D} + \zeta)$. Hence, the CRSS contribution of triangular voids is determined by [23]:

$$\tau_c^{tria} = \frac{\tau_c^{\text{rhom}}(\alpha_2, \beta_2, D, L) - \tau_c^{\text{rhom}}(\alpha_1, \beta_1, D/2, L)}{\ln(D/b) - \ln(D/2b)}\left[\ln\left(\frac{DL}{(D+L)b}\right) - \ln\left(\frac{D}{2b}\right)\right] + \tau_c^{\text{rhom}}(\alpha_1, \beta_1, D/2, L) \tag{4.18}$$

where $\tau_c^{\text{rhom}}(\alpha_i, \beta_i, D, L) = \frac{\mu b \sin\beta_i}{2\pi L \sin(\alpha_i+\beta_i)}\left(\cos^2\alpha_i + \frac{\sin^2\alpha_i}{1-v}\right)\left(\ln\frac{D}{b} + \zeta\right)$, $\zeta = 2.9$ by fitting the DD simulation data for $i = 1$ and 2 (red lines in Figure 4.6c). It is observed that the CRSS of the triangular voids with varying sizes falls near the three red lines (Figure 4.6c). This indicates the accuracy of the CRSS formula. Moreover, Figure 4.6c clearly displays the CRSS of triangular and rhombic voids.

By considering the spatial interaction between radiation defects (characteristic planes) and dislocations (specific slip systems), the plastic deformation behavior of irradiated metals and the evolution law of radiation defects are more precisely on quantity density and size of defects. For traditional alloys, such as BCC iron and FCC copper, corresponding irradiation crystal tensor models have been proposed for low temperature and low irradiation dose conditions [23,25]. These models capture the dislocation-loop hardening in BCC metals, and the spatial interaction between stacking fault tetrahedron and slip dislocation in FCC metals. They accurately predict the localized plastic deformation phenomena, post-yield softening, and the propagation/annihilation of irradiated defects. At high temperatures and high radiation doses, complex-shaped voids are produced in FCC HEAs [26]. To investigate the impact of polyhedral voids on the radiation-hardening behavior of FCC HEAs, it is necessary to consider probability-dependent spatial interactions between dislocations and polyhedral voids. Due to the anisotropy of single-crystal HEAs, irradiation-induced voids often exhibit polyhedral shapes, with cubic, octahedral, and cubic-octahedron voids being more common [27].

The 12 dislocation slip systems of FCC crystals are {111}<110>. As voids are symmetrical, the interaction between dislocations and voids in different slip systems is identical. Figure 4.7 illustrates the interaction mode and corresponding probabilities of (111)[$\bar{1}$10] dislocations with cubic, octahedral, and cubic-octahedron voids [23]. Due to the probability of dislocation cutting void at each position is the same, and the shape of the dislocation cutting through different positions of the void is different, the blocking effect of void on dislocation movement depends on the three-dimensional spatial shape. In Figure 4.7a, the shape and size of the dislocation cutting through different parts of the cubic void are different. From the perspective of [110], if the dislocation is evenly divided into n points between d and c, the probability of the dislocation cutting through each point is 1/n, and the void shape exhibits a triangular (first case, first point to the nth/3rd point), hexagon (the second case, n/3+1 point to the 2n/3 point), and triangular (the third case, 2n/3+1 point to the nth point) in Figure 4.7a. The CRSS of the dislocation cutting through the three-dimensional void is obtained by superimposing the two-dimensional plane CRSS of void:

$$\tau_d^{\alpha} = \frac{1}{n}\left(\sum_{i=1}^{n/3} \tau_f^i + \sum_{i=1+n/3}^{2n/3} \tau_s^i + \sum_{i=1+2n/3}^{n} \tau_t^i\right) \tag{4.19}$$

where $\tau_f^i = \tau_t^i = \frac{\tau_c^{\text{rhom}}(17^\circ, 60^\circ, D_i, L)}{\ln(2)} \ln\left(\frac{2L}{(D_i + L)}\right)$ is the CRSS provided by the triangular void (the first and third cases in Figure 4.7a) formed by the dislocation cutting through the ith point of the void. $\tau_s^i = \frac{\sin 60^\circ \mu b}{2 \sin 77^\circ \pi L}\left(\cos^2 17^\circ + \frac{\sin^2 17^\circ}{1-\nu}\right)$ $\left(\ln\left(\frac{D_i L}{(D_i + L)b}\right) + 2.9\right)$ is the CRSS provided by the hexagon void (the second case

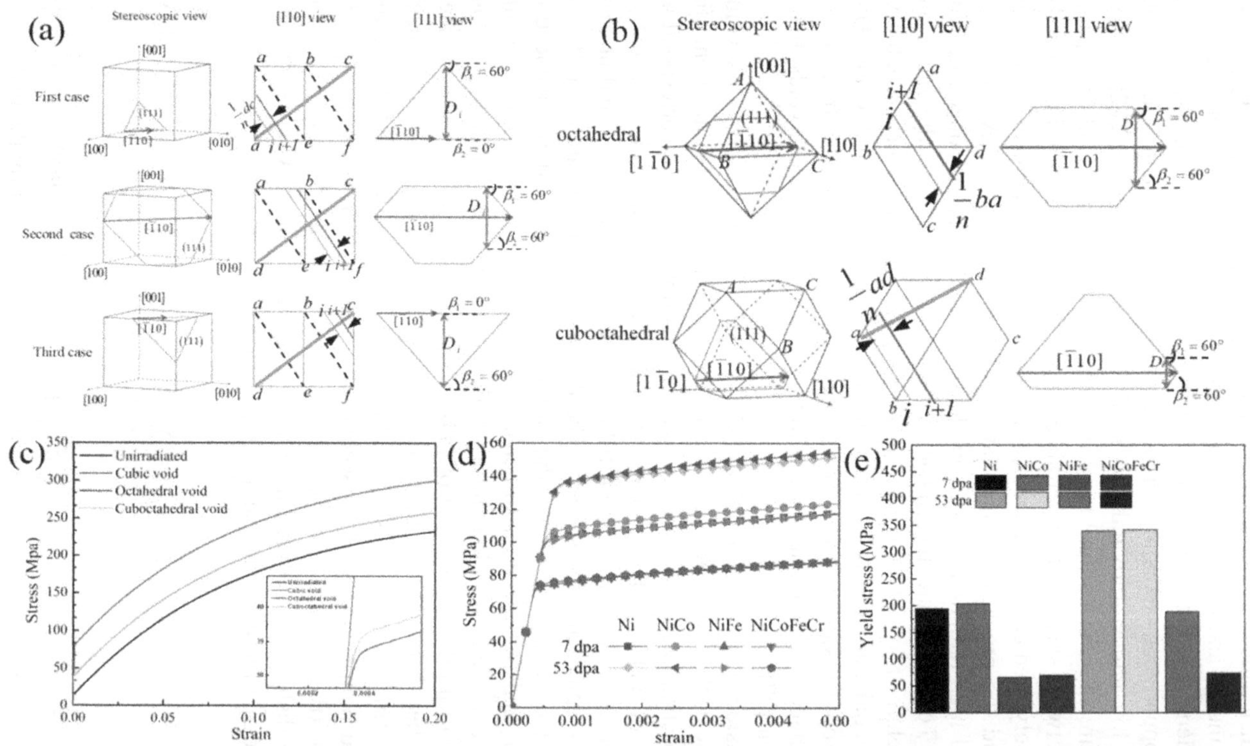

FIGURE 4.7 Schematic diagram illustrating the spatial interaction of dislocations and voids in FCC crystals: (a) cubic voids, (b) octahedral and cubic-octahedron voids. The void is depicted by the blue line, the red line illustrates the configuration of the dislocation, and the green arrow signifies the orientation of the dislocation. (c) stress-strain graphs for irradiation-induced voids. [23,25].

in Figure 4.7a caused by the dislocation cutting through the *i*th point of the void, and L is the spacing between voids [23].

In Figure 4.7b, there is only one case where the dislocation cuts through octahedral and cubic-octahedron voids:

$$\tau_d^{\alpha} = \frac{1}{n}\sum_{i=1}^{n} \tau^i \tag{4.20}$$

where $\tau_s^i = \frac{\sin 60°\mu b}{2\sin 77°\pi L}\left(\cos^2 17° + \frac{\sin^2 17°}{1-\nu}\right)\left(\ln\left(\frac{D_i L}{(D_i + L)b}\right) + 2.9\right)$ is the CRSS provided by the hexagon plane void caused by the dislocation cutting through the i-th point of the void [23].

Crystal plasticity theory combines with the CRSS formula to study the hardening effect of voids [28]. The results suggest that cubic voids provide a higher contribution to hardening than octahedral and cubic-octahedron voids, while octahedral and cubic-octahedron voids offer nearly identical hardening contributions. The dislocation cutting through the cube void will form a triangular cross-section and a hexagonal cross-section, while the dislocation cutting through the octahedron and the cubic-octahedron void only forms a hexagonal cross-section. The triangular cross-section has a larger void size under the same area, providing a higher CRSS [23].

Pure Ni, Ni-based alloys and FeNiCoCr HEAs will produce irradiation defects dominated by voids after high temperature and high dose irradiation [29]. In the irradiation conditions of 3MeV $Ni^{2+}$and 500°C, the parameters of voids for Ni, CoNi, FeNi and FeNiCoCr alloys at various irradiation doses are given in Table 4.3. The parameters for the crystal plasticity theoretical model include a Poisson's ratio

TABLE 4.3
Characteristics of Radiation-Induced Voids in Ni, NiCo, NiFe, and NiCoFeCr at Distinct Irradiation Doses During Irradiation Condition of 3MeV Ni^{2+} and 500°C

Alloy	Dose level (dpa)	Void size (nm)	Void number density (10^{20} m^{-3})	Shape
Ni	17	76.1	1.28	Octahedron
CoNi	17	21.9	7.21	Octahedron
FeNi	17	6.7	4.03	Octahedron
NiCoCrFe	17	8.2	3.37	Octahedron
Ni	53	131.6	0.596	1/2 Octahedron +1/2 Cube
CoNi	53	32.7	11.93	Octahedron
FeNi	53	44.9	0.54	1/2 Cube +1/2 Octahedron
NiCoCrFe	53	9.0	3.29	Octahedron

TABLE 4.4
Model Parameters for Crystal Plasticity of Irradiated Ni, CoNi, FeNi, and FeNiCoCr Alloys

Physical parameter	Explanation	Quantitative	Unit
K1	Kocks Mecking coefficient	2×10^9	m^{-1}
χ	Interaction coefficient	0.9	–
g	Normalized-activation energy	9×10^3	–
K	Boltzmann-constant	1.38×10^{-23}	$J\,K^{-1}$
D	Proportionality-constant	3.5×10^3	MPa
b	Burger-vector magnitude	0.257	nm
$\dot{\varepsilon}_0$	Reference strain rate	1×10^7	s^{-1}
$\dot{\gamma}_0$	Reference-plastic strain rate	1×10^{-3}	s^{-1}
m	Strain-rate sensitivity	0.05	–

of 0.34, shear modulus of 77 GPa, initial-dislocation density of $2 \times 10^{12}\ m^{-2}$, loading-strain rate of $1.2 \times 10^{-3}\ s^{-1}$, dislocation hardening coefficient of 2, and other parameters applied to the theoretical framework of crystal plasticity (Table 4.4). Only radiation hardening caused by voids is considered, and other hardening effects (such as grain boundary, lattice distortion, etc.) are not considered, to study the impact of radiation-induced defects on the radiation hardening of HEAs under varying irradiation doses.

By experimental results and void irradiation hardening model [29], single-crystal Ni, NiCo, NiFe alloy, and NiCoFeCr HEA are evaluated for their stress-strain profiles and corresponding yield strengths under various irradiation doses (Figure 4.7c-d). At a low irradiation dose of 7dpa, the irradiation hardening of single-crystal Ni and NiCo alloys is greater than that of NiFe alloys and NiCoFeCr HEAs. Meanwhile, the irradiation hardening of NiFe alloys and NiCoFeCr HEAs is identical, signifying that NiFe alloys and NiCoFeCr HEAs possess equally outstanding radiation resistance at low irradiation doses. As the irradiation dose raises to a high level of 53 dpa, the irradiation hardening of NiCoFeCr HEA remains unchanged, while the radiation hardening of NiFe alloy increases proportionally. This observation indicates that NiCoFeCr HEA has the superior radiation resistance of HEAs over traditional alloys. The high composition complexity and lattice distortion in NiCoFeCr HEA effectively hinder the migration of atomic defects [27]. Consequently, the voids produced by irradiation in HEAs are smaller, endowing HEAs with excellent radiation hardening resistance.

4.3.2 Dislocation Loop Hardening

The formation of dislocation loop under irradiation impedes the dislocation mobility within the slip system, leading to a radiation hardening effect on materials [30]. Dislocation loops are predicted to cause irradiation hardening. In Ni-based

TABLE 4.5
Relevant Parameters for the Elements Ni, Fe, Cr, and Mn [34]

	Ni	Fe	Mn	Cr
Atom radius (nm)	12.5	12.4	12.7	12.5
Shear modulus (GP)	76	82	81	115
Fraction percentage, at%	29.5	27.1	26.6	16.8

alloys (FeNi, FeNiCo, FeNiCoCr, and FeNiCoCrMn) exposed to 773 K and 35 dpa of Ni-ion irradiation [30], all the alloys produce dislocation loops. The alloy with a more complex composition leads to small-size and high-density dislocation loops. A high radiation resistance at high temperatures is indicated by the reduced radiation hardness of NiFeMnCr HEA as compared to room temperature [31]. The hardening effect of dislocation loops is studied using crystal plasticity theory. Considering the influence of lattice distortion in NiFeMnCr HEA under irradiation conditions (see Table 4.5 for physical parameters of each element), the hardening mechanism is primarily composed of dislocations, dislocation loops, and lattice distortion:

$$\tau_c^\alpha = \tau_s + \tau_n^\alpha + \tau_d^\alpha \tag{4.21}$$

The blocking effect of dislocation loop on dislocation is expressed by [32]:

$$\tau_d^\alpha(T) = b\mu(T)\sqrt{h_d(T)\sum_{\beta=1}^{N_d} \mathbf{N}^\alpha : \mathbf{H}^\beta} \tag{4.22}$$

Here, h_d means the hardening coefficient of the dislocation loop, $N_d = 4$ indicates the number of characteristic planes of the dislocation loop. $\mathbf{N}^\alpha = \mathbf{n}^\alpha \otimes \mathbf{n}^\alpha$ and $\mathbf{H}^\beta = \rho_1 \cdot 3d_1 \cdot \mathbf{M}^\beta$ are the rank-2 tensors used to describe dislocations and loops, where $\mathbf{M}^\beta = \mathbf{I}^{(2)} - \mathbf{n}^\beta \otimes \mathbf{n}^\beta$. $\mathbf{n}^\alpha$ and $\mathbf{n}^\beta$ represent the dislocation-slip-plane normal vectors and the characteristic plane of the dislocation loop. ρ_1 and d_1 denote the number density and the size of dislocation-loop, respectively. The dislocation loop will submerge due to interactions with dislocations, and its evolution follows the law described below:

$$\dot{\mathbf{H}}^\beta = -\eta \sum_{\alpha=1}^{N_S} (\mathbf{N}^\alpha : \mathbf{H}^\alpha)\mathbf{M}^\beta |\dot{\gamma}^\alpha| \tag{4.23}$$

where η is the annihilation coefficient of dislocation loop.

According to the material properties and relevant experimental data [28], the test conditions include a loading-strain rate of $2.8 \times 10^{-4}\ \text{s}^{-1}$, a mean-grain size of 35 μm, and an initial-dislocation density of $5 \times 10^{14}\ \text{m}^{-2}$. Other parameters related to the

TABLE 4.6
Crystal Plasticity Model Parameters of Irradiated NiFeMnCr HEA [34]

Parameter	Definition	Numerical	Unit
η	Annihilation parameter	10	–
K_1	Kocks Mecking coefficient	2×10^9	m^{-1}
χ	Interaction coefficient	0.9	–
g	Normalized-activation energy	9×10^3	–
K	Boltzmann coefficient	1.38×10^{-23}	$J\,K^{-1}$
D	Proportionality coefficient	10^4	MPa
b	Burger's vector coefficient	0.257	nm
$\dot{\varepsilon}_0$	Reference strain rate	1×10^7	s^{-1}
$\dot{\gamma}_0$	Reference-plastic strain rate	1×10^{-3}	s^{-1}
m	Strain-rate sensitivity	0.05	–
h_n	Dislocation-strength parameter	0.02	–
h_d	Dislocation-loop strength parameter	0.035	–

TABLE 4.7
Parameters Associated with Dislocation Loops in Irradiated NiFeMnCr HEA Specimens at Room Temperature

Dose Level (dpa)	Loop Size (nm)	Loop Number Density (m^{-3})
0.03	2.98	7.3×10^{22}
0.3	3.23	1.5×10^{23}

hardening model is listed in Table 4.6. Based on the observed dislocation loop data after irradiation experiments (Table 4.7), the number density and radius of dislocation loop increases with the increase in irradiation dose at room temperature. The theoretical calculation results align with the experimental data of nanoindentation [31,33,34]. Through the experimental-data analysis, the strength-coefficients of dislocation ($h_n = 0.02$) and the dislocation loop ($h_d = 0.036$) are determined by fitting the data obtained under different irradiation conditions. The yield strength and flow stress increase as the irradiation dose increases from 0 to 0.3 dpa (Figure 4.8a). The initial-dislocation density and lattice-distortion degree is assumed to be the same for the same material at different irradiation doses. It is observed that the transition time of NiFeMnCr HEA from elastic to plastic deformation is prolonged as the irradiation dose increases (Figure 4.8b). Figure 4.8c shows the low dislocation-growth rate at higher level of irradiation, due to the resistance of dislocation motion caused by irradiation-induced dislocation loops. The dislocation loops caused by irradiation leads to irradiation hardening in NiFeMnCr HEA and enhances its resistance to plastic deformation. The yield strength of NiFeMnCr HEA is determined using the hardening model (Figure 4.8a), which in line with experimental estimates at different levels of

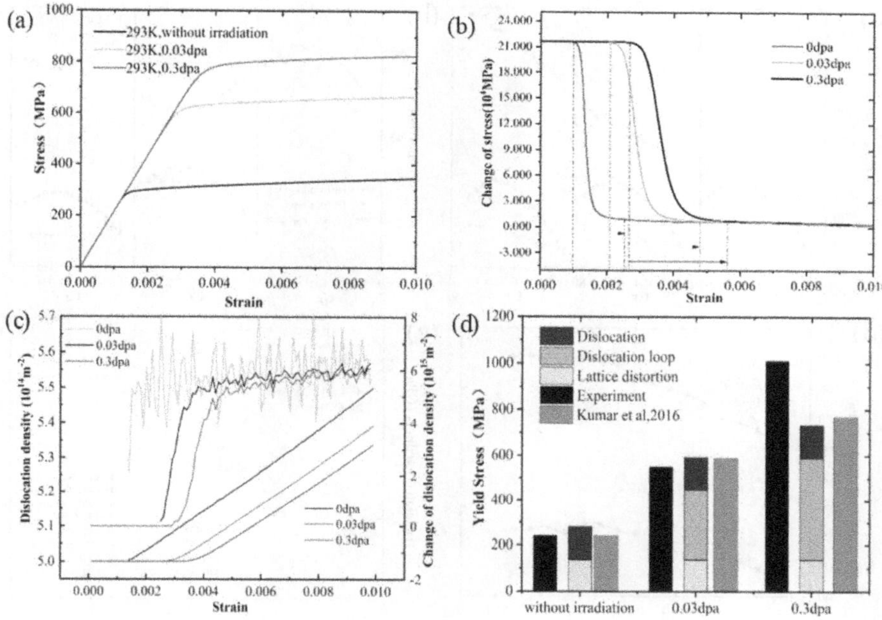

FIGURE 4.8 At 0, 0.03, and 0.3 dpa irradiation doses level at room temperature, (a) stress-strain relationships, (b) changes of stress per unit strain, (c) variations of dislocation density during plastic deformation on per unit strain, (d) the different-hardening-mechanisms contributions of yield stress in contrast to experimental and theoretical data [34].

irradiation [31]. At an irradiation dose of 0 dpa, the hardening mechanism of NiFeMnCr HEA results solely from lattice distortion and dislocation interaction (Figure 4.8d). However, the introduction of dislocation loops makes the irradiation hardening mechanism more intricate. A comparison of the impact of dislocation loops on the CRSS at 0, 0.03, and 0.3 dpa irradiation doses reveals that dislocation loops have a more significant effect at higher levels of irradiation, which explains why irradiation hardening is more prominent at higher doses.

The mechanical properties of NiFeMnCr HEA are significantly influenced by the element concentrations. This is because the lattice distortion effect in HEAs, resulting from variations in shear modulus and atomic radius, is closely linked to element concentration. To predict HEAs with superior mechanical properties, it is essential to estimate the concentration of each element. However, the elemental-concentration impact on irradiation-induced defects is often ignored in optimal-component prediction.

In $NiFeMnCr_x$ HEA, x represents the atomic fraction. Figure 4.9a illustrates the impacts of three hardening mechanisms on the yield strength at 293 K with irradiation doses of 0, 0.03, and 0.3 dpa. The solid-solution-strengthening contribution to the yield strength varies significantly as the Cr atomic fraction rises from 0.1 to 0.85. This is because the average-atom-size mismatch (δr^{ave}) and average-modulus mismatch ($\delta \mu^{ave}$) are noticeably altered with increasing Cr concentration. The maximum contribution occurs when the Cr fraction is 0.65. Cr has a higher shear

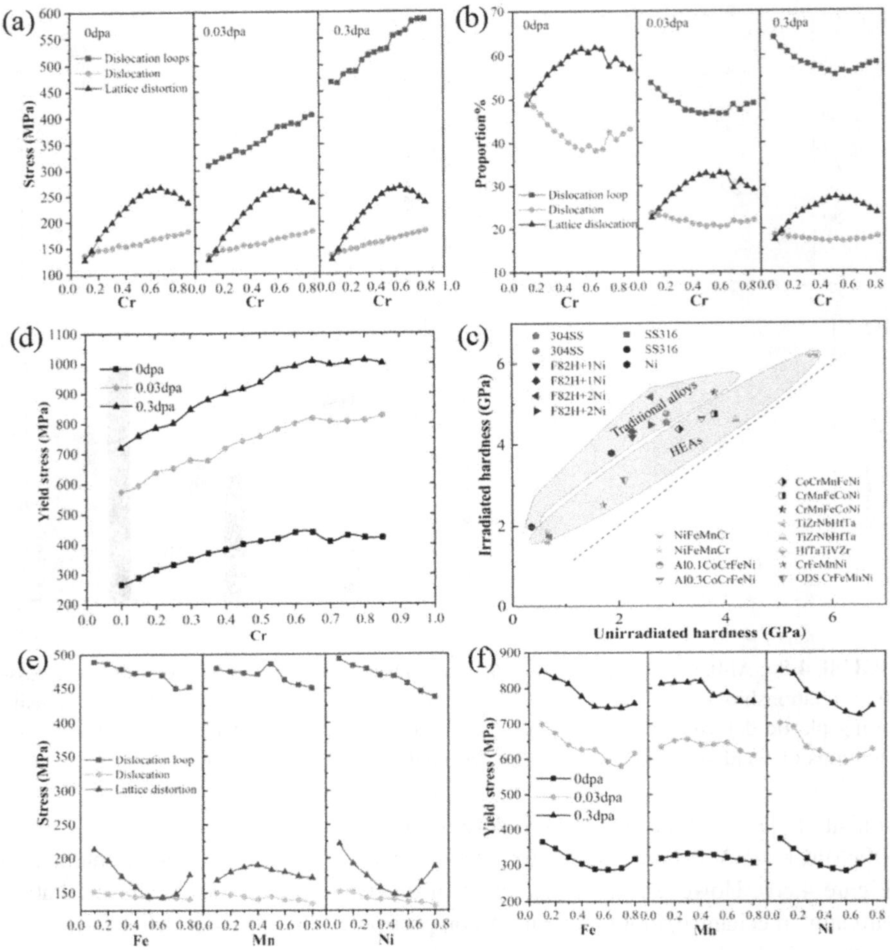

FIGURE 4.9 (a) The influences and (b) proportions of different hardening mechanisms to the yield strength of $NiFeMnCr_x$ HEA at 293 K temperature and 0, 0.03, and 0.3 dpa irradiation levels. (c) The relationship between yield strength and Cr atomic fraction for various irradiation levels is presented. (d) A comparison of irradiation effect on hardness in between HEAs and traditional alloys. Here, the dotted line represents the absence of any change in hardness following irradiation. (e) The strength caused by different hardening mechanisms for FexNiMnCr, FeNixMnCr, and FeNiMnxCr HEAs at 0.3 dpa irradiation levels. (f) The relationship between yield stress and element concentration at various irradiation levels [34].

modulus compared to other elements (Table 4.5). Hence, in $NiFeMnCr_x$ HEAs, the shear modulus is influenced by Cr concentration. The dislocation-hardening and dislocation-loop-hardening contributions rise slowly with increasing Cr concentration (Figure 4.9a). Note that this study does not consider the irradiation effect on dislocation and solid solution strengthening, and so their contributions remain approximately constant at different irradiation levels. Conversely, the dislocation-loop-strengthening contribution to yield strength in $NiFeMnCr_x$ HEAs is different because of the varying dislocation-loop number density and size at various irradiation levels. Figure 4.9b shows the three strengthening mechanisms' proportions on the yield strength of $NiFeMnCr_x$, indicating that the concentration of Cr causes dislocation strengthening to increase first and then decrease.

Figure 4.9c illustrates the corresponding variations in yield strength arising from the three distinct strengthening mechanisms. At a Cr fraction of 0.65, the combined effect of the three strengthening mechanisms on yield strength is maximized. At 0 and 0.03 dpa irradiation levels, the three hardening mechanisms exhibit an increasing trend. However, as the Cr concentration rises from 0.65 to 0.85, the lattice distortion effect slowly declines in NiFeMnCrx HEAs. When the Cr concentration surpasses 0.65, the yield stress has little variation, suggesting that lattice distortion has a greater influence on hardening compared to dislocation hardening and dislocation loop hardening. Comparing the hardness of irradiated HEAs to that of irradiated traditional alloys, it is revealed that HEAs possess higher irradiation hardening resistance than traditional alloys (Figure 4.9d) [35]. Therefore, the severe lattice distortion of HEAs plays a key role in improving irradiation tolerance compared to traditional alloys.

The influence of Ni, Mn, and Fe on the yield strength of NiFeMnCr HEA is also studied. Figure 4.9e describes three hardening mechanisms' contribution for FexNiMnCr, FeNixMnCr, and FeNiMnxCr HEAs at 0.3 dpa irradiation and 293 K temperature. The findings suggest that a decrease in shear modulus leads to a reduction in both dislocation hardening and dislocation loop hardening for all these HEAs. By comparing Figure 4.9c, f, the greatest impact on yield strength is associated with Cr atomic fraction, which has potential implications for designing a new excellent-irradiation-resistance HEA. Additionally, the highest yield strength for $NiFeMnCr_x$ HEAs at 0, 0.03, and 0.3 dpa irradiation levels is estimated by current theoretical approach of 439, 818, and 1008 MPa, when $x = 0.65$. The $NiFeMnCr_x$ HEA with $x = 0.65$ exhibits the greatest lattice distortion and highest dislocation loop hardening, contributing to its superior performance.

4.4 HIERARCHICAL MULTISCALE MODEL

4.4.1 Multiscale Model Building

The purpose in this section is to develop a hierarchical multiscale approach that links nanoscale, macroscale, and mesoscale. This methodology is used to explore the deformation behavior of HEAs by integrating molecular dynamics (MD), discrete dislocation dynamics (DDD), crystal plasticity finite element (CPFE) simulation techniques, and random-field theory (Figure 4.10). The AlxFeCoCrNi HEA has received extensive attention in the study of HEA systems because of its remarkable combination of high specific strength, resistance to radiation, and other excellent mechanical properties at low temperatures [36,37]. The elastic constants and details of the motion of single dislocations in HEAs are obtained through MD simulations at the nanoscale and subsequently incorporated into DDD simulation at larger scales. Using the random-field theory, the microscale

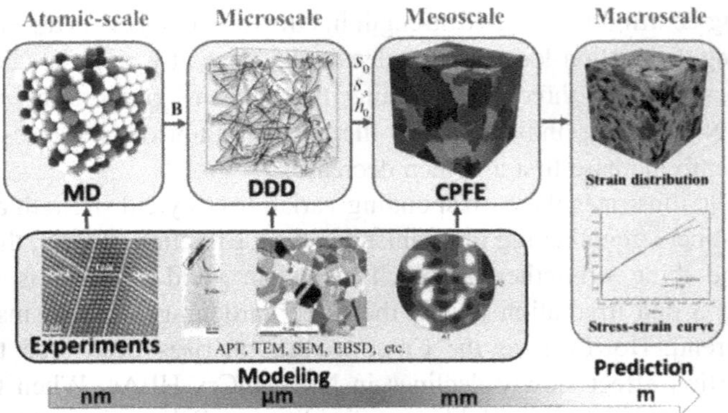

FIGURE 4.10 A hierarchical multiscale-modeling strategy. The mechanical behavior of polycrystalline material is projected across a range of length scales using coupled models of MD, DDD, and CPFE.

DDD simulations introduce the strain field induced by lattice distortion. The acquisition of hardening parameters for the crystal plasticity model requires the establishment of single-crystal DDD simulations with diverse crystal orientations of $Al_{0.1}$FeCoCrNi HEA. At the mesoscale, the mechanical behavior of a single crystal is forecasted, and the hardening coefficients for polycrystals are adjusted by matching them with the hardening coefficients of single crystals. By contrasting the anticipated mechanical reaction with experimental data obtained from uniaxial tension testing of the polycrystalline $Al_{0.1}$FeCoCrNi HEA at varying strain rates, the validity of the proposed multiscale approach is validated [38,39]. Under different strain rates, we analyze the distribution of stress and strain as well as the pole figures in the polycrystal $Al_{0.1}$FeCoCrNi HEA at various plasticity strains.

4.4.1.1 Molecular Dynamics

MD is founded upon the conventional theories of Newtonian mechanics and Hamiltonian equations. The overall energy of a system with N atoms is the summation of the potential and kinetic energy of each atom:

$$U_{total} = U_{ke} + U_{pe} = \sum_{i=1}^{n} (U_{ke,i} + U_{pe,i}) \tag{4.24}$$

where U_{pe} and U_{ke} are the potential energy and kinetic energy of the system, respectively. $U_{pe,i}$ and $U_{ke,i}$ are the potential energy and kinetic energy of atom i, respectively.

The potential-energy gradient gives rise to force acting on the atom i within the system:

$$\vec{F_i} = -\nabla_i U_{pe} = -\left(\vec{i}\frac{\partial}{\partial x_i} + \vec{j}\frac{\partial}{\partial y_i} + \vec{k}\frac{\partial}{\partial z_i}\right)U_{pe} \tag{4.25}$$

where $\vec{F_i} = m_i \vec{a_i}$. Therefore, the behavior of atoms follows Newton's law of dynamics.

4.4.1.2 Discrete Dislocation Dynamics

The dislocation is discrete into segments by nodes, and the force on each node, F_i^{node}, is computed by adding the impact of all dislocation segments that link to node, i:

$$F_i^{node} = \sum_j f_{ij} \tag{4.26}$$

here, the segment force, f_{ij}, is obtained by the following integral:

$$f_{ij} = \int_{C_j} N_i^j(s) f_{ij}^{\mathrm{PK}}(s)\, dL(s) \tag{4.27}$$

where, C_j is segment ij. $N_i^j(s) = s\ (0 \le s \le 1)$ represents the interpolation function, while s represents the index for location. For instance, when $s = 0$ and $s = 1$, the positions correspond to nodes j and i, respectively. The Peach-Koehler force at a given position s is influenced by the local stress:

$$f_{ij}^{PK}(s) = [\sigma^{ext}(s) + \sigma^{disl}(s) + \sigma^{mpea}(s)] \cdot \mathbf{b} \times \mathbf{t}_{ij} \tag{4.28}$$

here, $\sigma^{ext}(s)$ is the external load, and $\sigma^{disl}(s)$ is the stress caused by other dislocations. The stress, $\sigma^{mpea}(s)$, caused by the lattice distortion of HEAs is computed using a fractal function. The Burgers vector of the segment, ij, is represented by b_{ij}, and $\hat{t}_{ij}$ is the unit vector parallel to a dislocation line. Based on the force acting on the node, i, the velocity is determined by:

$$v_i = F_i^{node}/B \tag{4.29}$$

where B is the viscous drag coefficient.

The cross-slip rate of the energy barrier, is controlled by Escaig stress [40], and local stress, $\sigma(s)$. According to Arrhenius-type law, the cross-slip rate is:

$$\mathrm{R} = v_0 \exp\left(-\frac{E_b}{K_B T}\right) \tag{4.30}$$

where $v_0 = v_D(L/L_0)$ is the attempt frequency, K_B is the Boltzmann's constant, and T is the temperature. L represents the dislocation-segment length, v_D is the Debye frequency, and $L_0 = 1\ \mu\mathrm{m}$ represents a reference length. The energy barrier is determined by

$$E_b = E_a - V_a \Delta\tau_E \tag{4.31}$$

where $E_a = 0.4$ eV represents the energy of create a constriction point along a screw dislocation [41,42], $V_a = 20\ b^3$ is the activation volume [41], and $\Delta\tau_E$ represents the discrepancy between the Escaig stresses on the primary slip plane and the cross-slip plane.

Under an external load, the dislocation velocity is determined using Eq. 4.29. The plastic shear strain resulting from a dislocation segment with Burgers vector, b_n, on a slip plane area, A_n, over a distance, δ, is expressed as:

$$\delta\gamma_n^p = \frac{b_n \delta A_n}{V} \tag{4.32}$$

The macroscopic plastic strain rate tensor is determined as:

$$\delta\dot{\varepsilon}_{ij} = \sum_{\alpha=1}^{12} \frac{1}{2}\left(n_i^{(\alpha)} \otimes b_j^{(\alpha)} + b_j^{(\alpha)} \otimes n_i^{(\alpha)}\right)\delta\dot{\gamma}^{(\alpha)} \tag{4.33}$$

here, $b_j^{(\alpha)}$ is the Burgers vector, $n_i^{(\alpha)}$ is the unit vector of α-slip plane. The strain-hardening behavior resulting from the collective movement of dislocations in single crystals is obtained using DDD simulations, where the drag coefficient of a single dislocation is calculated by MD simulations.

4.4.1.3 Crystal Plasticity Framework

A crystal plasticity framework [43] is employed to model the constitutive behavior of the $Al_{0.1}$FeCoCrNi HEA. Plastic deformation is adapted by dislocation slip on the specific slip system. The total deformation gradient tensor F and velocity gradient tensor L are partitioned into elastic and plastic constituents, respectively:

$$\mathbf{F} = \mathbf{F^e F^p};\ \mathbf{L} = \mathbf{L^e} + \mathbf{L^p} \tag{4.34}$$

where $\mathbf{F^e}$ is elastic deformation and rigid body rotation of the crystal lattice, $\mathbf{F^p}$ denotes the plastic deformation of slip systems. The plastic part of the velocity gradient, $\mathbf{L^p}$, is determined as:

$$\mathbf{L^p} = \dot{\mathbf{F}}_\mathbf{p}\mathbf{F}_\mathbf{p}^{-1} = \sum_{\alpha=1}^{N} \dot{\gamma}^\alpha \mathbf{m}_0^\alpha \otimes \mathbf{n}_0^\alpha \tag{4.35}$$

here, $\dot{\gamma}^\alpha$ denotes the plastic shearing rate of αth slip system, $\mathbf{n}_0^\alpha$ denotes the slip plane normal, and $\mathbf{m}_0^\alpha$ denotes the slip direction. The integer α corresponds to the 12 slip systems in FCC materials. The Cauchy-Green elastic strain tensor, $\mathbf{E}^e$, is expressed as:

$$\mathbf{E}^e \equiv (1/2)(\mathbf{F}_e^T\mathbf{F}_e - \mathbf{I}),\quad \mathbf{T} \equiv \mathbb{C}\mathbf{E}^e \tag{4.36}$$

where $\mathbb{C}$ is the fourth-rank elastic tensor. The resolved shear stress τ acting on the α-slip system is expressed as:

$$\tau = (\mathbf{F}_e^T\mathbf{F}_e\mathbf{T})\cdot\mathbf{S}_0^\alpha \tag{4.37}$$

where $\mathbf{S}_0^\alpha = \mathbf{m}_0^\alpha \otimes \mathbf{n}_0^\alpha$ denotes the Schmid tensor. In the strain-rate-sensitivity crystal plasticity framework [44], the evolution of the plastic strain rate, $\dot{\gamma}^\alpha$, in α-slip system is written as:

$$\dot{\gamma}^\alpha = \dot{\gamma}_0 \left|\frac{\tau^\alpha}{\tau_{sr}^\alpha}\right|^{1/m} \operatorname{sign}(\tau^\alpha) \tag{4.38}$$

where $\dot{\gamma}_0$ represents the reference shear rate. τ_{sr}^α (> 0) is the deformation resistance stress of α-slip system, τ^α is the applied shear stress of α-slip system, and m is the sensitivity coefficient of the strain rate. The slip resistance, $\dot{\tau}_{sr}^\alpha$, of the α-slip system is expressed as:

$$\dot{\tau}_{sr}^\alpha = \sum_\beta |\dot{\gamma}^\beta| h^{\alpha\beta} \tag{4.39}$$

where $\dot{\gamma}^\beta$ is the shear rate of β-slip system, $\delta^{\alpha\beta}$ is the Kronecker delta, and $h^{\alpha\beta} = [q_0 + (1 - q_0)\delta^{\alpha\beta}]h^\beta$ represents both the self-hardening and latent hardening of slip systems. The expression for the self-hardening parameter, h^β, is given by:

$$h^\beta = h_0 \left|1 - \frac{\tau_{sr}^\beta}{s_s^\beta}\right|^{r^\beta} \tag{4.40}$$

where h_0 is the initial hardening modulus, s_s^β is the saturation slip resistance of the β-slip system, and r^β is the hardening exponent. The slip resistance parameters, including $\{s_0, r^\beta, s_s, h_0\}$, are assumed to be the same for all 12 slip systems for the CPFE simulations [45,46]. The hardening exponent and latent-hardening parameters used in CPFE simulations are $r = 2.25$ and $q_0 = 1.4$, respectively [47].

4.4.1.4 Coupling Methodology

In contrast to the traditional alloys, the strength of HEAs is improved due to the enhanced dislocation nucleation rate resulting from the severe lattice distortion effect, which strongly hinders dislocation movement and causes the local strain fields. In this section, the drag coefficient of a single dislocation and thermodynamic properties (such as activation volume and potential barrier) of the HEAs required for micro-scale DDD simulations are calculated using nanoscale MD simulations.

The plastic hardening and deformation of materials are caused by the cumulative motion of the dislocations. The strengthening correlation anticipated by the mathematical equation of the crystal plasticity (CP) model is linked to dislocation hardening. Therefore, it is possible to connect the results of DDD simulations with the CP model. The evolution of the plastic strain rate, $\dot{\gamma}^\alpha$, of CP constitutive model is represented as $\dot{\gamma}^\alpha = \dot{\gamma}_0 \left|\frac{\tau^\alpha}{\tau_{sr}^\alpha}\right|^{1/m} \text{sign}(\tau^\alpha)$. Using the thermally-activated slip theory, the reference shear rate, $\dot{\gamma}_0$, and the strain rate sensitivity parameter, m, are expressed as:

$$\dot{\gamma}_0 = b^2 \rho_m v_d \exp\left(\frac{-\Delta Q_s}{kT}\right), \quad \text{and } m = \frac{k_B T}{\alpha b^2 \mu r_0} \tag{4.41}$$

where ρ_m is the dislocation density, b is the Burgers vector, $k_B = 1.38 \times 10^{-23}\ \text{JK}^{-1}$ is the Boltzmann constant, v_d is the average velocity of dislocation, α is the forest strength parameter, T is the temperature, r_0 is the dislocation radius, μ is the shear modulus, and ΔQ_S is the slip-activation energy. At room temperatures, the strain rate sensitivity is 0.04, and r_0 is approximately $5b$. [48]. The applied shear stress, τ^α is decomposed into two components:

$$\tau^\alpha = \tau_{thermal}^\alpha + \tau_p^\alpha \tag{4.42}$$

where τ_p^α and $\tau_{thermal}^\alpha$ represent the athermal and thermal components of the slip resistance, respectively. In HEAs, τ_p^α is controlled by the interaction between dislocations, and $\tau_{thermal}^\alpha$ is mostly dominated by the lattice distortion effects and the solid solution strengthening [49,50]. The presence of solute atoms introduces stress fields that interact with dislocations, causing solid solution strengthening in alloys through elastic mechanisms. In comparison to conventional alloys, the HEAs exhibit more pronounced lattice distortion effects [50]. To account for the lattice distortion effect in DDD simulation, a 3D lattice strain field is incorporated using a fractal function.

The kinematics rule, $\dot{\gamma}^\alpha = \dot{\gamma}_0 \left|\frac{\tau^\alpha}{\tau_{sr}^\alpha}\right|^{1/m} \text{sign}(\tau^\alpha)$, is rewritten as

$$\dot{\gamma}^\alpha = \dot{\gamma}_0 \left|\frac{\tau_{thermal}^\alpha + \tau_p^\alpha}{\tau_{sr}^\alpha}\right|^{1/m} \text{sign}(\tau^\alpha) \tag{4.43}$$

here, $\dot{\gamma}^\alpha$ is computed from the deformation history of the DDD simulations. while the constants ($\dot{\gamma}_0$, m) are determined a priori using previous studies [48,51]. The remaining parameters, τ^α and $\dot{\gamma}^\alpha$, are acquired from DDD simulations since they are functions of time. As simulation progresses, only one unknown, τ_{sr}^α, in Eq. 4.43 evolves with the simulation time. Therefore, we could calculate the slip resistance, τ_{sr}^α, of each slip system in each time step using Eq. 4.43, and the DDD simulations can measure the rate of change in total slip resistance during simulation progress.

According to Eqs. 4.39 and 4.40, when slips appear on the β-slip system, the slip resistance of the α-th slip system is expressed as:

$$\dot{\tau}_{sr}^{\alpha} = \sum_{\beta} |\dot{\gamma}^{\beta}|[q_0 + (1 - q_0)\delta^{\alpha\beta}]h_0^{\beta}\left|1 - \frac{\tau_{sr}^{\beta}}{s_s}\right|^{r^{\beta}} \tag{4.44}$$

Since the total shear-strain rate across all slip systems is constant, the resistance to slip for the α-slip system is expressed as follows [51–53]:

$$\dot{\tau}_{sr}^{\alpha}(t) = \dot{\gamma}\sum_{\beta} [q_0 + (1 - q_0)\delta^{\alpha\beta}]h_0^{\beta}\left|1 - \frac{\tau_{sr}^{\beta}(t)}{s_s}\right|^{r} \tag{4.45}$$

here, $\dot{\tau}_{sr}^{\alpha}(t)$ is the time-dependent change rate of slip resistance of the α-slip system, $\dot{\gamma} = \Sigma|\dot{\gamma}^{\beta}|$, and $\tau_{sr}^{\beta}(t)$ denotes the change rate of slip resistance of the β-slip system. At each time node, 12 equations are associated with 12 slip systems, with each equation containing the identical unknowns of h_0 and s_s. A calibration procedure utilizing the Newton-Raphson method is constructed to obtain multiple hardening parameters. The residual, d^{α}, of the α-slip system, is considered by [54]:

$$d^{\alpha}(t) = \dot{\tau}_{sr}^{\alpha}(t) - \dot{\gamma}h_0\sum_{\beta} [q_0 + (1 - q_0)\delta^{\alpha\beta}]\left|1 - \frac{\tau_{sr}^{\beta}(t)}{s_s}\right|^{r} \tag{4.46}$$

Then, the sum of squares of the residual d^{α} is represented the function, $g(s_s, h_0)$:

$$g(s_s, h_0) = \sum_{\alpha} (d^{\alpha})^2 \tag{4.47}$$

The partial derivative of the function, $g(s_s, h_0)$, is computed to obtain the minimum value of the residual [54]:

$$f_1(s_s, h_0) = \frac{\partial g(s_s, h_0)}{\partial s_s} = 0; \; f_2(s_s, h_0) = \frac{\partial g(s_s, h_0)}{\partial h_0} = 0 \tag{4.48}$$

Solving two coupled nonlinear equations, (h_0, s_s), is achievable by minimizing the residuals. The Newton-Raphson program for nonlinear equations is utilized to calibrate the initial inferred value, resulting in the converged values of the hardening parameters, (h_0, s_s), at each time point. In DDD simulations, the slip resistance, $\tau_{sr}(t)$, changes as deformation progresses. To acquire the evolution of hardening parameters, (h_0, s_s), over time, the procedure mentioned before can be repeated for each time point of the deformation progress.

We can use the Taylor factor to determine the initial slip resistance, s_0, of the CPFE model, determined by the initial yielding observed in the stress-strain relationship. Typically, initial resistance to slip is calculated by dividing the experimentally determined macroscopic yielding stress by the Taylor factor [55]. The initial slip resistance of several single crystal directions is easily estimated through the DDD simulation results. From the stress-strain curve of DDD simulations, three values are extracted to find the CRSS: the yield stress at initial yielding, as well as the yield stress determined using 0.02% and 0.04% offset methods. These values are averaged and then divided by the Taylor factor to obtain the initial slip resistance.

To ensure that the CPFE hardening parameters remain constant after a certain time, the fluctuations in DDD simulations are smoothed and the parameter mean values are calibrated accordingly. Within a specific time frame, it is appropriate to compute the average value of the data, as long as the values are almost similar after the chosen time step. Using this averaging method can obscure the interactions between different dislocation types, much like coarse-graining over DDD simulations that include giant dislocation groups.

The atomic-scale modeling approach is utilized to understand the mechanism of single dislocation, specifically drag coefficient, which is used as input for DDD simulations at higher scales and strain rates ranging from 10^2~10^4 s^{-1} [56,57]. This is achieved through the calibration of dislocation mobility using MD simulation at a strain rate of 10^9 s^{-1}. To strike a balance between the computational burden and the prediction accuracy of results at higher scales [53], the strain rate effect is taken into account due to the large difference in time scales between DDD and CPFE. While concurrent multiscale models offer a potential solution to these problems [58], a parallel multiscale model that includes MD, DDD, and CPFE presents a challenge due to the issue of coupled boundary conditions [58].

MD simulations are leveraged to simulate the motion of a single dislocation and acquire mechanical characteristic parameters, which are then integrated into DDD simulations to model the hardening and plasticity of multiple dislocations. The coupling of DDD simulations with the CPFEM calculates hardening parameters, and hardening behavior is predicted at the mesoscale through CPFEM.

4.4.2 Parameter Transition and Simulation Setup

To obtain the dislocation-linear mobility ($V = F/B$) and the drag coefficient (B) for DDD simulations, MD simulations are utilized. Using microscale DDD simulations, the hardening parameters (s_0, h_0 and s_s) for different crystal directions are calibrated to obtain the single-crystal hardening parameters. The average values of single crystal parameters for various crystal directions are utilized to determine the parameters of the polycrystalline CPFE.

4.4.2.1 MD Simulations

The primary form of damping that controls edge dislocations is phonon drag or damping, which is quantified in atomic-scale MD simulations using a dislocation mobility. This parameter is an inherent material characteristic that is not affected by boundary conditions or strain rates. The dominant type of dislocation in the FCC alloy is edge dislocation [59]. In addition, edge and screw dislocations have different mobility characteristics, and they appear to have no significant impact on the strain hardening observed in DDD simulations [32,60]. For the present study, MD simulations are solely concerned with edge dislocations.

An open source Large-scale Atomic/Molecular Massively Parallel Simulator (LAMMPS) package [61] is utilized to carry out MD simulations, while an Ovito

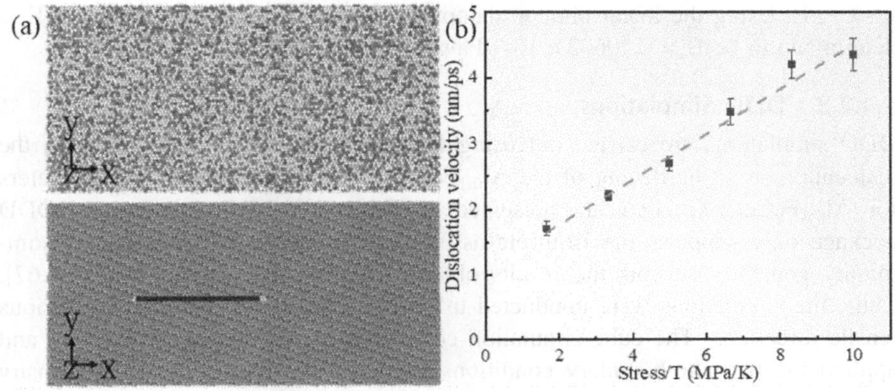

FIGURE 4.11 The simulation box used for MD simulations to determine the drag coefficient of edge dislocation slip on the (111) plane (a), the correlation between the velocity and the ratio of applied shear stress to temperature σ/B [54] (b).

software code [62] is employed to visualize the simulations. The embedded atom method potential [63] is utilized as the inter atomic potential in the simulations, as it is highly compatible with the mechanical behavior of $Al_{0.1}$FeCoCrNi HEA under deformation. As confirmed through simulations, the drag coefficient calculated is unaffected by the size of the simulation box, therefore, a cell size of $30.0 \times 14.0 \times 10.0$ nm^3 containing approximately 393,300 atoms is used.

The center of the simulation cell serves as the location for inserting a ½ $[1\bar{1}0]$ edge dislocation, with the slip direction of x = $[1\bar{1}0]$, the line direction of y = $[11\bar{2}]$, and a glide plane normal of z = [111]. The dislocation possesses a Burgers vector of $b = 2.52 \times 10^{-10}$ m. For the beginning 100 ps, the sample is equilibrated at the room temperature using the isothermal-isobaric ensemble, while ensuring that x- and z-direction pressures are maintained at zero, and periodic boundary conditions are imposed along the X and Z axes, while the Y axis has a fixed boundary condition (Figure 4.11a). The sample is initially relaxed for 100 ps under the canonical ensemble (NVT). The motion of an a/2 $[1\bar{1}0]$ edge dislocation is continuously monitored and it experiences significant local motion and slip under the increase of the shear load.

The relationship between the edge-dislocation velocity in the $Al_{0.1}$FeCoCrNi HEA and the ratio of various shear stresses to temperature are presented in Figure 4.11b. Within the extent of stress applied, there is an almost linear increase with σ/T for dislocation velocity, which is consistent with the phonon damping theory [59,64]:

$$v = v_0 + (b\sigma/T)\Gamma, \quad \Gamma = B/T \tag{4.49}$$

where drag coefficient is marked as B, and the shear stress is marked as σ. The MD simulations provide a dislocation drag coefficient that can serve as input for microscale DDD simulations, enabling predictions of synergistic behavior of numerous dislocations in HEA, and subsequently obtain hardening parameters for CPFE. Using the linear phonon damping law, the drag coefficient at 300 K is determined to be B = 2.20602×10^{-4} Pa·s (Figure 4.11b).

4.4.2.2 DDD Simulations

DDD simulations are carried out using the ParaDiS program [65] to obtain the dislocation strain hardening of the $Al_{0.1}FeCoCrNi$ HEA. The material parameters for $Al_{0.1}FeCoCrNi$ HEA are presented in Table 4.8. While the current DDD package only supports the isotropic assumption simulations, the materials community generally accepts the results obtained through this approach [53,66,67]. Thus, the simulations were conducted using the same shear modulus for various tensile directions. The cube simulation cell was set up with a size of 2 μm, and applied the periodic-boundary conditions [68,69]. The initial dislocation density was found to be approximately $5.5 \times 10^{12}\,m^{-2}$, which agrees with the range of values ($2 \times 10^{12}\,m^{-2}$ to $5 \times 10^{13}\,m^{-2}$) reported in previous experimental studies [70,71]. Subsequently, the dislocations were distributed uniformly across 12 slip systems and allowed to relax without applied any load. The crystallographic directions used in the simulation were [001], [101], [111], [112], [102], [212], and [213], with a strain rate of = $10^4\,s^{-1}$ [72].

The current model primarily takes into account the lattice distortion effect in the $Al_{0.1}FeCoCrNi$ HEA, given the difficulty in generating precipitation due to the phase structure remaining stable over a wide temperature range at low Al content in the FCC HEA [73]. To account for the impact of the non-uniform lattice strain field resulting from lattice distortion on the hardening behavior of HEA, a three-dimensional lattice strain field is created and integrated into DDD simulations.

The confirmation of the fractal characteristics of the lattice strain field in the HEA is based on experiments, leading us to utilize the generalized Weierstrass-Mandelbrot fractal function in the construction of the 3D lattice strain field [74,75].

$$\varepsilon(\mathbf{r}) = H^{D-3}\left(\frac{\ln \gamma}{M}\right)^{\frac{1}{2}} \sum_{m=1}^{M} \sum_{n=0}^{n_{\max}} \left[\cos(\phi_{m,n}) - \cos\left(\frac{2\pi\gamma^n \vec{n}_m \cdot \mathbf{r}}{L} + \phi_{m,n}\right)\right]\left(\frac{\gamma^n}{L}\right)^{D-4} \tag{4.50}$$

TABLE 4.8
Material Parameters of the $Al_{0.1}FeCoCrNi$ HEA in DDD Simulations [54]

Parameter	Symbol	Value
Shear modulus	μ	80.1 GPa
Poisson's ratio	v	0.3
Magnitude of Burgers vector	b	0.252 nm
Dislocation radius	r_0	5 b

TABLE 4.9
Strain Amplitude and Fractal Dimension in the $Al_{0.1}FeCoCrNi$ HEA of the Strain Components from Experimental [54,77]

Strain components	ε_{xx}	ε_{yy}	ε_{zz}	ε_{xy}	ε_{xz}	ε_{yz}
Fractal dimension, D_{ij}	D_{xx}	D_{yy}	D_{zz}	D_{xy}	D_{xz}	D_{yz}
	3.9065	3.8855	3.8961	3.9281	3.9281	3.9281
Strain amplitude, H_{ij} ($\times 10^{-4}$)	H_{xx}	H_{yy}	H_{zz}	H_{xy}	H_{xz}	H_{yz}
	5.539	5.539	5.539	3.2659	3.2659	3.2659

where $n_{max} = \text{int}\left[\frac{\log(L/L_s)}{\log\gamma}\right]$ is the upper limit, $k_0 = \frac{2\pi}{L}$ is a wavenumber, and $M = 50$ and $\gamma = 1.5$ are the number of superposed ridges and frequency density, respectively, $\vec{n}_m$ is unit vectors distributed in 3D hypersphere, $L_s = 4$ b is the cut-off size. and $\phi_{m,n}$ is a random phase. Using the structure function method [76] and experimental results of the lattice strain field of the $Al_{0.1}FeCoCrNi$ HEA [77], the fractal dimension, D, and strain amplitude, H, are presented in Table 4.9.

By applying the generalized Hooke's law, the stress field $\sigma^{mpea}(s)$ is derived from the lattice strain:

$$\sigma_{ij} = 2G\varepsilon_{ij} + \lambda\varepsilon_{kk}\delta_{ij} \tag{4.51}$$

Subtracting the overall average strain, $\bar{\varepsilon}$, of the whole DDD simulations from the 3D strain-field function can remove the non-equilibrium effects resulting from the random phase. A $100 \times 100 \times 100$ discrete point is used to calculate average strain, $\bar{\varepsilon} = \sum_k \sum_m \sum_n \varepsilon(x_k, y_m, z_n, H, D, \varphi_{m,n})/10^6$ (n, m, k = 1 ... 100), which is evenly distributed.

DDD simulations are conducted to examine the impact of screw and edge segment mobility on the mechanical behavior of $Al_{0.1}FeCoCrNi$ along the [001] direction. Figure 4.12a reveals that there is not a significant variation on the stress-strain curve for different dislocation mobilities. The calibrated hardening parameters are not significantly affected by the use of different mobility values in the DDD simulation [53]. Previous MD simulations have shown that the difference in drag coefficients between screw and edge dislocations is minor in the FCC FeNiCrCoCu HEA [78]. Thus, similar to other DDD simulation investigations on FCC crystals, the same dislocation mobility between edge and screw dislocation is taken into account [66,79,80]. The evolution of total plastic shear strain indicates that in $Al_{0.1}FeCoCrNi$ MPEA along the [212] direction, the plastic shear strain increases linearly with time (Figure 4.12b). Notably, the plastic shear strain rate is described by the slope, which is obtained from Figure 4.12b for the total plastic-shear strain rate ($9.84 \times 10^3\ s^{-1}$) and is similar to the load strain rate ($1 \times 10^4\ s^{-1}$). During plastic deformation, the relationship between the total plastic-shear strain and time is utilized to determine the total plastic-slip rate, which is in equilibrium with the applied strain rate.

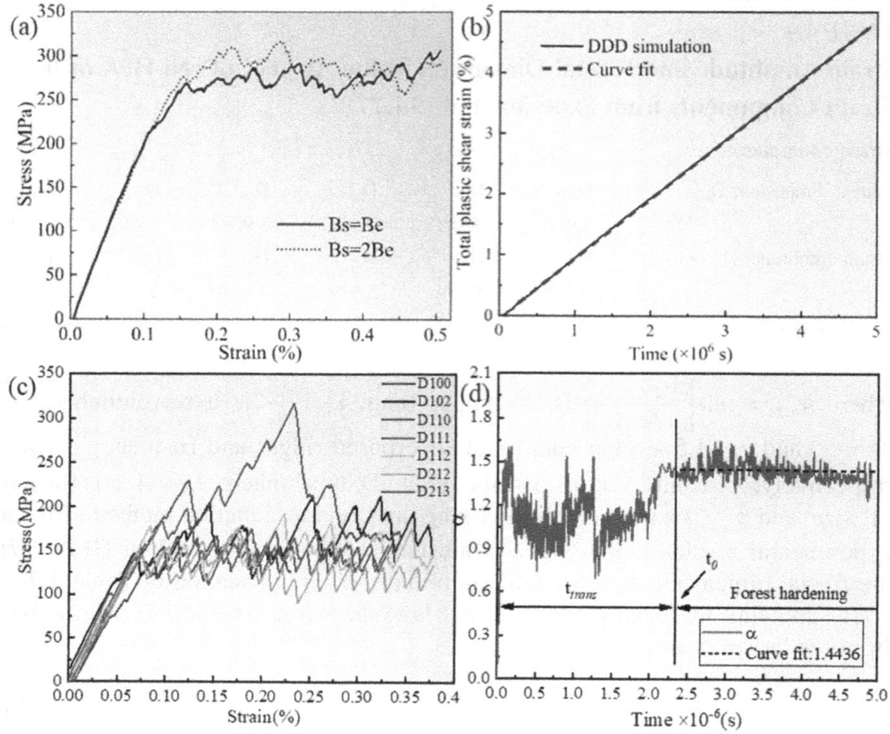

FIGURE 4.12 The stress-strain curve for $Al_{0.1}$CoFeCrNi single crystal along [001] direction obtained from DDD simulations under various dislocation mobilities (a). The correlation between the total plastic-shear strain and time in the $Al_{0.1}$FeCoCrNi HEA along the [212] direction (b). The mechanical response of the $Al_{0.1}$FeCrCoNi single crystal HEA of DDD simulations (c). The evolution of the dislocation hardening parameter as a function of time for single crystal along the [212] direction (d) [54].

From DDD simulations, Figure 4.12c illustrates the stress-strain curve of the FCC $Al_{0.1}$FeCoCrNi HEA along various crystal directions, including [001], [101], [111], [102], [112], [212], and [213]. The difference in yield stress of FCC $Al_{0.1}$FeCoCrNi HEA - under different load orientations is attributed to the different dislocation slip shear stresses caused by the correspondingSchmid factor. During the plastic deformation stage, the stress-strain curve for all the considered crystal orientations exhibits serrations (Figure 4.12c), which is attributed to the initial dislocation density not being able to meet the strain rate in deformation processes. The CRSS of DDD simulations is utilized to estimate the initial slip resistance, s_0, of various single crystal orientations. Since the stress-strain curve in Figure 4.12c exhibits significant fluctuations, the extracted values for each crystal orientation show the deviation of 0.02% and 0.04%. Table 4.10 presents the initial resolved slip resistance values of the CP model for the $Al_{0.1}$FeCrCoNi HEA single crystal under various loading orientations.

TABLE 4.10
Hardening Parameters for Various the Single-Crystal Tensile Directions of $Al_{0.1}$FeCoCrNi HEA Obtained Through DDD Simulations [54]

Orientation	[001]	[102]	[101]	[111]	[112]	[212]	[213]
s_0 (MPa)	60.37	71.56	59.34	57.53	62.02	55.65	80.05
h_0 (MPa)	359.74	303.87	304.08	402.34	343.45	312.51	285.05
s_s (MPa)	492.46	405.82	396.49	581.75	509.05	407.84	388.58

The hardening parameters of the CP model are obtained by outputting the resolved stresses of all active slip systems from DDD simulations. The forest hardening model is commonly employed to study strain hardening in alloys [60,81,82], and the Taylor law is expressed as:

$$\tau = \alpha \mu b \sqrt{\rho} \tag{4.52}$$

where α is the dislocation hardening coefficient, ρ is the dislocation density, b is the magnitude of the Burgers vector, and μ is the shear modulus. The dislocation hardening coefficient reflects the strength of the interaction between dislocations during deformation. For traditional alloys with dislocation density around 10^{12} m^{-2}, α usually between 0.25 and 0.45 in the forest hardening model [83]. Due to the lattice distortion effect, dislocation movement is hindered in HEAs, resulting in a higher resolved shear stress compared to traditional alloys with the same dislocation structure. As a result, the dislocation hardening coefficient in HEAs is higher than in traditional alloys (Figure 4.12d).

Since the data obtained from DDD simulations shows an intermittent fluctuation (Figure 4.12c-d), it is necessary to perform time averaging for the obtention of the average value of the hardening parameter for the CPFE model. Figure 4.12d shows the change in dislocation hardening coefficient over time along the [212] single crystal direction, which is divided into two distinct regions separated by a black solid line. In the first region, the value of α exhibits significant fluctuations since the initial dislocation structure is unable to accommodate the loading strain rate. As the strain increases, dislocations start to move, leading to a proliferation of dislocations. In the second region, the value of α fluctuates slightly around a constant value of 1.44 (a red dashed line in Figure 4.12d), indicating that the current dislocation structure, after the movement and multiplication of dislocations, meets the loading strain rate during the strain hardening stage.

Figure 4.13a displays the time-dependent evolution of the hardening parameters, s_s and h_0, over time, which ultimately converges to a constant. The hardening parameter mean values with respect to time with $t > t_0$ are extracted (represented by a red dotted line and a blue solid line parallel to the x-axis), and can be utilized in CP models to capture the mechanical response of the $Al_{0.1}$FeCrCoNi HEA at the mesoscale. Table 4.10 presents the hardening parameters for the $Al_{0.1}$FeCrCoNi HEA under various single crystal loading directions.

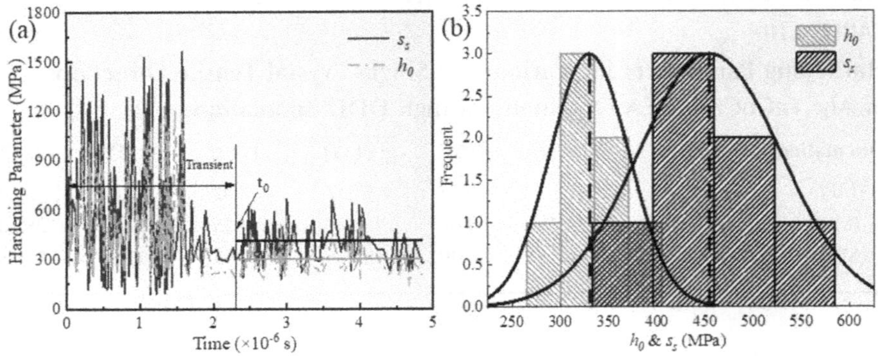

FIGURE 4.13 The time-dependent changes of hardening parameters, denoted as s_s and h_0 of single crystal $Al_{0.1}$FeCoCrNi HEA along the [212] direction (a). The frequency distribution histogram of hardening parameters in single crystals, with the average value indicated by a dotted line (b) [54].

The polycrystal plastic behavior is measured by taking the response of [100] and [111] single crystals multiplied by the corresponding Taylor factor in the traditional method. Nevertheless, the experimentally observed mechanical behavior of the polycrystalline material significantly differs from the predicted behavior at larger deformation. A new tactics involves calibrating the hardening parameters of the polycrystalline representative volume element (RVE) by averaging the hardening parameter values in seven crystallographic directions of single crystals, which can provide insights into the deformation mechanisms of polycrystals [53]. The boundary conditions of grains within a RVE are more intricate than those in a single crystal, making it impossible to characterize the CRSS of the polycrystal using an average of single crystal values. Due to the precise prediction of yield stress of HEAs with both FCC and BCC structures in our previous study [50], the calculation of CRSS for $Al_{0.1}$FeCoCrNi HEA is obtained from dividing the yield stress by Taylor factor, acquiring a s_0 of 62.5 MPa [54,55]. Table 4.10 displays the values of the hardening parameters, denoted as s_s and h_0, for seven distinct single crystal directions, as obtained through the hierarchical multiscale modeling approach. As a result, the hardening parameters s_s and h_0 are found to exhibit approximately normal distributions. The distribution histogram of hardening parameters is depicted in Figure 4.13b, where it can be observed that the mean values of s_s and h_0 are 454 MPa and 330 MPa, respectively.

4.4.2.3 CPFE Simulation

The PRISMS-Plasticity software [84,85] is employed to conduct CPFE simulations. The open-source software Paraview is utilized to visualize and display the outcomes [86]. Table 4.11 presents the values of the elastic parameters (C_{11}, C_{12}, and C_{44}) and the constitutive parameters of the CP model, which are used to simulate the mechanical behavior of $Al_{0.1}$FeCoCrNi HEA.

TABLE 4.11
Material Parameters Are Utilized to Model the Mechanical Behavior of $Al_{0.1}$FeCoCrNi HEAs [54]

Symbol	Description	Value
C_{11}	Elastic constants	275.31 GPa [39]
C_{12}	Elastic constants	107.02 GPa [39]
C_{44}	Elastic constants	106.59 GPa [39,87]
m	Strain-rate-sensitivity exponent	0.04
$\dot{\gamma}_0$	Reference shearing rate	10^{-3} s^{-1} [43]
q_0	Latent-hardening parameter	1.4 [88]
r	Hardening exponent	2.25 [88]
s_0	Initial slip resistance	From DDD simulations
h_0	Initial hardening modulus	From DDD simulations
s_s	Saturation stress	From DDD simulations

4.4.2.3.1 Single Crystals

To investigate the mechanical behavior of a single crystal $Al_{0.1}$FeCoCrNi HEA under uniaxial tension, a finite element model of $1{,}000 \times 1{,}000 \times 2{,}000$ (μm) is created using 3D-brick elements (Figure 4.14b). The model is discretized into $20 \times 20 \times 40$ (16,000) C3D8 hexahedral finite elements. To accurately simulate the mechanical behavior of polycrystalline $Al_{0.1}$FeCoCrNi HEA, seven distinct single-crystal orientations are selected from both the center and the edge of the stereographic triangle (Figure 4.14a). The reference frame z-axis is aligned with the directions of [001], [101], [102], [112], [111], [212], and [213] before the specimen is subjected to uniaxial tension of 10^{-3} s^{-1} strain rate at room temperature. To prevent the movement and rotation of rigid-body, two nodes at the bottom of the simulation box are fixed, applying variable positive displacement at the top of the box, maintaining a zero displacement for the bottom of the box along the z-axis during tensile loading.

4.4.2.3.2 Polycrystalline

Using DREAM.3D [89], the geometry of the finite element model is generated and meshed with the modules of PRISMS-Plasticity [84]. In contrast to Voronoi tessellation, the grains generated in DREAM.3D for the polycrystalline RVE are considered to be more realistic [90]. Figure 4.14c depicts an RVE consisting of 200 grains discretized by $32 \times 32 \times 32$ C3D8 finite elements. The grain-boundary morphology in HEAs is more elaborate than that of traditional alloys because of severe lattice distortions. Even though some crystal plasticity models have been developed to consider the complex grain boundary morphology, these models have not widely been accepted [46,91,92]. Consequently, the current crystal plastic model does not incorporate complex-grain-boundary-morphology influence. Hexahedral meshes are more efficient for post-processing results and visualizing them compared to cubic-octahedron

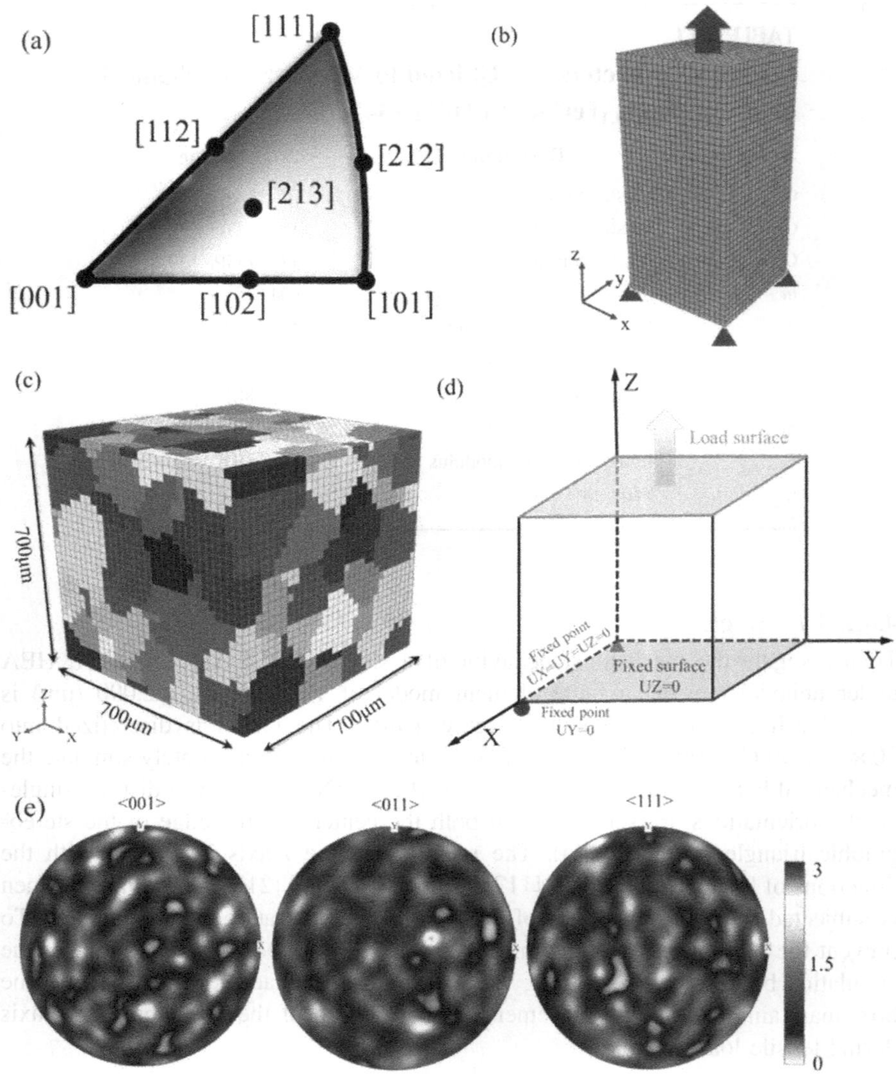

FIGURE 4.14 The schematic representation of the crystal orientations chosen from both the center and edge of the stereographic triangle (a). The model for a single crystalline CPFE simulation under tension (b). The model diagram of the RVE, which includes 200 randomly distributed grains (c). The boundary conditions for the polycrystal (d). The initial texture of $Al_{0.1}$FeCoCrNi HEA (e) [54].

meshes. We have verified that the number of grains and elements used to capture the mechanical behavior of a polycrystalline $Al_{0.1}$FeCoCrNi HEA is sufficient. This approach avoids the creation of stepped grain boundaries that leads to non-uniform stress field. The random orientations of the grains form an almost isotropic block, as the electron back-scattered diffraction measurement reveals no significant texture [93]. The pole figures of the polycrystalline HEAs are illustrated in Figure 4.14e.

Figure 4.14d illustrates the loading and boundary conditions imposed on the RVE. To simplify the representation, the degrees of freedom along the X, Y, and Z axes are denoted as UX, UY, and UZ. To avoid the displacement and rotation of rigid body during loading, the boundary conditions are set as follows: UZ = 0 at the bottom of the RVE, UX = UY = UZ = 0 at the coordinate origin, and UY = 0 at the outermost point of the X axis. The loading surface is positioned at the top of the RVE and subjected to positive displacements and quasi-static strain rates [94].

4.4.3 Result

4.4.3.1 Single Crystal Deformation Behavior

Here, the results of CPFE simulations predicting the mechanical properties of a monocrystal with various crystal orientations are presented (Figure 4.15). The strain hardening rates of [212] and [213] crystallographic orientations are more closely aligned with the [101] zone axis in the standard stereographic triangle, as opposed to the [001]-[111] symmetry axis. The yielding strengths for all crystal orientations, except for [111], fall in the 100–150 MPa range. The work hardening rates are reflected in strengthening coefficients determined through DDD simulations (Table 4.10). The [111] orientation exhibits the highest hardening coefficients, with values of 402.3 (h_0) and 581.7 MPa (ss). The strain hardening responses of these single crystals have been extensively previously discussed [95–97], but are not the focus of this study, which aims to successfully implement a hierarchical-multiscale theory.

Table 4.12 shows the Schmidt factors for different crystallographic directions, where the Schmidt factor for [111] is 0.27, significantly lower than other

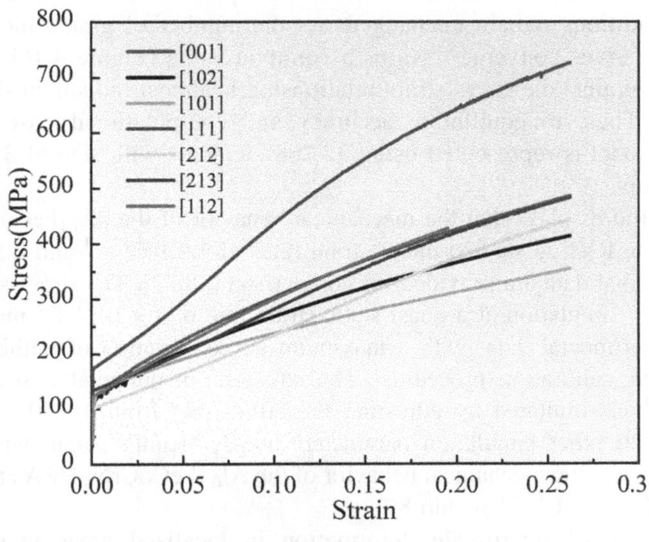

FIGURE 4.15 The stress-strain diagrams for the seven crystallographic directions [54].

TABLE 4.12
Schmid Coefficients of the $Al_{0.1}FeCoCrNi$ HEA for Various Loading Orientations [54]

	Loading direction						
Dislocation slip system	[001]	[102]	[101]	[111]	[112]	[212]	[213]
Schmid factor	0.41	0.49	0.49	0.27	0.41	0.41	0.47

orientations, resulting in stronger hardening behaviors. The highest Schmidt factor is 0.49 for crystallographic direction of [101] and [102], leading to flat stress-strain slopes (Figure 4.15). The difference of strengthening coefficients between crystal directions is because of the distinct Schmidt factors and the stress driving dislocation motion. Consequently, various dislocation multiplication rates result in diverse strain hardening rates among different crystallographic directions.

4.4.3.2 Polycrystalline Deformation Behavior

To improve simulation efficiency and completely represent the mechanical behavior of polycrystalline HEAs, the element-number and grain effect on computational findings should be explored. Figure 4.16c displays the mechanical response curves of RVEs with an equal number of grains (i.e., 200) and varying numbers of finite elements, demonstrating that the material behavior of RVEs with 32,768 ($32 \times 32 \times 32$) elements almost reaches convergence. RVEs containing different numbers of grains have been established, and their grain orientations have been randomly assigned in accordance with the previously explained method (Figure 4.16a). Uniaxial tension simulations are performed on four RVEs, while all other conditions remain unchanged. As the number of grains increases, the stress-strain curves converge towards a common value (Figure 4.16b). Between 200 and 400 grains, the stress-strain relationship is almost indistinguishable from each other. Thus, to equilibrate accuracy and computational cost, the polycrystalline model is represented using 32,768 elements with size of $32 \times 32 \times 32$ and 200 grains.

Figure 4.16d displays that the mechanical behavior of the $Al_{0.1}FeCoCrNi$ HEA polycrystalline RVE is studied under strain rates of 1×10^{-3} s^{-1} and 2.5×10^3 s^{-1}, and experimental data are provided for comparison [38,39]. The stress-strain curve obtained from simulation at a quasi-static strain rate of 1×10^{-3} s^{-1} matches well with the experimental data, with a maximum stress variation of within 26 MPa, validating our calibration procedure. The stress-strain curve at a strain rate of 2.5×10^3 s^{-1} is simulated by adjusting the strain rate from 1×10^{-3} s^{-1}, while maintaining all other simulation parameters and hardening parameters constant. It exactly predicts the mechanical behavior of the $Al_{0.1}FeCoCrNi$ HEA at a constant strain rate of 2.5×10^3 s^{-1} within 8%.

Identification of microscale deformation in localized areas is crucial for optimizing microstructures to significantly enhance mechanical properties.

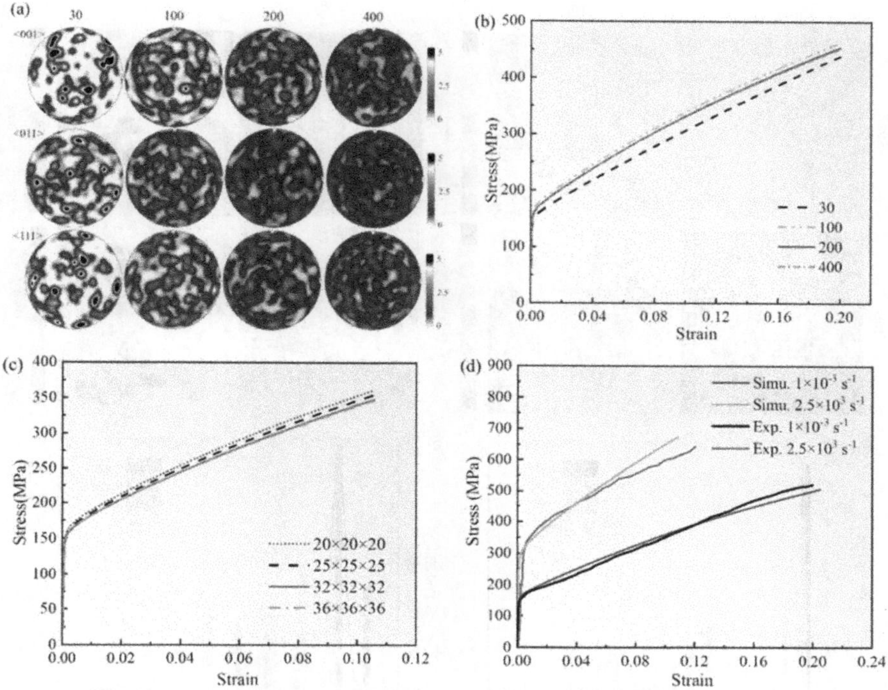

FIGURE 4.16 Display of the initial texture of $Al_{0.1}$FeCoCrNi HEA simulated at grain numbers 30, 100, 200, and 400 (a). Analysis of mechanical behavior of RVEs with varying grain numbers (size: 32 × 32 × 32) (b). Analysis of mechanical behavior of RVEs with varying numbers of elements (grain number: 200) (c). Analysis of predicted and experimental stress-strain responses of polycrystalline materials under strain rates of 1×10^{-3} s^{-1} and 2.5×10^3 s^{-1} (d) [54].

The strain distribution at the applied strains of 0.05%, 0.1%, 0.5%, and 1% is shown in Figure 4.17a-j. Because of crystallographic orientation-induced hardening behavior, the grain direction results in softening in certain grains that are sensitive to anisotropic strain responses, leading to extremely contrasting strain distributions in adjacent grains under the same loading conditions. In the process of yielding, the strain amplitude rises, and the deformation zone remains mostly unchanged (Figure 4.17b, f). In the stage of strain hardening, the strain amplitude further increases, and a few excessively high strain areas may appear in the original low strain region, with no predictable connection between strain and stress (Figure 4.17c, d, g, h). Notably, the areas of low strain exhibits high local stresses. Figure 4.17i, j shows the probability statistics of strain and stress distribution at various stages of strain, where the average values of strain and stress increase with increasing tensile deformation, and heterogeneity of high strain and stress areas increases at the corresponding stage as well. This suggests the existence of a strong stress/strain gradient with increasing applied loading, contributing significantly to material strength (Figure 4.17i, j). In-situ observation equipment can capture the strain field more easily, uncovering the mechanism of

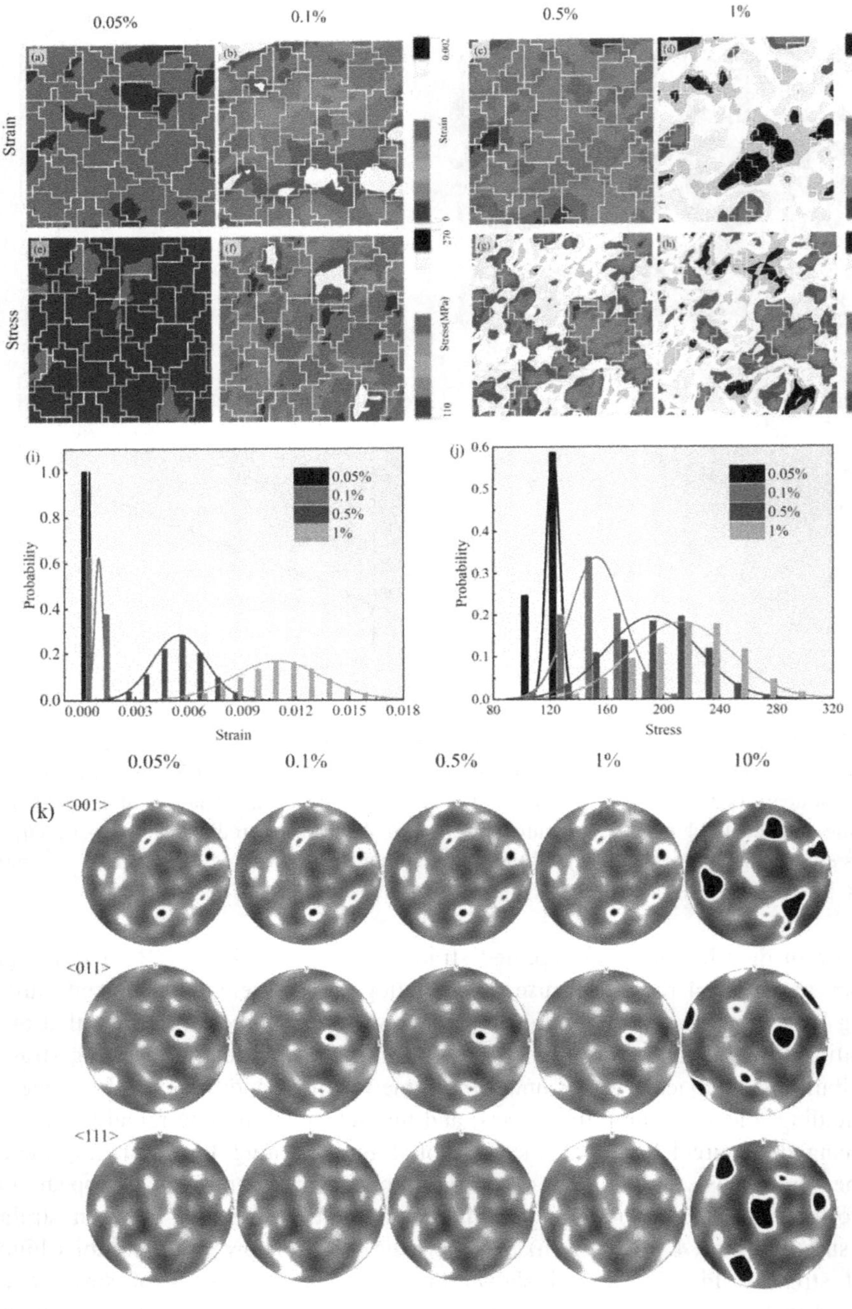

FIGURE 4.17 The distribution of strain during the initial elastic stage at a strain level of 0.05% (a), the yielding point at a strain level of 0.1% (b), and the strengthening process at strains of 0.5% and 1% (c, d). The distribution of stress during the initial elastic stage at a strain level of 0.05% (e), the plastic deformation stage at a strain of 0.1% (f), and the strain strengthening at strains of 0.5% and 1% (g, h). Probability analysis of stress and strain distribution at various stages of deformation (i, j). Pole figures at tension strains of 0.05%, 0.1%, 0.5%, 1%, and 10% (k) [54].

HEA strengthening through analysis of the strain field distribution. Apart from dislocation evolution at the nanoscale, non-uniform strain distribution may also play a key role in improving strength at the micron scale. The pole figures illustrate the grain orientations extracted from simulated data under increasing tension strain (Figure 4.17k). More newly-added local extreme points indicate significant changes in some grain orientations, potentially adjusting to inconsistent deformation behavior during the later stage of work hardening.

4.4.3.3 Effect of Strain Rate

The material behavior of FCC HEAs is significantly influenced by strain rate [39]. Metal materials exhibit distinct characteristics during loading at high strain rates as compared to static or quasi-static loading processes [98]. Furthermore, the motion of dislocations is distinctly altered under dynamic conditions due to the influence of viscous resistance [99]. Thus, the material behavior of $Al_{0.1}$FeCoCrNi HEA explored at varying strain rates is vital. The mechanical responses and stress-strain curve of the polycrystalline $Al_{0.1}$FeCoCrNi HEA are predicted at applied strain rates of 1×10^{-2} s^{-1}, 1×10^{-1} s^{-1}, 1×10^{0} s^{-1}, 1×10^{1} s^{-1}, and 1×10^{2} s^{-1}. The reference strain rate is 10^{-3} s^{-1} for CPFE simulations in this section. The stress-strain responses of the polycrystalline $Al_{0.1}$FeCoCrNi HEA at varying strain rates are illustrated in Figure 4.18k. As the strain rate increases, the yield stress monotonically increases at the same strain, and the slope of the stress-strain curve gradually increases, which is consistent with experimental observations [95,96].

Observations from Figure 4.18a-j indicate that at high strain rates, grains store large amounts of strain and stress. Moreover, as the strain rate increases, adjacent grains exhibit higher strain gradients (as depicted in Figure 4.18a-e). This intriguing phenomenon appears at stress fields (Figure 4.18f-j). In other words, extremely high deformation gradients can handle the energy brought about by higher strain rates in MPEA, revealing the fundamental reason for good energy absorption capacity.

4.4.4 Remark

In this section, a methodology for multiscale modeling is proposed that enables the consideration of the influence of the nano, micron, and mesoscale structures on the mechanical properties of HEAs. Extract the elastic parameters and dislocation resistance coefficients required for microscale DDD simulation from nano MD simulation, and then add the lattice distortion strain field caused by atomic mismatch to the DDD simulation of single crystals with different crystal orientations to quantify the interaction between dislocations and lattice distortion on dislocation motion. The dislocation collective behavior under the action of lattice distortion is applied to calibrate hardening parameters of single crystals for crystal plasticity model. The mechanical behavior under uniaxial tensile loading for different single crystal directions in $Al_{0.1}$FeCoCrNi HEA is predicted. To obtain the hardening parameter sets for polycrystals, the seven calibrated single-crystal hardening parameters are statistically averaged. The results of the CPEM

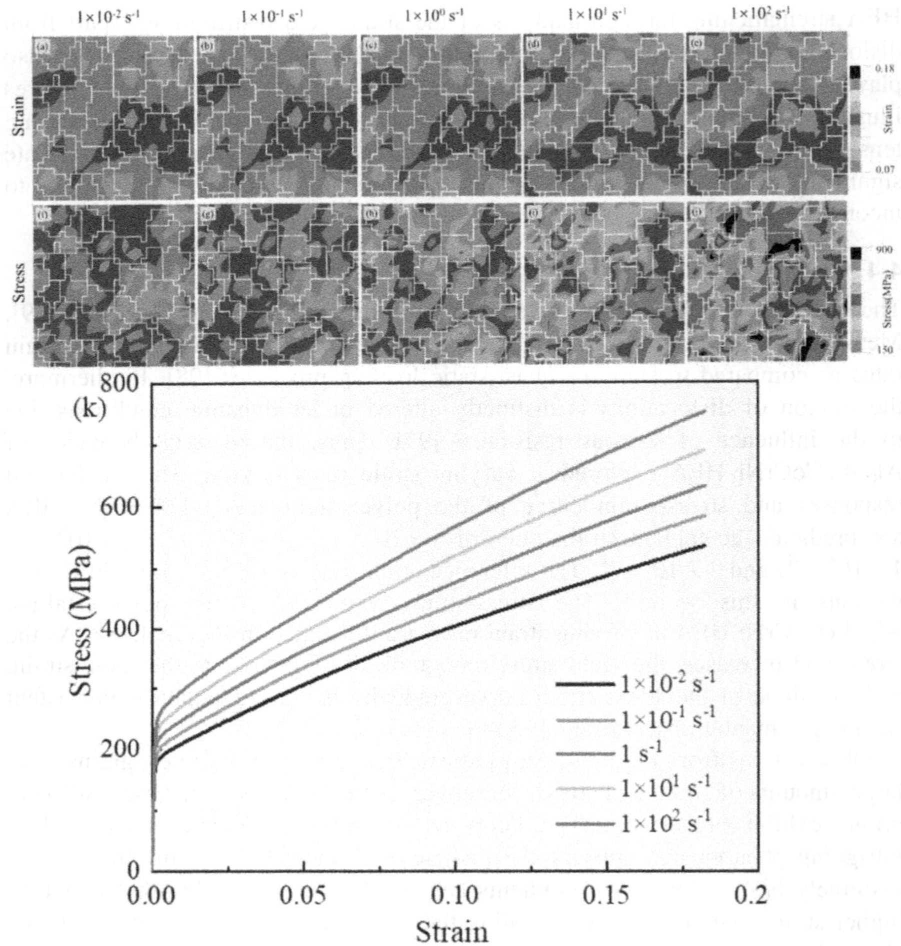

FIGURE 4.18 Strain and stress profiles at a strain of 10% for 0.01 s^{-1}, 0.1 s^{-1}, 1 s^{-1}, 10 s^{-1}, and 100 s^{-1} strain rates (a-j). The stress-strain responses of the polycrystals for various strain rates (k) [54].

simulation demonstrate that the established $Al_{0.1}FeCoCrNi$ HEA model can effectively predict the mechanical response of the polycrystal RVE on a meso-scale level.

REFERENCES

[1] Li W., Xie D., Li D., et al. 2021. Mechanical behavior of high-entropy alloys. *Progress in Materials Science*, 118: 100777.

[2] Song H., Tian F., Hu Q.M., Vitos L., Wang Y., Shen J., and Chen N. 2017. Local lattice distortion in high-entropy alloys. *Physical Review Materials*, 1(2): 023404.

[3] Tsai M.H. and Yeh J.W. 2014. High-entropy alloys: a critical review. *Materials Research Letters*, 2(3): 107–123.

[4] Zhang Z., Armstrong D.E., and Grant P.S. 2022. The effects of irradiation on CrMnFeCoNi high-entropy alloy and its derivatives. *Progress in Materials Science*, 123: 100807.

[5] Labusch R. 1970. A statistical theory of solid solution hardening. *Physica Status Solidi (b)*, 41(2): 659–669.

[6] Fleischer R.L. 1963. Substitutional solution hardening. *Acta Metallurgica*, 11(3): 203–209.

[7] Gypen L.A. and Deruyttere A. 1977. Multi-component solid solution hardening. *Journal of Materials Science*, 12(5): 1028–1033.

[8] Li L., Fang Q., Li J., et al. 2020. Lattice-distortion dependent yield strength in high entropy alloys. *Materials Science and Engineering a 784*.

[9] Senkov O.N., Scott J.M., Senkova S.V., et al. 2011. Microstructure and room temperature properties of a high-entropy TaNbHfZrTi alloy. *Journal of Alloys and Compounds*, 509(20): 6043–6048.

[10] Ma K.K., Wen H.M., Hu T., et al. 2014. Mechanical behavior and strengthening mechanisms in ultrafine grain precipitation-strengthened aluminum alloy. *Acta Materialia*, 62: 141–155.

[11] Li D.Y., Li C.X., Feng T., et al. 2017. High-entropy Al0.3CoCrFeNi alloy fibers with high tensile strength and ductility at ambient and cryogenic temperatures. *Acta Materialia*, 123: 285–294.

[12] Gwalani B., Soni V., Lee M., et al. 2017. Optimizing the coupled effects of Hall-Petch and precipitation strengthening in a Al0.3CoCrFeNi high entropy alloy. *Materials & Design*, 121: 254–260.

[13] Wu Z., Bei H., Pharr G.M., et al. 2014. Temperature dependence of the mechanical properties of equiatomic solid solution alloys with face-centered cubic crystal structures. *Acta Materialia*, 81: 428–441.

[14] Lin C.M., Juan C.C., Chang C.H., et al. 2015. Effect of Al addition on mechanical properties and microstructure of refractory AlxHfNbTaTiZr alloys. *Journal of Alloys and Compounds*, 624: 100–107.

[15] Yao H.W., Qiao J.W., Hawk J.A., et al. 2017. Mechanical properties of refractory high-entropy alloys: experiments and modeling. *Journal of Alloys and Compounds*, 696: 1139–1150.

[16] He J.Y., Liu W.H., Wang H., et al. 2014. Effects of Al addition on structural evolution and tensile properties of the FeCoNiCrMn high-entropy alloy system. *Acta Materialia*, 62: 105–113.

[17] Yeh J.W., Chang S.Y., Hong Y.D., et al. 2007. Anomalous decrease in X-ray diffraction intensities of Cu-Ni-Al-Co-Cr-Fe-Si alloy systems with multi-principal elements. *Materials Chemistry and Physics*, 103(1): 41–46.

[18] Owen L.R., Pickering E.J., Playford H.Y., et al. 2017. An assessment of the lattice strain in the CrMnFeCoNi high-entropy alloy. *Acta Materialia*, 122: 11–18.

[19] Coury F.G., Kaufman M. and Clarke A.J. 2019. Solid-solution strengthening in refractory high entropy alloys. *Acta Materialia*, 175: 66–81.

[20] Lan J., Shen X.J., Liu J., et al. 2019. Strengthening mechanisms of 2A14 aluminum alloy with cold deformation prior to artificial aging. *Materials Science and Engineering a-Structural Materials Properties Microstructure and Processing*, 745: 517–535.

[21] Osetsky Y.N., Bacon D.J., Serra A., et al. 2000. Stability and mobility of defect clusters and dislocation loops in metals. *Journal of Nuclear Materials*, 276(1–3): 65–77.

[22] Scattergood R.O. and Bacon D.J. 1982. The strengthening effect of voids. *Acta Metallurgica*, 30(8): 1665–1677.

[23] Chen Y., Liu Y., Fang Q., et al. 2020. An unified model for dislocations interacting with complex-shape voids in irradiated metals. *International Journal of Mechanical Sciences*, 185: 105689.

[24] Arsenlis A., Cai W., Tang M., et al. 2007. Enabling strain hardening simulations with dislocation dynamics. *Modelling and Simulation in Materials Science and Engineering*, 15(6): 553–595.

[25] Chen Y., Fang Q., Liu Y., et al. 2020. Void-shape dependent hardening model in irradiated face-centered-cubic metals. *Journal of Nuclear Materials*, 540: 152281.

[26] Yang L.X., Ge H.L., Zhang J., et al. 2019. High He-ion irradiation resistance of CrMnFeCoNi high-entropy alloy revealed by comparison study with Ni and 304SS. *Journal of Materials Science & Technology*, 35(3): 300–305.

[27] Lu C., Niu L., Chen N., et al. 2016. Enhancing radiation tolerance by controlling defect mobility and migration pathways in multicomponent single-phase alloys. *Nature Communications*, 7.

[28] Xiao X., Song D., Xue J., et al. 2015. A self-consistent plasticity theory for modeling the thermo-mechanical properties of irradiated FCC metallic polycrystals. *Journal of the Mechanics and Physics of Solids*, 78: 1–16.

[29] Yang T.-n., Lu C., Jin K., et al. 2017. The effect of injected interstitials on void formation in self-ion irradiated nickel containing concentrated solid solution alloys. *Journal of Nuclear Materials*, 488: 328–337.

[30] Lu C., Yang T., Jin K., et al. 2017. Radiation-induced segregation on defect clusters in single-phase concentrated solid-solution alloys. *Acta Materialia*, 127: 98–107.

[31] Kumar N.A.P.K., Li C., Leonard K.J., et al. 2016. Microstructural stability and mechanical behavior of FeNiMnCr high entropy alloy under ion irradiation. *Acta Materialia*, 113: 230–244.

[32] Ghosh S., Bai J. and Paquet D. 2009. Homogenization-based continuum plasticity-damage model for ductile failure of materials containing heterogeneities. *Journal of the Mechanics and Physics of Solids*, 57(7): 1017–1044.

[33] Wu Z. and Bei H. 2015. Microstructures and mechanical properties of compositionally complex Co-free FeNiMnCr18 FCC solid solution alloy. *Materials Science and Engineering: A*, 640: 217–224.

[34] Fang Q., Peng J., Chen Y., et al. 2021. Hardening behaviour in the irradiated high entropy alloy. *Mechanics of Materials*, 155: 103744.

[35] Li J., Fang Q. and Liaw P.K. 2021. Microstructures and properties of high-entropy materials: modeling, simulation, and experiments. *Advanced Engineering Materials*, 23(1): 2001044.

[36] Zhang Y., Zuo T.T., Tang Z., et al. 2014. Microstructures and properties of high-entropy alloys. *Progress in Materials Science*, 61: 1–93.

[37] Miracle D.B. and Senkov O.N. 2017. A critical review of high entropy alloys and related concepts. *Acta Materialia*, 122: 448–511.

[38] Wu S.W., Wang G., Yi J., et al. 2017. Strong grain-size effect on deformation twinning of an Al0. 1CoCrFeNi high-entropy alloy. *Materials Research Letters*, 5(4): 276–283.

[39] Jiang K., Ren T., Shan G., et al. 2020. Dynamic mechanical responses of the Al0· 1CoCrFeNi high entropy alloy at cryogenic temperature. *Materials Science and Engineering: A*, 797: 140125.

[40] Kuykendall W.P., Wang Y. and Cai W. 2020. Stress effects on the energy barrier and mechanisms of cross-slip in FCC nickel. *Journal of the Mechanics and Physics of Solids*, 144: 104105.

[41] Hussein A.M., Rao S.I., Uchic M.D., et al. 2015. Microstructurally based cross-slip mechanisms and their effects on dislocation microstructure evolution in fcc crystals. *Acta Materialia*, 85: 180–190.

[42] Rao S.I., Woodward C., Parthasarathy T.A., et al. 2017. Atomistic simulations of dislocation behavior in a model FCC multicomponent concentrated solid solution alloy. *Acta Materialia*, 134: 188–194.

[43] Kalidindi S.R. 1992. *Polycrystal plasticity: constitutive modeling and deformation processing*. Massachusetts Institute of Technology.

[44] Asaro R.J. and Needleman A. 1985. Overview no. 42 texture development and strain hardening in rate dependent polycrystals. *Acta Metallurgica*, 33(6): 923–953.

[45] Sundararaghavan V. and Zabaras N. 2008. A multi-length scale sensitivity analysis for the control of texture-dependent properties in deformation processing. *International Journal of Plasticity*, 24(9): 1581–1605.

[46] Lu X., Zhao J., Yu C., et al. 2020. Cyclic plasticity of an interstitial high-entropy alloy: experiments, crystal plasticity modeling, and simulations. *Journal of the Mechanics and Physics of Solids*, 142: 103971.

[47] Li Y., Zhu L., Liu Y., et al. 2013. On the strain hardening and texture evolution in high manganese steels: experiments and numerical investigation. *Journal of the Mechanics and Physics of Solids*, 61(12): 2588–2604.

[48] Li S. 2008. Orientation stability in equal channel angular extrusion. Part II: Hexagonal close-packed materials. *Acta Materialia*, 56(5): 1031–1043.

[49] Jiao Q., Sim G.-D., Komarasamy M., et al. 2018. Thermo-mechanical response of single-phase face-centered-cubic Al x CoCrFeNi high-entropy alloy microcrystals. *Materials Research Letters*, 6(5): 300–306.

[50] Li L., Fang Q., Li J., et al. 2020. Lattice-distortion dependent yield strength in high entropy alloys. *Materials Science and Engineering: A*, 784: 139323.

[51] Groh S., Marin E.B., Horstemeyer M.F., et al. 2009. Multiscale modeling of the plasticity in an aluminum single crystal. *International Journal of Plasticity*, 25(8): 1456–1473.

[52] Mecking H. and Kocks U.F. 1981. Kinetics of flow and strain-hardening. *Acta Metallurgica*, 29(11): 1865–1875.

[53] Chandra S., Samal M.K., Chavan V.M., et al. 2018. Hierarchical multiscale modeling of plasticity in copper: From single crystals to polycrystalline aggregates. *International Journal of Plasticity*, 101: 188–212.

[54] Fang Q., Lu W., Chen Y., et al. 2022. Hierarchical multiscale crystal plasticity framework for plasticity and strain hardening of multi-principal element alloys. *Journal of the Mechanics and Physics of Solids*, 105067.

[55] Zeng Z., Li X., Xu D., et al. 2016. Gradient plasticity in gradient nano-grained metals. *Extreme Mechanics Letters*, 8: 213–219.

[56] Chandra S., Samal M.K., Chavan V.M., et al. 2015. Multiscale modeling of plasticity in a copper single crystal deformed at high strain rates. *Plasticity and Mechanics of Defects*, 1(1).

[57] Li J., Chen Y., He Q., et al. 2022. Heterogeneous lattice strain strengthening in severely distorted crystalline solids. *Proceedings of the National Academy of Sciences*, 119(25): e2200607119.

[58] Chakraborty S. and Ghosh S. 2021. A concurrent atomistic-crystal plasticity multiscale model for crack propagation in crystalline metallic materials. *Computer Methods in Applied Mechanics and Engineering*, 379: 113748.

[59] Olmsted D.L., Hector L.G., Curtin W.A., et al. 2005. Atomistic simulations of dislocation mobility in Al, Ni and Al/Mg alloys. *Modelling and Simulation in Materials Science and Engineering*, 13(3): 371.

[60] Zhou C., Biner S.B. and LeSar R. 2010. Discrete dislocation dynamics simulations of plasticity at small scales. *Acta Materialia*, 58(5): 1565–1577.

[61] Plimpton S. 1995. Fast parallel algorithms for short-range molecular dynamics. *Journal of Computational Physics*, 117(1): 1–19.

[62] Stukowski A. 2009. Visualization and analysis of atomistic simulation data with OVITO–the Open Visualization Tool. *Modelling and Simulation in Materials Science and Engineering*, 18(1): 015012.
[63] Farkas D. and Caro A. 2020. Model interatomic potentials for Fe–Ni–Cr–Co–Al high-entropy alloys. *Journal of Materials Research*, 35(22): 3031–3040.
[64] Brailsford A.D. 1972. Anharmonicity contributions to dislocation drag. *Journal of Applied Physics*, 43(4): 1380–1393.
[65] Bulatov V. and Cai W. 2006. *Computer simulations of dislocations*. OUP Oxford.
[66] Rao S.I., Woodward C., Akdim B., et al. 2019. Large-scale dislocation dynamics simulations of strain hardening of Ni microcrystals under tensile loading. *Acta Materialia*, 164: 171–183.
[67] Wei D., Zaiser M., Feng Z., et al. 2019. Effects of twin boundary orientation on plasticity of bicrystalline copper micropillars: A discrete dislocation dynamics simulation study. *Acta Materialia*, 176: 289–296.
[68] Lehtinen A., Laurson L., Granberg F., et al. 2018. Effects of precipitates and dislocation loops on the yield stress of irradiated iron. *Science Reports*, 8(1): 6914.
[69] Arsenlis A., Rhee M., Hommes G., et al. 2012. A dislocation dynamics study of the transition from homogeneous to heterogeneous deformation in irradiated body-centered cubic iron. *Acta Materialia*, 60(9): 3748–3757.
[70] Dimiduk D.M., Uchic M.D. and Parthasarathy T.A. 2005. Size-affected single-slip behavior of pure nickel microcrystals. *Acta Materialia*, 53(15): 4065–4077.
[71] Norfleet D.M., Dimiduk D.M., Polasik S.J., et al. 2008. Dislocation structures and their relationship to strength in deformed nickel microcrystals. *Acta Materialia*, 56(13): 2988–3001.
[72] Fan H., Wang Q., El-Awady J.A., et al. 2021. Strain rate dependency of dislocation plasticity. *Nature Communications*, 12(1): 1–11.
[73] Gangireddy S., Whitaker D. and Mishra R.S. 2019. Significant contribution to strength enhancement from deformation twins in thermomechanically processed Al 0.1 CoCrFeNi microstructures. *Journal of Materials Engineering and Performance*, 28(3): 1661–1667.
[74] Ausloos M. and Berman D.H. 1985. A multivariate Weierstrass–Mandelbrot function. *Proceedings of the Royal Society of London. A. Mathematical and Physical Sciences*, 400(1819): 331–350.
[75] Yan W. and Komvopoulos K. 1998. Contact analysis of elastic-plastic fractal surfaces. *Journal of Applied Physics*, 84(7): 3617–3624.
[76] Wu J.-J. 2002. Analyses and simulation of anisotropic fractal surfaces. *Chaos, Solitons & Fractals*, 13(9): 1791–1806.
[77] Shao Y.-T., Yuan R., Hu Y., et al. 2019. The paracrystalline nature of lattice distortion in a high entropy alloy. *arXiv preprint arXiv:1903.04082*.
[78] Shen Y. and Spearot D.E. 2021. Mobility of dislocations in FeNiCrCoCu high entropy alloys. *Modelling and Simulation in Materials Science and Engineering*, 29(8): 085017.
[79] Lu S., Zhang B., Li X., et al. 2019. Grain boundary effect on nanoindentation: A multiscale discrete dislocation dynamics model. *Journal of the Mechanics and Physics of Solids*, 126: 117–135.
[80] Sills R.B., Bertin N., Aghaei A., et al. 2018. Dislocation networks and the microstructural origin of strain hardening. *Physical Review Letters*, 121(8): 085501.
[81] Devincre B., Kubin L. and Hoc T. 2006. Physical analyses of crystal plasticity by DD simulations. *Scripta Materialia*, 54(5): 741–746.
[82] Shehadeh M.A. 2012. Multiscale dislocation dynamics simulations of shock-induced plasticity in small volumes. *Philosophical Magazine*, 92(10): 1173–1197.

[83] Madec R. 2001. Dislocation interactions to plastic flow in fcc single crystals: a study by simulation of dislocation dynamics.
[84] Yaghoobi M., Ganesan S., Sundar S., et al. 2019. PRISMS-Plasticity: an open-source crystal plasticity finite element software. *Computational Materials Science*, 169: 109078.
[85] Bangerth W., Hartmann R. and Kanschat G. 2007. Deal. II—a general-purpose object-oriented finite element library. *ACM Transactions on Mathematical Software (TOMS)*, 33(4): 24-es.
[86] Ahrens J., Geveci B. and Law C. 2005. Paraview: an end-user tool for large data visualization. *The visualization handbook*, 717(8).
[87] Gnäupel-Herold T., Brand P.C. and Prask H.J. 1998. Calculation of single-crystal elastic constants for cubic crystal symmetry from powder diffraction data. *Journal of Applied Crystallography*, 31(6): 929–935.
[88] Anand L. and Kothari M. 1996. A computational procedure for rate-independent crystal plasticity. *Journal of the Mechanics and Physics of Solids*, 44(4): 525–558.
[89] Groeber M.A. and Jackson M.A. 2014. DREAM. 3D: a digital representation environment for the analysis of microstructure in 3D. *Integrating materials and manufacturing innovation*, 3(1): 56–72.
[90] Knezevic M., Drach B., Ardeljan M., et al. 2014. Three dimensional predictions of grain scale plasticity and grain boundaries using crystal plasticity finite element models. *Computer Methods in Applied Mechanics and Engineering*, 277: 239–259.
[91] Ganesan S., Yaghoobi M., Githens A., et al. 2021. The effects of heat treatment on the response of WE43 Mg alloy: crystal plasticity finite element simulation and SEM-DIC experiment. *International Journal of Plasticity*, 137: 102917.
[92] Lakshmanan A., Yaghoobi M., Stopka K.S., et al. 2022. Crystal plasticity finite element modeling of grain size and morphology effects on yield strength and extreme value fatigue response. *Journal of Materials Research and Technology*.
[93] Gangireddy S., Kaimiao L., Gwalani B., et al. 2018. Microstructural dependence of strain rate sensitivity in thermomechanically processed Al0.1CoCrFeNi high entropy alloy. *Materials Science and Engineering a-Structural Materials Properties Microstructure and Processing*, 727: 148–159.
[94] Stopka K.S., Yaghoobi M., Allison J.E., et al. 2021. Effects of boundary conditions on microstructure-sensitive fatigue crystal plasticity analysis. *Integrating materials and manufacturing innovation*, 1–20.
[95] Moon J., Jang M.J., Bae J.W., et al. 2018. Mechanical behavior and solid solution strengthening model for face-centered cubic single crystalline and polycrystalline high-entropy alloys. *Intermetallics*, 98: 89–94.
[96] Kireeva I.V., Chumlyakov Y.I., Vyrodova A.V., et al. 2020. Effect of twinning on the orientation dependence of mechanical behaviour and fracture in single crystals of the equiatomic CoCrFeMnNi high-entropy alloy at 77K. *Materials Science and Engineering: A*, 784: 139315.
[97] Kawamura M., Asakura M., Okamoto N.L., et al. 2021. Plastic deformation of single crystals of the equiatomic Cr–Mn–Fe–Co–Ni high-entropy alloy in tension and compression from 10 K to 1273 K. *Acta Materialia*, 203: 116454.
[98] Park J.M., Moon J., Bae J.W., et al. 2018. Strain rate effects of dynamic compressive deformation on mechanical properties and microstructure of CoCrFeMnNi high-entropy alloy. *Materials Science and Engineering a-Structural Materials Properties Microstructure and Processing*, 719: 155–163.
[99] Meyers M.A. 1994. *Dynamic behavior of materials*. John Wiley & Sons.

5 High-Entropy Amorphous Alloys

5.1 INTRODUCTION

Developed from amorphous alloys (metallic glass), amorphous high-entropy alloys (amorphous HEAs) that contain more constituent elements (≥ 5, equal or non-equal) have received extensive attention due to their superior strength, good elasticity and excellent corrosion resistance [1–3]. In this chapter, we provide a brief summary of the mechanical behavior and microstructural evolution related to amorphous HEAs, utilizing molecular dynamics (MD) simulations. Our focus is on examining the effects of alloying element, preparation process, microstructure, and load pattern [4–6]. Meanwhile, some theoretical models for predicting the mechanical properties of amorphous HEAs are developed successfully [6,7]. Finally, the deformation and properties of amorphous HEAs are summarized at the end of this chapter.

5.2 INDENTION-INDUCED DEFORMATION BEHAVIOR

In this section, $Cu_{29}Zr_{32}Ti_{15}Al_{5}Ni_{19}$ amorphous HEA and rigid spherical indenter are constructed to simulate the indentation process using MD simulation. The loading/unloading behavior for different contact depths is obtained to analyze the deformation mechanism of amorphous HEA. The deformation mechanisms of the targeted amorphous HEA revealed by MD simulations provide insight into the mechanical behaviors in the amorphous HEAs and assist further design in terms of superior mechanical properties [8].

5.2.1 Model

LAMMPS (Large-scale Atomic/Molecular Massively Parallel Simulator) is employed to construct the targeted model. The MD model of nanoindentation is observed by OVITO visualization software (Figure 5.1a). The nanoindentation model includes a virtual spherical indenter and amorphous HEA sample. By referencing the experimental preparation of amorphous HEA [9–12], the model of $Cu_{29}Zr_{32}Ti_{15}Al_{5}Ni_{19}$ amorphous HEA is obtained through the heat treatment processes of melting following by quenching, and the preparation flow chart of amorphous HEAs is summarized in Figure 5.1b. Thereafter, the models of amorphous HEA and the nanoindentation are designed to adopt periodic boundary conditions for avoiding the size effect. The relaxation of studied amorphous HEA is performed in microcanonical ensemble (NVE) at an initial temperature of

DOI: 10.1201/9781003225706-5

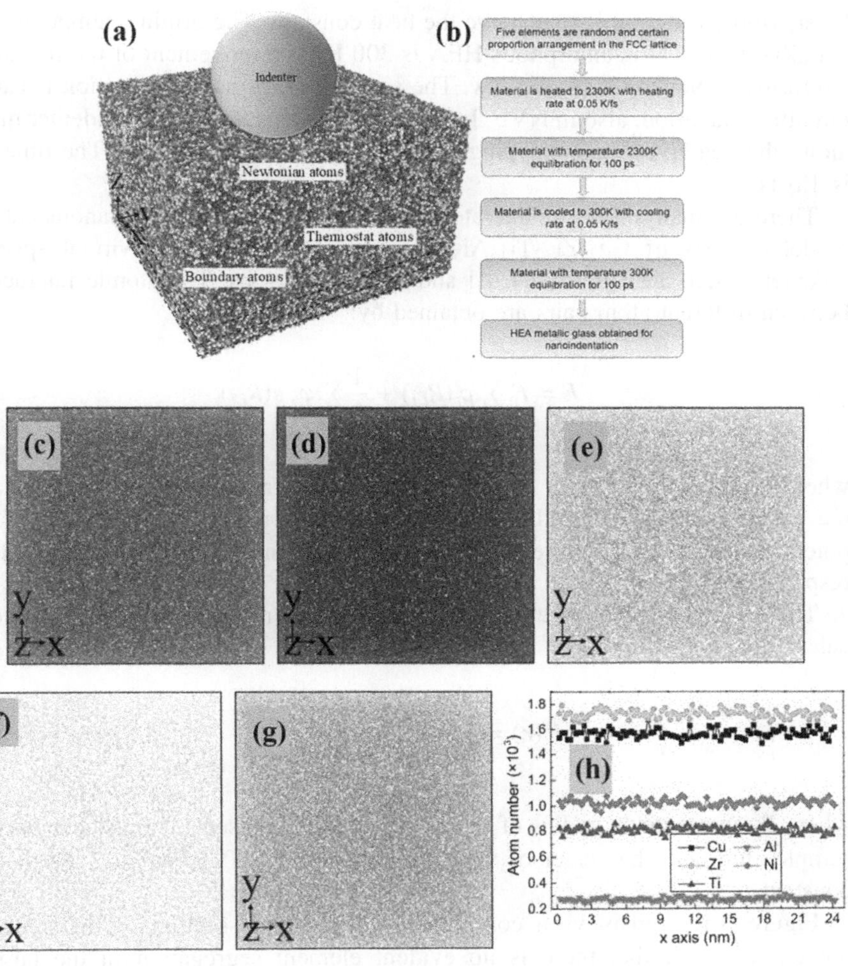

FIGURE 5.1 MD model (a) and the process of preparation (b) of $Cu_{29}Zr_{32}Ti_{15}Al_5Ni_{19}$ amorphous HEA. The alloying element distribution of amorphous HEA can be visualized through Z axis projection views: Cu (c), Zr (d), Ti (e), Al (f), and Ni (g). The distribution of alloying elements along the X axis (h) [8].

300 K. Then, the sample is rapidly heated to 2300 K at a rate of 0.05 K/fs and melted. After maintaining a melting temperature of 2300 K for 100 ps, the amorphous HEA is rapidly quenched to 300 K (0.05 K/fs). Finally, the crystalline mixture of $Cu_{29}Zr_{32}Ti_{15}Al_5Ni_{19}$ amorphous HEA forms a metallic glass structure by holding 100 ps at 300 K. The size of the studied amorphous HEA is 242 Å × 242 Å × 158 Å, containing 539,807 atoms. The size of virtual indenter is 50 Å. The sample consists of three layers with different atomic types: the Newtonian, the Thermostatic, and the Boundary. The boundary layer is set as the value of 10 Å to stabilize the sample during nanoindentation. The area with the size of 10 Å immediately above the boundary layer is regarded as the thermostatic

layer, whose function is to achieve the heat constant. The original temperature of $Cu_{29}Zr_{32}Ti_{15}Al_5Ni_{19}$ amorphous HEA is 300 K. The movement of the rest atoms conforms to Newton's second law. The simulation for nanoindentation is carried out after relaxation, also in NVE. In nanoindentation stage, virtual indenter moves along the negative Z axis keeping 10 m/s, and the depth is 24 Å. The time step is 1.0 fs.

There are many kinds of interatomic interactions in MD. The nanoindentation model consists of $Cu_{29}Zr_{32}Ti_{15}Al_5Ni_{19}$ amorphous HEA and virtual spherical indenter. There are Cu, Zr, Ti, Al and Ni atomic types. The atomic interactions between different atom pairs are obtained by:

$$E = F_\alpha \sum_{j\neq i} \rho_i(R_{i,j}) + \frac{1}{2}\sum_{j\neq i} \varphi_{\alpha,\beta}(R_{i,j}) \tag{5.1}$$

where E represents the total energy on atom i, and can be obtained by calculating the values of the F. In addition, φ and ρ correspond to the interaction of pair potential and the electron density, respectively. α and β stand for atoms i and j, respectively.

There is a repulsive force between virtual indenter and sample atoms, which is calculated by:

$$F(r) = \begin{cases} -K(R-r)^2 & r > R \\ 0 & r < R \end{cases} \tag{5.2}$$

where R represents the radius of virtual indenter, r represents the distance between sample atom and the center of virtual indenter. $K = 300eV/nm^2$ is a given force constant.

Figure 5.1c-g displays a comprehensive x-y cross-section of different elements, revealing that there is no evident element segregation in the targeted amorphous HEA. The uniform distribution of five elements in amorphous HEA can be observed, indicating that the studied amorphous HEA can be fabricated through fast quenching. Additionally, Figure 5.1h demonstrates the variation curve of alloying elements of $Cu_{29}Zr_{32}Ti_{15}Al_5Ni_{19}$ amorphous HEA along the X axis, further emphasizing the uniformity of the alloying elements. These results provide sufficient evidence that the fast melting and quenching method is an effective strategy to manufacture amorphous HEAs with consistent chemical composition.

5.2.2 Results and Discussion

The composition of alloying elements in this amorphous HEA shows a minor variation along the X axis, which confirms that the five alloying elements are present in equimolar amounts and there are no visible inconsistencies. The results of elemental distribution indicate that this amorphous HEA can be produced by the process of rapid cooling, according to prior research [13–15]. As shown in Figure 5.2a, the deformation

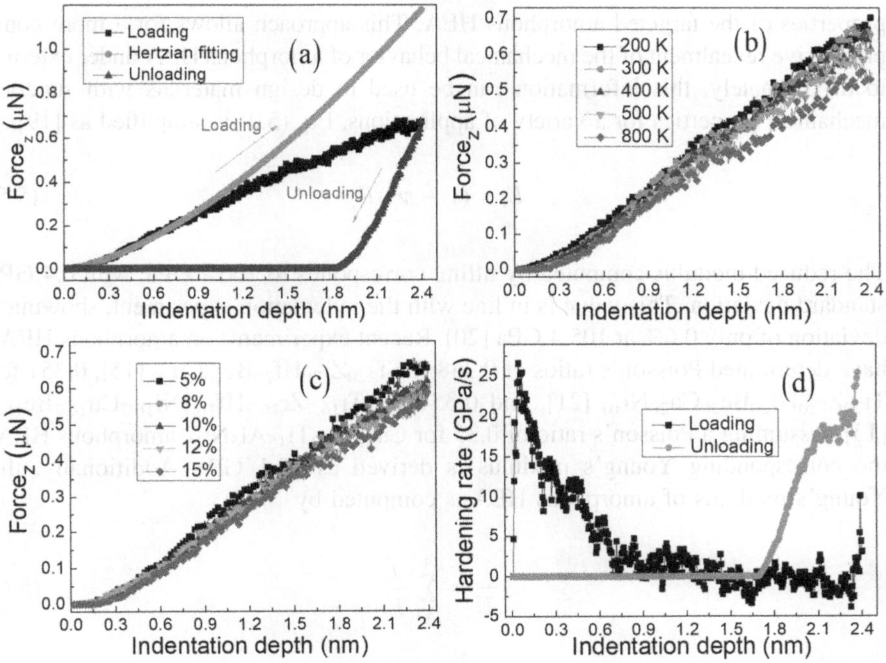

FIGURE 5.2 Load-displacement curve and its Hertzian fitting (a), the effects of temperature (b), Al concentration (c) and hardening rate on the indentation displacement [8].

behavior of $Cu_{29}Zr_{32}Ti_{15}Al_5Ni_{19}$ amorphous HEA involves an elastic stage followed by a plastic stage, resembling the behavior of this amorphous HEA under indentation as reported in previous studies [16,17]. Furthermore, the curve displays a sudden displacement. The Hertzian elastic theory [18] can be used to calculate the correlation between the applied load *P* and the depth of indentation *h* during the elastic deformation, which is given by the following equation:

$$P = 4E_* R^{1/2} h^{3/2} / 3 \tag{5.3}$$

where R and E* represent the size of indenter. The other parameters E* are calculated by

$$\frac{1}{E_*} = \frac{1 - \nu_i^2}{E_i} + \frac{1 - \nu_s^2}{E_s} \tag{5.4}$$

In the present MD simulation, a rigid indenter is constructed as a means of investigating the deformation behavior of the studied amorphous HEA. This method of analysis involves the application of Young's modulus and Poisson's ratio of the studied system. Specifically, the values assigned to m_i and E_i represent the Poisson's ratio and Young's modulus of the indenter, while m_s and E_s correspond to these same properties of the targeted amorphous HEA. This approach allows for a more comprehensive revealment of the mechanical behavior of amorphous HEA under external load. Ultimately, this information can be used to design materials with superior mechanical properties for a variety of applications. Eq. (5.2) is simplified as [19]

$$E_s = (1 - \nu_s^2)E_* \tag{5.5}$$

The reduced modulus computed by fitting corresponds to 106.1 GPa, with 0.4 GPa standard deviation. This value is in line with the indentation experiment, showing a deviation of only 0.6% at 105.4 GPa [20]. Recent experiments on amorphous HEAs have determined Poisson's ratios of 0.348 for $Ti_{20}Zr_{20}Hf_{20}Be_{20}Cu_{20}$ [15], 0.351 for $Ti_{20}Zr_{20}Hf_{20}Be_{20}Cu_{20}Ni_{10}$ [21], and 0.354 for $Ti_{16.7}Zr_{16.7}Hf_{16.7}Ni_{16.7}Cu_{16.7}Be_{16.7}$ [13]. Assuming a Poisson's ratio of 0.35 for $Cu_{29}Zr_{32}Ti_{15}Al_5Ni_{19}$ amorphous HEA, the corresponding Young's modulus is derived as 93.1 GPa. Additionally, the Young's modulus of amorphous HEA is computed by

$$\frac{1}{E} = \sum_{i=1}^{n} \frac{f_i}{E_i} \tag{5.6}$$

The Young's modulus of each element i within amorphous HEA is represented by E_i, with f_i indicating the atomic percentage of that element. Using this approach, the resulting Young's modulus of amorphous HEA has been regarded to be 101 GPa. In addition, the higher temperature during indention reduces the stiffness of amorphous HEA, as shown in Figure 5.2b. This observation is consistent with previous experiments, and suggests that the material softens at higher temperatures. Moreover, the maximum load required to achieve a given indentation depth also decreases at higher temperatures, providing further evidence for the decreased stiffness in amorphous HEA. These results indicate an important way for the research of amorphous HEA, as they highlight the effect of temperature on the mechanical behaviors. By understanding these factors, researchers can optimize the performance of amorphous HEA in various applications. These observations align with the findings of previous experiments [16,17].

The research related to HEA has established the significant effects of the Al element on HEA in terms of phase formation, strength, and ductility [22–24]. Consequently, this part investigates the effects of Al content by examining the mechanical response for amorphous HEA with varying Al content (Figure 5.2c). A lower indentation load can be observed when the Al content is increased, indicating a softening of the amorphous HEA due to the low Young's modulus induced by high Al content [25].

The ability of a material to bear the penetration of hard objects onto its surface, known as hardness H, is obtained by dividing the load P by the area of the thrust surface A_c:

$$H = P/A_c \tag{5.7}$$

where A_c can be calculated as 2πRh [26]. The hardness of $Cu_{29}Zr_{32}Ti_{15}Al_5Ni_{19}$ amorphous HEA is determined to be 8.8 GPa when indentation depth is 2.4 nm. The hardness of $Cu_{29}Zr_{32}Ti_{15}Al_5Ni_{19}$ amorphous HEA is found to be 7.45 GPa in a recent indentation experiment. However, this value is approximately 18% lower than the simulation results, which can be attributed to the strong nanosize effect.

The ratio of stiffness to elasticity modulus (*H*/*E* ratio) is a critical factor in determining wear resistance, as it reflects the depth of penetration [27,28]. In the case of $Cu_{29}Zr_{32}Ti_{15}Al_5Ni_{19}$ amorphous HEA, the *H*/*E* ratio is found to be 0.094. During deformation, the hardening rate shows a transition when the indention depth increases (Figure 5.2d). During unloading, the hardening rate consistently decreases at a fast rate.

In Figure 5.3, the allocation of shear strain in $Cu_{29}Zr_{32}Ti_{15}Al_5Ni_{19}$ amorphous HEA is displayed at various depths of indentation. The values of local shear strain for every atom induced by the indentation have been used to color them, with red

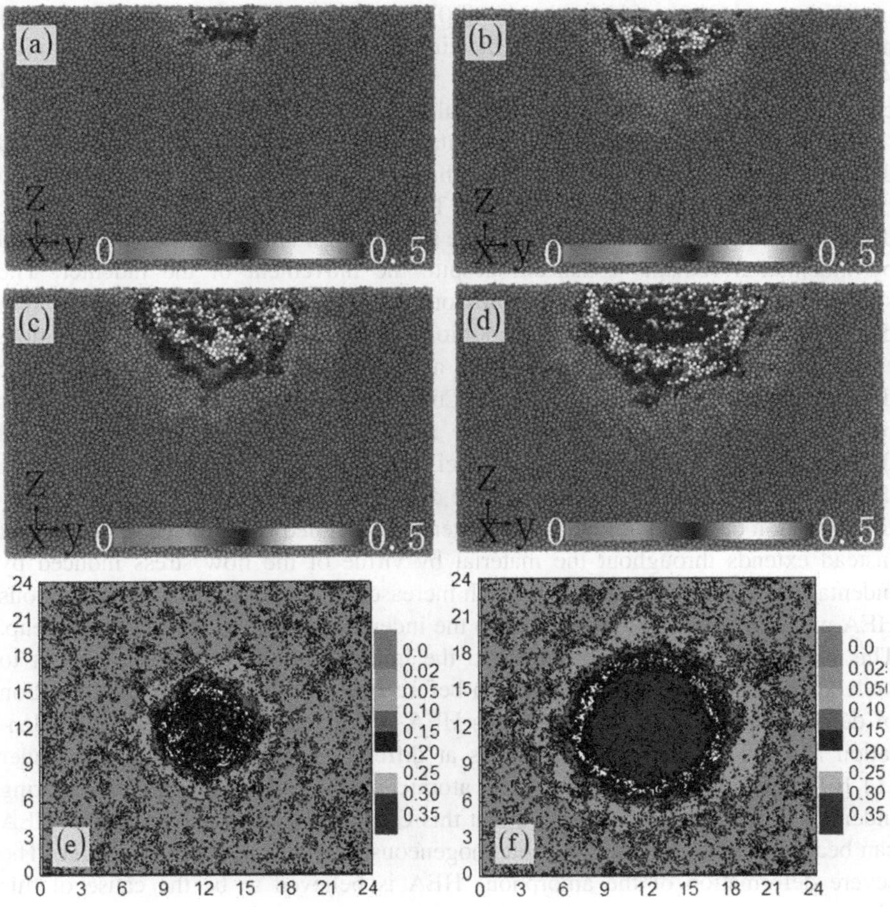

FIGURE 5.3 The distribution of shear strain (y-z plane cross-section) in the amorphous HEA induced by nanoindentation process at the indentation depths of (a) 0.6, (b) 1.2, (c) 1.8 and (d) 2.4 nm, respectively. The severest shear strain of (e) 1.2 and (f) 2.4 nm are displayed here [8].

indicating drastic shear strain and blue indicating relatively lower shear strain. These shear zones originate from the interface area that is located between the workpiece and the indenter before expanding inward. Interestingly, despite the symmetrical shape of the indenter, the high-entropy effect and significant level of disorder within the amorphous HEA result in an unsymmetric allocation of the shear zone. Furthermore, the strain is drastic at the center of the indenter and extremely low at the workpiece boundary, implying the inhomogeneity of plasticity. In other words, the area of the highest shear strain is primarily concentrated within the indentation region and its near surroundings at depths of 1.2 and 2.4 nm. These observations provide valuable insights into the deformation behavior of amorphous HEA under external load. This observation aligns with the Hertzian elastic theory [29] in a qualitative manner. To investigate mechanical behaviors near the indentation, surface morphology analysis is performed at different depths. The brittle deformation behavior observed at a depth of 0.6 nm aligns, which has been reported in prior experiments and simulations [20]. Above a depth of 2.4 nm, a mass of material pileups can be observed, indicating the deformation behavior of ductility. As the indentation load increases, shear stress flow leads to the displacement of contact atoms, influencing the mechanical properties of the sample [14,16].

The local displacement during indentation is characterized at typical loaded depths, where significant deformation within a local cluster can be observed through obvious local atom displacement [12,30]. The amorphous HEA experience isotropic displacement, which can be proved by the symmetry of atomic displacement caused by indentation remains consistent despite the movement of the indenter. The symmetry of atomic displacement is obvious in the contours observed at indentation depths (1.2 and 2.4 nm), providing additional evidence to support this observation.

For further understanding of the local atom displacement, the vector for atomic displacement caused by indentation in $Cu_{29}Zr_{32}Ti_{15}Al_5Ni_{19}$ amorphous HEA has been thoroughly investigated. Here, the arrows correspond to the direction and distance of atomic displacement, with their length providing information about the distance of atomic displacement (Figure 5.4). Interestingly, the significant curvilinear motion observed around the indenter is not limited to below the indenter, but instead extends throughout the material by virtue of the flow stress induced by indentation. When the indentation depth increases, the surface atoms of amorphous HEA move in the direction opposite to the indenter, leading to the material pileup. This phenomenon can be attributed to the inherent ductility, which allows it to undergo significant plastic deformation before fracture occurs. Thus, the transition of deformation behavior in amorphous HEA is influenced by the depth of indentation and the sliding of atom groups at different layers in an integral manner [14,16]. Drawing from the observed atomic displacement and direction during indentation, it has been determined that this $Cu_{29}Zr_{32}Ti_{15}Al_5Ni_{19}$ amorphous HEA can bear both homogeneous and inhomogeneous flow during the plastic stage. The severe deformation of the amorphous HEA is believed to be the cause of this phenomenon, which reduces the activation energy required to initiate shear deformation. Homogeneous flow is characterized by the uniform displacement of atoms in the material, whereas inhomogeneous flow is characterized by the displacement of atoms in a non-uniform manner. Interestingly, both forms of flow have been

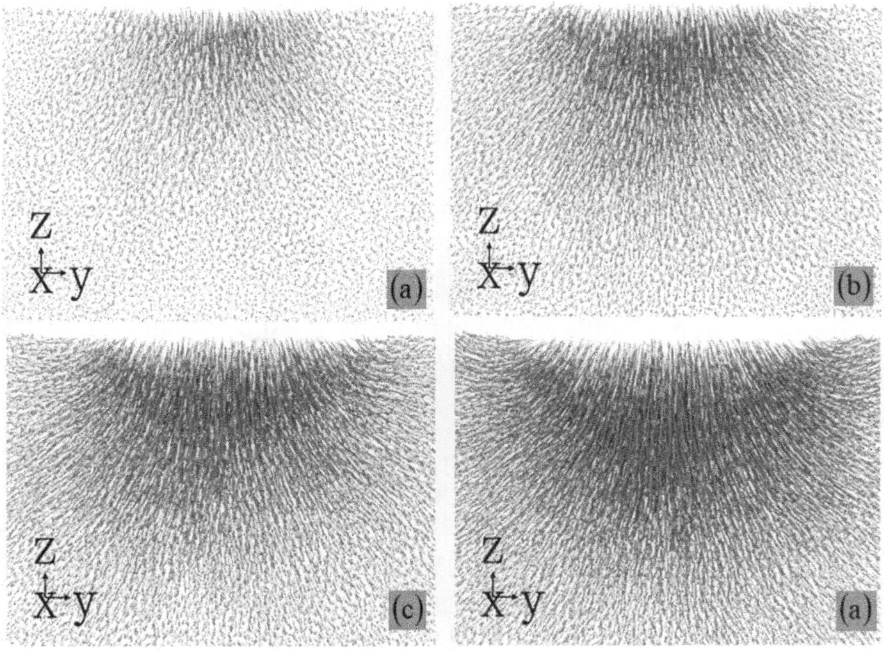

FIGURE 5.4 The vectors for atomic displacement of amorphous HEA can be viewed as cross-sectional snapshots of the y-z plane taken at indentation depths of (a) 0.6, (b) 1.2, (c) 1.8 and (d) 2.4 nm [8].

observed during plastic deformation, indicating that the studied amorphous HEA is capable of sustaining significant plastic deformation before fracture occurs.

Figure 5.5 displays the partial radial distribution function (RDF) for different atomic pairs. The RDF is essential in establishing the connection between intermolecular interactions and macroscopic thermodynamic properties [12,30]. The amorphous nature is reflected in the significant fluctuations observed in the first RDF peak when compared to the remaining peaks among the 15 partial RDFs analyzed here. The difference in atomic radius leads to a noticeable left or right shift in the first RDF peak of partial RDF for Al element. The atomic sizes of these alloying elements further emphasize the critical role that differences in radii between various elemental types play in glass formation ability (GFA) of amorphous alloys [31]. Earlier studies have indicated that the GFA depends on the content of elements and the critical pressure for melt-quenching [32,33]. In addition, the amorphous structure is confirmed by the room temperature RDF, which serves as a valuable tool for analyzing material structure. However, the first peak in the RDF is slightly reduced after indention, suggesting that the studied amorphous HEA can cause structural changes during plastic deformation. By analyzing the RDF of amorphous HEA at different temperatures and under various external loads, researchers can establish a more comprehensive understanding of the structural evolution and GFA. This information can be used to optimize the processing parameters for amorphous HEA and promote the potential applications. Overall, the RDF analysis provides valuable insights into

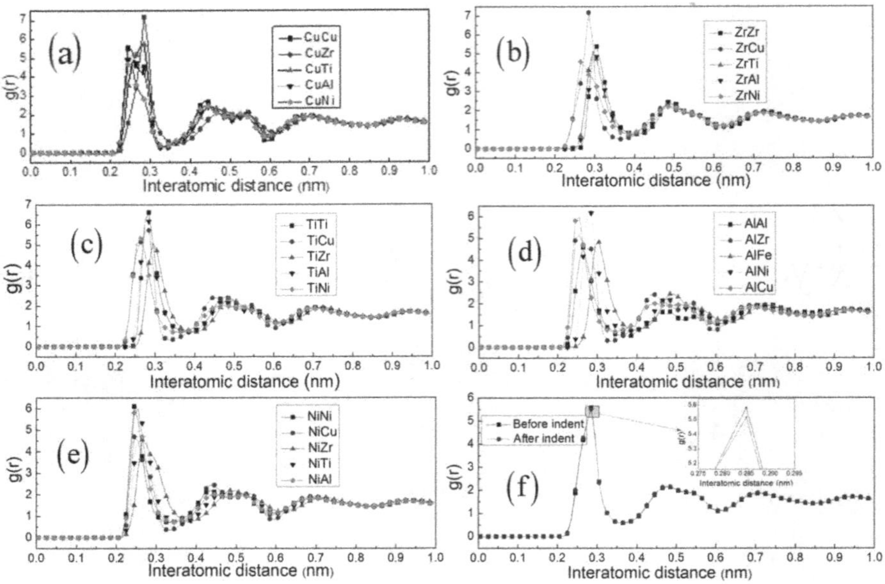

FIGURE 5.5 (a)-(e) display the correlation of partial RDF and interatomic distance after indentation for different atoms. The difference caused by indention is plotted in (f) [8].

the structural changes that occur in amorphous HEA during plastic deformation. By leveraging this information, researchers can work to optimize the mechanical properties and develop more effective methods for producing amorphous HEAs with consistent and predictable properties.

5.3 TENSION-INDUCED DEFORMATION BEHAVIOR

Although the amorphous HEAs possess many outstanding properties, the limited plasticity hinders their specific applications. The composite with amorphous component and crystalline component is one of effective strategies to promote their plasticity. Herein, the FeCoCrNi amorphous HEA is modified by developing a novel composite containing a corresponding crystalline structure. The MD simulation is applied to clarify the deformation behavior of the FeCrCoNi HEA composite, which is composed of amorphous HEA layers with different thicknesses and crystalline phases. This section focuses on analyzing the change in structure and interface, as well as the stress and shear strain, in order to determine how the size of the amorphous layer affects the deformation behavior. Figure 5.6 gives the MD models for targeted composite consisting of amorphous HEA and crystalline HEA [7].

5.3.1 Model

In this section, the sizes of amorphous structure are set as 0 Å, 30 Å, 60 Å, 90 Å, 120 Å and 155 Å. LAMMPS is used to construct the model of amorphous/crystalline FeCrCoNi HEA composites. The tensile model of amorphous/crystalline

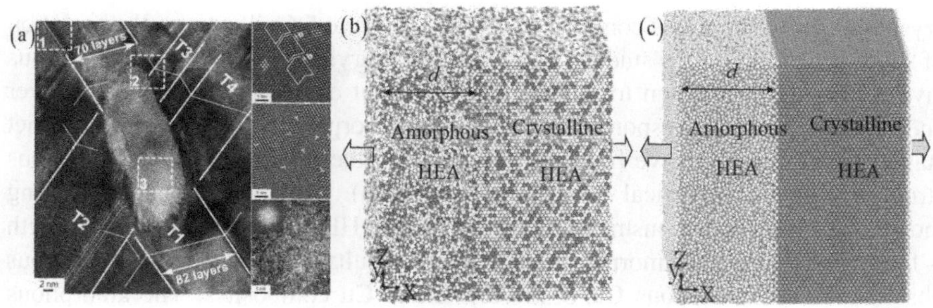

FIGURE 5.6 The microstructure images of the amorphous/crystalline FeCrCoNi HEA composites obtained from experiment (a). The constructed model for the studied composites (b, c). The spheres of red, blue, yellow and white represent Fe, Co, Cr, and Ni, respectively (b). Distinct structure types are reflected by varying colors assigned to the atoms (c) [7].

FeCrCoNi HEA visualized by OVITO is shown in Figures 5.6 (b) and (c), which consist of crystalline HEA and amorphous HEA. The metal atoms of alloying elements are randomly and evenly distributed in amorphous/crystalline HEA FeCoCrNi tensile model. The size of tensile model along x-[1 0 0], y-[0 1 0], z-[0 0 1] directions are 155 Å × 152 Å × 161 Å. In addition, the size of amorphous layer is changed from 0 nm to 15.5 nm in order to study the specific influence of amorphous HEA layers with different thickness on the deformation behavior of amorphous/crystalline FeCrCoNi HEA composites.

The interatomic interactions of amorphous/crystalline FeCrCoNi HEA composites are described by an EAM potential, which has been frequently applied to investigate the phase formation and deformation mechanism [30,34–36]. Previous studies have successfully used current EAM potential to study the phase composition, mechanical characteristics and deformation behavior of HEA with a single phase [34], HEA with complicated phases [37], HEA with nano-structure [38] and HEA in metallic glass state [5,39].

According to experiment [40], the process of obtaining the targeted composite is as follows: (a) At the initial temperature of 300 K, atoms of amorphous/crystalline HEA FeCoCrNi are arbitrarily distributed as the crystal structure; (b) to achieve amorphous HEA model, melting can first be performed in an isobaric-isothermal ensemble (NPT) at 2500 K temperature; (c) then cooled from 2500 K to 300 K at 10 K/ps [41–44]. The [011] crystal orientation is established along the X axis, Y axis and Z axis respectively for the crystalline HEA layer. Periodic boundary conditions are used in the X axis, Y axis and Z axis respectively. Before loading, the conjugate gradient method is employed to minimize the sample energy [37], and then all relax to reach equilibrium state. The strain rate is set as a constant value of 1×10^8 s^{-1} with the direction along the Xaxis [30,39] and the time step is 1 fs.

5.3.2 Results and Discussion

Earlier research has emphasized the importance of the thickness of crystal structure layers in determining the mechanical behaviors of composites that contain both crystalline and amorphous components [41–43]. Figure 5.6a illustrates the variation of stress and strain for the studied composite with varying thicknesses of amorphous layers, ranging from 0 nm to 15.5 nm. The content of the amorphous HEA layer and the composite corresponds to the size of amorphous structure. The distinct stages of deformation have been characterized by the influence of the amorphous structure on the mechanical behavior (Figure 5.6a). The highest strength among those composites is demonstrated by the crystalline HEA, while the lowest strength is found in the absolute amorphous HEA. These results are consistent with previous observations in amorphous Cu50Zr50/crystalline Cu composites. The amorphous Cu50Zr50 exhibits lower strength than metallic Cu [41]. Experimental results of amorphous Cu45Zr55/crystalline Cu composites are consistent with the observation that the strength of the high entropy composites (amorphous/crystalline) decreases when the size of the amorphous structure increases [43]. The variation of stress and strain for CuZr alloys, CuAg alloys, Mg alloys, and amorphous/crystalline Au nanowires all indicate that HEA with sample phase composition has greater mechanical properties than amorphous HEA. This significant size effect observed in the single crystal sample is believed to be responsible for this phenomenon [45–47]. In contrast, the formation of voids or cracks from grain boundary sliding at a lower critical stress leads to reduced strength in materials with complete structures with different grain sizes compared to their amorphous counterparts [48].

According to the mixed principle, the flow stress in this type of composites is calculated using $\sigma = \sigma_a t_a/(t_a + t_c) + \sigma_c t_c/(t_a + t_c)$, where σ_a, $\sigma_c t_a$ and t_c stand for the flow stress of the amorphous HEA, crystalline HEA and their sizes, respectively. The mechanical behaviors caused by tensile load are significantly influenced by the size of the amorphous structure. Corresponding to Figure 5.7a, flow stress declines with the amorphous HEA layer getting larger. Following by the yielding stage, the sample with an amorphous HEA layer of lower thickness possesses a significant decrease in stress (Figure 5.7a). Furthermore, a reduction in the size of the amorphous HEA promotes yielding strain (Figure 5.7b), suggesting the softening in advance. The general mechanical behavior caused by tensile load is largely dictated by the size of the amorphous HEA, highlighting its significant effects. (Figure 5.7).

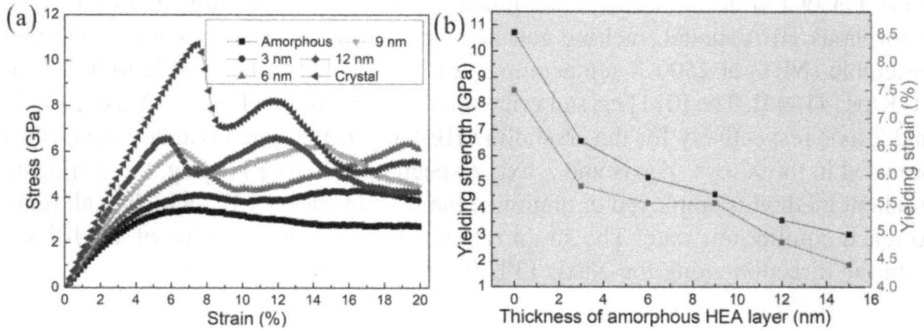

FIGURE 5.7 Strain-stress curves of amorphous HEA, crystal HEA, and amorphous/crystal HEA composites containing amorphous HEA layer with different thicknesses (a). The correlation between yielding strength and the size of the amorphous HEA layer (b) [7].

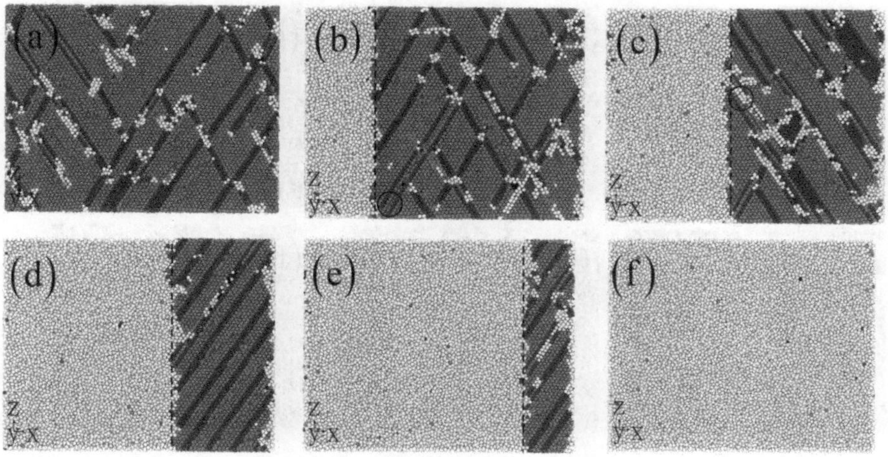

FIGURE 5.8 The evolution of microstructure at the strain of 20% for the crystalline HEA (a), the amorphous HEA layer with different thickness of 3 nm (b), 6 nm (c), 9 nm (d), and 12 nm (e), and the pure amorphous HEA (f). The interfaces between crystalline HEA and amorphous HEA are marked as dotted lines [7].

As plotted in Figure 5.8, the nucleation of Shockley partial dislocation is observed at the interface between amorphous structure and crystalline layers, with subsequent absorption of these dislocations at the opposite interface. Furthermore, the densities of dislocation vary in the range of 1.35×10^{17} m^{-2} to 9.62×10^{17} m^{-2}, including the pure amorphous HEA. The dynamic fluctuations in dislocation density can be observed when the size of amorphous HEA increases, suggesting the inhomogeneous plastic deformation in the crystalline structure, which is an intriguing result. In addition, the correlation between dislocation density and the deformation mechanism of the crystalline HEA has been widely recognized [49,50], which in turn relies on the size of the amorphous structure. The dominant factor contributing to the high strength of the studied amorphous/crystalline HEA composite is believed to be the effects of dislocation on crystalline structure. The plasticity of amorphous HEA layer is activated by the formation and movement of dislocations in the crystalline HEA [45–47,51]. Moreover, the mechanical behaviors of the amorphous/crystalline HEA composite are significantly influenced by the occurrence of deformation twins at the interface between the amorphous HEA and crystalline HEA [46]. The formation of dislocations results in a coarse interface structure, where the strip texture serves as the prefer position of nucleation for the shear band (Figure 5.9).

In order to further understand the effects of amorphous structure on the mechanical behaviors [41–43], the shear strain for the atom of the targeted amorphous/crystalline HEA composites is calculated following the tensile deformation. The pattern of shear strain in atom under a total strain of 20% deformation is depicted in Figures 5.9(a-d). To obtain the shear strain, the deformation gradient tensor is calculated by comparing the position of atom before and after deformation. In the component of amorphous HEA, a mass of shear bands can be observed as fine

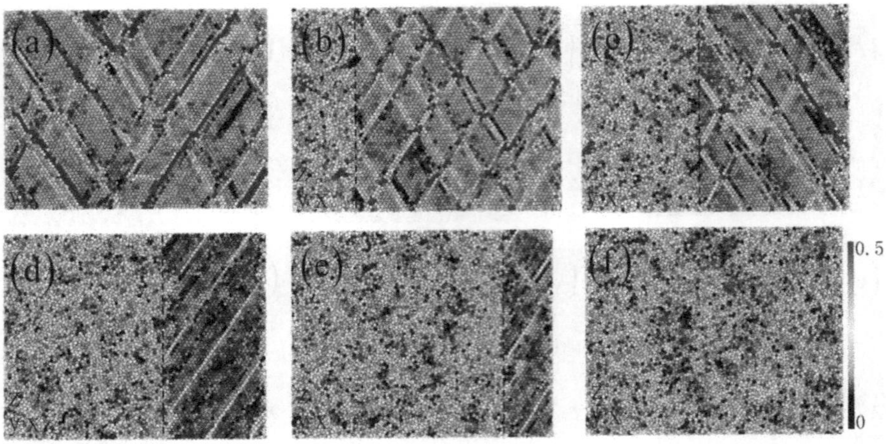

FIGURE 5.9 The distribution of shear strain for studied amorphous/crystalline HEA composites with different structure. a-e represent the amorphous HEA layers of 0 nm, 3 nm, 6 nm, 9 nm 12 nm, respectively. e gives the pure amorphous HEA [7].

morphology. Different with amorphous HEA, the shear bands are determined to be relatively larger and less in crystalline HEA. Additionally, the segregation of shear strain can be identified next to the interface of amorphous HEA layer. Unlike the findings of a previous study related to amorphous Cu50Zr50/crystalline Cu composites with amorphous structure, where the shear bands can pass through the interface between the crystal and amorphous structures [42], the current results indicates that the shear bands could not pass into another distinct structure due to the interface between them. Moreover, the drastic strain located at the interface suggests that dislocation can induce the nucleation of shear bands. Therefore, the plasticity of the amorphous HEA is accelerated by the dislocation slip mechanism in the studied amorphous/crystalline HEA composites.

The deformation behavior of the target HEA composite is significantly impacted by the presence of local stress concentrations, which ultimately determine the mechanical properties of the composite. In particular, the fields of stress in amorphous HEA components are relatively more complex and drastic, while it is more stable and slighter in the crystalline HEA components. As a result of this random distribution, the local tensile/compressive stresses exist in the characterized plane in turn. This is due to the stochastic position of alloying elements with inhomogeneous atomic radii in the amorphous HEA layer [37,52–56]. Hence, it is difficult for the shear bands to transfer to another component by breakdown interface. The variation of local stress also dominant the deformation mechanism and mechanical properties, particularly the softening. The movement of dislocations in the crystal structure and the nucleation of shear bands in the amorphous structure are influenced by stress fluctuations ranging from tensile to compressive that occur in the interface at the nanoscale. As a result of this trend, the plastic deformation is redistributed more evenly. In addition, the local yield of the amorphous structure is found and attributed to differences in atomic bonding, as evidenced by variations in the local

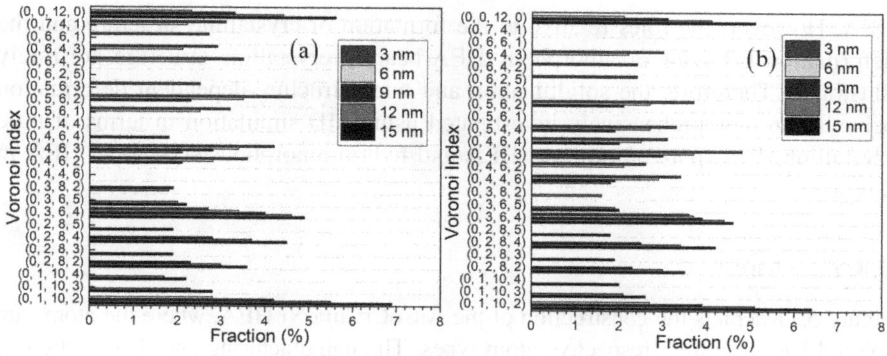

FIGURE 5.10 Arrangement and composition of major VPs for amorphous HEA layers with different thickness: (a) before deformation and (b) after deformation [7].

strain and stress. The local yield can affect the properties of the interface, which modify the yield behavior and hardening of amorphous/crystalline HEA composite at larger scale.

Following general method, the Voronoi polyhedron method (VPM) is carried out to determine the local microstructure of the deformed HEA [50–52]. The Voronoi polyhedron (VP) is represented by an index including n3 to ni, where ni is regarded as the number of specific faces with i-edge [57–59]. In the case where the quantity of atoms in the VP is lower than 1%, the vertex is not considered. Figure 5.10 shows different VPs and their fractions. As shown in Figure 5.10a, the top three VPs are <0, 1, 10, 2>, <0, 3, 6, 4>, and <0, 0, 12, 0> as the amorphous structure get larger (>9 nm) before deformation. After deformation, these top three VPs are maintained (Figure 5.10b). Under the relatively lower size of amorphous structure (<6 nm), the top two VPs of <0, 4, 6, 3> and <0, 5, 6, 2> are observed. As the amorphous HEA layer increases, it can be found that the VPs of <0, 3, 8, 2>, <0, 4, 4, 6>, <0, 4, 6, 2>, <0, 4, 6, 3>, <0, 4, 6, 4>, <0, 5, 4, 4>, <0, 5, 6, 1>, <0, 5, 6, 2>, <0, 5, 6, 3>, <0, 6, 2, 5>, <0, 6, 4, 2>, <0, 6, 4, 3>, <0, 6, 6, 1>, and <0, 7, 4, 1> increase, while VPs of <0, 1, 10, 2>, <0, 1, 10, 3>, <0, 1, 10, 4>, <0, 2, 8, 2>, <0, 2, 8, 3>, <0, 2, 8, 4>, <0, 2, 8, 5>, <0, 3, 6, 4>, and <0, 3, 6, 5> decrease. When the size of amorphous structure further reduces to 3 nm, new VPs of <0, 6, 6, 1> and <0, 7, 4, 1> are identified. The arrangement and composition of VPs after deformation is similar to that before deformation, with the inclusion of VPs of <0, 4, 4, 6>, <0, 5, 4, 4>, <0, 4, 4, 6>, and <0, 6, 2, 5>. Furthermore, some VPs can only be found in deformed HEA, while others can only be found in undeformed HEA. Therefore, the mechanical properties are regulated by the size of the amorphous structure, which in turn determines the evolution of VPs.

5.4 MICROSTRUCTURE-DEPENDENT DEFORMATION BEHAVIOR

The microstructure and mechanical properties of amorphous HEA can also be regulated by cooling rate. The formation of crystalline structure in the amorphous HEA is an expective method to improve the mechanical properties of amorphous

HEA. However, the roles relative to the formation of crystalline structure and the deformation behavior of amorphous HEA-bearing crystalline structure are rarely illustrated. Therefore, the solidification and microstructure-dependent deformation behavior are comprehensively investigated using MD simulation in terms of pure crystalline HEA, pure amorphous HEA and hybrid amorphous/crystalline HEA in this section [60].

5.4.1 Model

Figure 5.11 depicts the construction of the AlCoCrCuFeNi HEA, where the atoms are colored based on their respective atom types. The manufacturing and deformation of HEA samples are performed by the LAMMPS, which is utilized for MD simulations [61]. MD methods provide a powerful tool for understanding and displaying the nucleation and movement of dislocation or other type of crystal defects. Here, the evolution of dislocation in HEA is processed using visualization tool OVITO [62]. The size of independent HEA films along the x, y and z directions is 1079 Å × 36 Å × 1079 Å, containing 3.6 million atoms. The initial temperature of AlCoCrCuFeNi HEA is set at 300 K, where atoms spontaneously exist in face-centered cubic (FCC) phases. Then, the samples are heated under the rate of 1×10^{14} K/s to 3300 K and relax their speeds until AlCoCrCuFeNi HEA reaches equilibrium at 3300 K. Thereafter, the quenching stage is entered and the model is cooled rapidly at different cooling rates until the temperature is reduced to 300 K. After quenching, it relaxes to the equilibrium state at 300 K [30] and finally obtains AlCoCrCuFeNi HEA with different microstructure. The tensile simulation and relaxation are simultaneously performed in isothermal-isobaric (NPT) under zero pressure conditions. The MD model maintains periodic boundary conditions in different directions [63–65].

The formation of the amorphous structure in AlCoCrCuFeNi HEA are regulated by controlling the cooling rate during the solidification process. Specifically, high cooling rates are regarded as the positive factor for the formation of amorphous HEA, while low cooling rates (compared to that for amorphous HEA) are considered to promote

FIGURE 5.11 The constructed atomic model of the targeted AlCoCrCuFeNi HEA [60].

the generation of nanocrystalline HEA. At intermediate cooling rates, a hybrid HEA with amorphous and crystalline structure is produced. To gain a more comprehensive understanding of the solidification and deformation processes of HEAs with different structures, MD simulation is used to investigate the behavior of amorphous HEA, crystalline HEA, and hybrid amorphous/crystalline HEA. By leveraging this computational approach, researchers can gain insights into the underlying mechanisms that govern the mechanical properties of these amorphous HEAs [66].

The processes of melting and solidification in HEA are observed through changes in temperature, structure, and volume, as depicted in Figure 5.12. As the temperature rises above the melting point, the amorphous phases nucleate in FCC HEA matrix. At a temperature of 3,300 K, the solid phase transitions to liquid phase (as shown in Figure 5.12b). The liquid phase is subsequently cooled at a constant rate to 300 K, resulting in the formation of amorphous HEA. During the melting

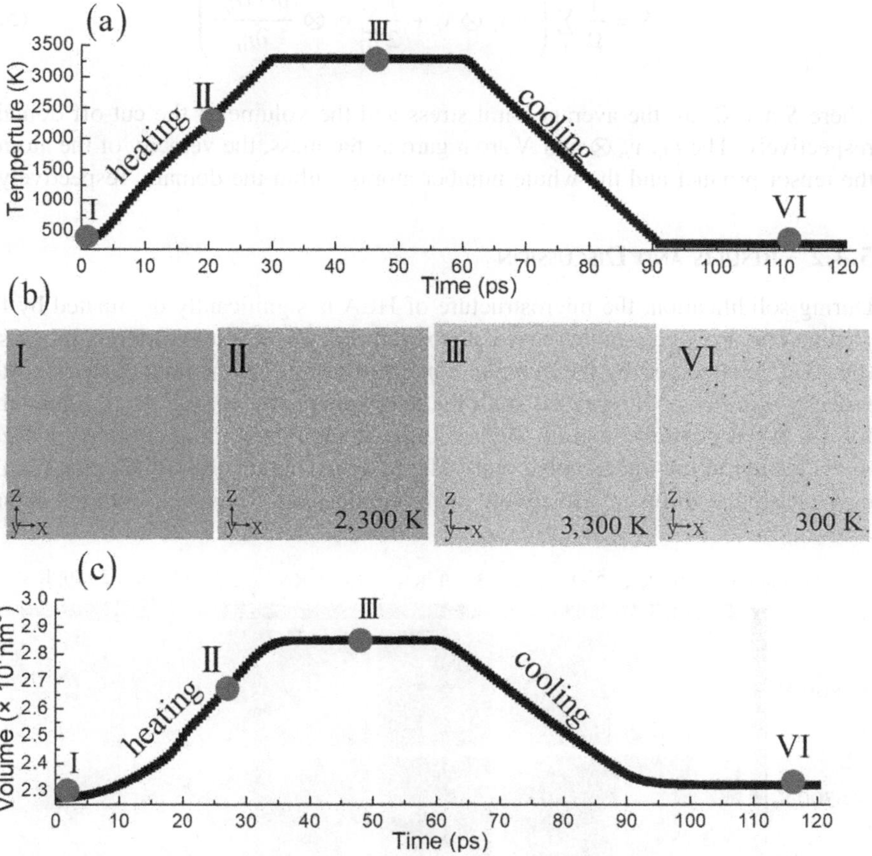

FIGURE 5.12 Schematic of the processes for heat-treatment (a), the atomic structure at different stages in (a). "*I*", "*II*", "*III*" and "*VI*" represent initial FCC structure, the microstructure at high temperature of 2,300 K, the liquid or melted state and amorphous structure, respectively. (c) plots the evolution of different microstructure correspond to (b) [60].

stage, the volume of FCC gradually increases in volume followed by a steady linear increase (Figure 5.12c). Conversely, linear volume decrease occurs during solidification due to the formation of the amorphous structure.

The interatomic interactions of elements in AlCoCrCuFeNi HEA are described using the embedded atom method (EAM) potential. Eq. (5.1) takes the form of a multibody potential energy function. As long as paired terms are adjusted accordingly, the EAM potential of a monatomic element remains unaffected by the transformation of the linear term of adding or subtracting electron density in the embedded energy function. As a result, many EAM functions for single atom developed in different research may look different but are actually similar when converted. The previous work is used to obtain the EAM [67–69], which is applied to understand the tensile deformation behavior and the preparation [39,70–72].

The average viral stress is calculated by [65].

$$S = \frac{1}{\Omega}\sum_{i}^{N}\left(m_i v_i \otimes v_i + \frac{1}{2}\sum_{i \neq j} r_{ij} \otimes \frac{\partial U(r_{ij})}{\partial r_{ij}}\right) \tag{5.8}$$

where S and Ω are the average viral stress and the volume of the cut-off domain, respectively. The m_i, v_i, $\otimes$ and N are regard as the mass, the velocity of the atom i, the tensor product and the whole number atoms within the domain, respectively.

5.4.2 Results and Discussion

During solidification, the microstructure of HEA is significantly dominated by the cooling rate. Previous studies have categorized cooling rates into different processes [66,73–75]. Hindered by the computational methods [76], this part focuses on the process with fast cooling rates. With the temperature decreasing, small grains that form at 800 K continue to grow (Figure 5.13). In addition, various cooling rates lead to nucleation in the undercooled melt. The body-centered cube (BCC) phase is the only phase that forms at a relatively lower cooling rate. However, both the crystal

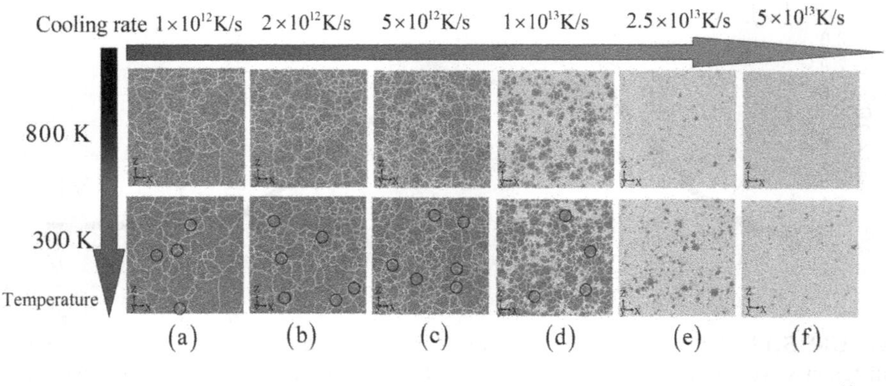

FIGURE 5.13 The correlation between microstructure and temperature/cooling rate: 1×10^{12} K/s (a), 2×10^{12} K/s (b), 5×10^{12} K/s (c), 1×10^{13} K/s (d), 2.5×10^{13} K/s (e), and 5×10^{13} K/s (f) [60].

FIGURE 5.14 Annealing twins are visible at the annealing temperature of 300 K and cooling rate of 2×10^{12} K/s, with the twin boundary represented by TB [60].

phase and the amorphous structure are found under the different cooling rates. At high cooling rates, amorphous phase can be observed. Cooling rates below 1×10^{13} K/s are associated with the presence of annealing twins (Figure 5.14), which is consistent with experimental results [77–79].

An examination is conducted on the arrangement of residual hydrostatic stress (Figure 5.15), represented by the equation $(\sigma_x + \sigma_y + \sigma_z)/3$. Grain boundaries are associated with elevated residual hydrostatic stress in the cooled samples studied. The mechanical behavior of the amorphous AlCoCrCuFeNi HEA may be affected by its intricate stress distribution, which differs from that of BCC HEA and hybrid-structured HEA. Prior research has emphasized the significant residual stress produced upon cooling from fabrication temperature to room temperature, in consistent with the current results [80–82]. The present results reveal the impact of cooling rates on residual stress levels at room temperature in HEAs.

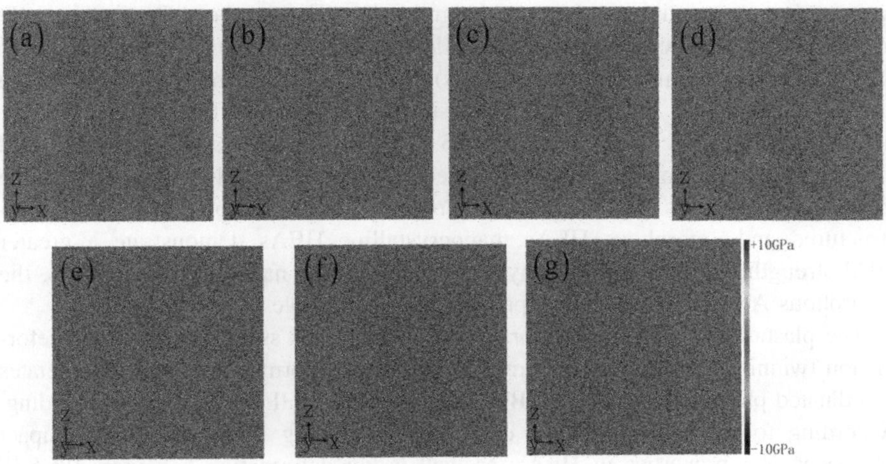

FIGURE 5.15 The residual hydrostatic stress at 300 K with cooling rates of: 1×10^{12} K/s (a), 2×10^{12} K/s (b), 5×10^{12} K/s (c), 1×10^{13} K/s (d), 2.5×10^{13} K/s (e), 5×10^{13} K/s (f), and 1×10^{14} K/s (g) [60].

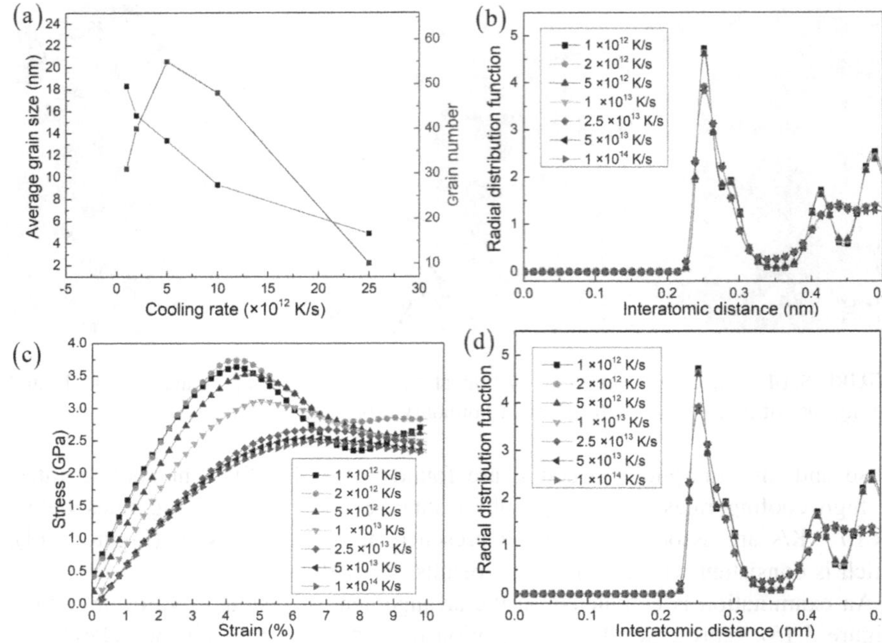

FIGURE 5.16 Variation of average grain size and grain number at different cooling rate (a). Relationship of RDF and cooling rate (b). Strain-stress curves at different cooling rate (c). Correlation between RDF and cooling rate after tension (d) [60].

The total grain size and RDF are shown in Figures 5.16a and b. RDF analysis can be applied to examine structural evolution. It is worthy of noting that only the grains larger than 0.5 nm are counted. As shown in Figure 5.16a, increasing the cooling rate results in a continuous decrease in grain size. The number of grains initially increases, then reduce when cooling rate decreases. The RDF curve corresponds to the observation of turning (5×10^{12} K/s) (Figure 5.16b). The emergence of an amorphous structure is suggested by a substantial peak shift in the RDF curve beyond 5×10^{12} K/s cooling rate. Figure 5.16c illustrates the variation of stress and strain observed in samples that are obtained using various cooling processes, while Figure 5.16d exhibits the RDF of AlCoCrCuFeNi HEA. In comparison to hybrid-structured and amorphous HEAs, nanocrystalline HEAs demonstrate a greater yield strength. Compared to the hybrid-structured and nanocrystalline HEAs, the amorphous AlCoCrCuFeNi HEA presents less noticeable softening effects.

The plastic deformation behavior of alloys with BCC structure, known as deformation twinning, is shown in Figure 5.17 [83–86]. Deformation twinning facilitates coordinated plastic deformation in BCC nanocrystalline HEAs under tensile loading. According to recent experiments, deformation twinning has a significant impact on mechanical properties in HEAs, as well as on deformation behaviors [86,87]. Anomalous grain growth is observed in nanocrystalline HEAs, which is consistent with prior experimental findings on grain growth in Ni-based alloys under deformation [88]. These phenomena have a significant impact on mechanical properties, as

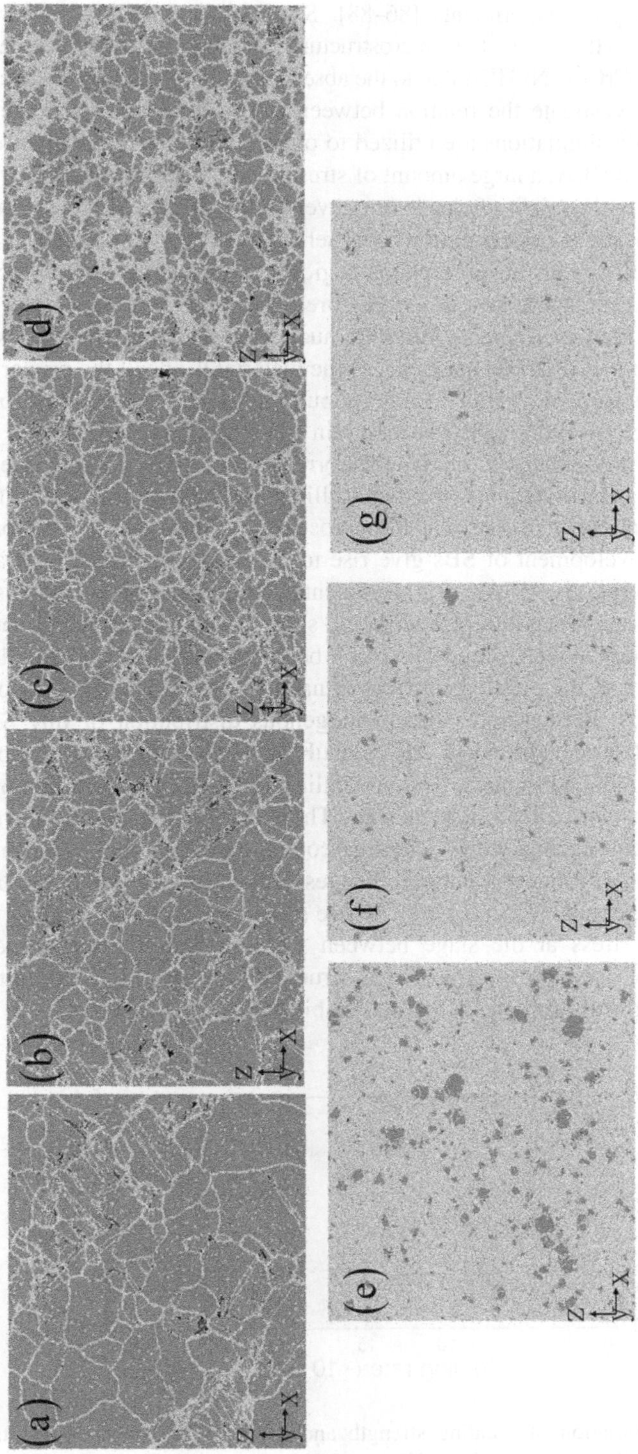

FIGURE 5.17 CNA-based atom coloring is used to illustrate microstructure evolution at a strain of 10% and various cooling rates: 1×10^{12} K/s (a), 2×10^{12} K/s (b), 5×10^{12} K/s (c), 1×10^{13} K/s (d), 2.5×10^{13} K/s (e), 5×10^{13} K/s (f), and 1×10^{14} K/s (g) [60].

demonstrated in prior experiments [86–88]. Small grain nucleation occurs in the hybrid-structured HEA, but the microstructure remains relatively unchanged in amorphous AlCoCrCuFeNi HEA due to the absence of significant higher stress region.

In order to investigate the relation between stress distribution and mechanical properties, atomic simulations are utilized to obtain the stress distribution of HEA. In nanocrystalline HEA, a large amount of stress is concentrated, especially at grain boundaries, due to the misfit of elastic or uneven flow of grains [89]. Conversely, no significant local stress concentration is observed in hybrid-structured and amorphous HEAs. HEA with an amorphous region exhibits more point-concentrated stresses, resulting in stable flow stress (Figure 5.11c).

Previous reports have indicated that the nucleation of shear bands (SBs) due to the relative sliding of different segments of the material is one of the primary plastic deformation mechanisms found in amorphous alloys and composites containing amorphous structures [90,91]. The shear strain configurations of HEAs at a strain of 10% have been investigated to clarify the correlation between cooling rate and the morphology of SBs, indicating that controlling the cooling rate can effectively modify the morphology of SBs due to microstructural differences. The formation, combine, and development of SBs give rise to mature shear bands in parallel orientation in amorphous HEAs [92]. Elemental heterogeneity in the disordered structure causes the formation of embryonic shear bands that do not penetrate the entire sample, deviating from the typical behavior observed in metallic glass [90–92]. In contrast, the plastic behaviors of nanocrystalline HEA mainly occurs in two primary SBs. Furthermore, an inhomogenous distribution of tiny SBs with high-density defects are found in AlCoCrCuFeNi HEA with a mixture of amorphous and crystalline structures. The crystalline HEA causes more homogeneous deformation behavior instead of local SBs. The yielding strength and lower stress exhibit a consistent decline with decreasing cooling rate (Figure 5.17), irrespective of any accompanying microstructural changes. The softening parameter δ can be calculated as the equation $(\sigma_y - \sigma_c)/\sigma_y$, where σ_y and σ_c represent yielding strength and the lowest stress at the stage between yielding and fracture, respectively. Additionally, the existence of amorphous structure can not only reduces the softening effect, but also promotes structural stability (Figure 5.18).

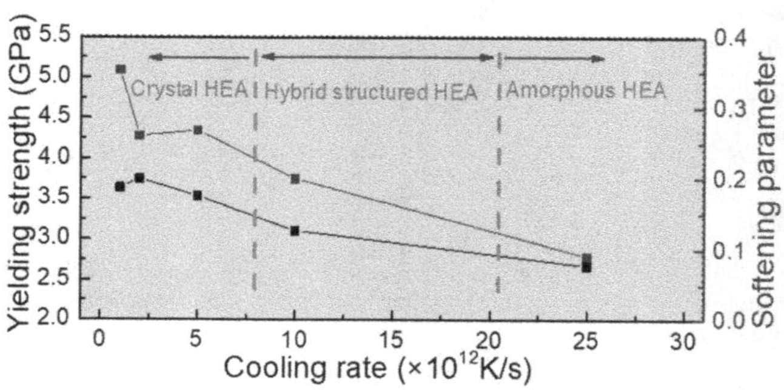

FIGURE 5.18 Variation of yielding strength and softening parameter of crystal HEA, hybrid structured HEA and amorphous HEA at different cooling rate [60].

5.5 CONCLUSIONS

In this chapter, the mechanical behaviors and properties of amorphous HEA with different microstructure under nanoindentation and tension have been clarified using MD simulation.

The effects of nanoindentation on the $Cu_{29}Zr_{32}Ti_{15}Al_5Ni_{19}$ amorphous have been investigated, with a focus on evaluating the shear strain, variation of load and displacement, surface morphology, and RDF. The results indicate that Young's modulus and the hardness of the amorphous HEA are 93.1 GPa and 8.8 GPa, respectively. A hardness-elasticity modulus ratio of 0.094 is determined by Hertzian fitting. These values are consistent with those obtained from atomic simulations and nanoindentation experiments. Atomic strain distribution analysis suggests that deformation localization can occur at loosely packed regions, which can be determined by the spatial arrangement of strain and indicate inhomogeneity of plasticity. The regular existence of atomic movement reveals homogeneous behavior in the amorphous HEA caused by indentation treatment. The drastic distortion resulted in a reduction of critical energy required for shear deformation, leading to the observation of a brittle to ductile transition. The Al atoms in the plastic area cause significant fluctuations in the first RDF peak, which is much higher than that of the other elements. New insights into the mechanical properties of amorphous HEA are provided by this section, which emphasizes the crucial effects of atomic misfit in the preparation of amorphous structures.

A systematic examination is conducted to investigate the correlation between tensile deformation behavior and the size of the amorphous structure. Dislocation formation in crystal structure and the creation of SBs in amorphous structure are triggered by plastic deformation. Consistent with prior experimental findings and the mixture principle, the strength of amorphous/crystal HEA composites declines as the size of the amorphous structure increased. Observable trends indicate that the reduction in size of the amorphous structure led to an increase in both softening stress and yielding strain. At the nanoscale, the interaction of plasticity and in crystal and glassy parts leads to an improved redistribution of plasticity, which accounts for the high plasticity observed in nanolaminated composites. MD simulation results demonstrate that modifying the size of amorphous structure could be efficient for adjusting the plasticity of composites containing amorphous structure. Insights into the mechanisms of yielding and adjustable plasticity, as revealed by these findings, may inform the development of amorphous HEA with excellent plasticity and high strength.

An investigation into HEA subjected to different cooling rates has led to a better understanding of its microstructures and mechanical properties. Experimental data can be well-matched by calculated RDF in the solid states. The formation of amorphous HEA occurs at high cooling rates, while crystallization is induced by low cooling rates. The initial solidification is dependent on cooling processes, which determines nucleation behavior. Additionally, nucleation from rapid cooled HEA is examined across different cooling rates, where size and number of grains are estimated for each rate. The HEA subjected to low cooling rate exhibit high strength with significant softening, whereas those subjected to high cooling rates had poor strength but stable flow stress.

REFERENCES

[1] Takeuchi A., Chen N., Wada T., et al. 2011. Pd20Pt20Cu20Ni20P20 high-entropy alloy as a bulk metallic glass in the centimeter. *Intermetallics*, 19(10): 1546–1554.

[2] Wei R., Tao J., Sun H., et al. 2017. Soft magnetic Fe26.7Co26.7Ni26.6Si9B11 high entropy metallic glass with good bending ductility. *Materials Letters*, 197: 87–89.

[3] Chen Y., Dai Z.W. and Jiang J.Z. 2021. High entropy metallic glasses: Glass formation, crystallization and properties. *Journal of Alloys and Compounds*, 866: 158852.

[4] Zhou X.Y., Wu H.H., Zhu J.H., et al. 2021. Plastic deformation mechanism in crystal-glass high entropy alloy composites studied via molecular dynamics simulations. *Composites Communications*, 24: 100658.

[5] Fang Q., Yi M., Li J., et al. 2018. Deformation behaviors of Cu29Zr32Ti15Al5Ni19 high entropy bulk metallic glass during nanoindentation. *Applied Surface Science*, 443: 122–130.

[6] Zhou Q., Du Y., Jia Q., et al. 2020. A nanoindentation study of Ti-based high entropy bulk metallic glasses at elevated temperatures Comment. *Journal of Non-Crystalline Solids*, 532: 119878.

[7] Li J., Chen H., Feng H., et al. 2020. Microstructure evolution and deformation mechanism of amorphous/crystalline high-entropy-alloy composites. *Journal of Materials Science & Technology*, 54: 14–19.

[8] Fang Q.H., Yi M., Li J., et al. 2018. Deformation behaviors of Cu29Zr32Ti15Al5Ni19 high entropy bulk metallic glass during nanoindentation. *Applied Surface Science*, 443: 122–130.

[9] Li J., Fang Q., Liu B., et al. 2016. Atomic-scale analysis of nanoindentation behavior of high-entropy alloy. *Journal of Micromechanics Molecular Physics*, 1(01): 1650001.

[10] Kao S.W., Yeh J.W. and Chin T.S. 2008. Rapidly solidified structure of alloys with up to eight equal-molar elements - A simulation by molecular dynamics. *Journal of Physics Condensed Matter,* 20(14): 145214.

[11] Wang Z., Li J., Fang Q., et al. 2017. Investigation into nanoscratching mechanical response of AlCrCuFeNi high-entropy alloys using atomic simulations. *Applied Surface Science*, 416: 470–481.

[12] Li J., Fang Q., Liu B., et al. 2016. Atomic-scale analysis of nanoindentation behavior of high-entropy alloy. *Journal of Micromechanics and Molecular Physics*, 01(01): 1650001.

[13] Ding H.Y., Shao Y., Gong P., et al. 2014. A senary TiZrHfCuNiBe high entropy bulk metallic glass with large glass-forming ability. *Materials Letters*, 125: 151–153.

[14] Wang W.H. 2014. High-entropy metallic glasses. *JOM*, 66(10): 2067–2077.

[15] Zhao S.F., Shao Y., Liu X., et al. 2015. Pseudo-quinary Ti20Zr20Hf20Be20(Cu20-xNix) high entropy bulk metallic glasses with large glass forming ability. *Materials Design,* 87: 625–631.

[16] Telford M. 2004. The case for bulk metallic glass. *Materials Today*, 7(3): 36–43.

[17] Wang W.H., Dong C. and Shek C.H. 2004. Bulk metallic glasses. *Materials Science and Engineering: R,* 44(2): 45–89.

[18] Zhang J.J., Sun T., Hartmaier A., et al. 2012. Atomistic simulation of the influence of nanomachining-induced deformation on subsequent nanoindentation. *Computational Materials Science*, 59: 14–21.

[19] Zhu P.Z. and Fang F.Z. 2012. Molecular dynamics simulations of nanoindentation of monocrystalline germanium. *Applied Physics A*, 108(2): 415–421.

[20] Pi J., Wang Z., He X., et al. 2016. Nanoindentation mechanical properties of glassy Cu29Zr32Ti15Al5Ni19. *Journal of Alloys and Compounds*, 657: 726–732.

[21] Gong P., Jin J., Deng L., et al. 2017. Room temperature nanoindentation creep behavior of TiZrHfBeCu(Ni) high entropy bulk metallic glasses. *Materials Science and Engineering: A*, 688: 174–179.

[22] Borkar T., Gwalani B., Choudhuri D., et al. 2016. A combinatorial assessment of AlxCrCuFeNi2 ($0 < x < 1.5$) complex concentrated alloys: Microstructure, microhardness, and magnetic properties. *Acta Materialia*, 116: 63–76.

[23] He J.Y., Liu W.H., Wang H., et al. 2014. Effects of Al addition on structural evolution and tensile properties of the FeCoNiCrMn high-entropy alloy system. *Acta Materialia*, 62: 105–113.

[24] Rao J.C., Diao H.Y., Ocelík V., et al. 2017. Secondary phases in AlxCoCrFeNi high-entropy alloys: An in-situ TEM heating study and thermodynamic appraisal. *Acta Materialia*, 131: 206–220.

[25] Wang W.H. 2012. The elastic properties, elastic models and elastic perspectives of metallic glasses. *Progress in Materials Science*, 57(3): 487–656.

[26] Li J., Guo J., Luo H., et al. 2016. Study of nanoindentation mechanical response of nanocrystalline structures using molecular dynamics simulations. *Applied Surface Science*, 364: 190–200.

[27] Cheng J.B., Liang X.B. and Xu B.S. 2014. Effect of Nb addition on the structure and mechanical behaviors of CoCrCuFeNi high-entropy alloy coatings. *Surface and Coatings Technology*, 240: 184–190.

[28] Nishikawa M. and Soyama H. 2011. Two-step method to evaluate equibiaxial residual stress of metal surface based on micro-indentation tests. *Materials Design*, 32(6): 3240–3247.

[29] Hertz H. 1896. *Miscellaneous papers*, Macmillan.

[30] Li J., Fang Q.H., Liu B., et al. 2016. Mechanical behaviors of AlCrFeCuNi high-entropy alloys under uniaxial tension via molecular dynamics simulation. *RSC Advances*, 6(80): 76409–76419.

[31] Guo S. and Liu C.T. 2011. Phase stability in high entropy alloys: Formation of solid-solution phase or amorphous phase. *Progress in Natural Science: Materials International*, 21(6): 433–446.

[32] Azumo S. and Nagayama K. 2006. Amorphous formation and magnetic properties of Nd-Fe-Co-Al alloys by gas flow type levitation process. *Materials Transactions*, 47(11): 2842–2845.

[33] Hu Z.Q., Ding B.Z., Zhang H.F., et al. 2001. Formation of non-equilibrium alloys by high pressure melt quenching. *Science and Technology of Advanced Materials*, 2(1): 41–48.

[34] Daw M.S. and Baskes M.I. 1984. Embedded-atom method: Derivation and application to impurities, surfaces, and other defects in metals. *Physical Review B*, 29(12): 6443–6453.

[35] Lu C.Y., Niu L.L., Chen N.J., et al. 2016. Enhancing radiation tolerance by controlling defect mobility and migration pathways in multicomponent single-phase alloys. *Nature Communications*, 7(1): 13564.

[36] Li W., Fan H., Tang J., et al. 2019. Effects of alloying on deformation twinning in high entropy alloys. *Materials Science and Engineering: A*, 763: 138143.

[37] Fang Q., Chen Y., Li J., et al. 2019. Probing the phase transformation and dislocation evolution in dual-phase high-entropy alloys. *International Journal of Plasticity*, 114: 161–173.

[38] Li J., Fang Q.H., Liu B., et al. 2018. Transformation induced softening and plasticity in high entropy alloys. *Acta Materialia*, 147: 35–41.

[39] Afkham Y., Bahramyan M., Mousavian R.T., et al. 2017. Tensile properties of AlCrCoFeCuNi glassy alloys: A molecular dynamics simulation study. *Materials Science and Engineering A*, 698: 143–151.

[40] Wu W.Q., Ni S., Liu Y., et al. 2017. Amorphization at twin-twin intersected region in FeCoCrNi high-entropy alloy subjected to high-pressure torsion. *Materials Characterization*, 127: 111–115.
[41] Song H.Y., Xu J.J., Zhang Y.G., et al. 2017. Molecular dynamics study of deformation behavior of crystalline Cu/amorphous Cu50Zr50 nanolaminates. *Materials Design*, 127: 173–182.
[42] Zhao L., Chan K.C. and Chen S.H. 2018. Atomistic deformation mechanisms of amorphous/polycrystalline metallic nanolaminates. *Intermetallics*, 95: 102–109.
[43] Sterwerf C., Kaub T., Deng C., et al. 2017. Deformation mode transitions in amorphous-Cu45Zr55/crystalline-Cu multilayers. *Thin Solid Films*, 626: 184–189.
[44] Song H.Y., Zhang K., An M.R., et al. 2019. Atomic simulation of interaction mechanism between basal/prismatic interface and amorphous/crystalline interface of dual-phase magnesium alloys. *Journal of Non-Crystalline Solids*, 521: 119550.
[45] Song H.Y., Zuo X.D., An M.R., et al. 2019. Superplastic dual-phase nanostructure Mg alloy: A molecular dynamics study. *Computational Materials Science*, 160: 295–300.
[46] Sun P., Peng C., Cheng Y., et al. 2019. Mechanical behavior of CuZr dual-phase nanocrystal-metallic glass composites. *Computational Materials Science*, 163: 290–300.
[47] Cui Y.N., Peng C.X., Cheng Y., et al. 2019. Deformation mechanism of amorphous/crystalline phase-separated alloys: A molecular dynamics study. *Journal of Non-Crystalline Solids*, 523: 119605.
[48] Meyers M.A., Mishra A. and Benson D.J. 2006. Mechanical properties of nanocrystalline materials. *Progress in Materials Science*, 51(4): 427–556.
[49] Guo W., Yao J., Jagle E.A., et al. 2015. Co-deformation of crystalline-amorphous nanolaminates. *Microscopy and Microanalysis*, 21(S3): 361–362.
[50] Kamaya M., Wilkinson A.J. and Titchmarsh J.M. 2005. Measurement of plastic strain of polycrystalline material by electron backscatter diffraction. *Nuclear Engineering and Design*, 235(6): 713–725.
[51] Wu C.D., Fang T.H. and Wu C.C. 2015. Atomistic simulations of nanowelding of single-crystal and amorphous gold nanowires. *Journal of Applied Physics*, 117(1): 014307.
[52] Zhang Y., Zuo T.T., Tang Z., et al. 2014. Microstructures and properties of high-entropy alloys. *Progress in Materials Science*, 61: 1–93.
[53] Shi P., Ren W., Zheng T., et al. 2019. Enhanced strength–ductility synergy in ultrafine-grained eutectic high-entropy alloys by inheriting microstructural lamellae. *Nature Communications*, 10(1): 489.
[54] Yeh J.W., Chen S.K., Lin S.J., et al. 2004. Nanostructured high-entropy alloys with multiple principal elements: Novel alloy design concepts and outcomes. *Advanced Engineering Materials*, 6(5): 299–303.
[55] Miracle D.B. and Senkov O.N. 2017. A critical review of high entropy alloys and related concepts. *Acta Materialia*, 122: 448–511.
[56] Li Z., Zhao S., Ritchie R.O., et al. 2019. Mechanical properties of high-entropy alloys with emphasis on face-centered cubic alloys. *Progress in Materials Science*, 102: 296–345.
[57] Wei Y.D., Peng P., Yan Z.Z., et al. 2016. A comparative study on local atomic configurations characterized by cluster-type-index method and Voronoi polyhedron method. *Computational Materials Science*, 123: 214–223.
[58] Hirata A., Kang L.J., Fujita T., et al. 2013. Geometric frustration of icosahedron in metallic glasses. *Science*, 341(6144): 376–379.
[59] Yu Q., Wang X.D., Lou H.B., et al. 2016. Atomic packing in Fe-based metallic glasses. *Acta Materialia*, 102: 116–124.

[60] Li J., Chen H., Li S., et al. 2019. Tuning the mechanical behavior of high-entropy alloys via controlling cooling rates. *Materials Science and Engineering: A*, 760: 359–365.
[61] Plimpton S.J. and Thompson A.P. 2012. Computational aspects of many-body potentials. *MRS Bulletin*, 37(5): 513–521.
[62] Stukowski A. 2010. Visualization and analysis of atomistic simulation data with OVITO-the Open Visualization Tool. *Modelling and Simulation in Materials Science and Engineering*, 18(1): 015012.
[63] Huang C., Peng X., Yang B., et al. 2019. Grain size dependence of tensile properties in nanocrystalline diamond. *Computational Materials Science*, 157: 67–74.
[64] Cui Y., Shibutani Y., Li S., et al. 2017. Plastic deformation behaviors of amorphous-Cu50Zr50/crystalline-Cu nanolaminated structures by molecular dynamics simulations. *Journal of Alloys and Compounds*, 693: 285–290.
[65] Li J., Liu B., Fang Q., et al. 2017. Atomic-scale strengthening mechanism of dislocation-obstacle interaction in silicon carbide particle-reinforced copper matrix nanocomposites. *Ceramics International*, 43(4): 3839–3846.
[66] Lin C.J. and Spaepen F. 1982. Fe-B glasses formed by picosecond pulsed laser quenching. *Applied Physics Letters*, 41(8): 721–723.
[67] Daw M.S., Foiles S.M. and Baskes M.I. 1993. The embedded-atom method: a review of theory and applications. *Materials Science Reports*, 9(7): 251–310.
[68] Zhou X.W., Johnson R.A. and Wadley H.N.G. 2004. Misfit-energy-increasing dislocations in vapor-deposited CoFe/NiFe multilayers. *Physical Review B*, 69(14): 144113.
[69] Lin Z., Johnson R.A. and Zhigilei L.V. 2008. Computational study of the generation of crystal defects in a bcc metal target irradiated by short laser pulses. *Physical Review B*, 77(21): 214108.
[70] Rao S.I., Varvenne C., Woodward C., et al. 2017. Atomistic simulations of dislocations in a model BCC multicomponent concentrated solid solution alloy. *Acta Materialia*, 125: 311–320.
[71] Xie L., Brault P., Thomann A.-L., et al. 2013. AlCoCrCuFeNi high entropy alloy cluster growth and annealing on silicon: A classical molecular dynamics simulation study. *Applied Surface Science*, 285: 810–816.
[72] Xie L., Brault P., Thomann A.L., et al. 2016. Molecular dynamics simulation of Al–Co–Cr–Cu–Fe–Ni high entropy alloy thin film growth. *Intermetallics*, 68: 78–86.
[73] Greer A.L. 1995. Metallic glasses. *Science*, 267(5206): 1947–1953.
[74] Davies H., Aucote J. and Hull J.J.N.P.S. 1973. Amorphous nickel produced by splat quenching. *Nature*, 246(149): 13–14.
[75] Zhong L., Wang J., Sheng H., et al. 2014. Formation of monatomic metallic glasses through ultrafast liquid quenching. *Nature*, 512(7513): 177–180.
[76] Mahata A. and Asle Zaeem M. 2019. Evolution of solidification defects in deformation of nano-polycrystalline aluminum. *Computational Materials Science*, 163: 176–185.
[77] Otto F., Dlouhý A., Somsen C., et al. 2013. The influences of temperature and microstructure on the tensile properties of a CoCrFeMnNi high-entropy alloy. *Acta Materialia*, 61(15): 5743–5755.
[78] Gali A. and George E.P. 2013. Tensile properties of high- and medium-entropy alloys. *Intermetallics*, 39: 74–78.
[79] Zaddach A.J., Niu C., Koch C.C., et al. 2013. Mechanical properties and stacking fault energies of NiFeCrCoMn high-entropy alloy. *JOM*, 65(12): 1780–1789.
[80] Das B., Nath A.K. and Bandyopadhyay P.P. 2018. Online monitoring of laser remelting of plasma sprayed coatings to study the effect of cooling rate on residual stress and mechanical properties. *Ceramics International*, 44(7): 7524–7534.

[81] Chobaut N., Carron D., Arsène S., et al. 2015. Quench induced residual stress prediction in heat treatable 7xxx aluminium alloy thick plates using Gleeble interrupted quench tests. *Journal of Materials Processing Technology*, 222: 373–380.
[82] Srolovitz D., Egami T. and Vitek V. 1981. Radial distribution function and structural relaxation in amorphous solids. *Physical Review B*, 24(12): 6936–6944.
[83] Tang Z., Yuan T., Tsai C.-W., et al. 2015. Fatigue behavior of a wrought Al0.5CoCrCuFeNi two-phase high-entropy alloy. *Acta Materialia*, 99: 247–258.
[84] Shi Y., Yang B., Xie X., et al. 2017. Corrosion of AlxCoCrFeNi high-entropy alloys: Al-content and potential scan-rate dependent pitting behavior. *Corrosion Science*, 119: 33–45.
[85] Li L., Fang Q., Li J., et al. 2019. Origin of strengthening-softening trade-off in gradient nanostructured body-centred cubic alloys. *Journal of Alloys and Compounds*, 775: 270–280.
[86] Lilensten L., Couzinié J.-P., Bourgon J., et al. 2017. Design and tensile properties of a bcc Ti-rich high-entropy alloy with transformation-induced plasticity. *Materials Research Letters*, 5(2): 110–116.
[87] Tsai C.-W., Chen Y.-L., Tsai M.-H., et al. 2009. Deformation and annealing behaviors of high-entropy alloy Al0.5CoCrCuFeNi. *Journal of Alloys and Compounds*, 486(1): 427–435.
[88] Miller V.M., Johnson A.E., Torbet C.J., et al. 2016. Recrystallization and the development of abnormally large grains after small strain deformation in a polycrystalline nickel-based superalloy. *Metallurgical and Materials Transactions A*, 47A(4): 1566–1574.
[89] Li J., Fang Q., Liu B., et al. 2017. Twinning-governed plastic deformation in a thin film of body-centred cubic nanocrystalline ternary alloys at low temperature. *Journal of Alloys and Compounds*, 727: 69–79.
[90] Zhao L., Chan K.C., Chen S.H., et al. 2019. Tunable tensile ductility of metallic glasses with partially rejuvenated amorphous structures. *Acta Materialia*, 169: 122–134.
[91] Bei H., Xie S. and George E.P. 2006. Softening caused by profuse shear banding in a bulk metallic glass. *Physical Review Letters*, 96(10): 105503.
[92] Cao A.J., Cheng Y.Q. and Ma E. 2009. Structural processes that initiate shear localization in metallic glass. *Acta Materialia*, 57(17): 5146–5155.

6 Nickel-Based Superalloys

6.1 INTRODUCTION

Nickel-based superalloys are widely employed to prepare hot-end components, such as gas turbines and aeroengines (Figure 6.1), due to their excellent high-temperature performances [1–10]. The distinguished high-temperature properties are essentially originated from the characteristics of precipitates with high-temperature stability [11–13]. The precipitates have an approximate lattice parameter in comparison to matrix in superalloys, and they show multi-modal and multi-scale distribution. The precipitates with different sizes hinder grain boundary and dislocation movement, which reduce dislocation migration rate, thereby improving creep life. Therefore, the multi-modal and multi-scale precipitate distribution is preferable to ensure their optimal comprehensive mechanical and service performance. Nickel-based superalloys are also denoted as the key materials for the nuclear application, because of the excellent creep and corrosion resistance [14–18].

Here, we study the mechanical properties (Section 6.2) and service performance (Section 6.3) of nickel-based superalloys through theoretical modeling and simulation. Specifically, considering the spatial uncertainty of the interaction between dislocations and precipitates, and the multi-scale statistical distribution of precipitates, a probability-dependent statistical precipitate strengthening model is established (Section 6.2). The relationship between processing technology and precipitate parameter is then established to obtain the optimal cooling rate range. Furthermore, combined with the dislocation dynamics method, the effect of precipitate enrichment on the strengthening mechanism is explored. A new statistical method of minimum spanning tree (MST) is proposed to describe precipitate enrichment, which accurately evaluates the quantitative strengthening from precipitate spatially randomly distribution. Based on the uncertainty of the interaction mechanism between dislocations and precipitates during the creep process, and co-existence of climbing mechanism, shearing mechanism and Orowan loop mechanism caused by multi-scale statistical distribution effect of precipitates, a theoretical model used in a wide temperature range is established (Section 6.3). The proposed model reveals the stress-controlled creep mechanism transformation, and quantitatively evaluates the occurrence probability of each mechanism and their contribution to the creep strength.

DOI: 10.1201/9781003225706-6

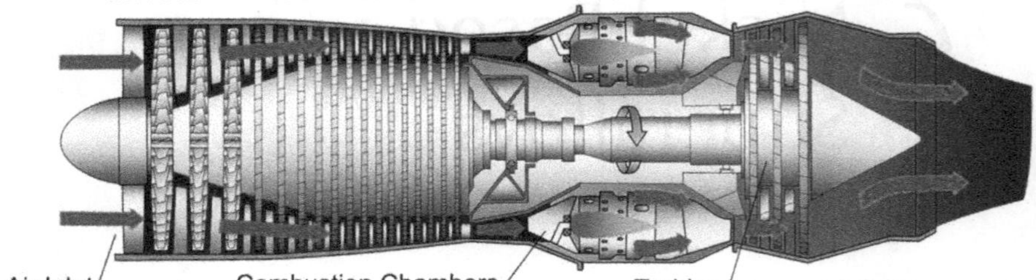

FIGURE 6.1 The schematic of aircraft engine [10].

6.2 MECHANICAL BEHAVIOR

6.2.1 Model and Process Optimization

6.2.1.1 Classical Precipitate Strengthening Model

The present weak-pair coupling mechanism and strong-pair coupling mechanism still have the following limitations: (1) For alloys with large precipitate volume fraction, the predicted stress from the strong-pair coupling model is much higher than the experimental result near the transition region between weak-pair coupling and strong-pair coupling. (2) The transformation from the weak-pair coupling mechanism to strong-pair coupling mechanism has an obvious stress mutation, which is extremely unreasonable.

The previous studies show that the dislocations propagating in matrix cannot enter the precipitate without the formation of anti-phase boundaries. Thus, the dislocation must bypass γ/γ' in pairs in nickel-based superalloys [15,19,20]. Here, the first dislocation enters γ' and forms the anti-phase boundary, while the second one eliminates anti-phase boundary generated by first one. The anti-phase boundary energy represents the obstacle need to be overcome if the cutting mechanism is effective. The transmission electron microscope (TEM) experiment shows that the value of anti-phase boundary energy is about 0.1 J/m^2. Therefore, the contributed stress from the cutting mechanism is about 400 MPa [21]. This value is considerable, indicating that ordered strengthening is very important in nickel-based superalloys. In fact, the difference in modulus, stacking fault energy, γ' precipitate interface energy is very small in nickel-based superalloys. The contribution from coherent strengthening and modulus strengthening of precipitate can be neglected, while the contribution of order strengthening far exceeds that of other mechanisms.

Figure 6.2a illustrates the correlation between the precipitate size and the distance between dislocation pairs, which is classified as two cases: weak-pair coupling mechanism and strong-pair coupling mechanism. The weak-pair coupling mechanism is effective when precipitate diameter is smaller than distance between two dislocations, indicating the presence of defective precipitates between the two pairs of dislocations. In the case where precipitate radius is smaller than r_w

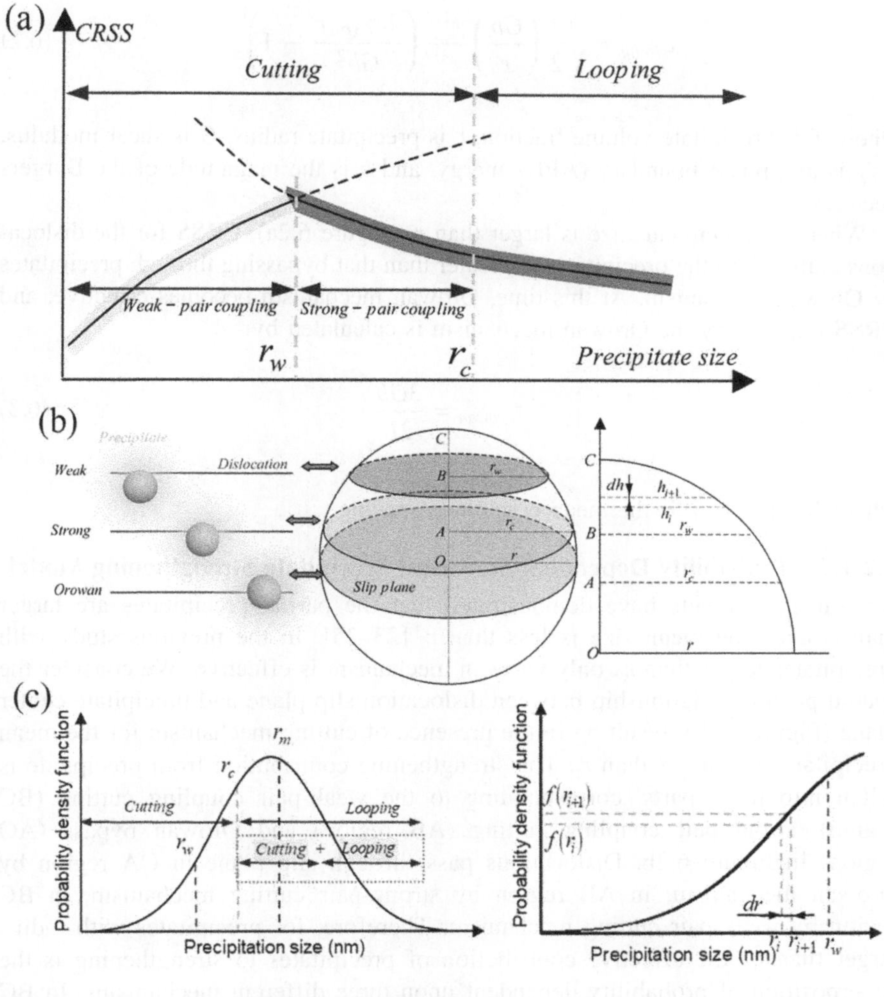

FIGURE 6.2 (a) The relationship between strengthening and precipitate size. (b) Three cases of the interaction between dislocation and a spherical precipitate. (c) The probability density function of the precipitate size and the schematic diagram of how the precipitate contributes to the yielding strength using the infinitesimal method [22].

(Figure 6.2a), the critical resolved shear stress (CRSS) for the weak-pair coupling mechanism is calculated by:

$$\tau_{weak} = \frac{\gamma_{APB}}{2b}\left[\left(\frac{6\gamma_{APB} f r}{2\pi T}\right)^{1/2} - f\right] \tag{6.1}$$

When the size of precipitate is larger than r_w but smaller than r_c (Figure 6.2a), the CRSS for strong-pair coupling mechanism is expressed as follows:

$$\tau_{strong} = \sqrt{\frac{3}{2}}\left(\frac{Gb}{r}\right)\frac{f^{1/2}}{\pi^{3/2}}\left(\frac{2\pi\gamma_{APB}r}{Gb^2} - 1\right)^{1/2} \tag{6.2}$$

where f is precipitate volume fraction, r is precipitate radius, G is shear modulus, γ_{APB} is anti-phase boundary (APB) energy, and b is the magnitude of the Burgers vector.

When the precipitate size is larger than r_c (Figure 6.2a), CRSS for the dislocations cutting into the precipitates is greater than that bypassing through precipitates by Orowan mechanism. At this time, Orowan mechanism becomes effective, and CRSS required by the Orowan mechanism is calculated by:

$$\tau_{orowan} = \frac{3Gb}{2L} \tag{6.3}$$

where $L = \left(\frac{2\pi}{3f}\right)^{1/2} r$ is the mean precipitate spacing.

6.2.1.2 Probability Dependent Statistical Precipitate Strengthening Model

Several experiments have demonstrated that the partial precipitates are larger than r_c even the mean size is less than r_c [23–25]. In the previous study with precipitates larger than r_c, only Orowan mechanism is effective. We consider the spatial position relationship between dislocation slip plane and precipitate center plane (Figure 6.2b), resulting in the presence of cutting mechanism for the mean precipitate size larger than r_c. The strengthening contribution from precipitate is fallen into three parts, corresponding to the weak-pair coupling cutting (BC region), strong-pair coupling cutting (AB region) and Orowan bypass (AO region) in Figure 6.2b. Dislocations pass through slip plane in OA region by Orowan mechanism, in AB region by strong-pair cutting mechanism, in BC region by weak-pair cutting mechanism. Therefore, for precipitates with radius larger than r_c, the effective contribution of precipitates to strengthening is the superposition of probability dependent upon three different mechanisms. In BC region, only the weak-pair cutting mechanism works. The probability for passing through the area of precipitate BC is obtained based on the correlation between the distance BC and precipitate radius:

$$p_1 = \frac{l_{BC}}{r} = 1 - \left(1 - \left(\frac{r_w}{r}\right)^2\right)^{1/2} \tag{6.4}$$

Similarly, the probability of dislocations passing through AB region of precipitate is dependent on the value of distance AB:

$$p_2 = \frac{l_{AB}}{T} = \left(1 - \left(\frac{r_w}{T}\right)^2\right)^{1/2} - \left(1 - \left(\frac{r_c}{T}\right)^2\right)^{1/2} \tag{6.5}$$

In the OA region, Orowan mechanism is a single effective mechanism, and its probability is:

$$p_3 = 1 - p_1 - p_2 \tag{6.6}$$

Then, the differential quadrature method is used to evaluate the contribution of precipitate strengthening. In Figure 6.2b, the BC, AB and OA regions are divided into n parts. Considering the probability effect, the CRSS of each mechanism is expressed as:

$$\begin{cases} \tau_{\text{weak}}^{P} = \dfrac{1}{n}\sum\limits_{i=1}^{n} \tau_{\text{weak}}(h_i),\ h_i \le r_w \\ \tau_{\text{strong}}^{P} = \dfrac{1}{n}\sum\limits_{j=1}^{n} \tau_{\text{strong}}(h_j),\ r_w < h_j < r_c \\ \tau_{\text{orowan}}^{P} = \dfrac{1}{n}\sum\limits_{\text{k}=1}^{n} \tau_{\text{orowan}}(L_{\text{k}}),\ r_{\text{c}} \le h_k \end{cases} \tag{6.7}$$

where for the τ_{weak}^{p}, $h_1 = r_w$, and h_{i+1} can be written as $h_{i+1} = \sqrt{r^2 - [(r^2 - h_i^2)^{1/2} + dh]^2}$. Here, $dh = l_{BC}/n$ (Figure 6.2b). As for τ_{strong}^{p}, $h_1 = r_c$, h_{j+1} can be expressed as $h_{j+1} = \sqrt{r^2 - [(r^2 - h_j^2)^{1/2} + dh]^2}$ and $dh = l_{AB}/n$. As for τ_{orowan}^{p}, $L_k = h_k(2\pi/3f)^{1/2}$, and $h_1 = r_c$, where h_{k+1} can be written as $h_{k+1} = \sqrt{r^2 - [(r^2 - h_k^2)^{1/2} - dh]^2}$, $dh = l_{OA}/n$.

Therefore, considering the probability dependent Orowan mechanism, CRSS for dislocations to pass through precipitates larger than r_c is expressed as:

$$\tau_{orowan}^{new} = p_1\tau_{weak}^{p} + p_2\tau_{strong}^{p} + p_3\tau_{orowan}^{p} \tag{6.8}$$

The spatial relationship between precipitate center plane and dislocation slip plane is considered. Furthermore, the statistical distribution effect of precipitate size is considered. The experiments show that the precipitate size distribution conforms to the lognormal distribution rule (Figure 6.2c) [7], in which the precipitate size is divided into three regions. The distribution of precipitate size is obtained by:

$$f(r) = \frac{1}{\sqrt{2\pi}\,\sigma r}\exp\left[-\frac{(\text{In}(r) - \mu)^2}{2\sigma^2}\right] \tag{6.9}$$

where μ and σ represent the mean precipitate size and deviation of precipitate size, respectively.

As shown in Figure 6.2c, even if the average precipitate size is larger than r_c, partial precipitates still have smaller sizes, leading to the presence of cutting mechanism. Similarly, when average precipitate size is smaller than r_c, the Orowan mechanism is still existing.

According to the statistical theory, each region ($r< r_w$, $r_w < r < r_c$ and $r_c < r$) in Figure 6.2c can be fallen into n parts, where precipitate strengthening contributed by the corresponding part is product of probability and precipitate strengthening contributed by corresponding precipitate size. The total precipitate strengthening is the sum of the precipitate strengthening contributed by the corresponding n parts, as illustrated in Figure 6.2c.

The statistically dependent CRSS of τ^s_{weak}, τ^s_{strong}, and τ^s_{orowan} is expressed as:

$$\begin{cases} \tau^s_{weak} = \sum_{i=1}^{n} \tau_{weak}(r_i) \int_{r_i}^{r_{i+1}} f(r)dr, & r \le r_w \\ \tau^s_{strong} = \sum_{i=1}^{n} \tau_{strong}(r_i) \int_{r_i}^{r_{i+1}} f(r)dr, & r_w < r < r_c \\ \tau^s_{orowan} = \sum_{i=1}^{n} \tau^{new}_{orowan}(r_i) \int_{r_i}^{r_{i+1}} f(r)dr, & r_c \le r \end{cases} \tag{6.10}$$

where τ^s_{weak} represents total precipitates with sizes less than r_w that contribute to precipitate strengthening, τ^s_{strong} represents total precipitates with sizes larger than r_w but less than r_c that contribute to precipitate strengthening, τ^s_{orowan} represents total precipitates with sizes larger than r_c that contribute to precipitate strengthening. $\tau_{weak}(r_i)$, $\tau_{strong}(r_i)$, and τ^s_{orowan} represent the CRSS of each mechanism, respectively. $\int_{r_i}^{r_{i+1}} f(r)dr$ is the probability of the radius r_i being in the corresponding part. $r_{i+1} = r_i + dr$, where dr is the increment of the infinitesimal.

Therefore, the modified precipitate strengthening is given by:

$$\sigma_{pre} = M(\tau^s_{weak} + \tau^s_{strong} + \tau^s_{orowan}) \tag{6.11}$$

where M is the Taylor coefficient.

The contribution of precipitates to the strength has been calculated above. In addition, there are grain boundary strengthening, solid solution strengthening, and dislocation strengthening contribute to the strength of nickel-based superalloys. Therefore, the tensile strength of nickel-based superalloy is expressed as:

$$\sigma_y = \sigma_{GB} + \sigma_{ss} + \sigma_{pre} + \sigma_{dis} \tag{6.12}$$

6.2.1.3 Model Validation

The nickel-based superalloy prepared through a series of vacuum induction melting, gas atomization, hot isostatic pressing, and hot extrusion processes. The nominal chemical composition (wt%) of prepared alloy is exhibited in Table 6.1. A dilatometer is employed to measure solution line temperature of the alloy, which is found to be about 1151 °C at a constant heating rate of 2 °C/s. The diameter of sample is determined as 4 mm, and length is determined as 10 mm. The alloy is

TABLE 6.1
The Nominal Chemical Composition of Nickel-Based Superalloy (wt.%) [22]

Ni	Co	Cr	Mo	W	Al	Ti	Nb	C	B	Zr	Hf
Bal.	26	13	4	4	3.2	3.7	0.95	0.05	0.025	0.05	0.2

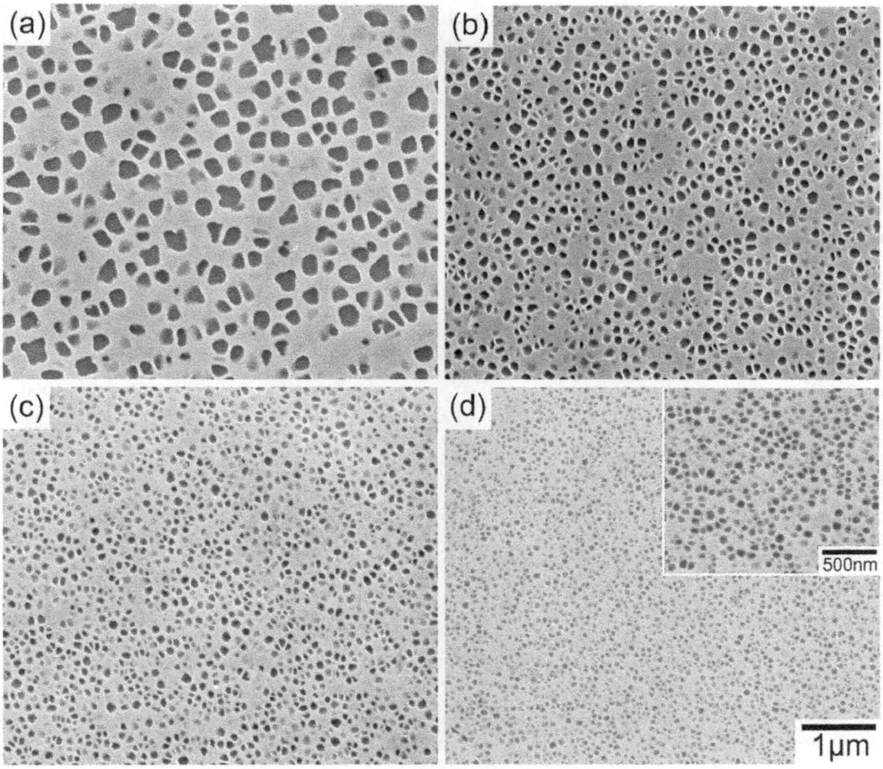

FIGURE 6.3 The SEM images of specimens with different cooling rates: (a) 20 °C/min,

solution treated at 1180 °C for 40 minutes, followed by six different cooling rates: 20 °C/min, 70 °C/min, 120 °C/min, 360 °C/min, 6000 °C/min, and 9000 °C/min.

The microstructure characterization shows that the morphology and microstructural parameter of the precipitates depend on cooling rate during preparation process (Figure 6.3). Surprisingly, the highest hardness value is not observed at fastest or slowest cooling rate, but at an intermediate cooling rate. The model is then verified by comparing the calculated strength with experimental data. The parameters in Table 6.2 are required for the calculation of precipitate strengthening. The parameters in Table 6.3 are required for the calculation of the tensile strength according to previous and current experiments.

TABLE 6.2
Microstructure Features for Specimens with Different Cooling Rates [22]

Cooling Rate (°C/min)	d_s (nm)	L_s (nm)	d_t (nm)	L_t (nm)	D (μm)
9,000	/	/	22.2 ± 5.2	14.8 ± 7.5	9.65
6,000	/	/	27.8 ± 3.9	22.7 ± 9.2	9.38
360	66.6 ± 16.0	95.8 ± 17.6	12.2 ± 1.3	6.2 ± 1.3	10.0
120	98.9 ± 28.5	117.5 ± 20.6	/	/	9.94
70	122.2 ± 41.7	177.0 ± 58.6	/	/	9.53
20	161.4 ± 69.4	236.0 ± 59.6	/	/	10.58

TABLE 6.3
Parameters Used in the Model [22]

Parameters	Symbol	Magnitude
Standard deviation	σ	0.25
Precipitate-size interval (nm)	dr	0.1
Maximum precipitate size (nm)	r_{max}	1000
Anti-phase boundary energy (J/m^2)	γ_{APB}	0.27
Magnitude of the Burger vector (nm)	b	0.248
Shear modulus (GPa)	G	80
Taylor factor	M	3.06
Empirical constant	α	0.33
Proportionality factor	ψ	0.2
Dynamic recovery constant	m	21.5
Dynamic recovery constant	k_{20}	21.6
Reference strain rate (s^{-1})	$\dot{\varepsilon}_0$	1

Figure 6.4a compares the strength obtained by the classical theory, the present probability-related model, and the experimental data. Compared with the classical model, the tensile strength predicted from the present probability-related model is in good agreement with experimental results. The deviation between classical theoretical model and the experimental data increases significantly as the precipitate size increases (Figure 6.4a). For example, for the precipitate size of 80.7 nm, the strength from the classical model is 1290 MPa with a deviation of 6.8% compared with 1385 MPa from the experiment, while the strength from the probability-related model is 1366 MPa with a deviation of 1.3%. In Figure 6.4a, the highest experimental strength is observed at a precipitate size of 33.3 nm with a strength of 1545 MPa. The calculated result of the present model is 1538 MPa with the deviation of 0.5%. Among them, 706 MPa is contributed from precipitate strengthening. Furthermore, the contribution from the weak-pair coupling mechanism is 223 MPa, from the strong-pair coupling mechanism is 440 MPa, from Orowan mechanism is

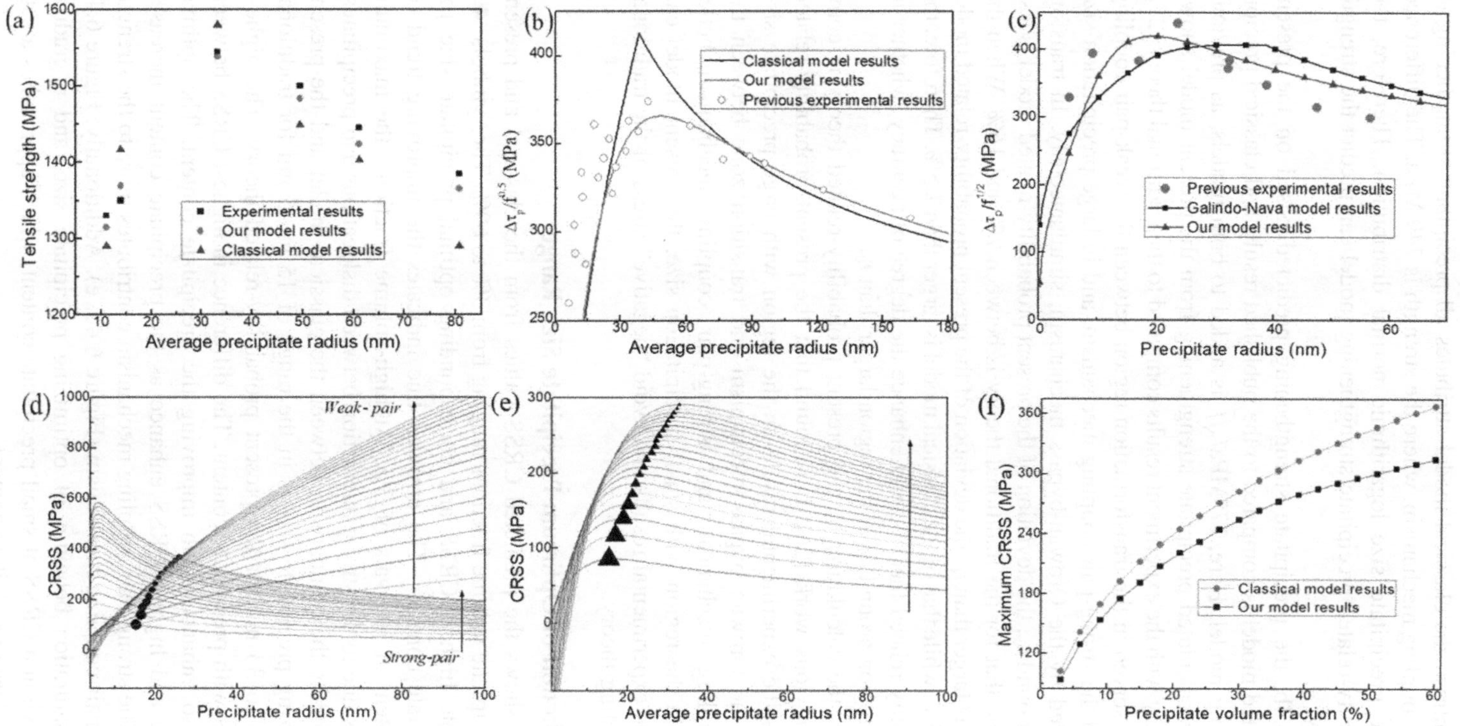

FIGURE 6.4 (a) The relationship between average precipitate size and tensile strength. (b, c) The predicted CRSS increment. (d) The predicted CRSS at various precipitate volume fractions for classical model, (e) for our model results. (f) The maximum CRSS as a function of the precipitate volume fraction [22].

43 MPa. In contrast, the classical model attributes all precipitate strengthening to the strong-pair coupling mechanism, where the strength is 749 MPa. The difference arises from the precipitate size logarithmic normal distribution. Therefore, the present probability-related precipitate strengthening model can predict the strength more accurately.

In Figure 6.4b, the precipitate strengthening predicted based on the present probability-related model is compared to the published results and classical precipitate strengthening model. Here, $175\text{MPa}/\sqrt{f}$ is added to both models, as previous study [26]. The calculated precipitate strengthening from the present model shows better consistency with the experiment results compared to the traditional theory [27]. This trend is observed in the transformation region between the weak-pair coupling mechanism and the strong-pair coupling mechanism, and in large precipitation size region dominated by the Orowan-bypass mechanism, simultaneously. In transformation region around r_w, the deviation of the present probability-related model is less than 2.5%, while that for the traditional theory is between 7% and 16%. When the precipitate size is larger than r_c, the deviation of the present probability-related model is less than 0.05%, while that of the classical model is larger than 2.5%. Therefore, the present probability-related theory could enhance the degree of accuracy, whether in transformation region around r_w or the region larger than r_c.

Furthermore, the calculation by the present probability-related theory is compared to the previous work [26]. It is found that the present probability-related theory captures the experimental results in the region with large precipitate size dominated by the Orowan bypass mechanism and transition zone between the weak-pair coupling mechanism and strong-pair coupling mechanism better (Figure 6.4c). In the region with small precipitation size, the present model can better match the experimental results and avoid "negative" stress of the traditional weak-pair coupling theory.

6.2.1.4 Prediction of Optimum Precipitate Size Range

Figure 6.4d, e shows the predicted CRSS values from the classical and present models for precipitate volume fractions ranging from 3% to 60%. The symbols '●' and '▲' denote highest CRSS and corresponding optimal precipitate size for various precipitate contents. The arrowed line indicates the improving trend of precipitate content at intervals of 3%. At high-volume fractions, the maximum CRSS increases due to the strong interaction between dislocations and precipitates. Figure 6.4f exhibits the comparison between the classical model and the present model. Optimizing precipitate size is in the range of 15 to 25 nm for traditional theory, and 16 to 33 nm for the present probability-related theory. The highest CRSS increases with precipitate content. The difference in highest CRSS between two models also enhances with improving the precipitate content. The optimal precipitate size and highest CRSS enhance as the precipitate content increases (Figure 6.4f). The strong-pair coupling mechanism contributes more to the strength than the weak-pair coupling mechanism (Figure 6.4d, e). Additionally, Figure 6.4f demonstrates saturation behavior of optimizing precipitate size, and a gradual decrease in maximum CRSS at small precipitate content, which approaches a saturation value at large precipitate content.

6.2.1.5 Prediction of Optimum Process

We have further established a relationship between processing technology and microstructural feature of precipitates in nickel-based superalloys, with the goal of guiding the design and preparation of high-performance materials (Figure 6.5a).

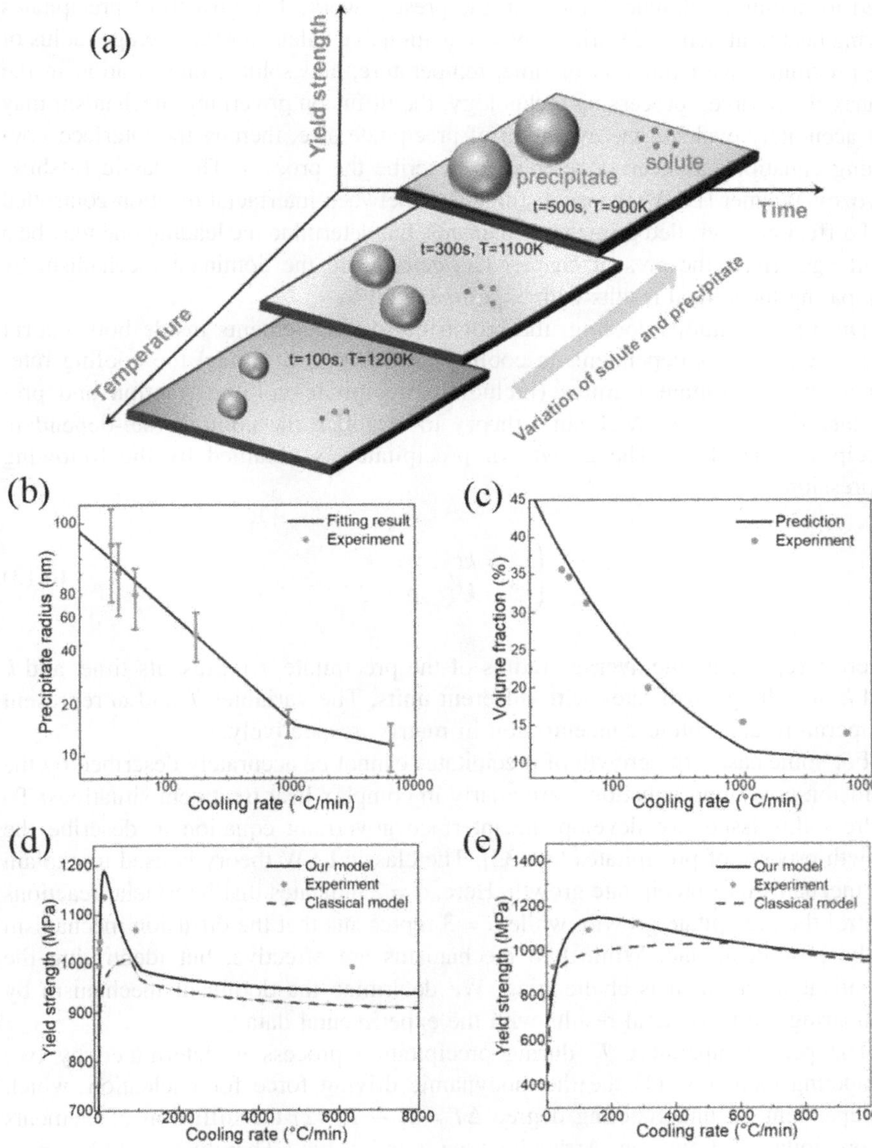

FIGURE 6.5 (a) The schematic diagram of the time- and temperature-dependent microstructural features of alloys. (b) The cooling rate dependent precipitate size: theory vs. experiments. (c) The cooling rate dependent precipitate volume fraction: theory vs. experiments. The cooling rate dependent yield strength of (d) the prepared alloy and (e) another alloy [28]: theory vs. experiment.

The cooling rate during heat treatment significantly affects the feature parameter and morphology of precipitates. The cooling rate is controlled by the correlation between temperature and time. To obtain the precipitates with various features, the prepared alloy is treated at different cooling rates. We have modeled the precipitate size and volume fraction at different processing technology. The McLean's theory has been used to obtain precipitate features in the present work. The growth of precipitates during heat treatment is described by an equation that calculates the average radius of the precipitate as a function of time, temperature, and solute concentration in the matrix. In complex processing technology, the diffusion governing mechanism may not accurately evaluate the evolution of precipitate size, thereby the interface governing equation has been developed to describe the process. The classic Lifshitz-Slyozov-Wagner (LSW) theory distinguishes between interfacial reaction-controlled and diffusion-controlled growth mechanisms, but determine the leading one may be a challenge. Here, the present chapter has determined the dominant mechanism by comparing theoretical results with experimental data.

During continuous cooling, the redistribution of elements inside both matrix and precipitates is dependent on cooling rate. Here, we model the cooling rate-dependent precipitate features (including precipitate volume fraction and precipitate size). We use McLean's theory to establish the cooling rate-dependent precipitate size [29]. The growth of precipitates is obtained by the following expression:

$$\begin{cases} r^{\alpha} = kt & \alpha = 3 \\ r^{\alpha} = k't & \alpha = 2 \end{cases} \tag{6.13}$$

where r represents the average radius of the precipitate, t represents time, and k and k' are the growth rates with different units. The variables T and ω represent temperature and solute concentration in matrix, respectively.

For some cases, the growth of precipitates cannot be accurately described by the diffusion governing equation, particularly in complex heat treatment situations. To address this issue, we develop the interface governing equation to describe the growth process of precipitates [30–32]. The classic LSW theory is used to explain the mechanism of precipitate growth. Here, $\alpha = 2$ indicates that interfacial reactions control the precipitate growth, while $\alpha = 3$ represents that the diffusion mechanism is the prominent one. While two mechanisms are effective, but identifying the dominant mechanism is challenging. We determine the dominant mechanism by comparing the theoretical results with the experimental data.

The peak temperature T_p during precipitation process is determined by two competing elements: (1) the thermodynamic driving force for nucleation, which is dependent on undercooling degree $\Delta T = T_e - T$; (2) the diffusion of elements in precipitate follows an Arrhenius-type relationship. Therefore, $K(T)$ can is obtained by:

$$K(T) \propto (T_e - T)\exp\left(-\frac{Q}{RT}\right) \tag{6.14}$$

where Q represents the activation energy, T_e represents the equilibrium transition temperature, and R represents the gas constant.

The total conversion rate is approximately written by:

$$K(T) = 1 - (1 - \tau)^2 = \tau(2 - \tau) \tag{6.15}$$

where τ is denoted as $\tau = \frac{T_e - T}{T_e - T_p}$.

Assuming the cooling rate is constant

$$\frac{d\tau}{dt} = \frac{c}{\Delta T^*} \Rightarrow \tau = \frac{ct}{\Delta T^*} \tag{6.16}$$

where $\Delta T^* = T_p - T_e$

Therefore, the conversion rate is approximately written as:

$$K(T) = \frac{ct}{\Delta T^*}\left(2 - \frac{ct}{\Delta T^*}\right) \tag{6.17}$$

where c is the cooling rate. Both k and K are sensitive to the temperature, it is expressed by:

$$r^\alpha = Kt \Leftrightarrow \dot{r} = \frac{1}{\alpha} K r^{1-\alpha} \tag{6.18}$$

where K is overall transformation rate. α is determined by precipitate growth and interfacial features. It is worth noting units of quantity are ineffective, so Eq. 6.18 is a numerical formulation [33].

By separating the variables in Eq. 6.18, then integrate, the following expression is obtained:

$$\int_0^r r^{\alpha-1} dr = \int_0^{t_{\max}} \frac{1}{\alpha} \frac{ct}{\Delta T^*}\left(2 - \frac{ct}{\Delta T^*}\right) dt \tag{6.19}$$

where $ct/\Delta T^* = 2$ represents that the precipitate growth terminates. Therefore, the precipitate radius is written by:

$$\frac{r^\alpha}{\alpha} = \frac{4}{3\alpha c} \Rightarrow r \propto c^{-1/\alpha} \tag{6.20}$$

Taking the logarithm of Eq. 6.20, the following expression is obtained:

$$\log r = -\frac{1}{\alpha} \log c + B \tag{6.21}$$

Subsequently, for the purpose of predicting precipitate volume fraction, precipitate number density would be determined. The number density of precipitates is evaluated based on the present experimental results. This relationship is expressed as following [34–36]:

$$N = Ec^F \tag{6.22}$$

By taking the logarithm, a linear relationship can be obtained. Therefore, the radius, volume fraction and number density of precipitate is related by [13,37,38]:

$$f = \frac{4\pi N r^3}{3} \tag{6.23}$$

Based on Eqs. 6.20, 6.22 and 6.23, the following expression is obtained:

$$f = \frac{4A^3 E\pi}{3} c^{F-3/\alpha} \tag{6.24}$$

where A represents $(4/3)^{1/\alpha}$.

Figure 6.5b exhibits the correlation of precipitate radius and cooling rate, which agree well with calculations. It can be concluded both diffusion and interfacial reactions controlled the precipitate features. The precipitate features are fallen into two different and continuous regions, including 0 -1000 °C/min and 1000–6346 °C/min to obtain a more accurate prediction, which have been reported in previous work [39–43]. The parameter in Eq. 6.20 lies between 0.5 and 0.33, reflecting that interfacial reaction plays a dominant role during precipitate growth process. The predicted precipitate volume fraction is compared with experimental data in Figure 6.5c. It can be seen that the calculated results of the present work capture experimental data very well. Figure 6.5d, e compares the predicted mechanical properties of the prepared nickel-based superalloy and published U720LI superalloy [44], respectively. Our model predicts that the highest strength occurs at 166 °C/min for the present prepared alloy (Figure 6.5d), and the predicted strength of the U720LI alloy agrees well with experimental result at intermediate cooling rates (Figure 6.5e).

6.2.2 Precipitate Random Distribution

The haphazard spatial arrangement of precipitates has a crucial impact on the mechanical properties of alloys. Nevertheless, the size and variance of precipitates have been overlooked, impeding the study of randomly dispersed precipitates in such alloys. Specifically, in nickel-based superalloys, the dissimilarity between the size and spacing of precipitates is minimal, resulting in a significant degree of size variation that cannot be disregarded [1]. To comprehensively understand the intricate interaction between dislocations and a large number of precipitates, some simulation methods can be used to reveal the precipitate-strengthening mechanism.

At the atomic level, the microstructural evolution of the interaction between dislocations and precipitates can be thoroughly investigated using molecular dynamics simulations [45]. Using discrete dislocation dynamics (DDD) simulation, the interaction between dislocations and a large number of precipitates can be observed in larger space and time scales [46]. In addition, the experimental observation shows that the precipitates are often obey the nonuniform size distribution [47,48]. Therefore, it will lead to the enrichment of the spatial distribution of precipitates, which will produce a significant impact on the mechanical characteristics. Building upon prior research, this section aims to elucidate the influence of precipitate size enrichment and spatial distribution on the mechanism of precipitate strengthening. Through DDD simulations, this section systematically elucidates the interaction between dislocations and precipitates with varying sizes and deviations in 3D space, and the strengthening mechanism mediated by the size enrichment and spatial distribution of precipitate is revealed. In addition, considering the size deviation and the enrichment of spatially randomly distributed precipitates, we introduce a statistical approach to assess the impact of precipitate distribution on strength, and validate the accuracy of this method through experimental data.

6.2.2.1 Derivation and Parameter Setting

We employ the open-source code ParaDiS [50] for all DDD simulations, with the simulation parameters listed in Table 6.4. To capture the crystal's overall characteristics, a 2 μm × 2 μm × 2 μm cube is selected as the simulation box [51], with an edge dislocation constructed on the plane. As pervious research [52], the strain rate is set to 1×10^3 s^{-1}. The precipitate radius follows a logarithmic normal distribution, as observed in experiments [7]. The probability density function (PDF) for the distribution of precipitate radius is derived by [53]

$$f_1(r) = \frac{1}{\sqrt{2\pi}\, r\phi} \exp\left[-\frac{(\ln(r) - \xi)^2}{2\phi^2}\right] \tag{6.25}$$

TABLE 6.4
DDD Simulation Parameters of Nickel-Based Superalloys [49]

Parameter	Symbol	Value
Burgers vector	b	0.248 nm
Dislocation radius	r_0	b
Shear modulus	μ	80 GPa
Poisson's ratio	v	0.31
Resistance coefficient	B	2.5×10^{-4} (Pa·s)
Anti-phase boundary energy	γ_{APB}	0.27 J/m^2
Size of precipitate	d	161.4 nm
Volume fraction of precipitate	f_v	0.245

The geometric mean can be expressed as:

$$\xi = \ln\left(\frac{E^2}{\sqrt{D + E^2}}\right) \tag{6.26}$$

Here, E is the average value of precipitate radius [54], and D represents the variance.

The geometric standard deviation is calculated using the following formula:

$$\phi = \sqrt{\ln\left(\frac{D}{E^2} + 1\right)} \tag{6.27}$$

Here, $\phi = 0$, 0.0392, 0.0554, 0.0783, 0.1105, 0.1558, 0.219 and 0.272 are used in the DDD simulation. The volume fraction is determined as 0.245 from the experiment [54]. The relationship between the precipitate radius, the precipitate distance, and precipitate volume fraction is $f_v = \frac{2\pi}{3}\left(\frac{\bar{r}}{L}\right)^2$.

Several mechanisms affect the precipitate strengthening, including the stacking faults, coherence and ordered strengthening [19,20,55]. The resistance from the APB prevented dislocations cutting precipitates [46,56]. The interface between precipitate and matrix is coherent, leading to negligible long-term stress field in nickel-based superalloy [15]. In addition, stacking fault strengthening and coherent strengthening effects can be ignored [19]. Therefore, the interaction between the dislocation and precipitate particle is selected as [57–59]:

$$\tau_{\text{obst}} = \begin{cases} -\gamma_{\text{APB}}/b & \text{inside a particle} \\ 0 & \text{outside} \end{cases} \tag{6.28}$$

where γ_{APB} represents the APB energy, b represents the burgers vector magnitude.

The mechanism of dislocation bypassing the precipitate is dependent on the precipitate size [28]. The CRSS of single dislocation cut through precipitate is given by [46]:

$$\tau_c^{\text{Shear}} = \frac{\gamma_{\text{APB}}}{b}\sqrt{\frac{3\pi\gamma_{\text{APB}} f_v r}{8T}} \tag{6.29}$$

Here, $T \approx 1/2\mu b^2$ represents the dislocation line tension. The CRSS for the Orowan mechanism is [55]:

$$\tau_c^{Orowan} = \frac{\mu b}{L_s} \tag{6.30}$$

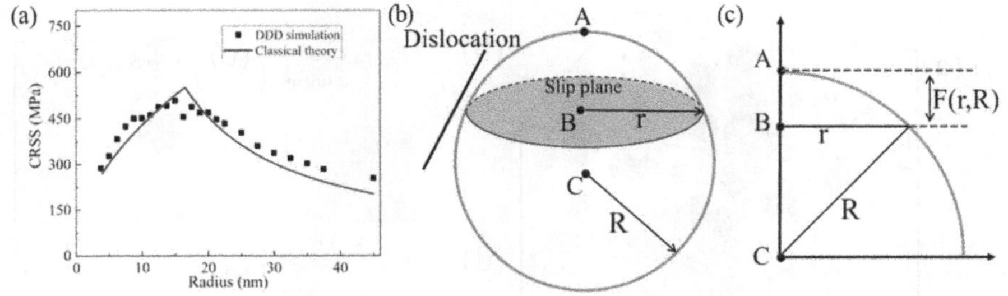

FIGURE 6.6 (a) CRSS of dislocation-precipitate interactions in nickel-based superalloy, calculated by DDD simulations and conventional theory. (b, c) Display of the cross-sectional radius CDF resulting from a dislocation-cut homogeneous precipitate from A to C [49].

where $L_s = L - 2r$ is the spacing between precipitate. The average precipitate spacing is $L = \left(\frac{2\pi}{3f}\right)^{1/2}\bar{r}$. The CRSS (see Table 6.4 for other simulation parameters) calculated through DDD simulations, at a precipitate volume fraction of 0.12, aligns with the outcomes predicted by traditional theory (Figure 6.6a).

The derivation process of the PDF has been demonstrated above. When the dislocation cuts the precipitate with radius R (Figure 6.6b, c), The cumulative distribution function (CDF) for the section is given by:

$$\mathrm{F}(r, R) = 1 - \sqrt{1 - \frac{r^2}{R^2}} \tag{6.31}$$

Based on the CDF, the PDF is:

$$f_s(r, R) = \frac{r}{R^2}\frac{1}{\sqrt{1 - r^2/R^2}} \tag{6.32}$$

The $f_2(r)$ PDF formed by that spherical precipitates cut by dislocation slip plane is related to the distribution $f_s(r, R)$ and radius $f_1(R, \mu, \sigma)$ of spherical precipitate:

$$f_2(r) = \int_{-\infty}^{+\infty} f_1(R, \mu, \sigma)\cdot f_s(r, R)dR \tag{6.33}$$

The distribution of spherical precipitate radius obeys logarithmic normal distribution.

6.2.2.2 Space Distribution Effect

As per previous experimental findings [47], there exists a notable difference in the spatial arrangement of precipitates with identical average size, leading to an important difference in the yield stress. Figure 6.7a illustrates the relationship between edge dislocation and randomly dispersed-spherical precipitates in DDD simulations. The precipitate diameter is arbitrarily chosen according to a

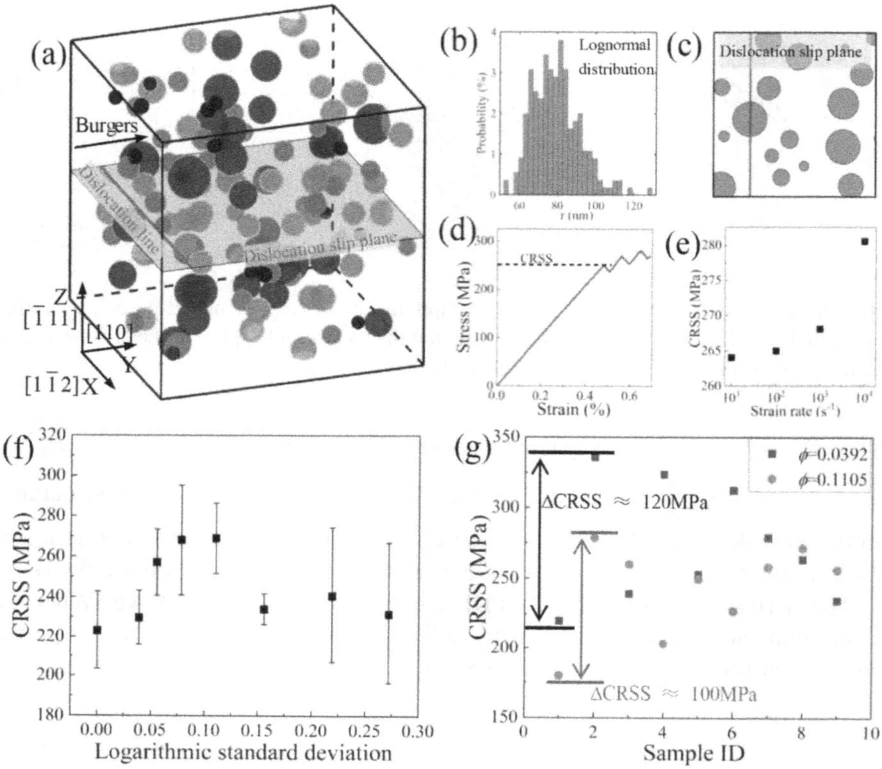

FIGURE 6.7 (a) The dislocation-precipitate interaction in DDD simulations. (b) The spherical-precipitation size distribution. (c) The precipitation spatial arrangement on the dislocation-slip surface. (d) The stress-strain relationship for alloys. (e) The association between CRSS and strain rate for alloys. (f) The association between CRSS and logarithmic-standard deviation of precipitate radii. (g) The association between CRSS and the ID with the equivalent logarithmic-standard deviation of precipitate radii [49].

lognormal distribution (Figure 6.7b) [7]. The dislocation-slip plane intersects with precipitation, resulting in a cross-section where precipitate size varies significantly (Figure 6.7(c)). The CRSS is identified as the first-drop in the stress-strain curve (Figure 6.7(c)) [52]. Figure 6.7(d) presents the mean CRSS across various precipitation deviations in the alloy. As the deviation in precipitate sizes increases, the CRSS increases firstly, and then decreases (Figure 6.7(d)), suggesting that appropriate size deviation of precipitate can improve the precipitate strengthening. In particular, Figure 6.7(e) shows that the CRSS has nearly twice the difference under the same precipitation radius deviation, showing that the spatial distribution has a significant impact on CRSS, which is consistent with experimental observations [47]. Therefore, it is necessary to reveal the mechanical mechanism of the impact of precipitation size variation and spatial arrangement on CRSS, so as to further improve the precipitate strengthening.

6.2.2.3 Size Distribution Effect

The precipitation radius on the dislocation-slip plane follows a logarithmic normal distribution, corresponding to the precipitation-radius distribution throughout the simulation cell. To account for the deviation of precipitate size, a PDF is proposed to characterize the dispersion of precipitate size in the precipitation cross-section, which aligns with the statistical outcomes obtained through DDD simulations (Figure 6.8a). Notably, precipitate-radius PDF on the precipitation cross-section is significantly different from that located at the dislocation-glide plane, indicating presence of a large number of small section radii that are often ignored in traditional precipitate strengthening theories [7,60].

Based on the results of cross-section-radius PDF, the small cross-section-radii likelihood rises with the relative growth in the number of precipitates to the average size deviation, leading to a growth in the number of precipitates on dislocation-glide plane (Figure 6.6b, c). It increases the likelihood of dislocation-precipitate interactions (Figure 6.8d) and improves the dislocation bending degree (Figure 6.8e) [57,61]. Longer dislocation lines require more energy to form, leading to higher CRSS [46,62]. As the size deviation of precipitates further

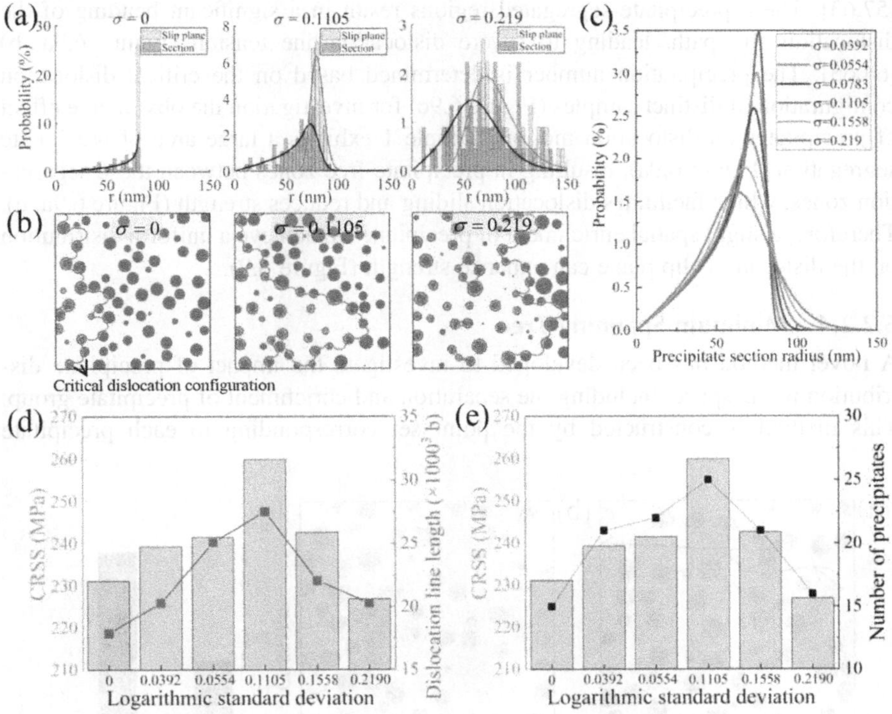

FIGURE 6.8 (a) Statistical results, (b) Critical dislocation configurations under different size deviations under CRSS, (c) The PDF of precipitate-section radius under different size deviations, (d) Critical dislocation configurations length (or CRSS) vs log standard deviation of precipitate size, (e) Precipitation number located on the critical-dislocation structure, and CRSS v.s. log standard deviation of precipitate size [49].

increases, the probability of the larger and smaller sections of the precipitates on the dislocation slip plane is higher. Larger cross sections will induce a small number of precipitates, while too small cross sections will cause little interference to dislocations [57], which can be ignored. Conversely, a large size deviation in precipitates will decrease the likelihood of dislocation-precipitate interactions, thereby reducing dislocation-segment length (Figure 6.8b, e) and resulting in a lower CRSS. Hence, the impact of precipitate-radius deviation on CRSS varies with dislocation line size.

The increase of precipitates-random-distribution degree in the alloy, rather than the equidistant, results in the formation of precipitate segregation zone [57]. Hence, it is crucial to investigate the impact of precipitate spatial distribution on CRSS. In areas where the spacing between precipitates is smaller than the average distance between dislocation pairs, the threshold-dislocation structure at CRSS is denoted by red curves. The light yellow, pink, and green curves represent the dislocation structure after the shear stress reaches CRSS, and the black arrow indicates the dislocation motion path. As a result, the dislocation motion path intersects the larger spacing between precipitate separation zones (the red arrow in Figure 6.9a, b), leading to 'Orowan Islands' formation around the precipitation-separation zones (Figure 6.9a, b) [57,63]. These precipitate-segregated regions result in a significant bending of the dislocation slip path, leading to severe dislocation-line tension (Figure 6.9a, b) [64,65]. The precipitation number is determined based on the critical dislocation configuration of distinct samples (Figure 6.9c) for investigation the obstructive effect of precipitates on dislocation motion. Sample 1 exhibits a large area of precipitate segregation (Figure 6.9a), resulting in precipitate-free zones between these segregation zones, which facilitates dislocation sliding and reduces strength (Figure 6.9a, c). Therefore, a slight spatial enrichment of precipitates to achieve a uniform distribution on the dislocation slip plane can enhance strength (Figure 6.9).

6.2.2.4 Minimum Spanning Tree

A novel method has been developed to investigate the impact of precipitate distribution in 3D space, including the separation and enrichment of precipitate group. This method is constructed by the point set corresponding to each precipitate

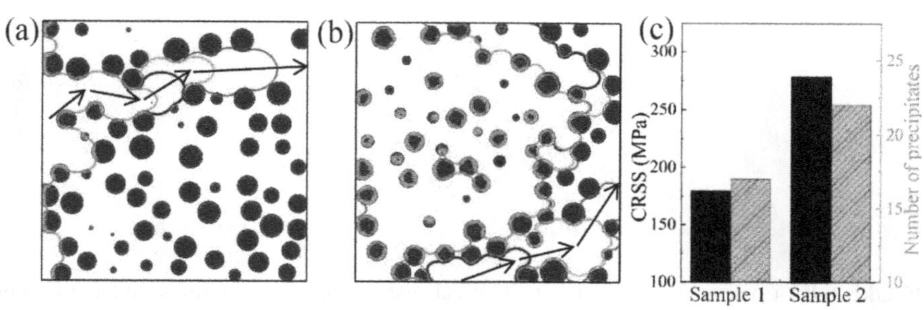

FIGURE 6.9 (a, b) The impact of precipitate spatial arrangement on dislocation movement in samples 1 and 2, which possess identical size deviation. (c) CRSS and number of precipitates on the threshold-dislocation structure for sample 1 and sample 2 [49].

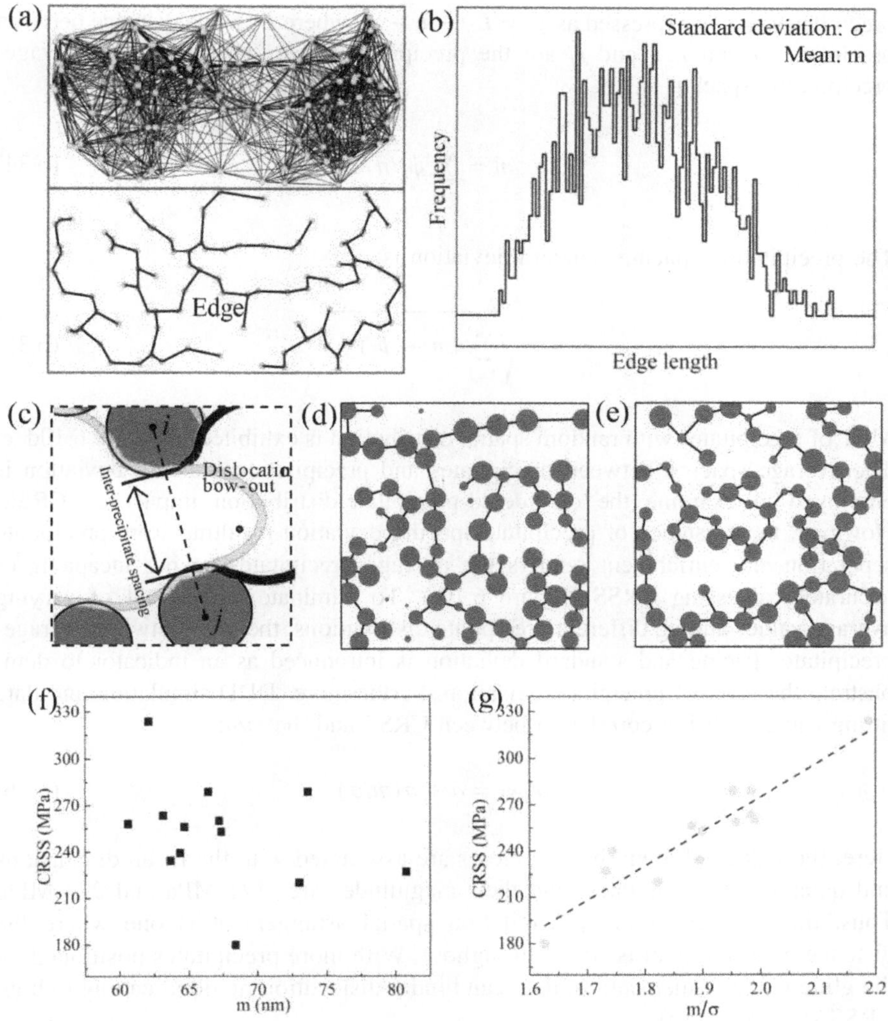

FIGURE 6.10 (a) Distribution of the MST with respect to the position of precipitates, which are treated as points. (b) The frequency distribution of the MST side length. (c) Partial expansion of the dislocation curve. (d, e) MST for sample 1 and sample 2 with the same precipitation size deviation in the precipitate distribution. (f) CRSS and average-precipitation spacing for a series of samples. (g) For a series of samples, the CRSS and the ratio of average-precipitation spacing to their standard deviation (m/σ) [49].

location based on the MST (Figure 6.10a, b). Here, MST will be used to evaluate the randomness of precipitate distribution, so as to reveal the influence of phase space arrangement on CRSS. Given that the radius of curvature for the bent critical dislocation structure is determined by distance between precipitates to assess dislocation line tension (Figure 6.10c), the MST of precipitate distribution is obtained by setting the distance between precipitates as the weight. The distance between precipitates can be expressed as $\psi_k = L_{ij} - r_i - r_j$, where L_{ij} is the distance between precipitates i and j, r_i and r_j are the precipitation size. Therefore, the average-precipitation spacing is:

$$m = \sum_{k=1}^{n} \psi_k / n \tag{6.34}$$

The precipitation-spacing-standard deviation is:

$$\sigma = \sqrt{\sum_{k=1}^{n} (m - \psi_k)^2 / n} \tag{6.35}$$

MST of precipitates with random spatial distribution is exhibited in Figure 6.10d, e. The average spacing between precipitates and precipitation-standard deviation is employed to examine the disordered-precipitate-distribution impact on CRSS. However, the existence of precipitate-spacing deviation resulting from precipitate separation and enrichment renders the average precipitate spacing incapable of accurately assessing CRSS (Figure 6.10f). To eliminate the impact of varying average values among different precipitate distributions, the ratio between average-precipitate spacing and standard deviation is introduced as an indicator to demonstrate the level of precipitate deviation. Furthermore, DDD simulation and data fitting can establish a correlation between CRSS and the m/σ:

$$\tau_{\mathrm{CRSS}} = a + b(m/\sigma) \tag{6.36}$$

Here, the values of a and b coefficients are associated with the mean dimensions and quantity of precipitates, and their magnitudes are -171 MPa and 223 MPa. Thus, the most favorable precipitation spatial arrangement is one where the distance between them is equal throughout. With more precipitates positioned on the glide plane of dislocation, they can hinder dislocation motion, leading to high CRSS (Figure 6.10g).

6.3 SERVICE PERFORMANCE

Increasing the service temperature and stress load of aircraft gas turbines can improve their efficiency. The high temperature and stress load can cause creep deformation and reduce service life of the gas turbines. Therefore, one of the main objectives in designing a new generation of nickel-based superalloys is to improve their creep resistance [66–75]. In addition, nickel-based superalloys are being considered as structural materials for the nuclear reactors. Based on the uncertainty of the interaction mechanism between dislocation and precipitate during the creep process in nickel-based superalloys, and cooccurrence of climbing mechanism, cutting mechanism, and Orowan mechanism arising from the multi-scale statistical distribution of precipitates, a theoretical model without adjustable parameter in a wide temperature range is developed to evaluate the creep rate. The calculated creep rates agree well with the experimental results, and the computational accuracy is significantly better than that of the previous work.

6.3.1 Spatial Distribution of Precipitate Size

It is generally believed that the effective mechanism of dislocation-precipitate interaction is dependent on the mean size of precipitates [15,76–79]. And the dislocations always pass through the center plane of the precipitate [15,76–79]. However, these features do not actually correspond to the real physical situation. The reality is that the slip plane of the dislocations is stationary, but precipitates are randomly distributed inside alloy. The different mechanisms of dislocation-precipitate interaction would exist even the size of the precipitates is the same. Therefore, the spatial distribution of the precipitates induces uncertainty of dislocation-precipitate interaction mechanism. Especially, the threshold stress comes from probability-dependent dislocation-precipitate interaction for large-size precipitate, including climbing mechanism, cutting mechanism, and Orowan mechanism. Taking the probability-dependent dislocation-precipitate interaction into account, the effective CRSS for creep resistance follows:

$$
\begin{aligned}
\tau_{weak}^{new}(r) &= p_0(r)\tau_{c\,\lim b} + \int_0^r p_1(r)\tau_{weak}(r)dr \quad r \le r_p \\
\tau_{strong}^{new}(r) &= p_0(r)\tau_{c\,\lim b} + \int_0^{r_p} p_1(r)\tau_{weak}(r)dr + \int_{r_p}^{r} p_2(r)\tau_{strong}(r)dr \quad r_p < r < r_c \\
\tau_{orowan}^{new}(r) &= p_0(r)\tau_{c\,\lim b} + \int_0^{r_p} p_1(r)\tau_{weak}(r)dr + \int_{r_p}^{r_c} p_2(r)\tau_{strong}(r)dr \\
&\quad + \int_{r_p}^{r} p_3(r)\tau_{Orowan}(r)dr \quad r_c \le r
\end{aligned}
\tag{6.37}
$$

where τ_{climb} is the CRSS of the climbing mechanism, $\tau_{weak}(r)$, $\tau_{strong}(r)$ and $\tau_{Orowan}(r)$ denote the CRSS of the weak-pair coupling mechanism, strong-pair coupling mechanism, and Orowan mechanism. $P_1(r)$, $P_2(r)$ and $P_3(r)$ can be denoted as $p_1(r) = 1 - (1 - (r_p/r)^2)^{1/2}$, $p_2(r) = (1 - (r_p/r)^2)^{1/2} - (1 - (r_c/r)^2)^{1/2}$, $p_3(r) = 1 - p_1(r) - p_2(r)$ respectively.

6.3.2 Statistical Distribution of Precipitate Size

Previous work has always supposed that "the interaction between dislocations and precipitate population" can be equated to "dislocations interacting with the precipitate of uniform average size" [15]. However, this assumption is not physically consistent with the actual situation. In fact, the precipitate sizes are dispersed in the experimental results, rather than a single size. Therefore, coupling statistical distribution of precipitate sizes could accurately predict the creep life.

$$
\sigma_{thr}(r) = M\left(\int_0^{r_p} \tau_{weak}^{new}(r)f(r)dr + \int_{r_p}^{r_c} \tau_{strong}^{new}(r)f(r)dr + \int_{r_c}^{r_{\max}} \tau_{orowan}^{new}(r)f(r)dr + \tau_{c\,\lim b}\right)
\tag{6.38}
$$

Most theoretical models are proposed to predict creep rates and their associated creep lifetimes based on the climbing mechanism and the Orowan bypassing mechanism [80–82]. Here, the creep rate is proposed as

$$\dot{\varepsilon}_{sec} = \frac{2\tau_L b c_L N}{M}\left(\frac{\sigma_{app} - \sigma_{thr}(r)}{\alpha MGb}\right)^3 \tag{6.39}$$

where τ_L is the dislocation line tension, which can be expressed as $\tau_L = 0.5Gb^2$, b the Burgers vector, c_L the strain hardening parameter, N the dislocation mobility, which depends on stress and temperature. M the Taylor factor, σ_{app} the applied stress, σ_{thr} the threshold stress, α the Taylor constant depending on the crystal structure, and R_g the shear modulus. Here, the new threshold stress dependent on the average precipitate size like the previous work, and statistical and spatial distribution of the precipitates. The present model in this section is valid for general precipitate-strengthened alloys.

6.3.3 Creep Rate

Figure 6.11a compares the experimental data in MAR-M247 high-temperature alloy with the prediction from the classical model and the current model [83]. The solid line denotes current model, dashed one denotes the classical model, both using the same parameters. The experimental data is represented by dots. The results from the present work match experimental data more closely than classical model at all temperatures and applied stresses. In particular, the slope of the creep rate curve varies with increasing stress at 1000 °C and 950 °C, because the transition of the effective creep mechanism from climbing mechanism to a cutting mechanism or an Orowan mechanism is captured by the current model. The degree of deviation between experimental and prediction is presented in Figure 6.11b. The solid dots indicate the predictions based on the current work, the hollow dots indicate the results of the traditional theory. At 800 °C, the average difference in creep rate for the current model is less than 300%, while that for classical model is 30,000%. At 900 °C, the average deviation of the present model is less than 50% compared to the experimental results, but that of the classical model is greater than 600%, indicating that the accuracy of present theory is obviously better compared with classical theory.

Figure 6.11c shows a comparison between the current and classical model predictions and experimental data in 718 high-temperature alloy [78]. The dots indicate the experimental results, the solid lines indicate calculation of current work, and the dashed one indicates that of classical model. When σ_{app} is below 100 MPa, the results show a linear relationship, which is consistent with creep behavior based on traditional theory (Figure 6.11c). Additionally, the transformation behavior of creep rate is exhibited in Figure 6.11a, c. Further analysis indicates that this trend originates from the transformation of the creep mechanism from climbing to the cutting mechanism induced by enhancement of the applied stress. Climbing dominates creep mechanism when σ_{app} is less than the

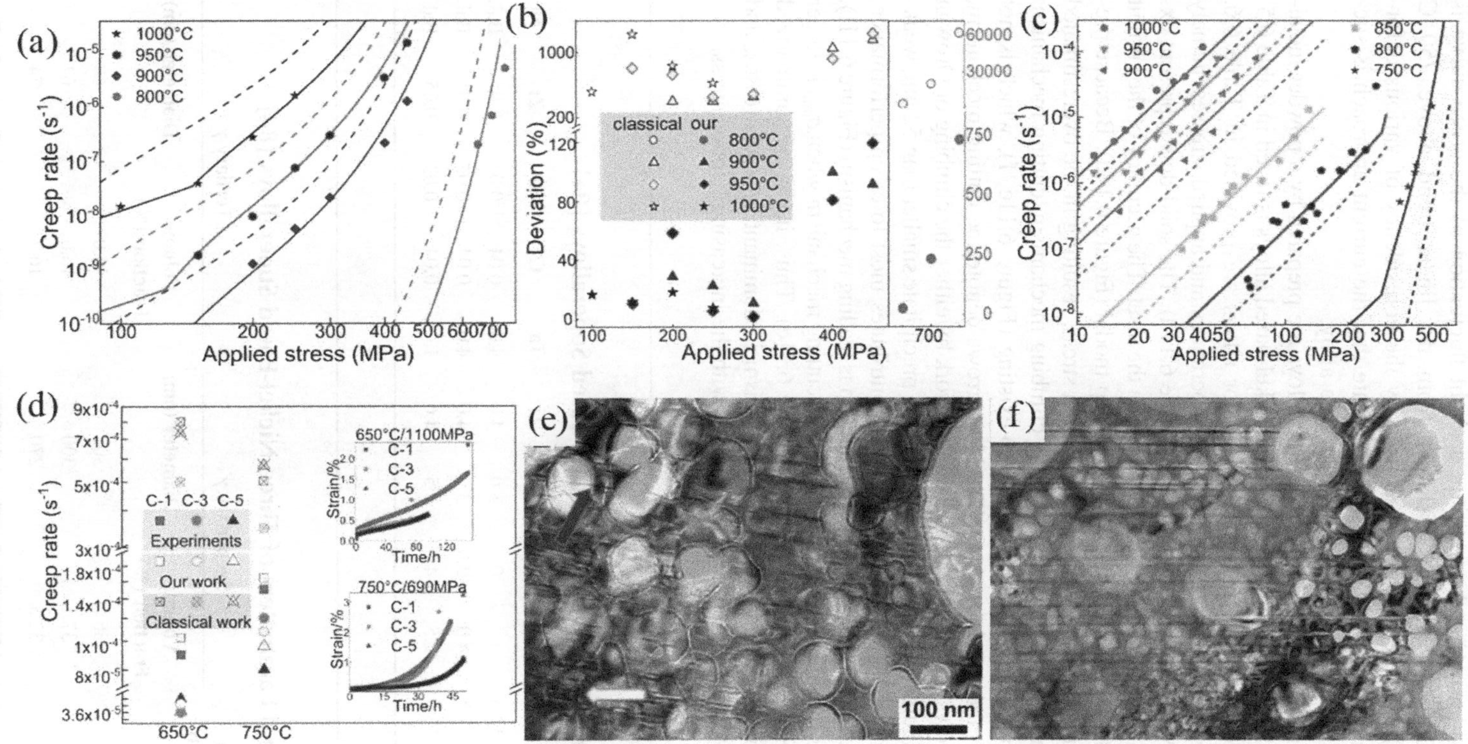

FIGURE 6.11 Comparison of current models with experiments and classical theory. (a) The log-log relationship between the creep rate of the MAR-M247 superalloy and the applied stress. (b) The deviation between model calculation and the experiment. (c) The creep rate of 718 superalloys. (d) Various creep mechanisms observed experimentally and compared with theory: Comparison between the model calculation and our experiments. (e) TEM results of C-5 alloy after creep deformation at 650 °C/1,100 MPa. (f) The coexistence of various mechanisms [84].

stress of transition point. In addition, a slight fluctuation of the creep curve is observed at 750 °C, which is different from the linear trend above 750 °C (Figure 6.11c). This result may be caused by the coresidency of various deformation mechanisms. These results demonstrate that the current approach is successfully applied to the 718 high-temperature alloy.

Three new nickel-based high-temperature alloys are prepared by a powder metallurgical route. The composition of the three nickel-based alloys is listed in Table 6.5. The initial microstructural parameters used for alloys are provided in Table 6.6. Creep tests are conducted at two different temperatures and applied stresses, namely 650 °C/1,100 MPa and 750 °C/690 MPa (Figure 6.11d). The solid dots represent experimental results obtained from calculations via the inset. The accuracy of the current model is significantly better than classical creep model (Figure 6.11d). Because previous studies underestimate the creep threshold stress, resulting in the overestimated creep rates. The experiments show that the climbing mechanism, cutting mechanism and Orowan bypassing mechanism are co-existing (Figure 6.11e, f), which is not consistent with the previous work. The blue arrows denote the cutting mechanism, and the red arrows indicate the dislocation network by either the climbing or Orowan mechanism (Figure 6.11e, f). In this case, for the precipitate smaller than 52 nm, weak-pair mechanism and strong-pair mechanism contributes most to creep performance. Inversely, the climbing/Orowan loop mechanism is leading mechanism (Figure 6.11f). Interestingly, the dominant creep mechanism is cutting mechanism when σ_{app} is larger than the CRSS for cutting mechanism (Figure 6.11e). Thus, it can be summarized that the climbing mechanism/Orowan mechanism is dominant at low stresses, and turning into two different cutting mechanisms with the increasing applied stress.

TABLE 6.5
Nominal Composition of Three Nickel-Based Superalloys [84]

	Co	Cr	Mo	W	Al	Ti	Ta	C	B	Zr	Ni
C-1	25.0	12.5	5	0	3.0	4.0	4.0	0.04	0.03	0.05	Bal.
C-3	25.0	12.5	4.0	2.0	3.0	4.0	4.0	0.04	0.03	0.05	Bal.
C-5	25.0	13.5	2.5	4.0	2.5	4.6	1.6	0.04	0.03	0.05	Bal.

TABLE 6.6
Microstructural Parameters of Three Nickel-Based Superalloys [84]

Alloy	Grain Size (μm)	Secondary γ'		Tertiary γ'	
		Volume Fraction (%)	Diameter (nm)	Volume Fraction (%)	Diameter (nm)
C-1	2.7	36.6	307	15.0	60
C-3	5.4	37.1	300	15.6	64
C-5	4.2	32.5	270	16.7	74

6.3.4 Quantitative Evaluation

Figure 6.12a, b shows the correlation between applied stresses and the effective creep mechanisms. Here, the change of applied stress (i.e., the red line in A moves up and down) results in the change of the occurrence probability and contribution of each creep mechanism (i.e., the pink line in B moves left and right). In zone I of Figure 6.12b, only the cutting mechanism is existing, because σ_{app} is greater than the threshold stress. In zone II, σ_{app} is not large enough for dislocations to shear into the precipitate, resulting in the appearance of climbing mechanism. However, the cutting mechanism exists due to the uncertainty of the precipitate-dislocation interaction, resulting in the coexistence of both mechanisms in region II. Therefore, in region II, the climbing mechanism and cutting mechanism coexist. In region III, σ_{app} is larger than the threshold stress, leading to the activation of Orowan bypass process. Due to the uncertainty of precipitate-dislocation interaction, the cutting and climbing mechanisms are still present. Therefore, the climbing mechanism, cutting mechanism and Orowan bypassing mechanism are present simultaneously in zone III. These results demonstrate that the present model reveals the cooccurrence of different mechanisms in the alloy, which is in keeping with the experiments.

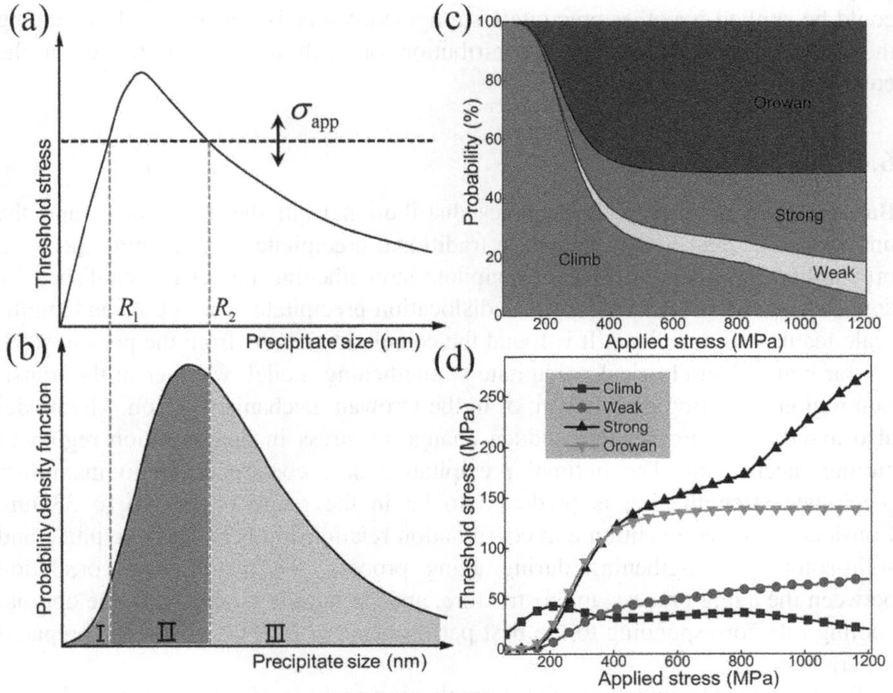

FIGURE 6.12 Calculation process of contributing creep resistance and probability for different mechanisms. (a) Relationship between σ_{app} and creep threshold stress. (b) the PDF for different mechanisms. (c) Appearance probability of each mechanism in C-5 superalloy depending on the σ_{app}. (d) Creep threshold stress from each mechanism in C-5 superalloy [84].

On the contrary, in previous theoretical models, only one creep mechanism was considered for a given service condition [82,85–90].

To enhance the service performance of the prepared superalloys by accurately regulating the size distribution of precipitates. Figure 6.12c quantitatively exhibits the appearance probability of various deformation mechanisms. When σ_{app} is less than 200 MPa, climbing leads creep process and makes the largest contribution for creep strength. As σ_{app} improves above 200 MPa, the occurrence probability of climbing decreases sharply, leading to an obvious transition. Under the intermediate stress region (larger than 300 MPa but less than 700 MPa), the contribution of strong-pair coupling to the creep resistance is comparable to that of the Orowan mechanism. Under the high stress region (larger than 700 MPa), the main contribution to creep resistance comes from the strong-pair coupling (Figure 6.12d). Interestingly, the cutting mechanism provides more subscription to the creep strength, despite the greater occurrence probability of the Orowan mechanism (Figure 6.12c, d). This finding is different from traditional perceive in which Orowan bypass mechanism provides more subscription to the creep strength. Furthermore, although Figure 6.12c, d presents for the current C-5 alloy at an experimental temperature of 750 °C, these results are typical for high temperature alloys owing to the small compositional fluctuations. Additionally, the current work could be applied for other precipitate-strengthened metals, quantitatively revealing the occurrence probability and contribution of each mechanism based on the corresponding features.

6.4 CONCLUSION

Based on the multi-scale precipitates distribution from the experiment, and the unreasonable stress abrupt change in traditional precipitate strengthening theory, a probability-dependent statistical precipitate strengthening model is established by coupling spatial uncertainty of the dislocation-precipitate interaction and multi-scale feature of precipitate. It is found the calculation results from the present work are far better than classical precipitate strengthening model, whether in the transition region of cutting mechanism, or in the Orowan mechanism region. The model also avoids the unreasonable sudden change of stress in the transition region of cutting mechanism. The optimal precipitation size corresponding to maximum precipitate strengthening is predicted to be in the range of 16 nm to 33 nm. Considering the competition and coordination relationship between precipitate and solid-solution strengthening during aging process, the quantitative correlation between the aging process, microstructure, and strength is established. The optimal cooling rate corresponding to the best performance is 166 °C/min for the prepared superalloys.

Based on DDD simulations, the strengthening analysis of random distribution for precipitates and the effect of size deviation in nickel-based superalloys is carried out. The interaction-probability for the dislocation-precipitate is controlled by changing the probability for size of cross section. It is found that the greatest precipitate strengthening appears when the mean size is equal to size variance. Slight spatial enrichment of precipitates corresponds to maximum strength, which prompts the precipitates to locate on the slip surface, which in turn results in serious dislocation bending. The influence of the random distribution of precipitate on the mechanical properties is quantitatively evaluated using the MST to describe change of precipitate spacing.

Considering the spatial uncertainty of the dislocations-precipitate interaction and coexistence of climb mechanism, cutting mechanism and Orowan mechanism during the deformation process in nickel-based superalloys, a unified creep rate prediction model without tuning parameters over a wide temperature range is established. The prediction of the proposed model captures the experimental results very well, and the calculation accuracy is obviously better than that of classical theory. Furthermore, appearance probability of individual mechanisms, and the weight of contribution from these mechanism to alloy's creep strength with increasing applied stress are obtained. It reveals that the climbing mechanism dominates the creep mechanism under low-stress service conditions, while the Orowan mechanism and cutting mechanism dominate under high-stress service conditions, which is different from the traditional view that only the Orowan mechanism is dominant one. These researches provide an important theoretical basis for the design and preparation of advanced nickel-based superalloys.

REFERENCES

[1] Pollock T.M. 2016. Alloy design for aircraft engines. *Nature Materials*, 15(8): 809–815.

[2] Kozar R.W., Suzuki A., Milligan W.W., et al. 2009. Strengthening mechanisms in polycrystalline multimodal nickel-base superalloys. *Metallurgical and Materials Transactions A*, 40a(7): 1588–1603.

[3] Meher S., Aagesen L.K., Carroll M.C., et al. 2018. The origin and stability of nanostructural hierarchy in crystalline solids. *Science Advances*, 4(11): eaao6051.

[4] Pollock T.M. and Tin S. 2006. Nickel-based superalloys for advanced turbine engines: Chemistry, microstructure, and properties. *Journal of Propulsion and Power*, 22(2): 361–374.

[5] Alabort E., Barba D., Sulzer S., et al. 2018. Grain boundary properties of a nickel-based superalloy: Characterisation and modelling. *Acta Materialia*, 151: 377–394.

[6] Anderson M.J., Schulz F., Lu Y., et al. 2020. On the modelling of precipitation kinetics in a turbine disc nickel based superalloy. *Acta Materialia*, 191: 81–100.

[7] Collins D.M. and Stone H.J. 2014. A modelling approach to yield strength optimisation in a nickel-base superalloy. *International Journal of Plasticity*, 54: 96–112.

[8] Xu Y.L., Jin Q.M., Xiao X.S., et al. 2011. Strengthening mechanisms of carbon in modified nickel-based superalloy Nimonic 80A. *Materials Science and Engineering A*, 528(13-14): 4600–4607.

[9] Huang S., Huang M.S. and Li Z.H. 2018. Effect of interfacial dislocation networks on the evolution of matrix dislocations in nickel-based superalloy. *International Journal of Plasticity*, 110: 1–18.

[10] Vaidya M., Pradeep K.G., Murty B.S., et al. 2017. Radioactive isotopes reveal a non sluggish kinetics of grain boundary diffusion in high entropy alloys. *Scientific Reports*, 7(1): 12293.

[11] Ebrahimi F., Bourne G.R., Kelly M.S., et al. 1999. Mechanical properties of nanocrystalline nickel produced by electrodeposition. *Nanostructured Materials*, 11(3): 343–350.

[12] Crudden D.J., Mottura A., Warnken N., et al. 2014. Modelling of the influence of alloy composition on flow stress in high-strength nickel-based superalloys. *Acta Materialia*, 75: 356–370.
[13] Cao W., Zhang F., Chen S.L., et al. 2016. Precipitation Modeling of Multi-Component Nickel-Based Alloys. *Journal of Phase Equilibria and Diffusion*, 37(4): 491–502.
[14] Blum W., Eisenlohr P. and Breutinger F. 2002. Understanding creep—a review. *Metallurgical and Materials Transactions A*, 33(2): 291–303.
[15] Reed R.C. 2008. *The superalloys: fundamentals and applications*. Cambridge university press.
[16] Courtney T.H. 2005. *Mechanical behavior of materials*. Waveland Press.
[17] Hall E.O. 1951. The deformation and ageing of mild steel: III discussion of results. *Proceedings of the Physical Society B*, 64(9): 747.
[18] Wei B.Q., Wu W.Q., Xie D.Y., et al. 2021. Strength, plasticity, thermal stability and strain rate sensitivity of nanograined nickel with amorphous ceramic grain boundaries. *Acta Materialia*, 212: 116918.
[19] Ardell A.J. 1985. Precipitation hardening. *Metallurgical Transactions A*, 16(12): 2131–2165.
[20] Nembach E. and Neite G. 1985. Precipitation hardening of superalloys by ordered γ'-particles. *Progress in Materials Science*, 29(3): 177–319.
[21] Kruml T., Conforto E., Lo Piccolo B., et al. 2002. From dislocation cores to strength and work-hardening: a study of binary Ni3Al. *Acta Materialia*, 50(20): 5091–5101.
[22] Fang Q.H., Li L., Li J., et al. 2019. A statistical theory of probability-dependent precipitation strengthening in metals and alloys. *Journal of the Mechanics and Physics of Solids*, 122: 177–189.
[23] Li D.F., Barrett R.A., O'Donoghue P.E., et al. 2017. A multi-scale crystal plasticity model for cyclic plasticity and low-cycle fatigue in a precipitate-strengthened steel at elevated temperature. *Journal of the Mechanics and Physics of Solids*, 101: 44–62.
[24] Li X.P., Ji G., Chen Z., et al. 2017. Selective laser melting of nano-TiB2 decorated AlSi10Mg alloy with high fracture strength and ductility. *Acta Materialia*, 129: 183–193.
[25] Ma K.K., Hu T., Yang H., et al. 2016. Coupling of dislocations and precipitates: Impact on the mechanical behavior of ultrafine grained Al-Zn-Mg alloys. *Acta Materialia*, 103: 153–164.
[26] Galindo-Nava E.I., Connor L.D. and Rae C.M.F. 2015. On the prediction of the yield stress of unimodal and multimodal gamma′ Nickel-base superalloys. *Acta Materialia*, 98: 377–390.
[27] Reppich B. 1982. Some new aspects concerning particle hardening mechanisms in γ'precipitating Ni-base alloys—I. Theoretical concept. *Acta Metallurgica*, 30(1): 87–94.
[28] Fang Q., Huang Z., Li L., et al. 2022. Modeling the competition between solid solution and precipitate strengthening of alloys in a 3D space. *International Journal of Plasticity*, 149: 103152.
[29] McLean D. 1984. Predicting growth of γ' in nickel alloys. *Metal Science*, 18(5): 249–256.
[30] Ardell A.J. and Ozolins V. 2005. Trans-interface diffusion-controlled coarsening. *Nature Materials*, 4(4): 309–316.
[31] Lifshitz I.M. and Slyozov V.V. 1961. The kinetics of precipitation from supersaturated solid solutions. *Journal of Physics and Chemistry of Solids*, 19(1-2): 35–50.

[32] Wagner C. 1961. Theorie der alterung von niederschlägen durch umlösen (Ostwald-reifung). *Zeitschrift für Elektrochemie, Berichte der Bunsengesellschaft für physikalische Chemie*, 65(7-8): 581–591.
[33] Papadaki C., Li W. and Korsunsky A.M. 2018. On the dependence of γ′ precipitate size in a nickel-based superalloy on the cooling rate from super-solvus temperature heat treatment. *Materials*, 11(9): 1528.
[34] Huang G.C., Liu G.Q., Feng M.N., et al. 2018. The effect of cooling rates from temperatures above the gamma′ solvus on the microstructure of a new nickel-based powder metallurgy superalloy. *Journal of Alloys and Compounds*, 747: 1062–1072.
[35] Semiatin S.L., Mahaffey D.W., Levkulich N.C., et al. 2018. The Effect of Cooling Rate on High-Temperature Precipitation in a Powder-Metallurgy, Gamma/Gamma-Prime Nickel-Base Superalloy. *Metallurgical and Materials Transactions A*, 49a (12): 6265–6276.
[36] Semiatin S.L., Kim S.L., Zhang F., et al. 2015. An Investigation of High-Temperature Precipitation in Powder-Metallurgy, Gamma/Gamma-Prime Nickel-Base Superalloys. *Metallurgical and Materials Transactions A*, 46a(4): 1715–1730.
[37] Guo H., Enomoto M. and Shang C.J. 2018. Simulation of bcc-Cu precipitation in ternary Fe-Cu-M alloys. *Computational Materials Science*, 141: 101–113.
[38] Herrnring J., Sundman B., Staron P., et al. 2021. Modeling precipitation kinetics for multi-phase and multi-component systems using particle size distributions via a moving grid technique. *Acta Materialia*, 215.
[39] Chiou M.S., Jian S.R., Yeh A.C., et al. 2016. High temperature creep properties of directionally solidified CM-247LC Ni-based superalloy. *Materials Science and Engineering A*, 655: 237–243.
[40] Pant N., Verma N., Ashkenazy Y., et al. 2021. Phase evolution in two-phase alloys during severe plastic deformation. *Acta Materialia*, 210: 116826.
[41] Smith T.M., Bonacuse P., Sosa J., et al. 2018. A quantifiable and automated volume fraction characterization technique for secondary and tertiary gamma′ precipitates in Ni-based superalloys. *Materials Characterization*, 140: 86–94.
[42] Wu H.Y., Huang Z.W., Zhou N., et al. 2019. A study of solution cooling rate on gamma′ precipitate and hardness of a polycrystalline Ni-based superalloy using a high-throughput methodology. *Materials Science and Engineering A*, 739: 473–479.
[43] Mitchell R.J., Preuss M., Tin S., et al. 2008. The influence of cooling rate from temperatures above the gamma′ solvus on morphology, mismatch and hardness in advanced polycrystalline nickel-base superalloys. *Materials Science and Engineering A*, 473(1–2): 158–165.
[44] Jackson M.P. and Reed R.C. 1999. Heat treatment of UDIMET 720Li: the effect of microstructure on properties. *Materials Science and Engineering A*, 259(1): 85–97.
[45] Krasnikov V.S., Mayer A.E. and Pogorelko V.V. 2020. Prediction of the shear strength of aluminum with θ phase inclusions based on precipitate statistics, dislocation and molecular dynamics. *International Journal of Plasticity*, 128: 102672.
[46] Chatterjee S., Li Y. and Po G. 2021. A discrete dislocation dynamics study of precipitate bypass mechanisms in nickel-based superalloys. *International Journal of Plasticity*, 145: 103062.
[47] Gao S., Hou J., Yang F., et al. 2017. Effect of Ta on microstructural evolution and mechanical properties of a solid-solution strengthening cast Ni-based alloy during long-term thermal exposure at 700° C. *Journal of Alloys and Compounds*, 729: 903–913.
[48] Jiao Z.B., Luan J.H., Miller M.K., et al. 2017. Co-precipitation of nanoscale particles in steels with ultra-high strength for a new era. *Materials Today*, 20(3): 142–154.

[49] Chen Y., Fang Q.H., Luo S.H., et al. 2022. Unraveling a novel precipitate enrichment dependent strengthening behaviour in nickel-based superalloy. *International Journal of Plasticity*, 155: 103333.
[50] Bulatov V.V., Hsiung L.L., Tang M., et al. 2006. Dislocation multi-junctions and strain hardening. *Nature*, 440(7088): 1174–1178.
[51] Lehtinen A., Laurson L., Granberg F., et al. 2018. Effects of precipitates and dislocation loops on the yield stress of irradiated iron. *Scientific Reports*, 8(1): 1–12.
[52] Santos-Guemes R., Bellon B., Esteban-Manzanares G., et al. 2020. Multiscale modelling of precipitation hardening in Al-Cu alloys: Dislocation dynamics simulations and experimental validation. *Acta Materialia*, 188: 475–485.
[53] Nan C.W. and Clarke D.R. 1996. The influence of particle size and particle fracture on the elastic/plastic deformation of metal matrix composites. *Acta Materialia*, 44(9): 3801–3811.
[54] Wu H., Li J., Liu F., et al. 2017. A high-throughput methodology search for the optimum cooling rate in an advanced polycrystalline nickel base superalloy. *Materials & Design*, 128: 176–181.
[55] Kelly A. 1963. Precipitation hardening. *Progress in Materials Science*, 10: 151–391.
[56] Lu S., Antonov S., Xue F., et al. 2021. Segregation-assisted phase transformation and anti-phase boundary formation during creep of a γ′-strengthened Co-based superalloy at high temperatures. *Acta Materialia*, 215: 117099.
[57] Mohles V. 2001. Simulations of dislocation glide in overaged precipitation-hardened crystals. *Philosophical Magazine A*, 81(4): 971–990.
[58] Wang X. and Xiong W. 2020. Uncertainty quantification and composition optimization for alloy additive manufacturing through a CALPHAD-based ICME framework. *npj Computational Materials*, 6(1): 1–11.
[59] Jiang M., Devincre B. and Monnet G. 2019. Effects of the grain size and shape on the flow stress: A dislocation dynamics study. *International Journal of Plasticity*, 113: 111–124.
[60] Santos-Guemes R., Capolungo L., Segurado J., et al. 2021. Dislocation dynamics prediction of the strength of Al-Cu alloys containing shearable θ" precipitates. *Journal of the Mechanics and Physics of Solids*, 151: 104375.
[61] Cui W., Cui Y. and Liu W. 2021. A statistical model of irradiation hardening induced by non-periodic irradiation defects. *Scripta Materialia*, 201: 113959.
[62] Li J., Chen H.T., Fang Q.H., et al. 2020. Unraveling the dislocation-precipitate interactions in high-entropy alloys. *International Journal of Plasticity*, 133: 102819.
[63] Aagesen L.K., Miao J., Allison J.E., et al. 2018. Prediction of Precipitation Strengthening in the Commercial Mg Alloy AZ91 using dislocation dynamics. *Metallurgical and Materials Transactions A*, 49(5): 1908–1915.
[64] Gilbert M.R., Schuck P., Sadigh B., et al. 2013. Free energy generalization of the Peierls potential in iron. *Physical Review Letters*, 111(9): 095502.
[65] Szajewski B.A., Pavia F. and Curtin W.A. 2015. Robust atomistic calculation of dislocation line tension. *Modelling and Simulation in Materials Science and Engineering*, 23(8): 085008.
[66] Birosca S., Liu G., Ding R.G., et al. 2019. The dislocation behaviour and GND development in a nickel based superalloy during creep. *International Journal of Plasticity*, 118: 252–268.
[67] Coakley J., Dye D. and Basoalto H. 2011. Creep and creep modelling of a multimodal nickel-base superalloy. *Acta Materialia*, 59(3): 854–863.
[68] Sass V., Glatzel U. and FellerKniepmeier M. 1996. Anisotropic creep properties of the nickel-base superalloy CMSX-4. *Acta Materialia*, 44(5): 1967–1977.

[69] Wu S., Song H.Y., Peng H.Z., et al. 2022. A microstructure-based creep model for additively manufactured nickel-based superalloys. *Acta Materialia*, 224: 117528.

[70] Yin W.M., Whang S.H., Mirshams R., et al. 2001. Creep behavior of nanocrystalline nickel at 290 and 373 K. *Materials Science and Engineering A*, 301(1): 18–22.

[71] Chen J.B., Chen J.Y., Wang Q.J., et al. 2022. Enhanced creep resistance induced by minor Ti additions to a second generation nickel-based single crystal superalloy. *Acta Materialia*, 232: 117938.

[72] Li K.S., Cheng L.Y., Xu Y.L., et al. 2022. A dual-scale modelling approach for creep-fatigue crack initiation life prediction of holed structure in a nickel-based superalloy. *International Journal of Fatigue*, 154.

[73] Sakane M. and Isobe N. 2022. Tension-torsion multiaxial creep-fatigue lives of the Nickel-based superalloy Alloy 738LC. *International Journal of Fatigue*, 155.

[74] Xu J.H., Li L.F., Liu X.G., et al. 2022. Fast characterization framework for creep microstructure of a nickel-based SX superalloy with high-throughput experiments and deep learning methods. *Materials Characterization*, 187.

[75] Zhang C.J., Wang P., Wen Z.X., et al. 2022. Study on creep properties of nickel-based superalloy blades based on microstructure characteristics. *Journal of Alloys and Compounds*, 890.

[76] Brown L.M. and Ham R.K. 1971. Strengthening methods in crystals. *Applied Science, London*, 9.

[77] Peng S.Y., Wei Y.J. and Gao H.J. 2020. Nanoscale precipitates as sustainable dislocation sources for enhanced ductility and high strength. *Proceedings of the National Academy of Sciences of the United States of America*, 117(10): 5204–5209.

[78] Drexler A., Fischersworring-Bunk A., Oberwinkler B., et al. 2018. A microstructural based creep model applied to alloy 718. *International Journal of Plasticity*, 105: 62–73.

[79] Mf A. 1975. Thermodynamics and kinetics of slip. *Progress in Materials Science*, 19: 1–281.

[80] Bai Z.T. and Fan Y. 2018. Abnormal Strain Rate Sensitivity Driven by a Unit Dislocation-Obstacle Interaction in bcc Fe. *Physical Review Letters*, 120(12).

[81] Kabir M., Lau T.T., Rodney D., et al. 2010. Predicting Dislocation Climb and Creep from Explicit Atomistic Details. *Physical Review Letters*, 105(9).

[82] Zhao J.F., Gong J.D., Saboo A., et al. 2018. Dislocation-based modeling of long-term creep behaviors of Grade 91 steels. *Acta Materialia*, 149: 19–28.

[83] Kvapilova M., Dvorak J., Kral P., et al. 2019. Creep behaviour and life assessment of a cast nickel - base superalloy MAR-M247. *High Temperature Materials and Processes*, 38: 590–600.

[84] Li L., Liu F., Tan L.M., et al. 2022. Uncertainty and statistics of dislocation-precipitate interactions on creep resistance. *Cell Reports Physical Science*, 3(1): 100704.

[85] Yang T., Zhao Y.L., Tong Y., et al. 2018. Multicomponent intermetallic nanoparticles and superb mechanical behaviors of complex alloys. *Science*, 362(6417): 933.

[86] Manonukul A., Dunne F.P.E. and Knowles D. 2002. Physically-based model for creep in nickel-base superalloy C263 both above and below the gamma solvus. *Acta Materialia*, 50(11): 2917–2931.

[87] Bensch M., Preussner J., Huttner R., et al. 2010. Modelling and analysis of the oxidation influence on creep behaviour of thin-walled structures of the single-crystal nickel-base superalloy Rene N5 at 980 degrees C. *Acta Materialia*, 58(5): 1607–1617.

[88] MacKay R.A., Gabb T.P. and Nathal M.V. 2013. Microstructure-sensitive creep models for nickel-base superalloy single crystals. *Materials Science and Engineering a-Structural Materials Properties Microstructure and Processing*, 582: 397–408.

[89] Smith T.M., Unocic R.R., Deutchman H., et al. 2016. Creep deformation mechanism mapping in nickel base disk superalloys. *Materials at High Temperatures*, 33(4-5): 372–383.

[90] Yang M., Zhang J., Wei H., et al. 2018. A phase-field model for creep behavior in nickel-base single-crystal superalloy: Coupled with creep damage. *Scripta Materialia*, 147: 16–20.

7 Light Alloys

7.1 INTRODUCTION

Generally, light alloys are defined as the alloy of density less than 3.5×10^3 kg m^{-3}, mainly including Al, Mg, Ti, Li alloys, etc. [1–3] Al alloys are frequently used in daily life and industrial production because of their excellent properties, like high strength, hardness, strong plasticity, good physical characteristics, and superior resistance to corrosion [1,2]. In comparison, Mg alloys are the lightest structural metals and possess excellent characteristics such as good strength, stiffness, electromagnetic interference resistance, vibration-damping performance, and cutting performance [3,4]. It has become an indispensable and important material in many fields such as aerospace, electronic communications, architectural decoration, mechanical and electrical, and petrochemical industry [2–5].

7.2 AL ALLOYS

7.2.1 Multilayer Nanotwin Effect

Al alloy contains a large amount of twin structure, caused by that substrate material is anti-stacked along [111] direction. The ideal stacking of (111) layers in face-centered cubic (FCC) represented by ABCABCABCABC is altered to ABCABC|A|CBACB. Here, letters A, B, and C refer to (111) closed-packed planes, and underline indicates twin position (Figure 7.1). Multilayer nanotwins in Al alloys exhibit a combination of strength, ductility, and thermal stability that make them attractive for various engineering applications. Therefore, the investigations for multilayer nanotwin effect in Al alloy are rather necessary and meaningful.

7.2.1.1 Model

The twin boundary energy (TBE) is calculated by determining the energy difference of per unit area between the nanotwin structure and a perfect structure:

$$\gamma_{Al-nt} = (E_{Al-nt} - E_{Al-perfect})/(2A) \tag{7.1}$$

where the perfect supercell energy is denoted by $E_{Al-perfect}$ ($E_{Al-perfect} = NE_{Al}$) and E_{Al-nt} is the energy of the nanotwins. The energy of single Al atom is E_{Al} and the number of Al atoms is N in the supercells. The area of the (111) plane is represented

DOI: 10.1201/9781003225706-7

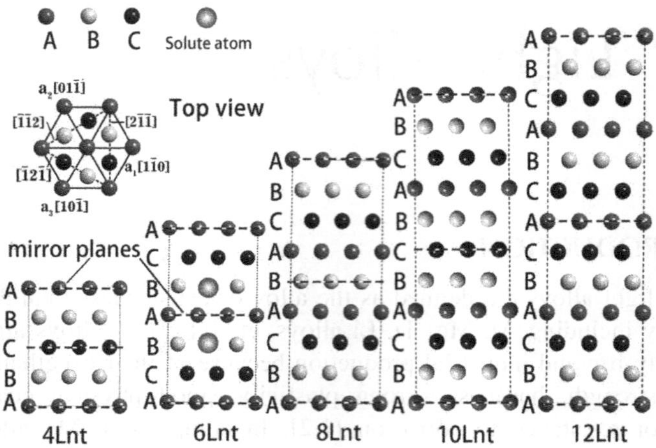

FIGURE 7.1 Schematic diagrams of multilayer nanotwins (4Lnt, 6Lnt, 8Lnt, 10Lnt, 12Lnt), with dash lines indicating the twin boundaries (TBs). It is worth noting that the TBs are conventionally described as the numbered 0 plane, which is a mirror plane, and the adjacent planes are sequentially numbered as 1, 2, 3 plane, and so on up to n plane. Due to the symmetry of nanotwin structures, only one site needs to be evaluated when adding solute atom in the same numeric nearest planes on both sides [6].

by A. The temperature-induced energy of nanotwin is calculated, when alloying atoms are introduced to various layers of the supercells [7]:

$$\gamma(T) = \gamma_{Al-nt} + \sum_n c_{nt-n}(T)E_{\text{int}-n}/A' \tag{7.2}$$

The area of (111) plane in unit cell is denoted by A'. The solute concentration of TB is represented by $c_{nt-n}(T)$ when solute atom is doped in the n-th layer, and the interaction energy $E_{\text{int}-n}$ between solute atoms and nanotwins is calculated [7]:

$$E_{int-n} = [E_{s-nt-n} - E_{s-\text{n}}] - [E_{Al-nt} - E_{Al-perfect}] \tag{7.3}$$

When doping a solute atom in the n-th layer of Al supercells, the total energy is denoted by E_{s-nt-n} for the existence of nanotwins and E_{s-n} for the inexistence of nanotwins. It is worth mentioning that the perfect supercells only have 3, 6, 9, or 12 atomic layers. Hence, we can calculate their energies for those perfect supercells with 4, 8, or 10 atomic layers as follow:

$$E_{s-n} = (N - 1)E_{Al} + E_{\text{solute}} + E_{enthalpy} \tag{7.4}$$

where the E_{solute} represents the energy of Al supercells with an individual solute atom, while $E_{enthalpy}$ refers to the solid solution enthalpy of corresponding solute concentration.

For investigating the impact of solute atoms on the nanotwins in Al alloys, two distribution models are employed: Fermi-Dirac distribution (FDD) and uniform distribution (UFD). According to FDD model, the solute concentration of nth layer, $c_{nt-n}(T)$, relative to the TB is [7]:

$$c_{nt-n} = \frac{1}{1 + \exp\left(\frac{E_{int-n}}{kT} - \ln\frac{c_0}{1-c_0}\right)} \tag{7.5}$$

where the alloying-atom concentration in a distant reference layer ($E_{\text{int}-n} = 0$) is denoted by c_0. The following equation is used to calculate the increments of the TBEs with respect to temperature and solute concentration:

$$\Delta\gamma(T) = \sum_n c_{nt-n} E_{int-n} / A' \tag{7.6}$$

The VASP (Vienna *ab* initio simulation package) is used to calculate TBEs of Al alloys [8]. Core-valence interactions are treated by the method of projector-augmented wave (PAW) [9]. Exchange-correlation functional is described by Perdew–Burke–Ernzerhof (PBE) of generalized-gradient approximation (GGA) [10].

7.2.1.2 Result

To investigate how solute atoms affect nanotwins, the TBE of nanotwin structures with 17, 18, or 19 atomic layers is first calculated (Figure 7.2a). To study the relationship between TBE and twin distance, 3 × 3 × L (L = 17, 18, or 19) supercells are used for calculating TBEs versus twin distance. The corresponding calculated results are shown by the line with triangular symbol in Figure 7.2a, the supercell with one stacking fault represents a distance of zero between twins (the first value in the line with triangular symbol). The TBEs of nanotwins range from 46–66 mJ·m^{-2} (Figure 7.2a), which consistent with previous results (44–66 mJ·m^{-2}) [11,12]. When two twins are formed in long-layers crystals,

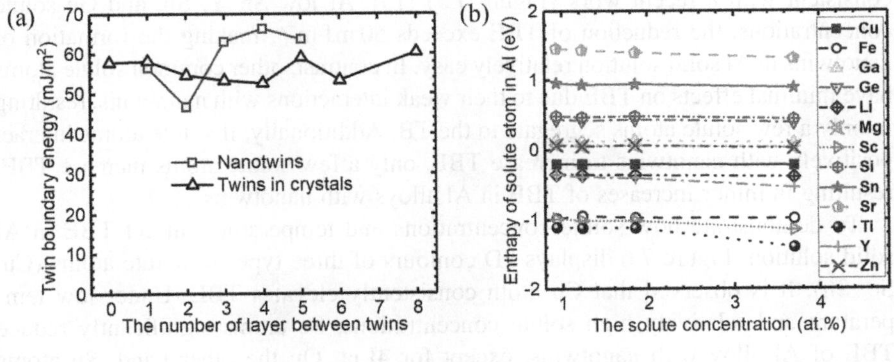

FIGURE 7.2 (a) TBEs of long crystals and nanotwin structures versus twin distance. (b) Formation enthalpies with respect to various solute concentrations [6].

TBEs slightly decrease first (~6 mJ·m^{-2}), and then increase to the maximum value as the distance raises between the two twins. In contrast, TBEs decrease from 56.5 mJ·m^{-2} to 46.7 mJ·m^{-2} when 4Lnt structure become 6Lnt structure, and then increase to a stable value of ~63 mJ·m^{-2}. TBE of 6Lnt structure in a single twin is the smallest, suggesting that 6Lnt structure in pure Al is more stable than other microstructures.

Figure 7.2b shows the calculated formation enthalpy when adding solute atoms in Al solid solution. The formation enthalpy remains nearly unchanged as the solute concentration increases, indicating that solute concentrations have a slight effect on the interaction mechanism. There exists a positive or negative formation enthalpy, which correlates with the ease of forming solid solution alloys. Solute atoms (Sr, Sn, Si, Ge) with positive formation enthalpy are difficult to add to Al alloys, while another atom (Sc, Ti, Li, Fe, Y, Cu) with negative formation enthalpy are easily incorporated in Al alloys. The formation enthalpy of solute atoms (Zn, Ga, Mg) close to zero, which indicates larger solubility [13].

The interaction energies between nanotwins and solute atoms are illustrated in Figure 7.3. Solute atoms are doped to only one part of the supercell, due to structural symmetry of TB. For most cases, as the distance between solute atoms and TBs increases, the interaction energy between them gradually decrease. The interaction energies of 12Lnt structure are almost zero. Solute atoms (Sr, Y, Sc, Sn, Si, Ge) have negative interaction energies, suggesting segregation in TBs, whereas solute atoms (Fe, Ti, Cu, Li, Mg, Ga, Zn) exhibit positive interaction energies, suggesting repulsion by TB. After a closer examination of the element characteristics, the atoms with more outermost electrons (≥3) are typically attracted to the TBs, with the exception of the Sr atom. Additionally, the strong interactions are produced for large atomic radius in these elements, hence Sr, Y, and Sc atoms have strong effects on the TBs.

For Sr, Y, Sn, and Ge solute atoms, an increase in solute concentration leads to a linear increase (positive or negative) in the increments of TBE (Figure 7.4) [14]. These solute atoms cause a decrease in TBEs of all nanotwins with increasing solute concentration, whereas Fe, Cu, Li, Zn, and Ti atoms cause an increase in TBE. Hence, Fe may not be the optimal solute atom for inducing nanotwins in Al alloys, which is consistent with a recent work (Figure 7.5) [15]. At low Sr, Y, Sn, and Ge solute concentrations, the reduction of TBE exceeds 50 mJ·m^{-2}, making the formation of nanotwins in Al solid solution relatively easy. In contrast, other common solute atoms have minimal effects on TBE due to their weak interactions with nanotwins, resulting in only a few solute atoms segregate to the TB. Additionally, if solute atoms interact positively with nanotwins to increase TBE, only a few solute atoms increase TBE, resulting in minor increases of TBE in Al alloys with nanotwins.

To demonstrate how solute concentrations and temperatures affect TBE in Al solid solution, Figure 7.6 displays 2D contours of three types of solute atoms (Cu, Sc, Sn). It is observed that Cu atom consistently elevates TBE. Under low temperature and relatively high solute concentrations, Sc atoms significantly reduce TBE of Al alloy with nanotwins, except for 4Lnt. On the other hand, Sn atoms induce a significant decrease of TBE for 4, 6, 8, and 10Lnts, but only slightly reduce it for 12Lnts.

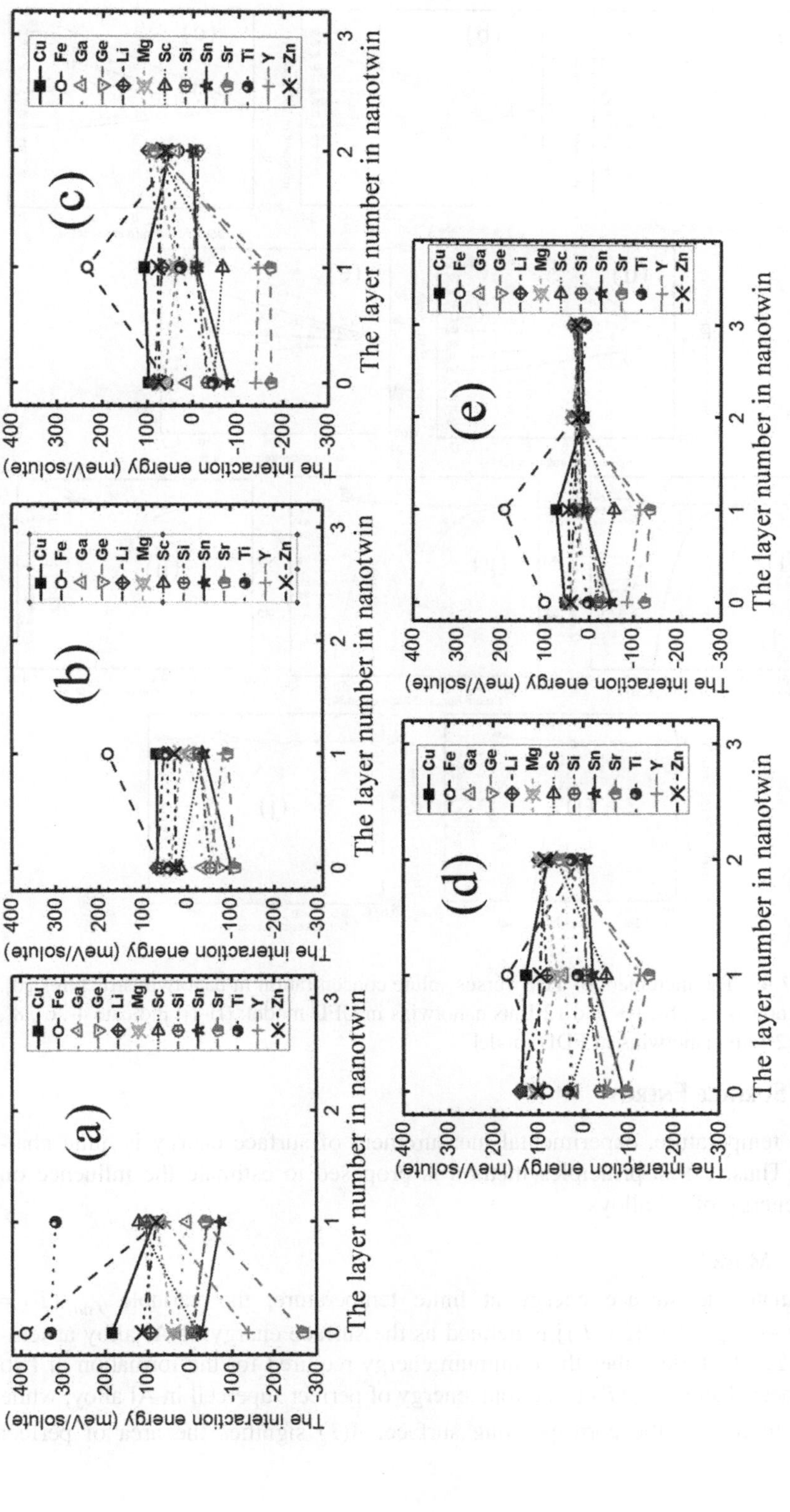

FIGURE 7.3 Interaction energies between nanotwins and solute atoms. (a)–(e) correspond to 4-, 6-, 8-, 10- and 12-Lnt structures [6].

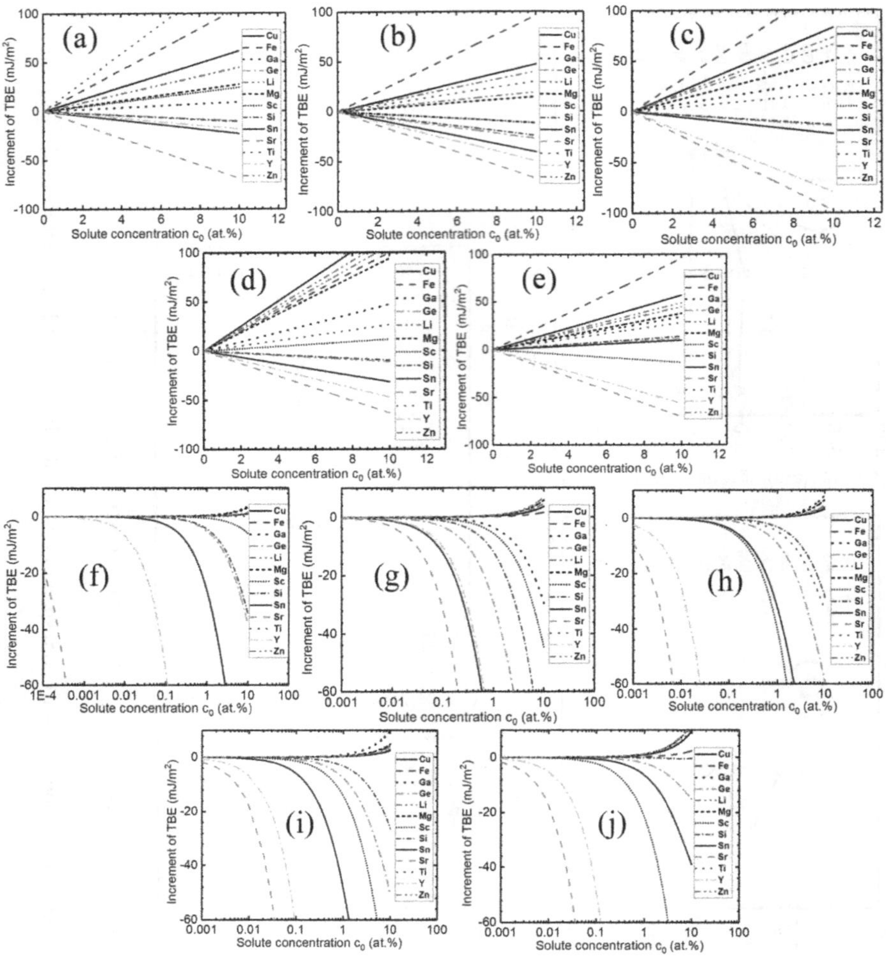

FIGURE 7.4 The increment of TBE versus solute concentration in nanotwin structures [6]. (a)–(e) denote 4-, 6-, 8-, 10- and 12-Lnts nanotwins in UFD model; (f)–(j) indicate 4-, 6-, 8-, 10- and 12-Lnts nanotwins in FDD model.

7.2.2 Surface Energy

At finite temperature, experimental measurement of surface energy is quite challenging. Thus, a first-principles method is proposed to estimate the influence on surface energy of Al alloys.

7.2.2.1 Model

To calculate the surface energy at finite temperature, the variable $\gamma_{surf}(T) = (E_{surf}(T) - E_{perf}(T))/[2A(T)]$ is defined as the surface energy of Al alloy at temperature T, which describes the minimum energy required for the formation of two free surfaces. Here, $\gamma_{surf}(T)$ is the total energy of perfect supercell in Al alloy, while $E_{surf}(T)$ represents the corresponding surface. $A(T)$ signifies the area of perfect

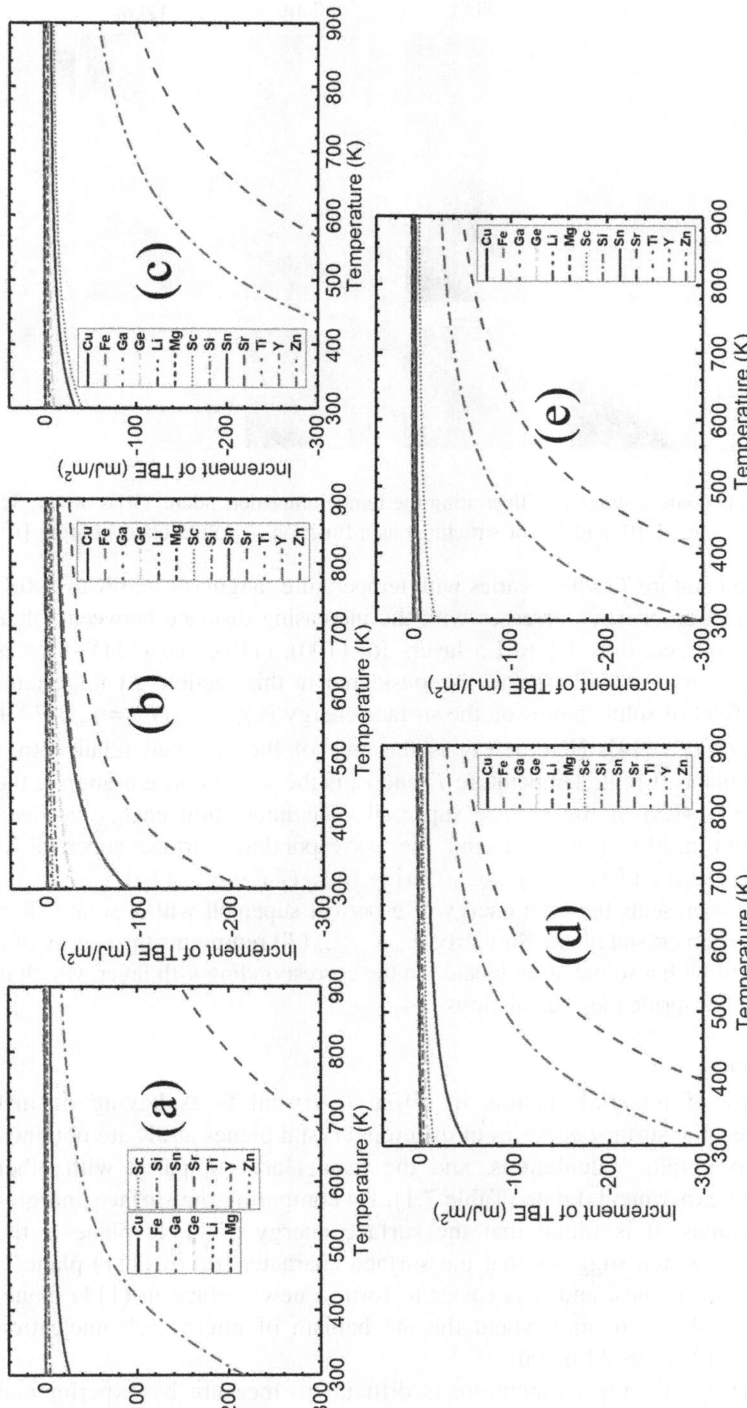

FIGURE 7.5 The increment of TBE in nanotwins versus temperature [6]. (a)–(e) denote 4-, 6-, 8-, 10- and 12-Lnts nanotwins in FDD model.

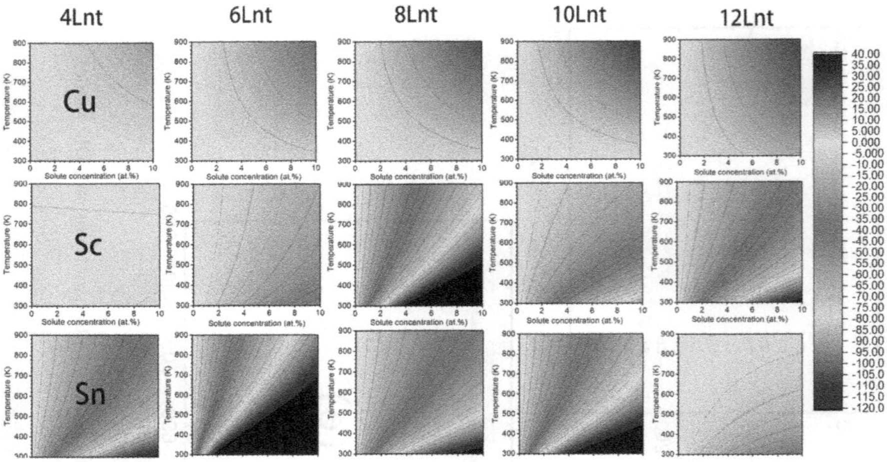

FIGURE 7.6 2D contour diagrams illustrating the temperature-dependent TBEs of Cu, Sc, and Sn atoms in 4, 6, 8, 10, and 12Lnt structures as a function of solute concentration [6].

supercell at temperature T, which varies with temperature ranging from 0K to 500K. Since the interaction energy decreases with the increasing distance between solute atoms and the surface, only L1 to L5 layers for (100), (110), and (111) surfaces and L1 to L8 layers for (112) surface is considered in this section. At finite temperature, the effect of solute atoms on the surface energy is $\gamma_{surf-sol}(T) = \gamma_{surf}(T) + \Sigma_n c_n E_{\text{int}-n}(T)/A'(T)$ [14]. Here, A'(T) is the area of the unit cell relative to a certain crystal plane at finite temperature T, and c_n is the solute concentration at the n-th layer ($n =$ L1~L8) of the surface supercell. The interaction energy between the solute atom in the n-th layer and the corresponding surface supercell is $E_{int-n}(T) = [E_{surf-sol-n}(T) - E_{perf-sol-Al}(T)] - [E_{surf}(T) - E_{perf}(T)]$. $E_{perf-sol-Al}(T)$ represents the total energy of a perfect supercell with a solute atom doped along certain crystal plane. Similarly, $E_{surf-sol-n}(T)$ represents the energy of a surface supercell with a solute atom located in the corresponding n-th layer, which is determined by first-principles calculations [14].

7.2.2.2 Result

The anisotropy of materials results in different crystal faces having distinct surface features, the surface energies of different crystal planes at 0K are obtained through first-principles calculations, and the results are compared with other calculation and experimental data (Table 7.1). By comparing the surface energies of different planes, it is found that the surface energy of (111) plane is the minimum value, which suggests that the surface characteristic of (111) plane is weaker than other planes, and it is easier to form a new surface in (111) plane. This result contributes to understand the mechanism of microcrack nucleation along the (111) plane in Al metal.

Surface energy at finite temperature is difficult to measure by experimental method, which hinders the study of high-temperature surface properties of Al alloys. First-principles calculations provide an effective tool for investigating

TABLE 7.1
Surface Energy (J/m^2) of Pure Al Is Compared with Theoretical Calculations and Experiments at 0K [16]

Surfaces	Present	Theoretical Calculations	Experiments
(100)	0.915	0.869 [17], 0.900 [18], 0.920 [19], 0.940 [20], 1.081 [21], 1.347 [22], 1.220 [23], 1.680 [24], 0.832 [25], 0.860 [26], 0.863 [27], 0.890 [28], 1.209 [29], 0.910 [30]	
(110)	1.031	1.006 [17], 0.972 [18], 1.020 [19], 1.000 [20], 1.090 [31], 1.271 [22], 1.300 [23], 1.840 [24], 0.954 [25], 0.930 [26], 0.942 [27], 0.960 [28], 1.286 [29], 1.060 [30]	
(111)	0.816	0.831 [17], 0.620 [18], 0.890 [19], 0.820 [20], 1.270 [33], 1.199 [22], 0.830 [34], 1.100 [23], 1.450 [24], 0.880 [35], 0.780 [25], 0.670 [26], 0.856 [27], 0.780 [28], 0.856 [29], 0.830 [30]	1.140 [32]
(112)	0.981	0.853 [18], 1.165 [29]	

their high-temperature surface properties. The investigation of finite temperature impacts on the surface energies for various crystal planes indicates that for (100), (110), (111), and (112) surfaces, surface characteristics of Al metal are minimally impacted by finite temperatures (<500K) (Figure 7.7a). These planes exhibit different sensitivities to temperature resulting in varying surface energy. Notably, compared to other planes, (111) plane exhibits the smallest impact from temperature due to its weaker atomic binding energy perpendicular to the plane (Figure 7.7a).

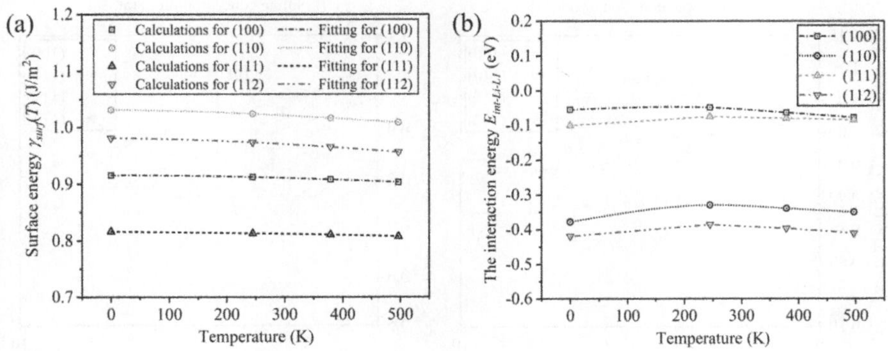

FIGURE 7.7 (a) Temperature-dependent surface energy $\gamma_{surf}(T)$ versus different crystal planes. (b) Temperature-dependent interaction energy $E_{int-Li-L1}(T)$ for Li atom added in the $L1$ layer of different surfaces [16].

$$\begin{cases} \gamma_{(100)}(T) = 0.91538 - 5.69861 \times 10^{-6}T - 8.35288 \times 10^{-8}T^2 \\ \quad + 4.72275 \times 10^{-11}T^3 \\ \gamma_{(110)}(T) = 1.03121 - 4.13749 \times 10^{-6}T - 1.14288 \times 10^{-7}T^2 \\ \quad + 6.42379 \times 10^{-11}T^3 \\ \gamma_{(111)}(T) = 0.81621 - 1.02953 \times 10^{-5}T - 5.48259 \times 10^{-10}T^2 \\ \quad - 2.57465 \times 10^{-11}T^3 \\ \gamma_{(112)}(T) = 0.98099 - 6.14421 \times 10^{-6}T - 1.05091 \times 10^{-7}T^2 \\ \quad + 3.58659 \times 10^{-11}T^3 \end{cases} \tag{7.7}$$

The expansion volume-temperature relationships are utilized to calculate the interaction energies at various temperatures according to previous work [36,37] (Figure 7.7b). At finite temperatures, the negative interaction energies cause the segregation of Li atoms in the surface of each crystal plane (Figure 7.7b). Specifically, the (110) and (112) planes exhibit higher interaction energies than the (100) and (111) planes. However, in general, finite temperatures (<500K) have a negligible effect on the interaction energies. Figure 7.8 shows the

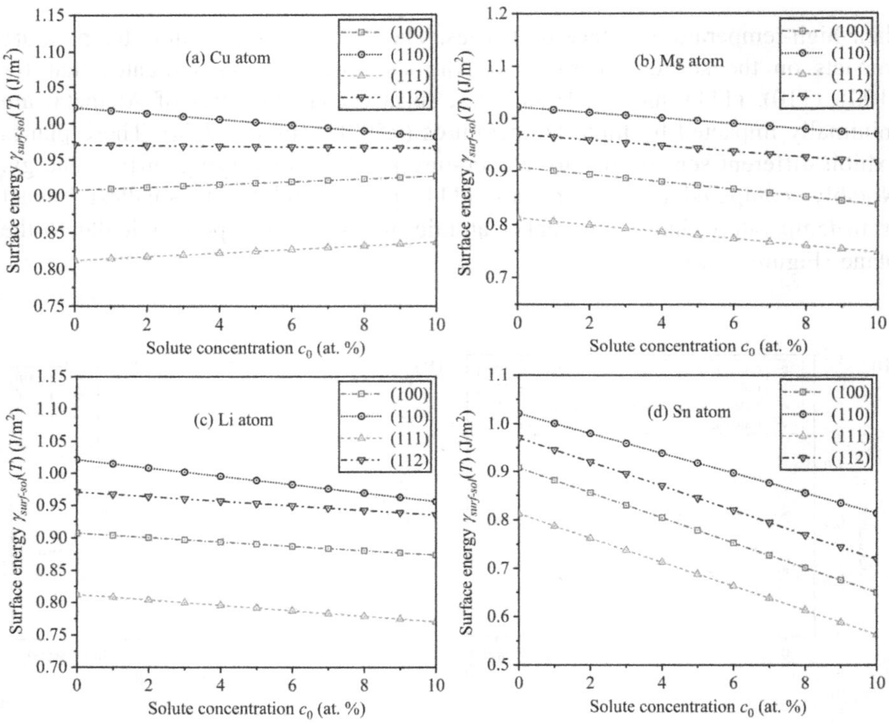

FIGURE 7.8 Surface energy $\gamma_{surf-sol}(T)$ versus solute concentration c_0 at 300K for Cu, Mg, Li, and Sn atoms [16].

TABLE 7.2
Surface Energy (J/m²) Is Compared with Other Results in Al_xMg (x = 0~10%) Alloys at 0K [16,30]

x (at. %)	(100)		(110)		(111)		(112)
	Present	Ref. [30]	Present	Ref. [30]	Present	Ref. [30]	Present
0	0.915	0.913	1.031	1.061	0.816	0.826	0.981
2	0.901	0.915	1.021	1.056	0.803	0.819	0.970
4	0.887	0.908	1.011	1.037	0.790	0.809	0.959
6	0.873	0.902	1.000	1.022	0.777	0.787	0.948
8	0.859	0.893	0.990	1.022	0.765	0.767	0.937
10	0.845	0.870	0.979	0.986	0.752	0.745	0.926

investigation of Cu, Mg, Li, and Sn atoms impacts on the surface energy at 300K. Among the (100), (110), (111) and (112) planes, the (111) plane exhibits the lowest surface energy at finite solute concentration (c_0 < 10%). Figures 7.8a-d indicate that the surface energies of all four surfaces decrease upon the addition of Mg, Li, and Sn atoms. However, Sn atom significantly reduces the surface energy of the four surfaces, owing to stronger interaction between Sn solute atoms and the various surfaces compared to Mg and Li solute atoms, as well as the lower surface energy, which makes it easier to form a new surface. A minor rise in the surface energies of the (100) and (111) planes occur with an increase in Cu solute concentration, while the ones of (110) and (112) planes decrease. These results indicate that the probability of (111) microcrack nucleation is higher when doped with solute atoms compared to other surfaces. Furthermore, the promotion of microcrack nucleation varies among different solute atoms, with Cu atom inhibiting it and Sn atom having the most substantial promoting effect. Table 7.2 displays the computed findings of Al-Mg alloys at 0K, which are in good accord with previous investigations [30].

To investigate the impact of finite temperatures and solute concentrations on surface energies, the 2D contour diagrams demonstrate the relationships between surface energies and solute concentration/finite temperature for four types of solute atoms: Cu, Mg, Li, and Sn (Figure 7.9). The results indicate that at high solute concentrations, Sn atoms cause a substantial reduction in surface energies for all surfaces, and as temperature increases, doping solute atoms reduces surface energies. High temperatures and Sn concentration decrease the atomic-binding energy in Al alloys, potentially resulting in the formation of a new surface, like a microcrack. At the same solute concentration and temperature, when solute atoms are doped, the (111) plane consistently exhibits the lowest surface energy among the (100), (110), and (112) planes. Additionally, the atomic-binding energy in the perpendicular direction of (111) plane is lower than that of the (100), (110), and (112) planes, which makes the (111) surface the preferred site for microcrack nucleation.

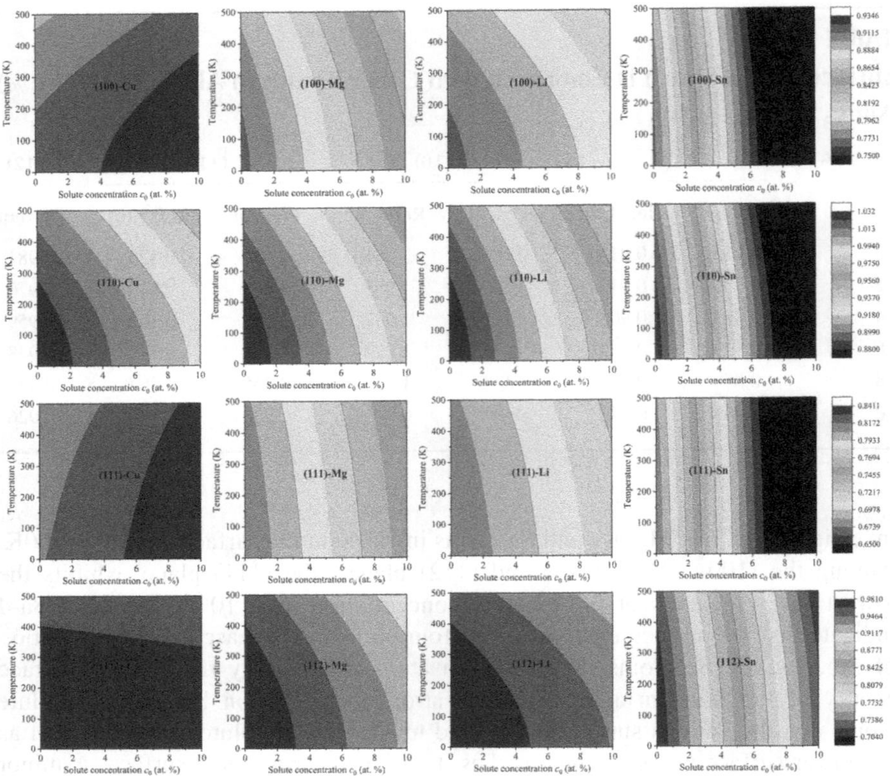

FIGURE 7.9 2D contour plots of surface energy versus finite temperature and solute concentration for four types of solute atoms (Cu, Mg, Li, and Sn) in different surfaces [16].

7.2.3 9R Phase Stabilization

The stability of the 9R phase remains a critical concern, as it contributes significantly to the exceptional mechanical properties of Al alloys. The effect of solute atoms on the stability of the 9R phase in Al alloys is thoroughly investigated in the following part.

7.2.3.1 Model

All energy terms obtained through first-principles calculations indicate the ground-state properties of the crystal ($T = 0$ K) in this study. At finite temperature ($T <$ 900 K), for Al alloys with a 9R phase structure, the temperature-dependent energy difference for each area caused by the 9R phase as relative to the perfect Al supercell: $\gamma_{Al-9R}(T) = [E_{Al-9R}(T) - E_{Al-perfect}(T)]/A(T)$. Here, $E_{Al-9R}(T)$ is the total energy of the 9R phase supercell, $E_{\mathrm{Al-perfect}}(T)$ denotes one of the perfect supercells in Al alloy, and $A(T)$ is the area of the (111) supercell at temperature T. The temperature-dependent intrinsic stacking fault energy (ISFE) $\gamma(T)$ in the 9R phase structure doped with solute atoms is redefined as: $\gamma(T) = \frac{1}{3}\gamma_{Al-9R}(T) + \sum_n c_{9R-n} E_{int-n}(T)/A'(T)$, where $A'(T)$ represents the area of the (111) unit cell at temperature T, c_{9R-n} is the solute concentration in the n-th layer (n = $L1$, $L2$) of the 9R phase structure. $E_{int-n}(T)$ denotes the interaction energy between the solute atom in the n-th layer and the 9R phase at temperature T, which is expressed as: $E_{int-n}(T) = [E_{s-9R-n}(T) - E_{s-Al}(T)] - [E_{Al-9R}(T) - E_{Al-perfect}(T)]$, where $E_{s-9R-n}(T)$ is the total energy of the 9R phase structure involving one solute atom in the n-th layer, and $E_{s-Al}(T) = (N-1)E_{Al}(T) + E_{solute}(T) + E_{enthalpy}(T)$ is one of the perfect Al supercells with one solute atom. Here, N is the number of Al atoms in the supercell, $E_{Al}(T)$ and $E_{solute}(T)$ represent the energy of a single Al atom and solute atom in the bulks, respectively, and $E_{enthalpy}(T)$ is the formation enthalpy of solute atoms in Al alloys at temperature T. The solute concentration c_{9R-n} in the FDD model is: $c_{9R-n} = \frac{1}{1+\exp\left(\frac{E_{int-n}(T)}{kT} - \ln\frac{c_0}{1-c_0}\right)}$, c_0 is the solute concentration when $E_{int-n}(T) = 0$, and k is the Boltzmann constant. For a given solute concentration and finite temperature, the increment of ISFE in the 9R phase structure, $\Delta\gamma(T)$, is calculated as: $\Delta\gamma(T) = \sum_n c_{9R-n}E_{int-n}(T)/A'(T)$.

7.2.3.2 Result

The formation enthalpy of solute atoms is used to measure their solubility in a solid solution. According to Figure 7.10a, Ga, Ge, Si, Sn, and Sr atoms have positive formation enthalpies, indicating their low solubility in Al alloys. Conversely, Cu, Fe, Li, Sc, Ti, and Y atoms have significantly negative formation enthalpies, suggesting their high solubility in Al alloys. The formation enthalpies of Mg and Zn atoms are nearly zero, implying their considerable solubility in Al alloys. Figure 7.10b shows the interaction energies between 9R phase and solute atoms, with the solute atoms located at the n-th ($n = L1,\ L2$) atomic layer in 9R phase. The interaction energies between these atoms (Cu, Fe, Li, Mg, Ti, Zn) and 9R phase are positive, implying repulsion between them. In contrast, the interaction energies between other atoms (Ga, Ge, Sc, Si, Sn, Sr, Y) with 9R phase are negative, indicating their inclination to aggregate in a 9R structure. Due to the lower interaction energy between the $L1$ stacking fault layer and the $L2$ stacking fault layer, they are likely to separate on the stacking fault plane of the 9R phase structure. Among them, Sr has the lowest interaction energy, and is most effective in promoting 9R phase stabilization. The solute atoms' capacity for promoting the stabilization of 9R structure is sorted as: Sr > Y > Sn > Ge > Si > Sc > Ga.

Since the effect of finite temperature (<900 K) on the stacking fault energy (SFE) of Al alloys is minimal, we compare the effects of solute concentration on the increment of ISFE in the 9R phase structure at T = 300K between the UFD and FDD models (Figure 7.10c-f). Based on the UFD model, Figure 7.10c shows the linear relationship between solute concentrations and ISFE increments. Ga, Ge, Sc, Si, Sn, Sr, and Y solute atoms are found to be effective in decreasing the increments of ISFE and improving the stabilization of 9R structure. The Sr and Y are the most effective solute atoms to promote the stability of the 9R phase, which is consistent with FDD model calculation results. Although the high-temperature diffusion of solute atoms inhibits the stability of the 9R phase, increasing the solute concentration (Ga, Ge, Ge, Sc, Si, Sn, Sr, and Y) at a certain temperature can promote the stability of the 9R phase (Figure 7.10d) [39–41].

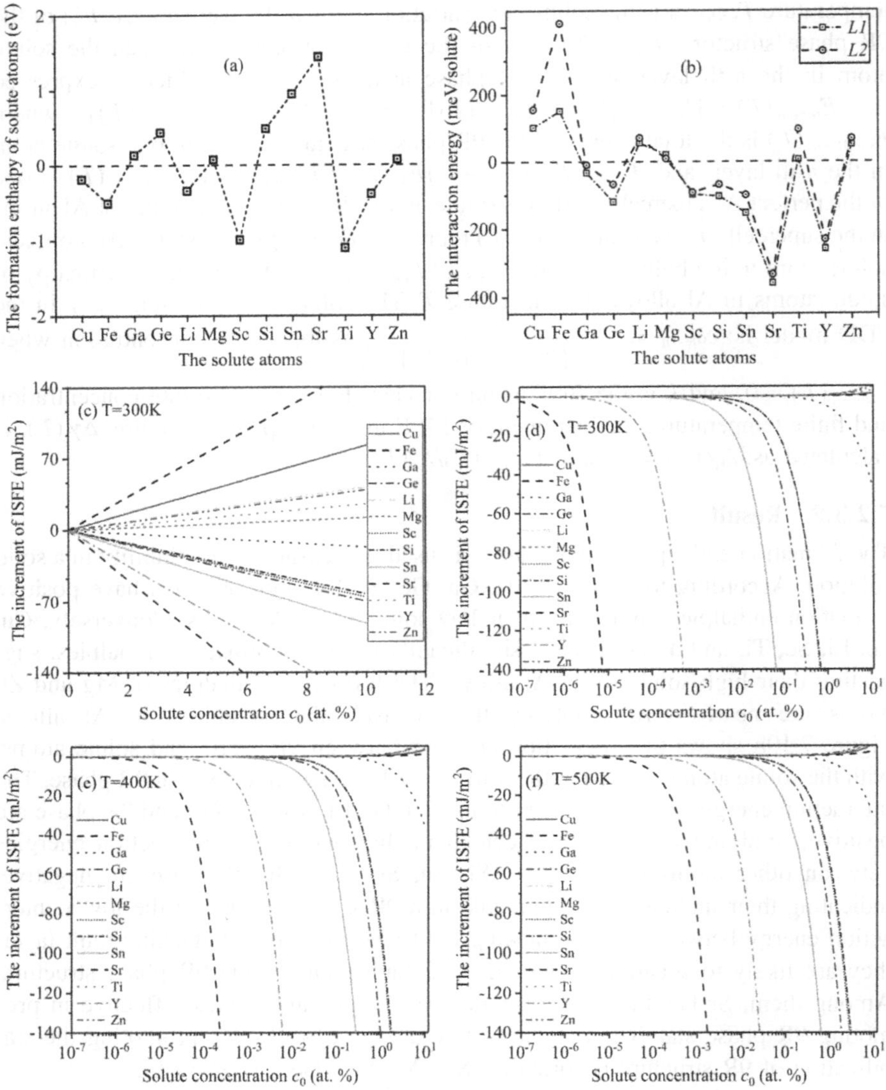

FIGURE 7.10 (a-b) Schematic diagrams of the formation enthalpy and interaction energy versus different solute atoms. (c-f) The relationship between solute concentration and the increments of ISFE at certain temperature in 9R phase structure [38].

Figure 7.11a-d depicts the relationship between ISFE increments and finite temperatures versus different solute concentrations based on the FDD model. The results indicate that, regardless of solute concentration, the increments of ISFE increase as the temperature rises, implying that the 9R structure stabilization is disturbed by high-temperature diffusion. However, Sr and Y atoms have a beneficial impact on ISFE increments, even at high temperatures, promoting the

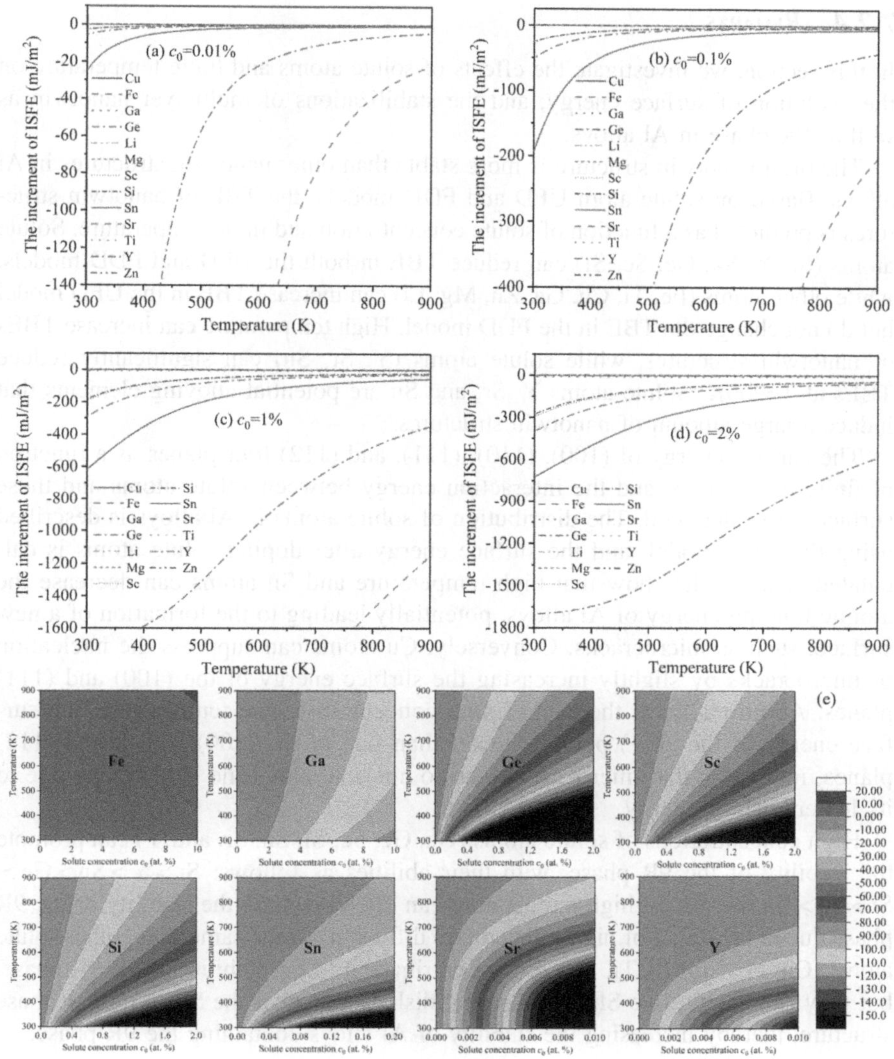

FIGURE 7.11 (a)-(d) Relationship between ISFE increments and finite temperatures at certain solute concentration in 9R phase structure. (e) 2D contours diagrams of ISFE increments versus different solute concentrations and finite temperatures in 9R phase structure [38].

stabilization of 9R structure and improves the mechanical properties. Figure 7.11e illustrates the impact of temperature and solute concentrations on ISFE increments through 2D contours of these solute atoms (Fe, Ga, Ge, Sc, Si, Sn, Sr, Y). At finite temperatures, Fe consistently raises the ISFE increments of 9R phase, while other solute atoms decrease them to varying degrees. It is worth noting that the increments of ISFE are significantly reduced by Sr and Y atoms both low solute concentration and high temperature.

7.2.4 Remarks

In this section, we investigate the effects of solute atoms and finite temperature on the evolution of surface energy, and the stabilizations of multilayer nanotwin as well as 9R phase in Al alloys.

The 6Lnt nanotwin structure is more stable than other nanotwin structures in Al alloys. Based on solute atom UFD and FDD models, the TBE of nanotwin structures is predicted as a function of solute concentration and finite temperature. Solute atoms (Sr, Y, Sn, Ge, Sc, Si) can reduce TBE in both the UFD and FDD models, while other atoms (Fe, Ti, Cu, Ga, Zn, Mg, Li) can increase TBE in the UFD model but do not change the TBE in the FDD model. High temperature can increase TBEs of nanotwin structures, while solute atoms (Y, Sr, Sn) can significantly reduce TBEs at T=900K. Solute atoms Y, Sr, and Sn are potential alloying elements that induce a large amount of nanotwin structures.

The surface energy of (100), (110), (111), and (112) four planes as a function of finite temperature and the interaction energy between solute atoms and these surfaces are calculated. The distribution of solute atoms in Al alloys is described using the UFD model, and the surface energy after doping solute atoms is calculated. The results show that high temperature and Sn atoms can decrease the atomic binding energy of Al alloys, potentially leading to the formation of a new surface, such as microcracks. Conversely, Cu atoms can suppress the nucleation of microcracks by slightly increasing the surface energy of the (100) and (111) planes. Additionally, at the same solute concentration and temperature, the surface energy of the (111) plane is lower than that of the (100), (110), and (112) planes, indicating that microcracks tend to nucleate along the (111) plane due to its lower surface energy.

High concentrations of solute atoms (Ga, Ge, Sc, Si, Sn, Sr, and Y) can promote the stability of the 9R phase, with their abilities as follows: Sr > Y > Sn > Ge > Si > Sc > Ga. However, high temperature can often decrease the stability of the 9R phase due to the effect of high-temperature diffusion. At the same time, other solute atoms (Cu, Fe, Li, Mg, Ti, and Zn) are detrimental to the stability of the 9R phase, but they can anchor two Shockley partial dislocations near the SF in the 9R phase structure, thereby increasing the stability of Al alloys containing the 9R phase.

7.3 MG ALLOYS

7.3.1 Transverse Propagation of Deformation Twinning

The thickening mechanism of deformation twins (DTs) has been extensively studied in many research studies, and their lateral propagation has begun to attract attention. Recently, some researchers have reported that the twin front of the $\{10\bar{1}2\}$ mode in Mg is composed of a conjugate twin plane and a prism/basal (PB) interface, and the combination of these planes' migration rates determines the overall kinetics of twin propagation. Based on this, a phase field model has been proposed to study the equilibrium shape of stretched twins and the dynamics of twin fronts in this section. To describe the orientation-dependent properties of TBs, a new form of surface-free energy is introduced in the model. The simulation reproduces the

PB interface well, and the results show that anisotropic surface energy dominates the formation of irregular small facets on the twin plane. To study the equilibrium and migration rate of TBs, the generalized energy-momentum tensor of shear loads is derived and analyzed.

7.3.1.1 Crystallography

The dominant deformation mechanism of hexagonal close-packed (HCP) metals at low temperatures and strain rate is $\{10\bar{1}2\}\langle\bar{1}011\rangle$ type twinning. In classic crystallographic theory about twinning, a macroscopic twinning is characterized by the parameters $\{\mathbf{K}_1, \mathbf{K}_2, \eta_1, \eta_2\}$ referred to as the "twinning element". Figure 7.12 illustrates the crystal geometric relationships related to the twinning element. In Figure 7.12b, the orthogonal axes e_1^c, e_2^c and e_3^c are aligned with the standard orthogonal directions of the HCP crystal. The homogeneous simple shear is expressed by an affine deformation [42,43]:

$$\mathbf{v} = \mathbf{S}\cdot\mathbf{u}; \quad \text{and } \mathbf{S} = 1 + \gamma_0 \hat{\mathbf{m}} \otimes \hat{\mathbf{n}} \tag{7.8}$$

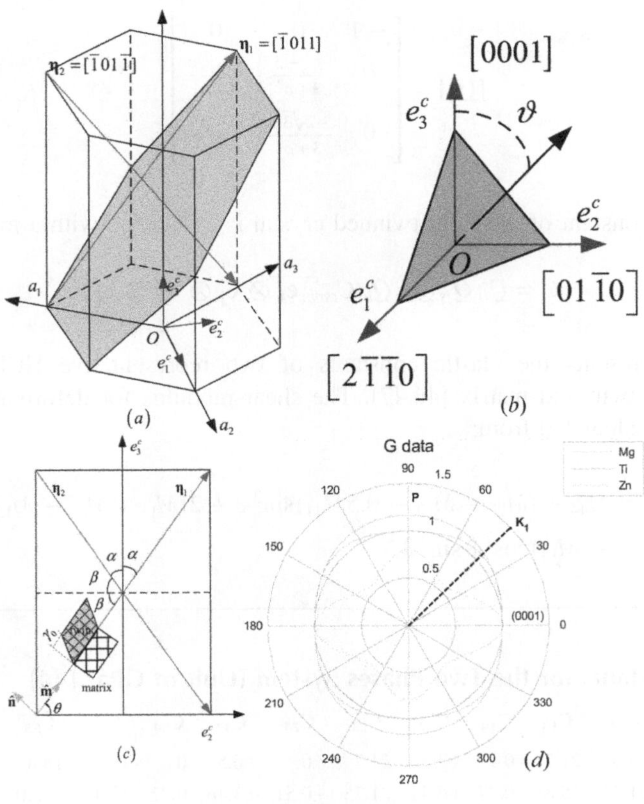

FIGURE 7.12 (a) $\{10\bar{1}2\}\langle\bar{1}011\rangle$ twinning model. (b) Relationship among the e_1^c, e_2^c and e_3^c axes and crystal direction. (c) The schematic diagram of the shear plane. (d)The relationship between the three HCP metals (Mg, Ti, Zn) and their shear modulus G [44].

where **S** is a second-rank tensor, and **u** and **v** represent the corresponding crystal lattice vectors of the twin and parent, respectively. Both $\hat{\mathbf{m}}$ and $\hat{\mathbf{n}}$ lie on the **P** plane, and $\hat{\mathbf{m}}$ represents the direction parallel to the shear direction, and $\hat{\mathbf{n}}$ denotes the unit normal vector of the twin interface $\mathbf{K}_1$ (Figure 7.12). The twin shear is written as $\gamma_0 = (3 - \rho^2)/(\sqrt{3}\rho)$, ρ is the lattice parameter ratio (c/a).

A re-orientation tensor is used to describe the crystallographic elements of $\{10\bar{1}2\}$ [42,45]:

$$\mathbf{Q} = 2\hat{\mathbf{m}} \otimes \hat{\mathbf{n}} - \mathbf{1} \tag{7.9}$$

where "**1**" denotes the second unit tensor. Here, $1 = \delta_{ij}\mathbf{e_i} \otimes \mathbf{e_j}$. In the e_i^c coordinates, the $\hat{\mathbf{m}}$ and $\hat{\mathbf{n}}$ is expressed as:

$$\hat{\mathbf{m}} = (0, \sqrt{3}a, c)/\sqrt{3a^2 + c^2}; \quad \hat{\mathbf{n}} = (0, -c, \sqrt{3}a)/\sqrt{3a^2 + c^2} \tag{7.10}$$

The matrix form of the tensor for re-orientation is:

$$[Q_{ij}] = \begin{bmatrix} -1 & 0 & 0 \\ 0 & \frac{\rho^2 - 3}{3 + \rho^2} & \frac{-2\sqrt{3}\rho}{3 + \rho^2} \\ 0 & \frac{-2\sqrt{3}\rho}{3 + \rho^2} & \frac{3 - \rho^2}{3 + \rho^2} \end{bmatrix} \tag{7.11}$$

The elastic constant of the fully twinned crystal is associated with a matrix:

$$C_* = Q_{ir} Q_{js} Q_{km} Q_{ln} C_{rsmn} \mathbf{e_i} \otimes \mathbf{e_j} \otimes \mathbf{e_k} \otimes \mathbf{e_l} \tag{7.12}$$

Table 7.3 presents the elastic constants of two representative HCP materials between the twin and matrix [46,47]. The shear modulus for deformation in any direction is calculated from:

$$G = M_{44} + (M_{11} - M_{12} - 0.5M_{44})\sin^2\vartheta + 2(M_{11} + M_{33} - M_{13} - M_{14})\cos^2\vartheta\sin^2\vartheta \tag{7.13}$$

TABLE 7.3
Elastic Constants for the Two Phases System (Unit of GPa) [44]

	C_{11}	C_{12}	C_{13}	C_{14}	C_{22}	C_{23}	C_{24}	C_{33}	C_{34}	C_{44}	C_{55}	C_{56}	C_{66}
Parent	63.5	25.9	21.7	0	63.5	21.7	0	66.5	0	18.4	18.4	0	18.8
Twin	63.5	21.72	25.88	0.27	66.43	21.75	–0.51	63.46	0.32	18.45	18.80	0.026	18.40
ΔC	0	–4.18	4.18	0.27	2.93	0.054	–0.51	–3.04	0.32	0.054	0.40	0.026	–0.40
C^{eff}	63.5	23.81	23.79	0.14	64.97	21.73	–0.26	64.98	0.16	18.43	18.60	0.013	18.60

where M_{ij} is the compliance constant of the selected metals. To compare the anisotropy of elasticity, the shear modulus of each plane is normalized (Figure 7.12d), where $\Delta\mathbf{C} = \mathbf{C}^{twin} - \mathbf{C}^{parent}$ is the separation of the elastic in the phases. $\mathbf{C}^{eff} = (\mathbf{C}^{twin} + \mathbf{C}^{parent})/2$ represent the effective modulus. The ratio of c/a for Mg is 1.6235.

7.3.1.2 Phase Field Model

7.3.1.2.1 Elastic Strain Energy

$\phi(\chi, t)$ is an order parameter to distinguish the different phases: twinning phase for $\phi = 1$ and parent phase for $\phi = 0$, and twin boundary (TB) is the interfaces between the two phases. The plastic shear is calculated according to the classic theory of twinning nucleation and propagation [48]:

$$\varepsilon^{p} = \frac{1}{2}\gamma_0 \varphi(\phi)(\hat{\mathbf{m}} \otimes \hat{\mathbf{n}} + \hat{\mathbf{n}} \otimes \hat{\mathbf{m}}) \tag{7.14}$$

where $\varphi(\phi)$ is a polynomial function obeying $\varphi(0) = 0$, $\varphi(1) = 1$ and $\varphi'(0) = \varphi'(1) = 0$. The polynomial function with monotonically increasing type is written as: $\varphi(\phi) = \phi^3(10 - 15\phi + 6\phi^2)$ [49,50].

To accurately assess the non-uniform impacts of twinning, the re-orientation of elastic constants must be taken into account. The elastic stiffness coefficients at different phases are given by the following equation:

$$\mathbf{C}(\phi) = \mathbf{C}^{eff} + [\mathbf{C}(1) - \mathbf{C}(0)]\psi(\phi) \tag{7.15}$$

where $\mathbf{C}^{eff} = (\mathbf{C}(0) + \mathbf{C}(1))/2$ represents the effective modulus for a cell, and $\mathbf{C}(0) = \mathbf{C}^{matix}$, $\mathbf{C}(1) = \mathbf{C}^{twin}$, $\psi(\phi) = \varphi(\phi) - 0.5$ [49,50].

Therefore, the elastic stress of both the matrix and the twin phase is deduced according to Hooke's law:

$$\boldsymbol{\sigma} = \mathbf{C}(\phi) : \varepsilon^{el} = \mathbf{C}(\phi) : \frac{1}{2}[\mathbf{u} \otimes \nabla + \nabla \otimes \mathbf{u} - \gamma_0 \varphi(\hat{\mathbf{m}} \otimes \hat{\mathbf{n}} + \hat{\mathbf{n}} \otimes \hat{\mathbf{m}})] \tag{7.16}$$

The stress equilibrium equation under periodic constraints is solved to obtain the elastic strain energy:

$$\nabla \cdot \boldsymbol{\sigma} = 0 \text{ in } \Omega \tag{7.17}$$

Since the stress tensor $\boldsymbol{\sigma}$ and the elastic modulus $\mathbf{C}(\phi)$ are periodic on Ω in a periodic microstructure system, the displacement field $\mathbf{u}(\chi)$ is expressed as:

$$\mathbf{u} = \bar{\mathbf{E}} \cdot \chi + \mathbf{u}^{*} \tag{7.18}$$

where $\mathbf{u}^{*}$ represents the periodic displacement field. The $\bar{\mathbf{E}}$ is the homogenous strain tensor associated with the mean strain of the cell Ω. The mean strain tensor in a represent volume element (RVE) is:

$$\bar{\mathbf{E}} = \langle \{\boldsymbol{\varepsilon}\} \rangle = \frac{1}{V} \int_{\Omega} \{\boldsymbol{\varepsilon}\} d\Omega \quad (7.19)$$

where $\langle \{\cdot\} \rangle$ means the mean quantity operation, V denotes the volume of the RVE. According to Eq. 7.18 the periodic strain field is derived as:

$$\boldsymbol{\varepsilon} = \bar{\mathbf{E}} + \boldsymbol{\varepsilon}^* \quad (7.20)$$

where $\boldsymbol{\varepsilon}^* = 0.5(\nabla \otimes \mathbf{u}^* + \mathbf{u}^* \otimes \nabla)$ denotes the periodic strain tensor, which satisfies the constraint of $\langle \{\boldsymbol{\varepsilon}^*\} \rangle = 0$ according to the homogenization theory [51].

$$W^{el} = \frac{1}{2}\varepsilon^{el}\colon \mathbf{C}(\phi)\colon \varepsilon^{el} = \frac{1}{2}(\bar{\mathbf{E}} + \varepsilon^* - \varepsilon^p)\colon \mathbf{C}(\phi)\colon (\bar{\mathbf{E}} + \varepsilon^* - \varepsilon^p) \quad (7.21)$$

Due to $\partial\varphi/\partial\phi = \partial\psi/\partial\phi$, the elastic chemical potential is written as:

$$\mu^{el} = \frac{\delta W^{el}}{\delta\phi} = (W^{inh} - \tau_{mn}\gamma_0)\frac{\partial\varphi}{\partial\phi} \quad (7.22)$$

where $\tau_{mn} = \sigma\colon (\hat{\mathbf{m}} \otimes \hat{\mathbf{n}})$ denotes the resolved shear stress (RSS) applied to the corresponding plane, and $W^{inh} = \boldsymbol{\varepsilon}^{el}\colon \Delta\mathbf{C}\colon \boldsymbol{\varepsilon}^{el}/2$ is the free energy arising from non-uniform elasticity.

7.3.1.2.2 Orientation-Dependent Interface Free Energy

The surface energy is dependent on the angle θ, which represents the angle between the $[10\bar{1}1]$ direction and its interface [52]. For the growth of $\{10\bar{1}2\}$ twin crystals, the prismatic/basal (PB) interfaces exhibit low energy and cause a cusp in the orientation-dependent interface free energy (ODIE). Thus, utilizing a smooth four-peak function, that is related to angle θ describing the ODIE, appears to be appropriate and precise. In Figure 7.13a, the four peak function's valleys appear when the twin interface is parallel to the $\mathbf{K_1}$ and $\mathbf{K_2}$ planes and the PB interfaces, and PB interfaces show a higher energy than the $\mathbf{K_1}$ plane and $\mathbf{K_2}$ plane.

Here, $\Gamma(\theta)$ represents a periodic function:

$$\Gamma(\theta) = \Gamma(\theta + \pi) \quad (7.23)$$

A Fourier function is constructed to characterize this surface anisotropy.

$$\Gamma(\theta) = a_0 + \sum_{i=1}^{n} (a_i \cos(iw\theta) + b_i \sin(iw\theta)) \quad (7.24)$$

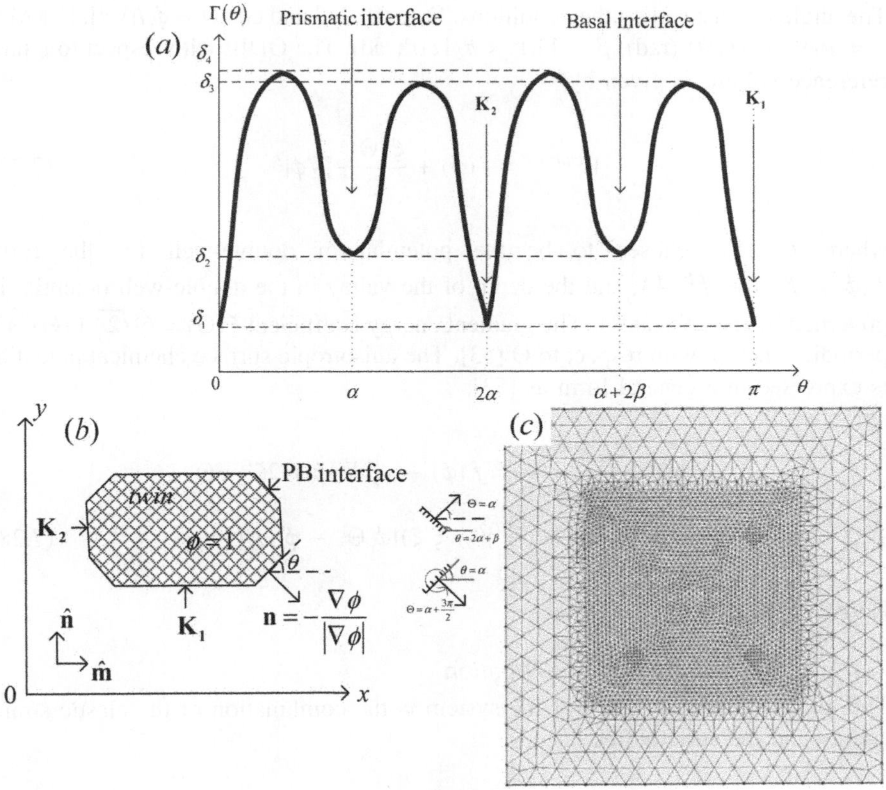

FIGURE 7.13 The dependence of the ODIE on the angle between the twin direction and the TB (a). The relationship between Θ and θ in geometry (b). Partition of the finite cell grid (c) [44].

Θ is the angle between the twinning direction $\hat{\mathbf{m}}$ and the normal vector $\mathbf{n} = -\nabla\phi/|\nabla\phi|$ of the TB. Figure 7.13b shows the geometric relationship between Θ and θ. Based on this, and incorporating the periodic condition from Eq. (7.23), a surface energy function is obtained with respect to Θ:

$$\Gamma(\Theta) = \Gamma\left(\theta - \frac{\pi}{2}\right) \tag{7.25}$$

A rectangular coordinate system is established in Figure 7.13b, in which the twinning direction $\hat{\mathbf{m}} = [0 \quad 1]^{T}$ is aligned with the X axis and $\hat{\mathbf{n}} = [0 \quad 1]^{T}$ is aligned with the Y axis. In this coordinate, the RSS is expressed as $\tau_{mn} = \boldsymbol{\sigma}\colon (\hat{\mathbf{m}} \otimes \hat{\mathbf{n}})$, and the angle Θ is formulated as:

$$\Theta = \arctan\left(\frac{\phi_y}{\phi_x}\right) \tag{7.26}$$

The angle Θ must satisfy the conditions: $\Theta \in [0, 2\pi]$ and $\cos \Theta = \phi_x/|\nabla\phi|$. For Mg, $\alpha = 46.9° \times \pi/180°$(rad), $\beta = 43.1° \times \pi/180°$(rad). The ODIE with respect to a unit reference volume is given by:

$$W^{surface} = f(\phi) + \frac{\xi(\Theta)^2}{2}|\nabla\phi|^2 \tag{7.27}$$

where $f(\phi)$ is chosen to be the potential of double-well for the form $f(\phi) = \phi^2(1 - \phi^2)/4A$, and the depth of the valley in the double-well potential is governed by the constant A. The gradient energy coefficient $\xi(\Theta) = 6\sqrt{2A}\,\Gamma(\Theta)$ is a periodic function with respect to Θ [53]. The anisotropic surface chemical potential is expressed in a general form as [53]:

$$\mu^{ch} = \frac{\delta W^{surface}}{\delta\phi} = f(\phi) - \{\xi^2\nabla^2\phi + 2\xi'\xi\nabla\Theta \cdot\nabla\phi + ([\xi']^2 + \xi'\xi)[\phi_x\Theta_y - \phi_y\Theta_x]\} \tag{7.28}$$

Here $\xi' = d\xi/d\Theta$, $f'(\phi) = \partial f/\partial\phi$.

7.3.1.2.3 GinzburgeLandau Equation

The total Gibbs free energy of the system is the combination of the elastic strain energy and the surface free energy:

$$G = W^{surface} + W^{el} \tag{7.29}$$

The order parameter that characterizes the deformation twinning (DT) is non-conserved, and the rate of evolution is assumed to vary linearly. Therefore, the equilibrium of the phase field is defined by [51,54]:

$$\frac{\partial\phi}{\partial t} = -L\frac{\delta G}{\delta\phi} = -L\left(\frac{\delta G}{\delta\phi} - \mathrm{div}\frac{\delta G}{\delta\nabla\phi}\right) = -L(\mu^{ch} + \mu^{el}) \tag{7.30}$$

$$\frac{1}{L}\dot{\phi} = \nabla\cdot\begin{pmatrix}\xi^2\phi_x - \xi\xi'\phi_y \\ \xi^2\phi_y - \xi\xi'\phi_x\end{pmatrix} - f'(\phi) + (\tau_{xy}\gamma_0 - W^{inh})\varphi(\phi) \tag{7.31}$$

The boundary condition is: $\phi^+ = \phi^-$ (the symbols + and − indicate the opposite boundary). The elastic chemical potential is obtained by solving the equation for the equilibrium of the periodic stress. Therefore, $\mathbf{t}^+ = -\mathbf{t}^-$ and $\mathbf{u}^{*+} = \mathbf{u}^{*-}$, where $\mathbf{t}$ is the boundary traction.

Eqs. 7.17 and 7.31 are solved by the finite element method [55]. The mesh consists of 6166 linear triangular elements with the cell size of 50 nm × 50 nm, and the nodes of boundary on opposite sides are symmetric (Figure 7.13c).

The application of an increasing external strain to the cell results in the production of pure shear stress in the simulation cell:

$$\varepsilon_{ij}^{appl} = (\delta_{i1}\delta_{j2} + \delta_{i2}\delta_{j1})\dot{\varepsilon}_0 t + \varepsilon_{ij}^{cr} \quad (7.32)$$

where $\dot{\varepsilon}_0$ denotes the loading strain rate, t is the loading time. ε_{ij}^{cr} is the critical value of shear strain in the initial twin nucleus, which is obtained by using classical nucleation theory [48,56]. The $\{10\bar{1}2\}$ twin embryos reach the minimum equilibrium thickness of 17 atomic layers, which is approximately 3.32 nm [57,58]. Therefore, using Eshelby's eigenstrain method, the ε_{ij}^{cr} is estimated for a circular twin nucleus with an initial radius of 3 nm:

$$\varepsilon_{ij}^{cr} = \frac{\tau_{cr}^{xy}}{2G} = \frac{\gamma_0}{16(1-\nu)} + \frac{\Gamma}{2G\gamma_0 r} \quad (7.33)$$

where Γ denotes the energy of the $\{10\bar{1}2\}$ plane, and τ_{cr}^{xy} represents the critical resolved shear stress of twin nucleation. The value of simulation parameters are chosen as follows: For elastic properties of Mg, shear modulus $G = 19.1$ GPa, Lame constant $\lambda = 24$ GPa; For Fourier function, $a_0 = 192.9$ mJ/m^2, $a_1 = 192.9$ mJ/m^2, $a_2 = 192.9$ mJ/m^2, $b_1 = 192.9$ mJ/m^2, w = 4; For double-well potential, $A = \chi/(46\Gamma) = 0.178$ GPa^{-1}, and the equilibrium thickness of the interface is chosen as $\chi = 1$ nm [57]; kinetic coefficient $L = 50$ m^3/s/J; critical shear strain is $\varepsilon_{ij}^{cr} = 0.0191$, and the loading strain rate is $\dot{\varepsilon}_0 = 0.2$ns^{-1}.

7.3.1.3 Result

7.3.1.3.1 The Topology of the Twin Front

Figure 7.14a shows the square domain containing initial twin nuclei. In Figure 7.14b, the TB consists of three interfaces: the PB interfaces and a pair of conjugate twinning plane $\mathbf{K_2}$. The clear PB interfaces configuration existed on the TB indicates the accuracy and efficiency of the ODIE. The geometry of the twin nucleus has little impact on the formation and growth of PB interfaces. The shear strain of the twin is equal to the plastic shear strain under the small strain limit (Figure 7.14c). According to Figure 7.14d, the twin exhibits lower shear stress compared to the average stress of the cell, and the PB interfaces show higher shear stress compared to the $\mathbf{K_2}$ and $\mathbf{K_1}$ planes. With the pure shear loading applied, the prismatic planes are tension stressed, and the basal planes are compression stressed (Figure 7.14e-f). The two pair of stress are nearly same and the rest of twin surfaces ($\mathbf{K_2}$ and $\mathbf{K_1}$) do not exist significant normal stress. The growth of the twin plays a crucial role in dominating the deformation of the cell, and the stress within the twin region is significantly different from that of the parent phase [59,60].

7.3.1.3.2 The Stability of PB Interfaces and K_2 Plane

Micromechanics theory suggests that the stability for a given twin plane is determined by capillary forces (the numeric value is equal to the surface energy)

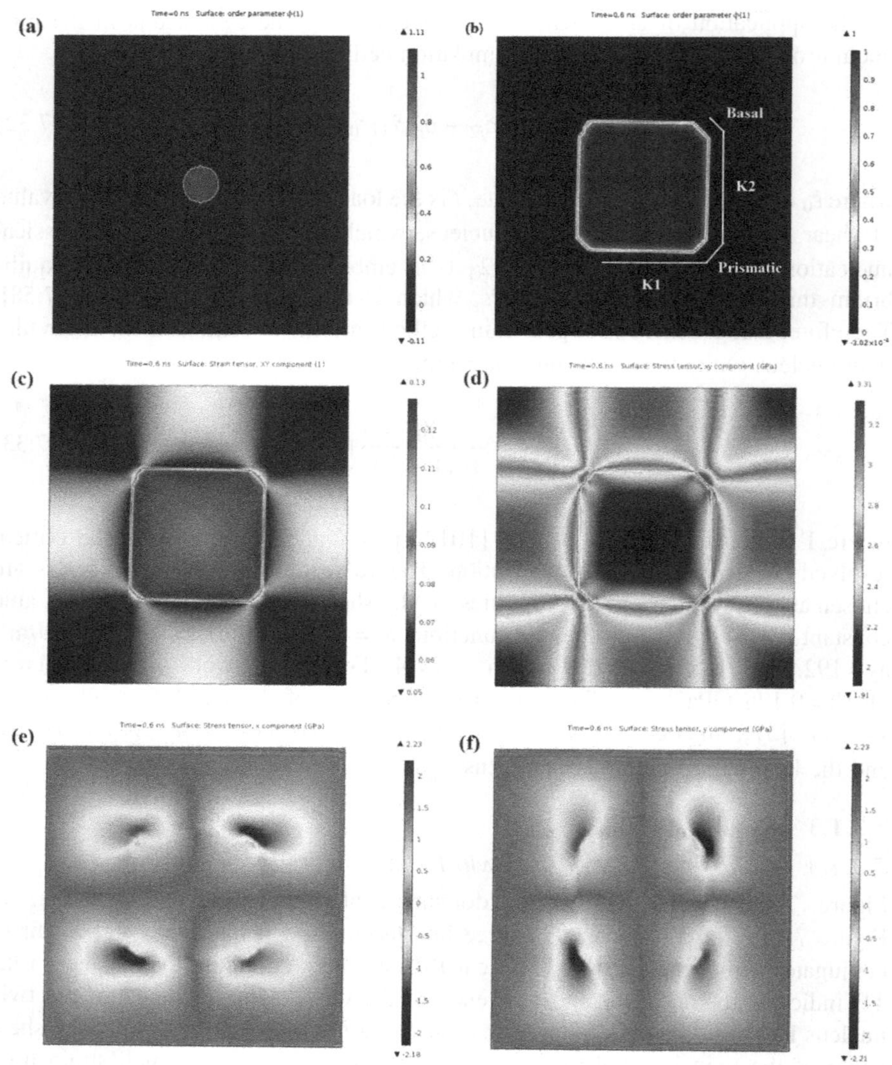

FIGURE 7.14 (a) The image of twin initial embryo (b) configuration at the front of the twinning; (c, d) the shear strain $\boldsymbol{\varepsilon}_{xy}$ and shear stress τ_{xy} upon twinning. (e, f) the normal stress σ_x and σ_y upon the twinning [44].

and the elastic force across TBs is discontinuity. To clarify the stability of PB interfaces, the balance of capillary forces at the point of intersection of TBs was achieved (Figure 7.15a-b) [61]. When the Γ_{PB} is smaller/larger than $\sqrt{2}\,\Gamma_{K1}$ under the condition of $\Gamma_{K1} = \Gamma_{K2}$, the PB interface exhibits stable/unstable (Figure 7.15a). However, when the Γ_{PB} exceeds a critical value, the PB interfaces transform to the $\mathbf{K_1}$ and the $\mathbf{K_2}$ planes, even if Γ_{PB} is still in the valley region of the Fourier function. However, simulation results indicate that the PB interfaces

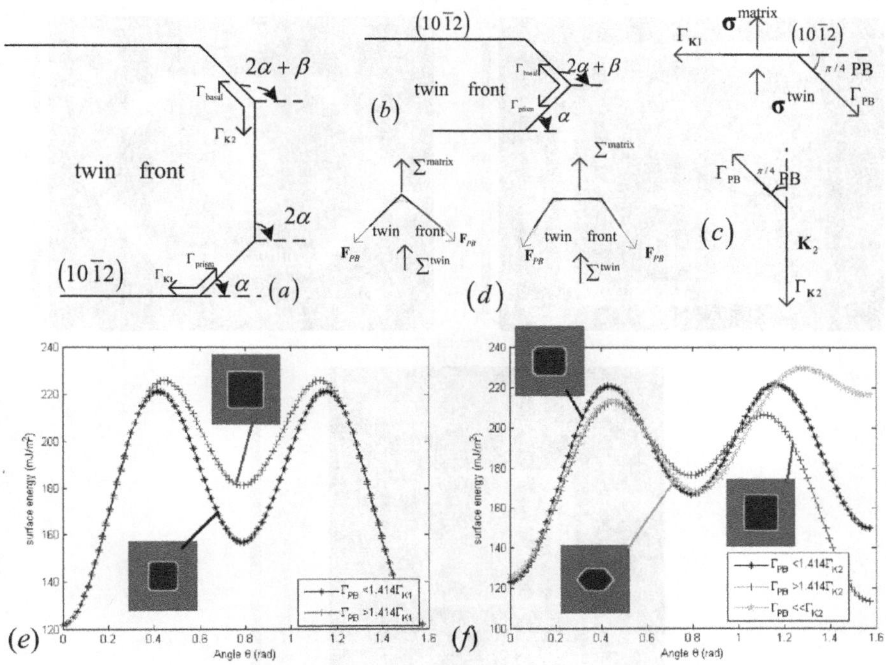

FIGURE 7.15 The geometry and capillary forces of twin kind of twin fronts (a) stable (b) unstable $\mathbf{K_2}$ plane (c) the capillary forces balance on the junction (d) the configurational forces balance on the twin front. The stability of (e) PB interfaces and (f) $\mathbf{K_2}$ plane [44].

can remain stable even when the value of Γ_{PB} is below $\sqrt{2}\,\Gamma_{K1}$ estimated from capillary force balance.

The simulations results examine the stability of the $\mathbf{K_2}$ plane with the condition of $\Gamma_{K1} \neq \Gamma_{K2}$ (Figure 7.15). The black line mark in Figure 7.15f indicates that the PB interfaces and the $\mathbf{K_2}$ plane can remain stable within the twin front when the value of Γ_{K2} exceeds the value of $\Gamma_{PB}/\sqrt{2}$, that also suggests that it is possible to achieve a larger aspect ratio using current phase field model especially when the Γ_{K1} is set to a very small value relative to Γ_{K2}. The red line marked in Figure 7.15f demonstrates that the $\mathbf{K_2}$ plane remains relatively stable even the PB interfaces transform to $\mathbf{K_2}$ plane, when the value of Γ_{K2} is smaller than both the $\Gamma_{PB}/\sqrt{2}$ and the Γ_{K1}. Otherwise, even if the Γ_{K2} greatly exceeds the Γ_{PB}, the $\mathbf{K_2}$ plane is unstable and will shrink to the PB interfaces shown as the blue line. This trend leads to the formation of a cuspidal twin front, which is consistent with the previous molecular dynamic (MD) simulations.

The direction of configurational force, which are parallel to each other but opposed to the propagation direction of the TBs, is shown in Figure 7.16. The configurational force is regarded as a cohesive force that must be overcome during the twinning process [62]. The configurational forces acting on the PB interfaces are smaller than those on the $\mathbf{K_1}$ and $\mathbf{K_2}$ planes, suggesting that the resistance to

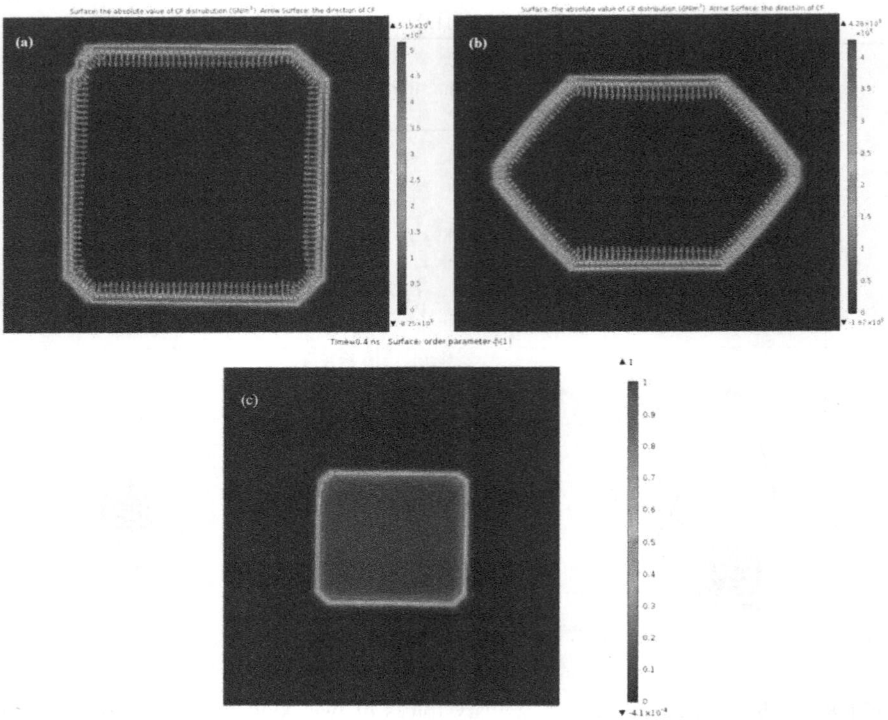

FIGURE 7.16 The distribution and direction of configurational force on a twin with (a) stable and (b) unstable $\mathbf{K_2}$ plane. (c) A twin computed by the path-dependent function with $\ell = 4$ [44].

twinning is lower on PB interfaces than on other twin planes. The twin expansion process usually occurs through the deposition of twinning dislocations on the PB interfaces, with further growth requiring the nucleation of new twinning dislocations on the habit plane. The conclusion drawn here is consistent with the classical theory of crystal growth and the observation of previous MD work for the transverse propagation of $\{10\bar{1}2\}$ twins [63].

7.3.1.3.3 The Transverse Propagation Mechanism of Deformation Twinning

It is widely recognized that twins typically exhibit a high aspect ratio shape, as their length is much greater than their thickness. In terms of free energy considerations, the primary factor contributing to the high aspect ratio of twins is not the minor difference in TBEs between interfaces oriented parallel and perpendicular to the direction of twinning. Twins tend to propagate in the transverse direction rather than thicken. It may be caused by the dislocation dissociation or atomic shuffling between the parent and twin region, or the anisotropic barrier energy induced by the different mechanisms of thickening and transverse propagation. In this section, it is assumed that when a twin starts to thicken, the energy barrier of new generated twinning dislocation at the habit plane is greater than that involved in driving an existing twinning dislocation to glide. As a result, the rate of transverse propagation is constrained by the lattice resistance that is associated with the movement of twinning dislocations on the twin front, while thickening rate is limited by the nucleation of new twinning dislocations on the twinning plane. To accommodate the assumption, we have transformed the mobility parameter L into a function that is dependent on the path:

$$L(\Theta) = \begin{cases} L_{th} & if\ |\sin \Theta| > \text{tolerance} \\ L_{tp} & else \end{cases} \tag{7.34}$$

where L_{th} is a constant related to the thickening speed, L_{tp} is related to the transverse propagation speed, the ratio $l = L_{th}/L_{tp}$ dominates the aspect ratio of a twin. The test of $l = 4$ is chosen. The computed results indicate the hypothesis is rational, and the current method is expected to be more accurate in reproducing the high aspect ratio twins compared to the previous phase field approach (Figure 7.16c).

7.3.2 Nucleation and Growth Mechanisms of Nanoscale Deformation Twins

Twinning is a common deformation mode in hcp metals, but there are very few theoretical models on twin nucleation and growth mechanisms compared to FCC metals. The nucleation and growth of TBs involve the sliding of partial dislocations on the twin plane. This section aims to establish a theoretical model for the nucleation and migration mechanism of nanoscale deformation twins in HCP Mg.

7.3.2.1 Model

7.3.2.1.1 Twinning Dislocations from Dislocation Dissociations

Twinning dislocations behave differently from slip dislocations in that they shear materials only in one direction on their twin plane and along the Burgers vector. The orientation, Burgers vector, and glide plane of twins depend on their type and c/a ratio [65]. When dealing with Mg with a c/a ratio of 1.623, twinning dislocations glide along the $[10\bar{1}1]$ direction on the $\{\bar{1}012\}$ plane but not along the $[10\bar{1}1]$ direction (Figure 7.17a). Thus, twin nucleation modeling focuses on $\{\bar{1}012\}[10\bar{1}1]$ twins in this section.

Many Shockley partial dislocations are generated by the dislocation separation of a slip dislocation [66]. The configuration of dislocations in dissociation is usually non-planar, consisting of twin dislocations and stair-rod dislocations between the original dislocation slip plane and twin plane (Figure 7.17c). Two rules must be followed for these dissociations [67]: (1) the position of the dislocation line has to coincident with the intersection of the twin plane and the slip plane; (2) The Burgers vector of the twin dislocation has to produce the necessary shear force for twinning. The first rule indicates that the formation of twin nuclei is inherently probabilistic. The second rule indicates that at least one of the reaction products is a twin dislocation located on its twin plane.

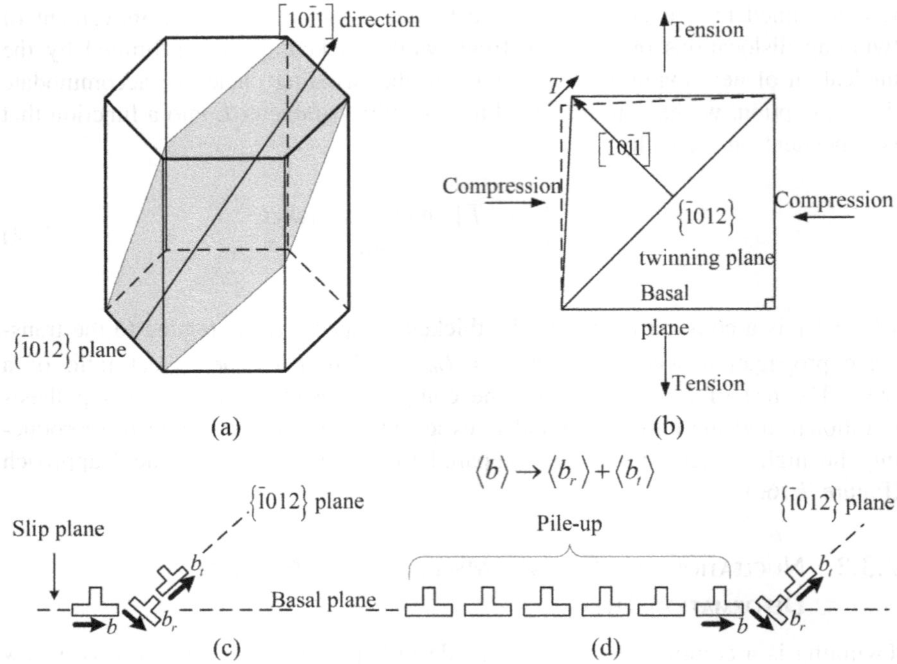

FIGURE 7.17 The twinning mode of Mg is illustrated: (a) the twinning plane and direction are described; (b) the shape change resulting from twinning is shown; (c) a single dislocation from (a) is displayed; (d) the process of dislocation pile-up and dissociation is revealed [64].

Many dissociation reactions meet Mendelson's two conditions, however, only dissociations from slip dislocations with Burgers vector of $\langle a\rangle = 1/3\langle\bar{2}110\rangle$ from the basal plane in this section, as the Mg deforms predominantly by basal sliding. The single dissociation mechanism of an ideal basal dislocation on plane $\{\bar{1}012\}$ (Figure 7.17c) is [67]:

$$\langle b_{ini}\rangle \rightarrow \langle b_r\rangle + \langle b_t\rangle \tag{7.35}$$

$$\langle a\rangle \rightarrow [(\langle a\rangle - 3s\langle d\rangle - s\langle c\rangle) + s(3\langle d\rangle + \langle c\rangle)] \tag{7.36}$$

The left-hand side of Eq. (7.35) represents the Burgers vector of the initial slip dislocation b, while the right-hand side lists the Burgers vector of the stair-rod dislocation b_r and the Burgers vector of the twin dislocation b_t. In Eq. (7.36), $\langle a\rangle = 1/3\langle\bar{2}110\rangle$, $\langle c\rangle = \langle 0001\rangle$ and $\langle d\rangle = 1/3\langle\bar{1}010\rangle$. The value of s is determined by $s = Ns' - m$. Here, $s' = 1/[\rho/3 + (\sigma^2/4\rho)(c/a)^2]$ for $\{\bar{1}012\}$ plane represents the needed shear to form the fault layers of N. The corresponding values of ρ and σ for the twinning plane $\{\bar{1}012\}$ are $\rho = 3$ and $\sigma = 2$, respectively. Furthermore, m is chosen to make the absolute value of s less than 1. The dissociation of individual $\langle a\rangle$ glide dislocations is not a feasible mechanism for generating twin dislocations [67]. Dislocation pile-ups with relatively small slip can lead to dissociation of the head (5 to 10) (Figure 7.17c).

7.3.2.1.2 The Glide of Twinning Dislocations

According to the assumption in this section, nanoscale deformation twinning is produced by the sliding of glissile Shockley partial dislocations (twinning dislocations) that arise from non-planar dissociation of $\langle a \rangle$ glide dislocations on the $\{111\}$ planes. Therefore, the mechanism of nanoscale deformation twinning is as follows: first, a pile-up of pre-existing perfect dislocation is located on the basal plane of the deformed Mg HCP metal (Figure 7.18a). These dislocations are created during the preceding plastic deformation, therefore and they are non-equilibrium. The head of pile-up perfect dislocation then dissociates into Shockley partial dislocations (twin dislocations) and a stair-rob located between the sliding plane of the pile-up perfect dislocation and the twin plane (Figure 7.18a). The pile-up can simultaneously drive this dissociation further and provide repulsive forces to drive the twin dislocations away from the center of the core (Figure 7.18b). Lastly, twin dislocations slip on each slip plane within the nanoscale region, leading to the formation and growth of deformation twins at the nanoscale (Figure 7.18c-d).

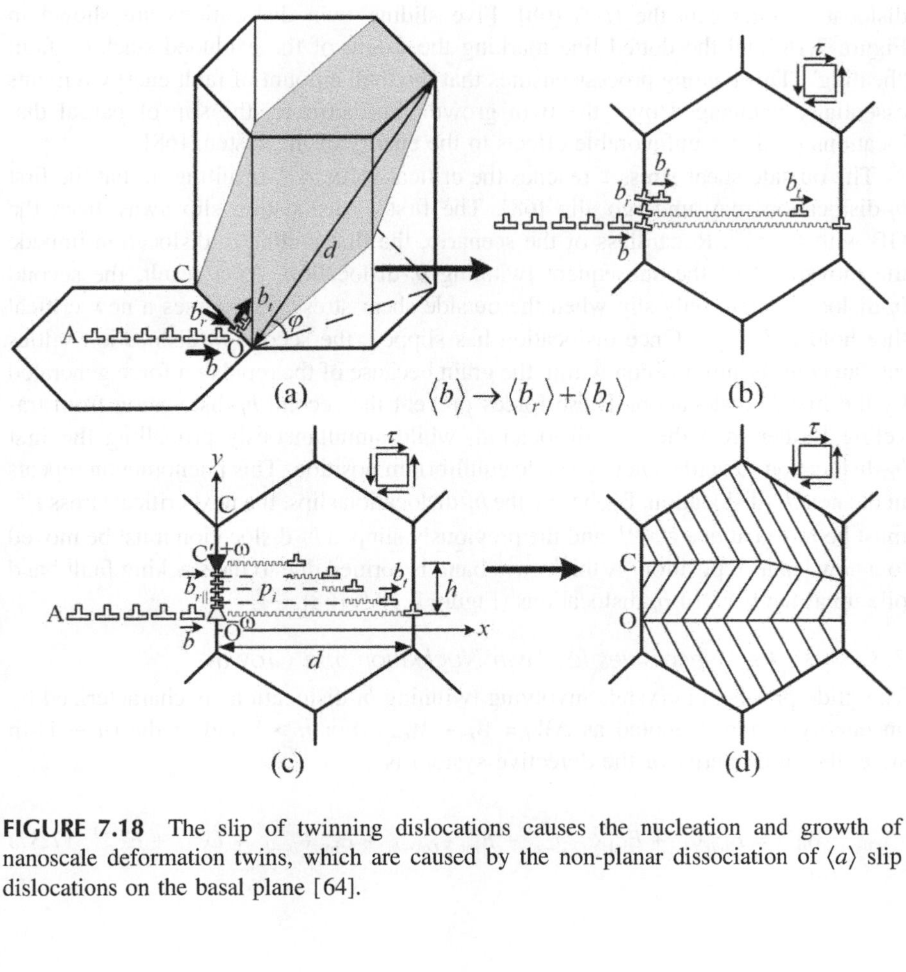

FIGURE 7.18 The slip of twinning dislocations causes the nucleation and growth of nanoscale deformation twins, which are caused by the non-planar dissociation of $\langle a \rangle$ slip dislocations on the basal plane [64].

The geometrical characteristics of nanoscale deformation twinning induced by dislocation dissociation and twinning dislocation slip in deformed Mg need to be investigated. A pile-up of dislocations *OA* is impeded by the crystal grain boundary (GB), which is located on the basal plane (Figure 7.18a). These *b*-dislocations undergo deformation when subjected to a shear stress τ. Thus, the pile-up of dislocation develops through dissociations of its head dislocations, which causes stair rod dislocations at the GB and twinning dislocations with Burgers vectors b_r and b_t respectively. These stair rod dislocations create an OC' wall-like structure on the *OC* GB, while twin dislocations slide into adjacent grains.

Under the assumption of a continuum approach, the possibility of twin dislocations slipping into grains is considered, and the space between twin dislocations and the stair rod dislocation is consist of the stacking fault band of width p_i (where *i* represents the number of sliding b_t-dislocations) with unit area surface energy γ_{SF}. φ is the angle among the basal plane and twin plane, and the spacing δ among twinning dislocation depends on the crystallography of the metal. Thus, if the number of sliding twin dislocations *i* exceeds 2, then the p_i slip distance of each twin dislocations eliminates the stacking fault at the back of the dislocation of the (*i*-1) segment until the *i*-th dislocation intersects the (*i*-2) [68]. Five sliding twin dislocations are shown in Figure 7.18, and the dotted line marking the extent of the excluded stacking fault "healing". This healing process ensures that the total amount of fault energy remains essentially unchanged over the twin growth. Furthermore, the slip of partial dislocations produces unfavorable effects to the energy of the system [68].

The outside shear stress τ reaches the critical value $\tau_c^{(1)}$, resulting in that the first b_t-dislocation may undergo slip [68]. The first b_t-dislocation slip away from the GB with $\tau \geq \tau_c^{(1)}$. Regardless of the scenario, the first sliding b_t-dislocation impede the movement of the subsequent twinning b_t-dislocation. As a result, the second b_t-dislocation can only slip when the outside shear stress τ surpasses a new critical threshold $\tau_c^{(2)} > \tau_c^{(1)}$. Once dislocation has slipped, the second b_t-dislocation settles into an equilibrium position within the grain because of the repulsion force generated by the first b_t-dislocation. These forces prevent the second b_t-dislocation from traveling further than the first dislocation, while simultaneously propelling the first b_t-dislocation towards a newly stable equilibrium position. This phenomenon repeats in the next b_t-dislocation. Each time the b_t-dislocation slips, the next critical stress $\tau_c^{(n)}$ must be raised above $\tau_c^{(n-1)}$, and the previously slipped b_t-dislocation must be moved to a new balance position. A thick twin band is formed due to the stacking fault band pile up at the back of b_t-dislocations (Figure 7.18d) [68].

7.3.2.1.3 Key Necessities for Twin Nucleation and Growth

The glide process in crystals involving twinning b_t-dislocations is characterized by an energy change denoted as $\Delta W_n = W_n - W_{n-1}$. For $n > 1$ and at the $(n-1)$-th state, the sum energy of the defective system is:

$$W_{n-1} = E_{self\sum b_t}^{n-1} + E_{\text{int}\sum OC'-b_t}^{n-1} + E_{\text{int}\sum OA-b_t}^{n-1} + E_{\text{int}\sum b_t-b_t}^{n-1} + E_{\sum\gamma}^{n-1} + E_{\sum\tau}^{n-1} \quad (7.37)$$

where $E_{self\Sigma b_t}^{n-1}$ represents the sum self-energy of the $(n-1)$ twin b_t-dislocations, $E_{\text{int}\,\Sigma OC'-b_t}^{n-1}$ represents the total elastic interaction energy among the stair-rod dislocation wall OC' of and the $(n-1)$ twin b_t-dislocations. $E_{\text{int}\,\Sigma OA-b_t}^{n-1}$ indicates the overall interaction energy among the site pile-up OA b-dislocations and the $(n-1)$ twin b_t-dislocations, $E_{\text{int}\,\Sigma b_t-b_t}^{n-1}$ means the total elastic interaction energy among all twin b_t-dislocations. $E_{\Sigma\gamma}^{n-1}$ signifies the total energy of the new nanoscale TBs and adjacent pile-up fault segments, and $E_{\Sigma\tau}^{n-1}$ represents the total interaction energy between the shear stress τ and the $(n-1)$ twin b_t-dislocations.

In the case of the elastically isotropic solid defects, the total $E_{self\Sigma b_t}^{n-1}$ self-energy of $(n-1)$ twin b_t-dislocations are given by [69]:

$$E_{self\Sigma b_t}^{n-1} = \frac{Db_t^2}{2}\sum_{i=1}^{n-1}\left(1 + \ln\frac{R}{r_{b_t c}}\right) \tag{7.38}$$

where $D = G/(2\pi(1-\nu))$. G is the shear modulus, and ν is the Poisson's ratio. R is the cut-off radius of the dislocation stress field, and $r_{b_t c} \approx b_t$. In accordance with the theory of solid defects [70], the stresses produced by GB $b_{r\perp}$- dislocation walls are effectively represented as a dipole of GB wedge disclinations marked by disclination strengths $\pm\,\omega$ ($\omega = 2arctan(b_{r\perp}/2\delta)$). The energy $E_{\text{int}\,\Sigma OC'-b_t}^{n-1}$ is decided by the below formula [71,72]:

$$\begin{aligned} E_{\text{int}\,\Sigma OC\prime-b_t}^{n-1} = {} & \frac{Db_t\omega}{2}\sum_{i=1}^{n-1}\left((y_i - L)\ln\frac{R^2+L^2+y_i^2-2Ly_i}{p_i^2+L^2+y_i^2-2Ly_i} - y_i\ln\frac{R^2+y_i^2}{p_i^2+y_i^2}\right) \\ & -\frac{Db_tb_{r\parallel}}{2}\sum_{i=1}^{n-1}\sum_{j=1}^{n-1}\left(\ln\left(1+\frac{R^2-2Rp_i}{p_i^2+y_{ij}^2}\right) - \frac{2y_{ij}^2(R^2-2Rp_i)}{(p_i^2+y_{ij}^2)(p_i^2+y_{ij}^2+R^2-2Rp_i)}\right) \end{aligned} \tag{7.39}$$

where $L = (n-2)\delta$ is the distance among the dipoles, $y_{ij} = h_j - h_i$ and $h_i = i\delta$.

The interaction energy $E_{\text{int}\,\Sigma OA-b_t}^{n-1}$ among the pile-up OA of b-dislocations and $(n-1)$ twinning b_t-dislocations is [71]:

$$E_{\text{int}\,\Sigma OA-b_t}^{n-1} = -Db_tb\sum_{i=1}^{n-1}\sum_{j=1}^{n_c}\left(\begin{aligned} & \frac{\cos\varphi}{2}\ln\frac{p_i^2+x_j^2+y_i^2+R^2+2p_i\cos\varphi(R+x_j)+2Rx_j}{p_i^2+x_j^2+y_i^2+2p_ix_j\cos\varphi} + \\ & \frac{(R+x_j+p_i\cos\varphi)\sqrt{y_i^2+p_i^2\sin^2\varphi}\,\sin\varphi+(y_i^2+p_i^2\sin^2\varphi)\cos\varphi}{p_i^2+x_j^2+y_i^2+R^2+2p_i\cos\varphi(R+x_j)+2Rx_j} \\ & -\frac{(x_j+p_i\cos\varphi)\sqrt{y_i^2+p_i^2\sin^2\varphi}\,\sin\varphi+(y_i^2+p_i^2\sin^2\varphi)\cos\varphi}{p_i^2+x_j^2+y_i^2+2p_ix_j\cos\varphi} \end{aligned}\right) \tag{7.40}$$

where n_c and x_j respectively represent the quantity and location coordinates of the balance positions of the dislocations. $x_j = Gb\pi(j-1)^2/(16n_c\tau(1-\nu))$, $y_i = h_i = i\delta$ with the lower index being $i = 1, 2, 3 \ldots n-1$ [73].

The overall interaction energy $E^{n-1}_{\text{int}\ \Sigma b_t - b_t}$ of the twin dislocations is expressed as [71]:

$$E^{n-1}_{\text{int}\ \Sigma b_t - b_t} = \frac{Db_t^{\,2}}{2}\sum_{i=1}^{n-2}\sum_{j=i+1}^{n-1}\left(\ln\left(1+\frac{R^2-2Rp_{ij}}{p_{ij}^{\,2}+y_{ij}^{\,2}}\right) - \frac{2y_{ij}^{\,2}(R^2-2Rp_{ij})}{(p_{ij}^{\,2}+y_{ij}^{\,2})(p_{ij}^{\,2}+y_{ij}^{\,2}+R^2-2Rp_{ij})}\right) \quad (7.41)$$

where $p_{ij} = p_j - p_i$. The energy $E^{n-1}_{\Sigma\gamma}$ associated with the interaction among twinning b_t-dislocations and stair rod b_r-dislocations is transformed into the energy of the TBs. Additionally, the energy of the beside segment of stacking fault with a length of $(p_1 - p_2)$ is:

$$E^{n-1}_{\Sigma\gamma} = \begin{cases} \gamma_{SF}p_1 & n-1=1 \\ \gamma_{SF}(p_1-p_2) + 2\gamma_{TB}p_2 & n-1 \geq 2 \end{cases}. \quad (7.42)$$

where γ_{TB} represents the particular energies associated with the new nanotwin interface located on the $\{\bar{1}012\}$ plane.

The amount of work needed to push a twinning b_t-dislocation a distance p_i affected by the outside shear stress τ is represented by the energy E^i_τ. The overall interplay energy $E^{n-1}_{\Sigma\gamma}$ between the outside shear stress τ and (n-1) twinning b_t-dislocations is:

$$E^{n-1}_{\Sigma\tau} = -b_t\tau\sum_{i=1}^{n-1}p_i \quad (7.43)$$

The overall energy of the defective system in the n-th state shows as follows:

$$W_n = E^n_{self\Sigma b_t} + E^n_{\text{int}\ \Sigma OC'-b_t} + E^n_{\text{int}\ \Sigma OA-b_t} + E^n_{\text{int}\ \Sigma b_t-b_t} + E^n_{\Sigma\gamma} + E^n_{\Sigma\tau} \quad (7.44)$$

For $n = 1$, the overall energy W_1 is obtained by substituting $n = 1$ and $E^1_{\text{int}\ \Sigma b_t-b_t} \equiv 0$ in Eq. (7.44), and all terms involved in the energies W_{n-1} and W_n is calculated. The energy variation $\Delta W_n = W_n - W_{n-1}$ describes the transition of the defective system from the $(n-1)$-th state to the n-th state. The previously glided b_t-dis-dislocations have stable equilibrium positions, known as $\tilde{p}_i$, which correspond to the points of minimum energy of $\Delta W_n(p_i)$. These positions are determined by $\partial\Delta W_n/\partial p_i = 0$ [68]. The steady balance position of the first glided b_t-dislocation denoted by $\tilde{p}_1$ is found by $\partial\Delta W_1/\partial p_1 = 0$ with $\Delta W_1 = W_1$. $\tilde{p}_1$ is set to a fixed value, then, the equilibrium position $\tilde{p}_2$ for the second glided b_t-dislocation is calculated by $\partial \Delta W_2/\partial p_2|_{p_1=\tilde{p}_1} = 0$ with $\Delta W_2 = W_2 - W_1$. The resulting value of $\tilde{p}_2$ and the equation $\partial \Delta W_2/\partial p_1|_{p_2=\tilde{p}_2} = 0$ is used to correct a new value of $\tilde{p}_1$. This corrected value of $\tilde{p}_1$ is then used to adjust the value of $\tilde{p}_2$ by $\partial \Delta W_2/\partial p_2|_{p_1=\tilde{p}_1} = 0$. The iterative correction process was applied to calculate a newly revised value of $\tilde{p}_1$ based on the newly corrected value of $\tilde{p}_2$, and this process continued until the values of $\tilde{p}_1$ and $\tilde{p}_2$ converged to a constant value. These constant values are considered the steady balance positions of the first and second twinning b_t-dislocations (for $n = 2$). By using this process, the steady balance positions $\tilde{p}_i$ of n twinning b_t-dislocations with a high degree of accuracy are determined [68].

7.3.2.2 Result

7.3.2.2.1 Nucleation Conditions for Deformation Twinning

The parameter values for HCP Mg: Poisson's ratio $\nu = 0.35$, shear modulus $G = 40.06$ GPa [67], lattice parameter $c \approx 0.52$ nm, $a \approx 0.32$ nm, $c/a = 1.623$ [74], $\varphi \approx 42°$, stacking fault energy $\gamma_{SF} = 14$ mJ/m^2, and TBE $\gamma_{TB} = 189$ mJ/m^2 on the $\{\bar{1}012\}$ plane [67,75]. Figure 7.19 clearly illustrates the dependence of the transform in system energy ΔW_1, on the twin length p_1, at different levels of external shear stress τ. At $d = 100$ nm, $n_c = 5$ in Figure 7.19a, and $n_c = 10$ in Figure 7.19b. It is observed that before reaching the minimum value, the energy change ΔW_1 decreases sharply, and then slightly increases with the increase of twin length p_1. This behavior indicates that twin dislocations are advantageous in occupying balanced state within the grain. As the applied shear stress τ increases, balanced position changes towards the opposite GB.

In addition, there exists a critical shear stress $\tau_c^{(1)}$, such that for any p ($p \le d$) and $\tau > \tau_c^{(1)}$, the energy alters ΔW_1 reduces, which is beneficial for twinning nucleation. This critical shear stress is determined by $\Delta W_1(p = p*) = 0$, with $p* = 1$ nm, $\Delta W_1|_{p_1 > p^*} < 0$, and $\partial \Delta W_1/\partial p_1|_{p_1 < p^*} \le 0$. The choice of $p* = 1$ nm is due to the fact that it is the physical shortest length for any twin. The results show that $\tau_c^{(1)} \approx 519.51$ MPa for stacking $n_c = 5$ and $\tau_c^{(1)} \approx 466.97$ MPa for stacking $n_c = 10$.

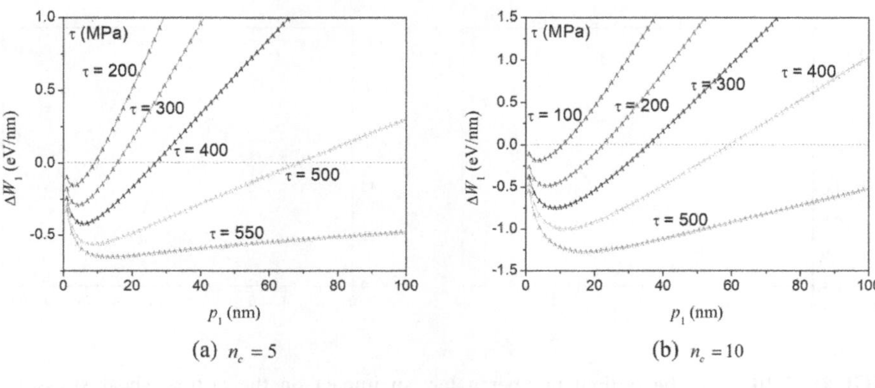

FIGURE 7.19 For different levels of applied shear stresses, the relationship between energy change ΔW_1, and twin length p_1, is examined [64].

The nucleation stresses for these deformed twins are markedly elevated compared to the previously reported critical shear stress for $\{\bar{1}012\}$ deformed twins [76]. However, these stresses are calculated based on a nanoscale model of deformed twins, whereas previous values are obtained according to bulk-averaged stresses, macroscopic or microscopic stresses.

However, the stress required for deformation twin nucleation is lower than the stress required for initiation, which are determined through in situ measurements of much smaller samples (800 MPa) and through MD simulations of single-crystal Mg tensile tests (800 ± 200 MPa) [77]. This is due to the nucleation stress for deformed twins in this section are acquired based on a theoretical model. In an ideal situation and assuming a pre-existing dislocation pile-up of 5~10, whereas the shear stress used on accommodate the dislocation pile-up was already included in the experimental and simulation results. Hence, if the shear stress exerted on the pre-existing dislocation pile-up is considered ($\tau_{n_c} \approx 313.88$ MPa for pile-up $n_c = 5$ and $\tau_{n_c} \approx 627.77$ MPa for pile-up $n_c = 10$ with $\tau_{n_c} = Gn_c b/(\pi(1-\nu)d)$), the nucleation stresses for deformed twins is between ≈833.39 ~ 1094.74 MPa, which is consistent with previous research [77].

An increase in dislocation pile-up n_c results in a reduce in the critical shear stress (Figure 7.19a,b). This indicates that the dislocation pile-up can promote twin formation by generating stress concentrations and appropriate directed stress fields. It is vital to note that the dislocation pile-up serves two roles, which is constant with previous research [67]. Before dissociation, it provides the necessary stress for dissociation. After dissociation, it assists the twinning dislocations in overcoming attractive forces and reaching a stable extension far enough away from the reaction center.

7.3.2.2.2 Growth Conditions for Deformation Twinning

The critical shear stress $\tau_c^{(n)}$, for the n-th twin b_t-dislocations during slip is calculated using $\Delta W_n = 0$. The stable position of the slip twinning b_t-dislocations is calculated by combining this equation with the previously described algorithm. The growth of nanoscale deformation twins increases the critical shear stress $\tau_c^{(n)}$ (Figure 7.20a). However, when the thickness of the twinning reaches a certain

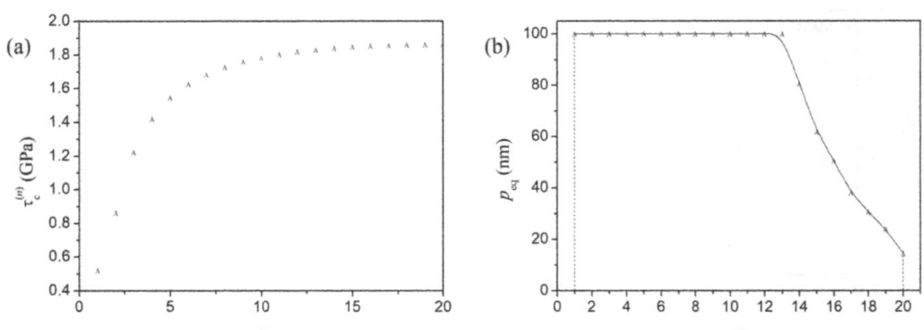

FIGURE 7.20 (a) The ordinal number n has an impact on the critical shear stress $\tau_c^{(n)}$ required for the glide of twinning dislocations. (b) When n is equal to 20, the balance position p_{eq} of each twinning dislocation is calculated [64].

value, the rate of increase is much slower than the initial stages of twinning growth. The reason for this is that as the number of twinning dislocations increases, the promoting effect of pile-ups and step barriers on twinning growth becomes significantly greater. The equilibrium positions for each twinning dislocation are calculated as the amount of twinning dislocations n is equal to 20 (Figure 7.20b). The thickness $h = (n - 1)\delta$ of the twin layer near its bottom is affected by the amount of twinning dislocations n.

7.3.3 Transition of Dynamic Recrystallization Mechanism

Due to the HCP structure of Mg alloys at room temperature, their formability is significantly low. However, some slip modes and dynamic recrystallization (DRX) that rarely occur at room temperature are activated at intermediate and high temperatures, greatly enhancing plasticity. The DRX behavior significantly affects the microstructure and texture, determining the macroscopic mechanical properties of Mg alloys. The previous research has presented various simulation methods for DRX and effectively captured the impacts of subgrain development (SD), particle-stimulated nucleation (PSN), and GB bulging dynamic recrystallization (GBBDRX) on microstructures, textures, flow stress, and macro-mechanical performance of alloys during hot deformation. The lack of models that can accurately quantify twin-induced dynamic recrystallization (TDRX) nucleation and growth during the early stages of hot deformation contributes to an insufficient understanding of the effects of TDRX and DRX transition on hot deformation. Additionally, there is a need for a physical explanation for the transition between TDRX and GBBDRX in as-cast Mg alloys.

7.3.3.1 Model

7.3.3.1.1 A Theoretical Dynamic Recrystallization Transition Criterion

The nucleation and the growth of twin are explained by the competition between the strain energy arising from local stress concentrations and the occurrence of dislocation slip in the material [78]. The twin transformation strain plays a crucial role in determining the stress state of the parent, the twin, and the surrounding environment [79]. A novel dislocation-based strain nucleus model and Green's function method are utilized to compute the local stress field that arises from deformation twinning [80]. The findings indicate that incident twin growth is strongly influenced by the local transformed strains in barrier twins.

Figure 7.21a shows a schematic diagram of the double twinning (DTW) model. The green region depicts the contraction twinning (CTW) with primary $\{10\bar{1}1\}$, while the blue illustrates the extension twinning (ETW) with secondary $\{10\bar{1}2\}$ growing within the primary CTW. both x_1 and x_2 axes are parallel to the $\{10\bar{1}0\}$ direction and c-axis, respectively. The d and h_{PriCtw} are the length and thickness of the primary twin band. $\theta = 61.9°$ indicates the angle between the basal plane of the CTW and the twinning plane, $\varphi = 18.75°$ is the angle between twinning plane of the CTW and that of the ETW, and the value of χ signifies the distance coefficient between the starting point and the origin of the secondary ETW. The ratio of the

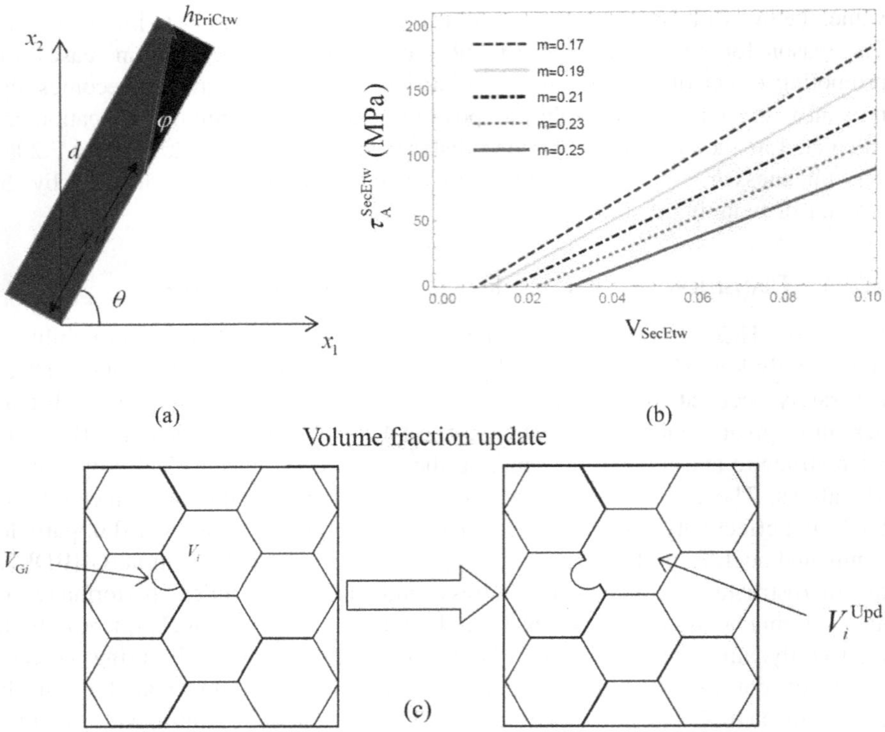

FIGURE 7.21 (a) A DTW model. (b) Applied shear stress of the secondary ETW versus the volume ratio of the secondary ETW to primary CTW. (c) Schematics of the volume fraction update by strain-induced GB bulging DRX [81].

volume of the secondary ETW to that of the primary CTW is defined as $V_{\text{SecEtw}} = \frac{h_{\text{SecEtw}}}{d \sin \varphi}$, and the ratio of the thickness to the length of the primary CTW is $m = h_{\text{PriCtw}}/d$.

Figure 7.21b shows that the local stress of the primary CTW increases with an increase in the ratio m, which also acts as a driving force for the growth of the secondary ETW. Additionally, it is observed that the applied shear stress $\tau_{\text{A}}^{\text{SecEtw}}$ on the secondary twinning is proportional to the volume ratio v_{SecEtw}. The assumption is made that the DRX occurs solely as a result of the primary ETW and $\{10\bar{1}1\}$-$\{10\bar{1}2\}$ DTW. The resolved shear stress $\tau_{\text{A}}^{\text{SecEtw}}$ varies in direct proportion to the compression stress, and the stress-strain relationship is expressed as $|\sigma_{\text{A}}| = \sigma_0 + h_0 |\varepsilon^p|^D$ where σ_0 is the initial yield stress, h_0 denotes the strength coefficient, and D signifies hardening coefficient. It is reasonable to assume that there exists an equilibrium relationship between the volume ratio v_{SecEtw} and equivalent strain $\bar{\varepsilon}^p$ [81]:

$$\text{v}_{\text{SecEtw}} = A + B\,[\bar{\varepsilon}^P(k)]^D \tag{7.45}$$

The initial volume fraction of the secondary ETW is expressed as:

$$A = \frac{-m_0\sigma_0 - \tau_r + \dfrac{\left[\int \dfrac{\chi d \quad \cos\theta + \xi d \quad \cos(\varphi+\theta) \frac{-\tau^{\mathrm{SecTw}}_{\mathrm{pri}\to\mathrm{sec}}}{\cos(\pi-\theta-\varphi)}}{\chi d \quad \cos\theta} dx\right]}{\xi d}}{\beta\left[\dfrac{E(k_0)}{k'_0} + \dfrac{v}{1-v}\dfrac{1}{k'_0}\dfrac{E(k_0) - k'^2_0 K(k_0)}{k_0^2}\right]\mu e^{\mathrm{T}}_{\mathrm{Sec}} \sin\varphi},$$

and the volume growth rate from plastic deformation, $B = \frac{-m_0 h_0}{\beta\left[\frac{E(k_0)}{k'_0} + \frac{v}{1-v}\frac{1}{k'_0}\frac{E(k_0) - k'^2_0 K(k_0)}{k_0^2}\right]\mu e^{\mathrm{T}}_{\mathrm{Sec}} \sin\varphi}$, where $m_0 = \tau_{\mathrm{A}}^{\mathrm{SecEtw}}$ is the ratio of the resolved shear stress to the compression stress. It is evident that A and B are significantly influenced by various material constants and microstructural features.

As the strain increases, the saturation state of the secondary ETW $v_{\mathrm{SecEtw}} = 1$ is achieved in two ways: either the primary CTW is completely retwinned due to the combined effects of external applied stress and local stress, or the grain diameter significantly decreases. By incorporating the saturation condition $v_{\mathrm{SecEtw}} = 1$ into Eq. (7.45), it is possible to determine the saturation strain for nucleation of TDRX. When the volume ratio exceeds 1 ($v_{\mathrm{SecEtw}} = 1$), the TDRX is no longer viable due to the exhaustion of nucleation sites. In order to relieve the accumulated strain, TDRX is replaced by GBBDRX, which occurs through bulging of boundaries of recrystallized grains that have evolved from twins [82].

7.3.3.1.2 VPSC Simulation Incorporating DRX Schemes

To simulate Mg crystal hot deformation, DRX and texture evolution, the visco-plastic self-consistent (VPSC) model incorporating double DRX schemes is applied. In the VPSC model, every grain is considered as an ellipsoidal visco-plastic inclusion that is surrounded by an effective visco-plastic medium. The properties of both the inclusion and medium are fully anisotropic. The polycrystal is modeled using weighted orientations, where the orientations represent individual grains and the weights represent the respective volume fractions. By employing the Eshelby inclusion formalism and linearization scheme, the stress and strain rate of a single crystal are self-consistently related to the those of the effective medium. This allows for accurate prediction of the overall behavior of the polycrystal under deformation.

7.3.3.1.3 VPSC Simulation

The local viscoplastic constitutive response (in a specific orientation) is characterized using a non-linear rate-sensitivity equation [83]:

$$\dot{\varepsilon}_{ij} = \dot{\gamma}_0 \sum {}_s m^s_{ij} \left(\frac{m^s_{kl}\sigma_{kl}}{\tau^s_0}\right)^n \tag{7.46}$$

where $\dot{\gamma}_0$ indicates a coefficient for normalization, $\dot{\varepsilon}_{ij}$ represents the deviatoric strain rate, σ_{kl} represents the stress, n signifies the reciprocal of the rate-sensitivity exponent, and m_{ij}^s denotes the symmetric Schmid tensor and τ_0^s is the threshold stress that is related to the slip or twinning systems. In this section, a random texture is employed as the initial texture for the hot compression of the as-cast Mg alloys. Additionally, the VPSC code achieves the Secant linearization scheme [84], which is correlated with the interaction between the grain and matrix.

The threshold stress τ_C^s represents the minimum stress required to activate the deformation modes, and is typically increase with increasing deformation. The Voce hardening model is employed to characterize the variation of the threshold stress with the accumulated shear strain in each grain:

$$\tau_C^s = \tau_0^s + (\tau_1^s + \theta_1^s \Gamma)\left[1 - \exp\left(-\Gamma\left|\frac{\theta_0^s}{\tau_1^s}\right|\right)\right] \tag{7.47}$$

where Γ indicates the shear accumulated in the grain, τ_0^s is the initial critical resolved shear stress (CRSS), θ_0^s represents the initial hardening rate, θ_1^s is the asymptotic hardening rate and $\tau_0^s + \tau_1^s$ signifies the back-extrapolated CRSS of the deformation mode.

The first-order derivative of with respect to Γ is expressed as follows:

$$\frac{d\tau_C^s}{d\Gamma} = \theta_1^s + \left(\left|\frac{\theta_0^s}{\tau_1^s}\right|\tau_1^s - \theta_1^s\right)\exp\left(-\Gamma\left|\frac{\theta_0^s}{\tau_1^s}\right|\right) + \left|\frac{\theta_0^s}{\tau_1^s}\right|\theta_1^s\Gamma\exp\left(-\Gamma\left|\frac{\theta_0^s}{\tau_1^s}\right|\right) \tag{7.48}$$

As the obstacle effects on slip and twinning are insignificant at the temperature of 573 K [85], the latent hardening component is not taken into account in this model.

7.3.3.1.4 TDRX Model

At the beginning of the deformation process, when the strain is low, the primary mechanism of deformation in coarse grains is twinning. TDRX can lead to the formation of a necklace-like structure composed of many fine particles, which subdivides the primary particles [86]. The dominant twinning modes that are widely distributed within the coarse grains are the $\{10\bar{1}2\}$ ETW, the $\{10\bar{1}1\}$ CTW, and the $\{10\bar{1}1\}$-$\{10\bar{1}2\}$ DTW. As the strain increases, the $\{10\bar{1}2\}$ ETW rapidly spreads throughout the primary twinning and matrix grains, leading to the abundant formation of $\{10\bar{1}1\}$-$\{10\bar{1}2\}$ DTW and twin intersections. The forming twins are favorably oriented for slip, and thus can accumulate a higher density of dislocations compared to the matrix, resulting in the storage of significant strain energy. This energy accumulation triggers DRX and induces the formation of a large number of twin-walled grains with smooth boundaries [87]. It is hypothesized that the twin-walled grains formed through TDRX roughly maintain the c-axis orientation of the parent twins, with a random rotation of less than 30° around the c-axis of the twin host [86,88].

As the twin-walled grain nuclei gradually increase in number with increasing plastic strain during the low-strain stage, it is possible to define a probability model of TDRX nucleation within each orientation:

$$P_{\mathrm{NTDRX}} = \begin{cases} 0 & \text{If } \bar{\varepsilon}^{\mathrm{P}} \le E_c \\ \mathrm{Exp}\left[-C\left(\frac{E_s - \bar{\varepsilon}^{\mathrm{P}}(k)}{\bar{\varepsilon}^{\mathrm{P}}(k) - E_c}\right)\right] & \text{If } E_c \ge \bar{\varepsilon}^{\mathrm{P}} > \text{If } \bar{\varepsilon}^{\mathrm{P}} \le E_c \\ 0 & \text{If } \bar{\varepsilon}^{\mathrm{P}} > E_s \end{cases} \quad (7.49)$$

where P_{NTDRX} represents the probability of nucleation of secondary and primary ETW each system, E_c is the strain threshold required for the nucleation of TDRX within each twin system, the value of E_s is obtained by satisfying $v_{\mathrm{SecEtw}} = 1$, which denotes the saturation strain for TDRX nucleation, and the value of the fitting parameter C is between 0 and 1.

Apart from oriented nucleation, the growth process of TDRX nuclei is also influenced by a stochastic factor of the misorientation angle between the grown grains and the twin host. The $\varphi_2^{\mathrm{DrxGro}}$ of the grown DRX grains should be $\varphi_2^{\mathrm{DrxNuc}}$ plus a random value of [−30°, 30°], and the selected growth probability P_{GTDRX}. In each system, the volume fraction of the grown DRX grains V_{DRX} is expressed as:

$$P_{\mathrm{GTDRX}} = 1 - \mathrm{Exp}\left[-b\left(\frac{\mathrm{Abs}[\mathrm{Random}[-30°, 30°]]}{\theta_c}\right)^C\right] \quad (7.50)$$

in grain k, the volume fraction of the primary ETW-induced DRX is:

$$V_{\mathrm{DRX1}} = V_{\mathrm{EtwSym}}(1 - aV_{\mathrm{SecEtw}})P_{\mathrm{NTDRX}}P_{\mathrm{GTDRX}} \quad (7.51)$$

and the volume fraction of the secondary ETW-induced DRX is:

$$V_{\mathrm{DRX2}} = aV_{\mathrm{EtwSym}}V_{\mathrm{SecEtw}}P_{\mathrm{NTDRX}}P_{\mathrm{GTDRX}} \quad (7.52)$$

where V_{EtwSym} is the volume fraction of the ETW system in grain k, which has been defined in the VPSC program. The a is the ratio of the CTW system volume to the ETW system, $\theta_c = 15°$ is the critical angle value that serves as a boundary between high-angle and low-angle grain boundaries. Abs[] represents the absolute value function, Romdom[] indicates the range of a random numbers, b is a positive constant and c denotes the power law exponent.

The TDRX approach is a useful method for improving the grain orientation and increasing slip. This model involves using the weighted mean method to calculate the orientations between the matrix and DRX grains after each deformation step, and updating the orientation of each grain accordingly:

$$\varphi_1^{\mathrm{MGCurr}} = \varphi_1^{\mathrm{MGDef}}\left(1 - \sum_{i=1}^{6} V_{\mathrm{DRX1}i} - \sum_{j=1}^{6} V_{\mathrm{DRX2}j}\right) + \sum_{i=1}^{6} \varphi_{1i}^{\mathrm{DrxNuc}} V_{\mathrm{DRX1}i} + \sum_{j=1}^{6} \varphi_{1j}^{\mathrm{DrxNuc}} V_{\mathrm{DRX2}j} \tag{7.53}$$

$$\phi^{\mathrm{MGCurr}} = \phi^{\mathrm{MGDef}}\left(1 - \sum_{i=1}^{6} V_{\mathrm{DRX1}i} - \sum_{j=1}^{6} V_{\mathrm{DRX2}j}\right) + \sum_{i}^{6} \phi_i^{\mathrm{DrxNuc}} V_{\mathrm{DRX1}i} + \sum_{j=1}^{6} \phi_j^{\mathrm{DrxNuc}} V_{\mathrm{DRX2}j} \tag{7.54}$$

$$\varphi_2^{\mathrm{MGCurr}} = \varphi_2^{\mathrm{MGDef}}\left(1 - \sum_{i=1}^{6} V_{\mathrm{DRX1}i} - \sum_{j=1}^{6} V_{\mathrm{DRX2}j}\right) + \sum_{i=1}^{n} \left(\varphi_{2i}^{\mathrm{DrxNuc}} + \mathrm{Random}[-30°, 30°]\right) V_{\mathrm{DRX1}i} + \sum_{j=1}^{n} \left(\varphi_{2j}^{\mathrm{DrxNuc}} + \mathrm{Random}[-30°, 30°]\right) V_{\mathrm{DRX2}j} \tag{7.55}$$

where $\varphi_1^{\mathrm{MGDef}}$, ϕ^{MGDef} and $\varphi_2^{\mathrm{MGDef}}$ represent the Euler angles that describe the orientation of the deformed material in the current step, $\varphi_1^{\mathrm{MGCurr}}$, ϕ^{MGCurr} and $\varphi_2^{\mathrm{MGCurr}}$ refer to the Euler angles that describe the orientation of the grain after applying the TDRX method, and i and j are the index of the primary and secondary ETW systems, respectively.

7.3.3.1.5 Model for Grain Boundary Bulging

During the high-strain stage, the occurrence of TBs within the refined grains becomes infrequent, and the previous TDRX consumes a large number of TBs. As a result, the TDRX mechanism becomes less frequent, and the DRX process switches to the strain-induced GBBDRX mechanism. In contrast to the TDRX, the migration of grain boundaries from the low dislocation density side to the high dislocation density side results in serrated grain boundaries, and the bulges can develop into new grains with lower dislocation densities [86,89]. These DRX grains inherit the orientations of their parents coarsely.

The DRX process as plastic strain increases are characterized by the saturation state for TDRX nucleation, which marks the transition point. GBBDRX nucleation is modeled by a probability model with two critical conditions based on the equivalent plastic strain and grain orientations [90]:

$$P_{\mathrm{NGBB}i} = \begin{cases} 0 & \text{If } V_{\mathrm{SecEtw}} \le 1 \text{ and } F(i) > F_c \\ \mathrm{Exp}\left[-\left(\dfrac{F_i - F_{Min}}{F_c - F_{Min}}\right)^{dg}\right] & \text{If } V_{\mathrm{SecEtw}} > 1 \text{ and } F(i) \le F_c \end{cases} \tag{7.56}$$

where P_{NGBBi} is the probability of nucleation for the ith grain, $F(i)$ is the Taylor factor for the ith grain which is used to estimate the dislocation density of the grain, F_c is the exceeding critical value of Taylor factor, which strongly restricts the nucleation, F_{Min} is the minimum value of the Taylor factor, d_g denotes a gaussian exponent, and $V_{SecEtw} > 1$ represents a post-saturation state of the secondary ETW in which the TDRX is no longer feasible as the nucleation sites for TDRX are depleted. Instead, the GBBDRX is activated due to the accumulation of local strain within the refined grains.

In this model, the volume fraction of the DRX nucleus for a given parent orientation is expressed as:

$$V_{Ni} = P_{NGBBi} V_i \tag{7.57}$$

where V_i is the volume fraction of orientation i. P^{L}_{GGBBij} represents the average local growth probabilities:

$$P^{T}_{GGBBij} = \sum {}_j P^{L}_{GGBBij} V_j \tag{7.58}$$

where V_j is the volume fraction of the j-th grain, and local growth probabilities is: $P^{L}_{GGBBij} = \begin{cases} 1 & \text{If } misorientation\ \theta_{ij} \geq \theta_c \\ 0 & \text{If } misorientation\ \theta_{ij} < \theta_c \end{cases}$. The total volume of the grown grains from GBBDRX is expressed as:

$$V_{Gi} = c_g \left(P^{T}_{GGBBi}\right)^{P} V_{Ni} \tag{7.59}$$

where c_g is a normalization constant that influences the steepness of the flow softening portion and P is an exponent that quantifies the relative contributions of nucleation and growth processes, in terms of their respective importance.

Ultimately, it is necessary to interpose the volume fraction of each orientation and the normalization total volume fraction (Figure 7.21c). Contrary to TDRX, the flow softening is governed by the volume fraction modification rather than the orientation modification:

$$V_i^{Upd} = (V_i + V_{Gi}) / \sum {}_{i=1}(V_i + V_{Gi}) \tag{7.60}$$

7.3.3.2 Result

The VPSC model is employed, incorporating DRX schemes, enables the simulation of authentic stress-strain curves, slip and twinning activity, and texture. The Voce hardening parameters (Table 7.4) and DRX parameters (Table 7.5a, b) is fitted according to experimental results [81]. The simulation includes basal ⟨a⟩, prismatic ⟨a⟩, pyramidal ⟨c + a⟩, $\{10\bar{1}2\}$ ETW, and $\{10\bar{1}1\}$ CTW. The rate-sensitivity exponent is established at 0.05.

TABLE 7.4
Voce Hardening Parameters for Hot Compression [81]

Deformation Modes for Hot Compression	Voce Parameters			
	τ_0^s (MPa)	τ_1^s (MPa)	θ_0^s (MPa)	θ_1^s(MPa)
Basal ⟨a⟩	1	11	60	0
Prismatic ⟨a⟩	1.5	15	320	0
Pyramidal ⟨c + a⟩	23	73.5	700	0
{10$\bar{1}$2} extension twinning	3	5	1	0
{10$\bar{1}$1} contraction twinning	5	0	0	0

TABLE 7.5A
DRX Parameters for the TDRX Model [81]

Secondary Twinning			DRX Nucleation			DRX Growth		
A	B	D	C	E_c	E_a	a	b	c
0	2.5	1	0.14	0.0	0.4	1	5	4

TABLE 7.5B
DRX Parameters for the GBBDRX Model [81]

DRX Nucleation			DRX Growth	
F_{min}	F_c	d_g	C_g	p
2	3.1	6	0.016	2

7.3.3.2.1 Voce Hardening

Figure 7.22a shows that $\theta_0^s \geq \theta_1^s \geq 0$ and $\tau_1^s \geq 0$ indicates the yield stress increases and the hardening rate decreases, respectively. Additionally, the hardening rate gradually approaches θ_1^s starting from an initial value of θ_0^s. This law has a limit case known as linear hardening for $\tau_1^s = 0$. On the other hand, the situation of rigid-perfectly-plastic hardening meets $\theta_0^s = \theta_1^s = \tau_1^s = 0$. The stress-strain curves of Mg alloy without softening effect typically exhibits a monotonic increase in threshold stress and a decrease in hardening rate, as Γ increases. An evolution law is established by utilizing a group of parameters, namely $\tau_0^s > 0$, $\tau_1^s > 0$, $\theta_0^s > \theta_1^s \geq 0$, to describe this behavior.

Initially, the experimental results [92] and initial yield stress are used to establish the τ_0^s parameters for five deformation modes. Secondly, the θ_0^s parameter is established by evaluating the slope of the initial hardening section of the experimental

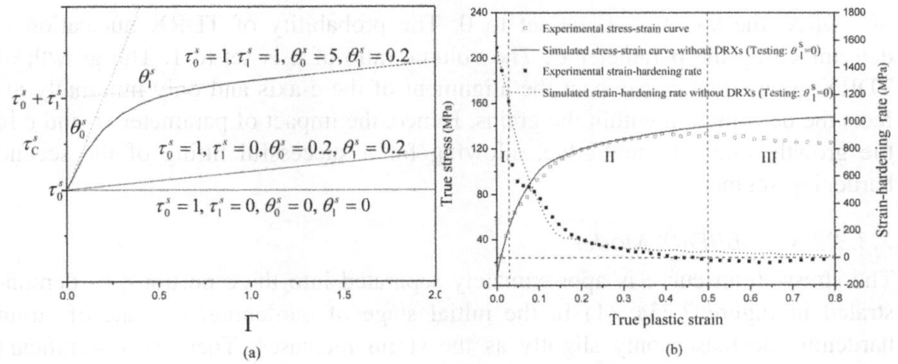

FIGURE 7.22 (a) Variation of τ_C^s versus **** [91]; (b) A fitting test is conducted to determine the Voce hardening parameters. Specifically, the first hardening segment within the strain interval of 0 to 0.03 was represented by parameter I, while the second hardening segment within the strain interval of 0.03 to 0.5 was represented by parameter II, and the third hardening segment within the strain interval of 0.5 to 0.8 was represented by parameter III [81].

curve (strain: 0–0.03), as the hardening rate shows only a slight decrease with strain in this segment. The third point is that the reduction in hardening rate as strain increases is achieved by numerical calculation, as θ_1^s is set to 0. Nevertheless, even if τ_1^s is set to an appropriate value, the stress-strain curve obtained from fitting fails to reproduce the second hardening region of the experimental curve (strain range: 0.03–0.5). Moreover, there is a disparity between the simulated and experimental hardening rates, particularly at the commencement of the second hardening region, where the simulated rate significantly exceeds the experimental rate. This dissimilarity is particularly evident in the initial stages of the second hardening region. The softening segment (strain: 0.5–0.8), as well as the second hardening segment, cannot be accurately captured. The simulated hardening rate exhibits a positive trend, whereas the experimental rate demonstrates a negative trend. As a result, the TDRX and following GBBDRX models must be incorporated.

7.3.3.2.2 TDRX Model

The DRX parameters of the TDRX model are shown in Table 7.5a. The $D = 1$ is determined according to the near-linear hardening behavior that is observed during the initial hardening stage. Through numerical analysis using Eq. (7.45), it has been observed that the volume fraction of secondary ETW A and the growth rate with plastic deformation B both exhibit a significant increase ranging from 0 to 10^{-2} and 10^{-1} to 10^{0} orders, respectively, when the grain size d decreases from 250 μm to 5 μm and the primary twin thickness remains constant. Due to the VPSC model's inability to account for the real grain size, A is designated as 0, while B is set to 2.5. The determination of E_s solution is achieved by taking into account the critical DRX transition condition with $V_{SecEtw} = 1$. Furthermore, taking into account that the initiation of twinning growth in the as-cast Mg alloy having an initial grain size of 245.5 μm occurs at a true plastic strain of 0, followed by the nucleation and growth of TDRX, the value of E_c is set to 0. The probability of TDRX nucleation is determined by the parameter C. The volume ratio of a is set to 1. The growth of TDRX grains does not impact the alignment of the c-axis and only minimally affects the deformation within the grains. Hence, the impact of parameters b and c in the growth model is negligible, allowing for a successful fitting of the second hardening segment.

7.3.3.2.3 GBBDRX Model

The stress-strain curve is approximately separated into three portions, as demonstrated in Figure 7.23a: (1) In the initial stage of hardening, the rate of strain hardening decreases only slightly as the strain increases. There is a significant increase in twinning observed in the larger grains (Figure 7.23b,c). (2) The second stage of hardening has a rapid reduce in the rate of strain hardening. The stress-strain curves that are simulated without TDRX fail to depict the second hardening phase, whereas the curves simulated using double DRXs are able to accurately capture it. In this particular section, the abundance of dislocations concentrated in

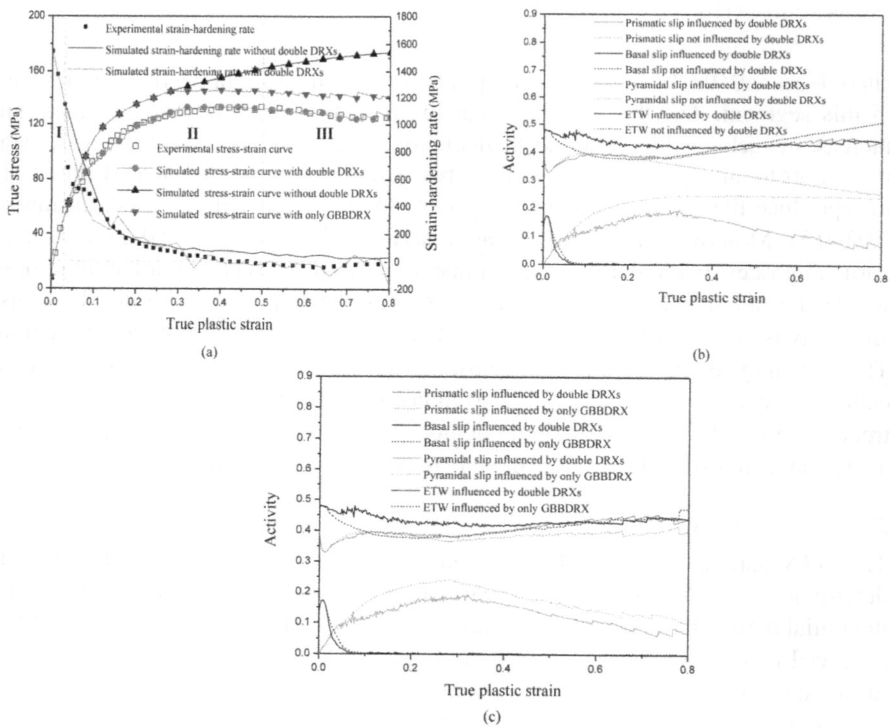

FIGURE 7.23 (a) The stress-strain behavior of the material is divided into three distinct segments. The first segment, denoted as I, corresponds to hardening within the strain range of 0 to 0.03. The second segment, denoted as II, corresponds to hardening within the strain range of 0.03 to 0.5. The third segment, denoted as III, corresponds to softening within the strain range of 0.5 to 0.8. (b) Affected and unaffected of double DRX. (c) The activity of twinning and slip is influenced by both double DRX, as well as by GBBDRX only [81].

twins that are favorably oriented results in a significant occurrence of TDRX. This occurrence leads to a drastic reduction in the rate of strain-hardening, causing it to continually decrease below the simulated rate that excludes double DRXs. As a result, in certain grains, the combination of significant reduction in grain size, heightened applied stress, and localized stress can lead to the attainment of a post-saturation state for secondary ETW where $V_{SecEtw} = 1$, and depletion of TDRX nucleation sites occurs. Inevitably, the serration of grain boundaries in re-crystallized grains that developed from twins occurs as a means of alleviating accumulated strain. (3) The softening region is characterized by a decrease in flow stress as strain increases. Only the simulation that incorporates DRX is able to precisely model this softening region. During this stage, the dominant GBBDRX supplants TDRX, resulting in flow softening and uninterrupted plastic deformation.

Based on the findings presented in Figure 7.22a, it is observed that the hardening rates obtained through simulations that solely utilize the GBBDRX mechanism and exclude double DRXs exhibit a considerably higher value compared to the experimental rate and the simulated rate incorporating TDRX, at the commencement of the second hardening segment. The GB migration according to GBBDRX rate is significantly limited in the coarse-grained domain during low-strain phase [93]. Consequently, the initial coarse grains with average diameter of 245.5 μm in the as-cast Mg alloys is challenging to refine during the low-strain stage without TDRX. Numerous twinning has the potential to propagate extensively within coarse-grained materials, leading to a sustained larger in strain-hardening rate. The intersections of twins and the junctions between twins and grain boundaries induces significant stress concentrations, ultimately resulting in the formation of numerous microcracks [80]. As a result, the occurrence of brittle fracture may occur in situations where the intense stress concentration cannot be adequately alleviated by the TDRX [94]. In the case of as-cast Mg alloys featuring coarse grains, the TDRX plays a pivotal role in alleviating stress concentrations, refining grains, facilitating the DRX transition to the GBBDRX, and preventing the onset of brittle fracture resulting from rapid strain hardening.

During the low-strain stage, the proliferation of twinning results in an elevated rate of strain hardening. During hot compression, a significant number of twins are consumed by the TDRX process. This results in twin-walled grains inheriting the c-axis orientation of the parent twins, thereby weakening the deformation basal texture. The application of TDRX may result in a weakening of the deformation texture observed in a coarse-grained Mg-2Zn-2Nd Mg alloy subjected to hot compression [95]. The promotion of basal slip activity and the reduction of ETW activity results in a substantial decline in the strain-hardening rate (Figure 7.23).

As the strain increases, fine grains exhibit a decreased tendency at TBs, which are predominantly consumed by prior TDRX. The bulging phenomenon, resulting from the migration of grain boundaries from the low dislocation density region towards the high dislocation density region, gradually replaces the TDRX and dominates the behavior of DRX. The formation of new low-dislocation-density zones presents increased opportunities for slip, thereby promoting the activity of prismatic slip (Figure 7.23b). It results in a significant flow-softening effect and prolonged plastic deformation (Figure 7.23a). Additionally, the GBBDRX causes a

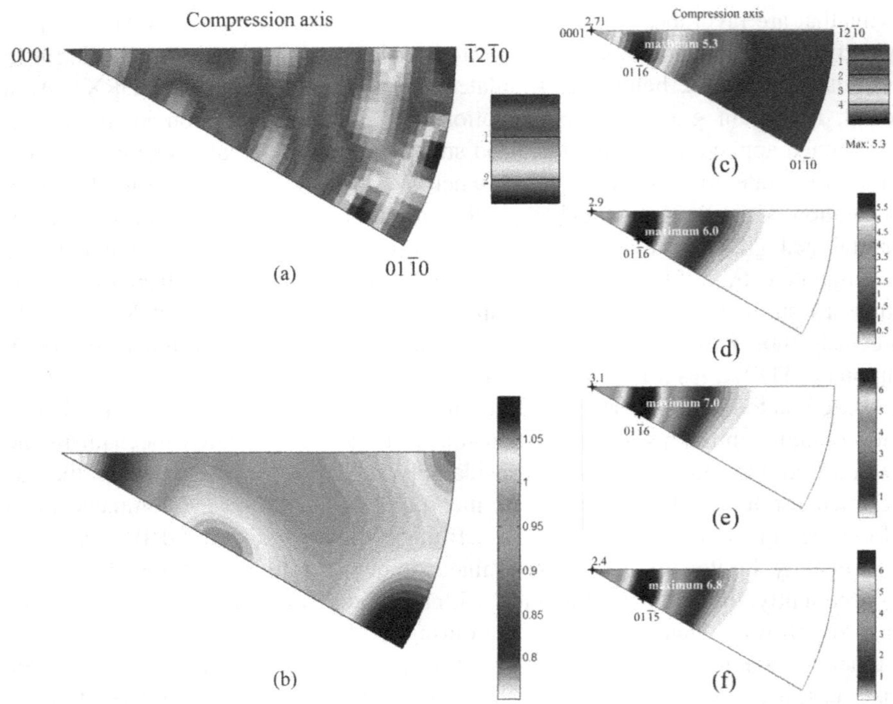

FIGURE 7.24 Inverse pole figure (IPF) comparison in-between (a) experimental initial texture; and (b) simulated random texture. IPFs acquired from c(a) the hot compression experiment; d(b) the VPSC simulation incorporating double DRX schemes; e(c) the VPSC simulation only incorporating GBBDRX scheme; and f(d) the VPSC simulation not incorporating DRX schemes [81].

reduction in pyramidal slip activity (Figure 7.23b), which strengthens the basal texture of the material (Figures 7.24e, f).

The experimental intensity of the strongest orientation is merely 2.57, exhibiting only a marginal improvement over other orientation. The experimental orientations are randomly dispersed (Figure 7.24), resembling a dispersed texture with no apparent strengthening. The inverse pole figure (IPF) map of the compression axis of the specimen reveals that the basal plane normal are inclined at an average angle of approximately 7.1° with respect to the compression direction along the z-axis (Figure 7.24a-c). The $\langle 01\bar{1}6\rangle$ plane also exhibited maximum intensities of 6.0 and 7.0, respectively (Figure 7.24d,e). Figure 7.24f demonstrates that the $\langle 01\bar{1}5\rangle$ plane exhibits the highest intensity at 6.8. Furthermore, the introduction of the TDRX effect results in a less pronounced basal texture, as illustrated in Figure 7.24d in comparison to Figure 7.24e. The introduction of the GBBDRX effect results in a more pronounced basal texture, as depicted in both Figure 7.24d and Figure 7.24e, in contrast to Figure 7.24f. The orientation of the c-axis in relation to the compression axis, exhibits the highest degree of similarity to the experimental outcome (Figure 7.24c,d). The results presented above demonstrate that the basal texture evolution is influenced by both the TDRX-induced weakening effect and the GBBDRX-induced strengthening effect.

7.3.4 Remarks

In this section, we investigate transverse propagation, nucleation, growth mechanisms of twinning, and the transition of DRX mechanisms in HCP Mg alloys.

A continuous phase-field model is presented to investigate the most frequently seen twinning mode. The effect of heterogeneous elasticity between the parent phase and the twin phase on the twin is negligible. In contrast to the elastic forces, the direction-dependent surface energy has a dominant effect on the formation of the PB interface at the twin front during twin growth. The PB interface is a high normal stress concentration and a high shear stress as well. Configuration force is a type of cohesive force. The distribution of this force on the TB prevents the growth of the twin, while the stress value on the PB interface is lower than that on the other twin planes, which means that the lateral propagation of DT is in favor of the creation of the PB interface. It is crucial to understand that the anisotropic energy of TBs is not accountable for the high aspect ratio characteristic of twins, and the interplay between thickening and transverse propagation mechanisms of the DT may control the dynamic equilibrium shape of twins.

A dislocation-mediated mechanism is described in theory for efficient nucleation and growth of nanoscale deformation twins of HCP metallic Mg. The increase in pile-up can decrease critical shear stress, which means that pile-up can propagate twin faults and assist in the formation of twin faults by generating stress concentration and appropriate directional stress fields. The critical shear stress increases as the nanoscale deformed twin grows. After a certain value of twin thickness is reached, however, the rising rate is much lower than that in the initial stage of twin formation. The thickness of the twin layer near its bottom is finite, which is determined by the number of twinning dislocations. The longitudinal cross-section of the twins is a nearly rectangular shape.

The transformation of the DRX mechanism of Mg alloy during hot compression is modeled theoretically. The evolutionary process is emulated by VPSC, which includes the continuous TDRX and GBBDRX schemes. Twinning within the coarse grains of as-cast Mg alloy is very prevalent in the low-strain stage. As the strain increases, ETW spreads quickly within the primary twinned grains and the matrix grains, resulting in a large number of DTW and twinned grains intersecting. The smaller twin-walled grains that inherit the c-axis direction of the matrix twin can weaken the deformed substrate texture, enhance the substrate slip behavior, and decrease ETW activity. However, the GBBDRX mechanism can reduce the density of partial dislocations, enhances the substrate texture by limiting pyramidal slip. Higher strain hardening rates due to massive twinning is significantly reduced due to the enhanced substrate slip behavior and reduced ETW activity generated by TDRX at the low strain stage and at the intermediate temperature of 573 K. The variety of DRX influences on texture and properties evolution is evident. This leads to a significant pragmatic result that it is the potential to quantitatively manage the textural evolution and mechanical properties of Mg alloys during plastic processing by governing the DRX transition. Initial specimens with traditional strong C-axis textures are generally less plastic under C-axis compression than specimens with random textures.

REFERENCES

[1] Sun W., Zhu Y., Marceau R., et al. 2019. Precipitation strengthening of aluminum alloys by room-temperature cyclic plasticity. *Science*, 363(6430): 972–975.

[2] Zhao H., Chakraborty P., Ponge D., et al. 2022. Hydrogen trapping and embrittlement in high-strength Al alloys. *Nature*, 602(7897): 437–441.

[3] Xu W., Birbilis N., Sha G., et al. 2015. A high-specific-strength and corrosion-resistant magnesium alloy. *Nature Materials*, 14(12): 1229–1235.

[4] Wu G., Chan K.-C., Zhu L., et al. 2017. Dual-phase nanostructuring as a route to high-strength magnesium alloys. *Nature*, 545(7652): 80–83.

[5] Esmaily M., Svensson J.E., Fajardo S., et al. 2017. Fundamentals and advances in magnesium alloy corrosion. *Progress in Materials Science*, 89: 92–193.

[6] Fan T.W., Wang Z.P., Lin J.J., et al. 2019. First-principles predictions for stabilizations of multilayer nanotwins in Al alloys at finite temperatures. *Journal of Alloys and Compounds*, 783: 765–771.

[7] Shao Q., Liu L., Fan T., et al. 2017. Effects of solute concentration on the stacking fault energy in copper alloys at finite temperatures. *Journal of Alloys and Compounds*, 726: 601–607.

[8] Kresse G. and Furthmüller J.J.P.r.B. 1996. Efficient iterative schemes for ab initio total-energy calculations using a plane-wave basis set. *Physical review B*, 54(16): 11169.

[9] Blöchl P.E.J.P.r.B. 1994. Projector augmented-wave method. *Physical review B*, 50(24): 17953.

[10] John P., Perdew K. and Burke M. 1997. Ernzerhof, Generalized Gradient Approximation Made Simple. *Phys. Rev. Lett*, 77: 3865–3868.

[11] Kibey S., Liu J., Johnson D., et al. 2007. Predicting twinning stress in fcc metals: Linking twin-energy pathways to twin nucleation. *Acta Materialia*, 55(20): 6843–6851.

[12] Liu L., Wang R., Wu X., et al. 2014. Temperature effects on the generalized planar fault energies and twinnabilities of Al, Ni and Cu: First principles calculations. *Computational Materials Science*, 88: 124–130.

[13] Soer W., De Hosson J.T.M., Minor A., et al. 2004. Effects of solute Mg on grain boundary and dislocation dynamics during nanoindentation of Al–Mg thin films. *Acta Materialia*, 52(20): 5783–5790.

[14] Fan T., Wei L., Tang B., et al. 2014. Effect of temperature-induced solute distribution on stacking fault energy in Mg–X (X= Li, Cu, Zn, Al, Y and Zr) solid solution: A first-principles study. *Philosophical Magazine*, 94(14): 1578–1587.

[15] Li Q., Xue S., Wang J., et al. 2018. High-Strength Nanotwinned Al Alloys with 9R Phase. *Advanced Materials*, 30(11): 1704629.

[16] Wang Z., Chen D., Fang Q., et al. 2019. Effects of finite temperature on the surface energy in Al alloys from first-principles calculations. *Applied Surface Science*, 479: 499–505.

[17] Kim S.P., Lee S.C., Lee K.R., et al. 2008. Asymmetric surface intermixing during thin-film growth in the Co-Al system: Role of local acceleration of the deposited atoms. *Acta Materialia*, 56: 1011–1017.

[18] Zhang J.M., Ma F. and Xu K.W. 2004. Calculation of the surface energy of FCC metals with modified embedded-atom method. *Applied Surface Science*, 229: 34–42.

[19] Wang X.C., Jia Y., Yao Q.K., et al. 2004. The calculation of the surface energy of high-index surfaces in metals at zero temperature. *Surface Science*, 551: 179–188.
[20] J. L. F. D. S. 2005. All-electron first-principles calculations of clean surface properties of low-Miller-index Al surfaces. *Physical Review B*, 71: 195416.
[21] Bohnen K.P. and Ho K.M. 1988. First principles calculation of lattice relaxation and surface phonons on Al(100). *Surface Science*, 207: 105–117.
[22] Vitos L., Ruban A.V., Skriver H.L., et al. 1998. The surface energy of metals. *Surface Science*, 411(1–2): 186–202.
[23] Galanakis I., Papanikolaou N. and Dederichs P.H. 2002. Applicability of the broken-bond rule to the surface energy of the fcc metals. *Surface Science*, 511: 1–12.
[24] Jiang Q., Lu H.M. and Zhao M. 2004. Modelling of surface energies of elemental crystals. *Journal of Physics: Condensed Matter*, 16: 521–530.
[25] Kim S.P., Lee K.R., Chung Y.C., et al. 2009. Molecular dynamics simulation study of deposition and annealing behaviors of Al atoms on Cu surface. *Journal of Applied Physics*, 105: 114312.
[26] Singh-Miller N.E. and Marzari N. 2009. Surface energies, work functions, and surface relaxations of low-index metallic surfaces from first principles. *Physical Review B*, 80: 235407.
[27] Kong Y., Shen L.M., Proust G., et al. 2011. Al–Pd interatomic potential and its application to nanoscale multilayer thin films. *Materials Science & Engineering A*, 530: 73–86.
[28] Jelinek B., Groh S., Horstemeyer M.F., et al. 2012. Modified embedded atom method potential for Al, Si, Mg, Cu, and Fe alloys. *Physical Review B*, 85: 245102.
[29] Rodríguez A.M., Bozzolo G. and Ferrante J. 1993. Multilayer relaxation and surface energies of fcc and bcc metals using equivalent crystal theory. *Surface Science*, 289: 100–126.
[30] Bernstein N., Goswami R. and Holtz R.L. 2012. Surface and Interface Energies of Complex Crystal Structure Aluminum Magnesium Alloys. *Metallurgical & Materials Transactions A*, 43: 2166–2176.
[31] Ho K.M. and Bohnen K.P. 1985. Investigation of multilayer relaxation on Al(110) with the use of self-consistent total-energy calculations. *Physical Review B*, 32(6): 3446–3450.
[32] Tyson W.R. and Miller W.A. 1977. Surface free energies of solid metals: Estimation from liquid surface tension measurements. *Surface Science*, 62: 267–276.
[33] Skriver H.L. and Rosengaard N.M. 1992. Surface energy and work function of elemental metals. *Physical Review B*, 46(11): 7157–7168.
[34] Polatoglou H.M., Methfessel M. and Scheffler M. 1993. Vacancy-formation energies at the (111) surface and in bulk Al, Cu, Ag, and Rh. *Physical Review B*, 48(3): 1877–1883.
[35] Silva J.L.F.D., Stampfl C. and Scheffler M. 2006. Converged properties of clean metal surfaces by all-electron first-principles calculations. *Surface Science*, 600: 703–715.
[36] Touloukian Y.S., Kirby R.K., Taylor R.E., et al. 1975. Thermal expansion: Metallic elements and alloys. *Thermophysical Properties of Matter*, 12.
[37] Liu L.H., Chen J.H., Fan T.W., et al. 2015. The possibilities to lower the stacking fault energies of aluminum materials investigated by first-principles energy calculations. *Computational Materials Science*, 108: 136–146.
[38] Wang Z., Fang Q., Fan T., et al. 2019. Effects of solute atoms on 9R phase stabilization in high-performance Al alloys: a first-principles study. *JOM*, 71: 2047–2053.

[39] Fan T.W., Luo L.G., Ma L., et al. 2013. Effects of Zn atoms on the basal dislocation in magnesium solution from Peierls-Nabarro model. *Materials Science and Engineering: A*, 582(2): 299–304.
[40] Xue S.C., Fan Z., Lawal O.B., et al. 2017. High-velocity projectile impact induced 9R phase in ultrafine-grained aluminium. *Nature Communications*, 8(1653): 1–9.
[41] Li Q., Xue S.C., Wang J., et al. 2018. High-Strength Nanotwinned Al Alloys with 9R Phase. *Advanced Materials*, 30(11): 1704629.
[42] Christian J.W. and Mahajan S. 1995. Deformation twinning. *Progress in Materials Science*, 39(1–2): 1–157.
[43] Pandey A., Kabirian F., Hwang J.-H., et al. 2015. Mechanical responses and deformation mechanisms of an AZ31 Mg alloy sheet under dynamic and simple shear deformations. *International Journal of Plasticity*, 68: 111–131.
[44] Pi Z.P., Fang Q.H., Liu B., et al. 2016. A phase field study focuses on the transverse propagation of deformation twinning for hexagonal-closed packed crystals. *International Journal of Plasticity*, 76: 130–146.
[45] Staroselsky A. and Anand L. 2003. A constitutive model for hcp materials deforming by slip and twinning: application to magnesium alloy AZ31B. *International Journal of Plasticity*, 19(10): 1843–1864.
[46] Hearmon R.F.S. 1984. The elastic constants of crystals and other anisotropic materials. *Landolt-Bornstein Tables*, III(18), 1154.
[47] Slutsky L.J. and Garland C.W. 1957. Elastic constants of magnesium from 4.2 K to 300 K. *Physical Review*, 107(4): 972.
[48] Lebensohn R.A. and Tomé C.N. 1993. A study of the stress state associated with twin nucleation and propagation in anisotropic materials. *Philosophical Magazine A*, 67(1): 187–206.
[49] Wang S.L., Sekerka R.F., Wheeler A.A., et al. 1993. Thermodynamically-consistent phase-field models for solidification. *Physica D: Nonlinear Phenomena*, 69(1–2): 189–200.
[50] Gururajan M.P. and Abinandanan T.A. 2007. Phase field study of precipitate rafting under a uniaxial stress. *Acta Materialia*, 55(15): 5015–5026.
[51] Hu S.Y. and Chen L.Q. 2001. A phase-field model for evolving microstructures with strong elastic inhomogeneity. *Acta Materialia*, 49(11): 1879–1890.
[52] Xu B., Capolungo L. and Rodney D. 2013. On the importance of prismatic/basal interfaces in the growth of (1¯ 012) twins in hexagonal close packed crystals. *Scripta Materialia*, 68(11): 901–904.
[53] McFadden G.B., Wheeler A.A., Braun R.J., et al. 1993. Phase-field models for anisotropic interfaces. *Physical Review E*, 48(3): 2016.
[54] Hu S.Y., Li Y.L., Zheng Y.X., et al. 2004. Effect of solutes on dislocation motion—a phase-field simulation. *International Journal of Plasticity*, 20(3): 403–425.
[55] Zhou K. 2012. Elastic field and effective moduli of periodic composites with arbitrary inhomogeneity distribution. *Acta Mechanica*, 223(2): 293–308.
[56] Yoo M.H. and Lee J.K. 1991. Deformation twinning in hcp metals and alloys. *Philosophical Magazine A*, 63(5): 987–1000.
[57] Clayton J.D. and Knap J. 2011. A phase field model of deformation twinning: nonlinear theory and numerical simulations. *Physica D: Nonlinear Phenomena*, 240(9-10): 841–858.
[58] Wang J., Beyerlein I.J. and Tomé C.N. 2010. An atomic and probabilistic perspective on twin nucleation in Mg. *Scripta Materialia*, 63(7): 741–746.
[59] Abdolvand H., Majkut M., Oddershede J., et al. 2015. On the deformation twinning of Mg AZ31B: A three-dimensional synchrotron X-ray diffraction experiment and crystal plasticity finite element model. *International Journal of Plasticity*, 70: 77–97.

[60] Bieler T.R., Wang L., Beaudoin A.J., et al. 2014. In situ characterization of twin nucleation in pure Ti using 3D-XRD. *Metallurgical and Materials Transactions A*, 45: 109–122.

[61] Ostapovets A. and Gröger R. 2014. Twinning disconnections and basal–prismatic twin boundary in magnesium. *Modelling and Simulation in Materials Science and Engineering*, 22(2): 025015.

[62] Kuhn C. and Müller R. 2010. A continuum phase field model for fracture. *Engineering Fracture Mechanics*, 77(18): 3625–3634.

[63] Leclercq L., Capolungo L. and Rodney D. 2014. Atomic-scale comparison between twin growth mechanisms in magnesium. *Materials Research Letters*, 2(3): 152–159.

[64] Feng H., Fang Q.H., Liu B., et al. 2017. Nucleation and growth mechanisms of nanoscale deformation twins in hexagonal-close-packed metal magnesium. *Mechanics of Materials*, 109: 26–33.

[65] Kumar M.A., Kanjarla A.K., Niezgoda S.R., et al. 2015. Numerical study of the stress state of a deformation twin in magnesium. *Acta Materialia*, 84: 349–358.

[66] Mendelson S. 1969. Zonal dislocations and twin lamellae in hcp metals. *Materials Science and Engineering*, 4(4): 231–242.

[67] Capolungo L. and Beyerlein I.J. 2008. Nucleation and stability of twins in hcp metals. *Physical Review B*, 78(2): 024117.

[68] Gutkin M.Y., Ovid'ko I.A. and Skiba N.V. 2007. Mechanism of deformation-twin formation in nanocrystalline metals. *Physics of the Solid State*, 49(5): 874–882.

[69] Gutkin M.Y., Ovid'Ko I.A. and Skiba N.V. 2003. Crossover from grain boundary sliding to rotational deformation in nanocrystalline materials. *Acta Materialia*, 51(14): 4059–4071.

[70] Vladimirov V.I. and Romanov A.E.e.. 1986. Disclinations in crystals. *Leningrad Izdatel Nauka*.

[71] Gutkin M.Y., Ovid'ko I.A. and Skiba N.V. 2005. Emission of partial dislocations from triple junctions of grain boundaries in nanocrystalline materials. *Journal of Physics D: Applied Physics*, 38(21): 3921.

[72] Gutkin M.Y., Ovid'Ko I.A. and Skiba N.V. 2004. Emission of partial dislocations by grain boundaries in nanocrystalline metals. *Physics of the Solid State*, 46: 2042–2052.

[73] Hirth J.P., Lothe J. and Mura T. 1983. Theory of dislocations. *Journal of Applied Mechanics*, 50(2): 476.

[74] Partridge P.G. 1967. The crystallography and deformation modes of hexagonal close-packed metals. *Metallurgical Reviews*, 12(1): 169–194.

[75] Serra A. and Bacon D.J. 1991. Computer simulation of twinning dislocation in magnesium using a many-body potential. *Philosophical Magazine A*, 63(5): 1001–1012.

[76] Koike J. 2005. Enhanced deformation mechanisms by anisotropic plasticity in polycrystalline Mg alloys at room temperature. *Metallurgical and Materials Transactions A*, 36(7): 1689–1696.

[77] Yu Q., Qi L., Chen K., et al. 2012. The nanostructured origin of deformation twinning. *Nano Letters*, 12(2): 887–892.

[78] Galindo-Nava E.I. and Rivera-Díaz-del-Castillo P.E.J. 2014. Thermostastitical modelling of deformation twinning in HCP metals. *International Journal of Plasticity*, 55: 25–42.

[79] Abdolvand H. and Wilkinson A.J. 2016. On the effects of reorientation and shear transfer during twin formation: Comparison between high resolution electron backscatter diffraction experiments and a crystal plasticity finite element model. *International Journal of Plasticity*, 84: 160–182.

[80] Xie C., Fang Q.H., Liu X., et al. 2016. Theoretical study on the $\{\bar{1}012\}$ deformation twinning and cracking in coarse-grained magnesium alloys. *International Journal of Plasticity*, 82: 44–61.

[81] Xie C., He J.M., Zhu B.W., et al. 2018. Transition of dynamic recrystallization mechanisms of as-cast AZ31 Mg alloys during hot compression. *International Journal of Plasticity*, 111: 211–233.

[82] Sitdikov O. and Kaibyshev R. 2001. Dynamic recrystallization in pure magnesium. *Materials Transactions*, 42(9): 1928–1937.

[83] Lebensohn R.A. and Tomé C.N. 1993. A self-consistent anisotropic approach for the simulation of plastic deformation and texture development of polycrystals: application to zirconium alloys. *Acta metallurgica et materialia*, 41(9): 2611–2624.

[84] Hutchinson J.W. 1976. Bounds and self-consistent estimates for creep of polycrystalline materials. *Proceedings of the Royal Society of London. A. Mathematical and Physical Sciences*, 348(1652): 101–127.

[85] Kuang J., Low T.S.E., Niezgoda S.R., et al. 2016. Abnormal texture development in magnesium alloy Mg–3Al–1Zn during large strain electroplastic rolling: Effect of pulsed electric current. *International Journal of Plasticity*, 87: 86–99.

[86] Zhang J., Chen B. and Liu C. 2014. An investigation of dynamic recrystallization behavior of ZK60-Er magnesium alloy. *Materials Science and Engineering: A*, 612: 253–266.

[87] Basu I. and Al-Samman T. 2015. Twin recrystallization mechanisms in magnesium-rare earth alloys. *Acta Materialia*, 96: 111–132.

[88] Al-Samman T., Molodov K.D., Molodov D.A., et al. 2012. Softening and dynamic recrystallization in magnesium single crystals during c-axis compression. *Acta Materialia*, 60(2): 537–545.

[89] Yu S., Liu C., Gao Y., et al. 2017. Dynamic recrystallization mechanism of Mg-8.5 Gd-2.5 Y-0.4 Zr alloy during hot ring rolling. *Materials Characterization*, 131: 135–139.

[90] Hildenbrand A., Tóth L.S., Molinari A., et al. 1999. Self-consistent polycrystal modelling of dynamic recrystallization during the shear deformation of a Ti IF steel. *Acta Materialia*, 47(2): 447–460.

[91] Tome C., Canova G.R., Kocks U.F., et al. 1984. The relation between macroscopic and microscopic strain hardening in FCC polycrystals. *Acta Metallurgica*, 32(10): 1637–1653.

[92] Barnett M. 2003. A Taylor model based description of the proof stress of magnesium AZ31 during hot working.

[93] Zhou G., Li Z., Li D., et al. 2017. A polycrystal plasticity based discontinuous dynamic recrystallization simulation method and its application to copper. *International Journal of Plasticity*, 91: 48–76.

[94] Somekawa H., Singh A. and Mukai T. 2009. Fracture mechanism of a coarse-grained magnesium alloy during fracture toughness testing. *Philosophical Magazine Letters*, 89(1): 2–10.

[95] Wang T., Jonas J.J. and Yue S. 2017. Dynamic recrystallization behavior of a coarse-grained Mg-2Zn-2Nd magnesium alloy. *Metallurgical and Materials Transactions A*, 48: 594–600.

8 Chemomechanical Modeling of Lithiation

8.1 INTRODUCTION

Sustainable energy would be the great challenge for meeting the needs of humans in the future [1–3]. Due to rapid global industrialization, some serious problems have appeared, such as energy shortage and environmental pollution. In order to solve these issues, renewable energy sources are used to provide highly effective storage and mobile energy for meeting the extensive power needs of equipment. The electrochemical energy storage achieves the transformation of energy between chemical energy and electrical energy based on migration of electrons and has been used for the past hundreds of years. The lithium-ion battery (LIB) shows a high rate capability, cycle performance, and energy density, in comparison with electrochemical capacitor and fuel cell [1–5].

LIBs have developed by leaps and bounds in large-scale usage in automotive and mobile electronic devices [6–8]. For the charging, Li-ion is firstly extracted from a negative electrode, and then diffuses through the electrolyte towards a positive electrode; for the discharging, Li-ion moves from a positive electrode to electrolyte, and finally to a negative electrode [9,10]. Unfortunately, the long cycling process of LIB results in the damage inside the electrode, and reduce performance or even failure. One of the reasons is the diffusion-induced stress (DIS) to cause the microstructure evolution of the electrode materials. In recent years, the influence of some key factors on DIS is investigated using the modeling and experiment [8–10]. For example, the small size and large aspect ratio reduce DIS of the ellipsoidal electrode [11]. The modulus and tension of surface relieve DIS of the nanoscale electrode [12]. The increase of temperature reduces the Li-ion concentration gradient [13]. The DIS leads to the nucleation and growth of crack in a nanowire electrode [14]. The impacts of the dislocations on DIS are reported in a spherical electrode [15]. During the potentiostatic and galvanostatic processes, the Li-ion diffusion cause the dislocations for suppressing the crack nucleation in a cylindrical electrode [16]. In the above works, various influence factors lead to DIS, which are separately studied based on experiment, simulation and model. However, the coupled effect on DIS under Li-ion diffusion is rarely considered.

This chapter develops a chemomechanical modeling of lithiation in the spherical and cylindrical electrolyte considering the coupled effect during LIB charging at different operations, and analyses the Li-ion concentration and DIS. The stresses generated from the surface, dislocation and strain energy are computed. In addition,

DOI: 10.1201/9781003225706-8

a theoretical method is used to provide some useful information for prolong life of LIB in the nanoscale electrode.

8.2 BASIC MODELING

8.2.1 Model of Cylindrical Electrolyte

The electrodes are separated by a solid or a liquid electrolyte (Figure 8.1a), and Li-ions are transported from a positive to negative electrode during discharging. Figure 8.1b shows a cylindrical LIB, a simple structure, in which two electrodes are connected to a current collector.

Using a cylindrical coordinate system, the relationship of stress and strain is expressed by [11]:

$$\varepsilon_r = \frac{1}{E}[\sigma_r - \nu(\sigma_\theta + \sigma_z)] + \frac{1}{3}\Omega C(r, t) \tag{8.1}$$

$$\varepsilon_\theta = \frac{1}{E}[\sigma_\theta - \nu(\sigma_r + \sigma_z)] + \frac{1}{3}\Omega C(r, t) \tag{8.2}$$

$$\varepsilon_z = \frac{1}{E}[\sigma_z - \nu(\sigma_r + \sigma_\theta)] + \frac{1}{3}\Omega C(r, t) \tag{8.3}$$

where ε_r, ε_θ and ε_z denote the strains along the radial, tangential and axial direction. σ_r, σ_θ and σ_z denote the stress along the radial, tangential and axial direction. E is Young's modulus, ν is Poisson ratio, Ω is the partial molar volume of solute, and $C(r, t)$ is the Li-ion concentration.

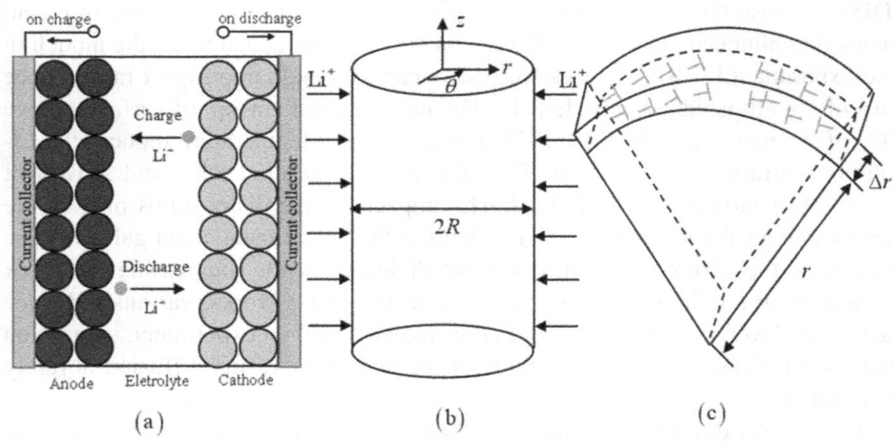

FIGURE 8.1 (a) Schematic composition of a LIB. (b) Li-ion diffusion in a cylindrical electrode of the diameter 2 R during insertion and extraction, where the diameter of cylindrical electrode is 2 R. (c) The edge dislocations nucleated from an element of diffusion layer in cylindrical electrode [16].

For the infinitesimal deformation, the radial and tangential strains are given by:

$$\varepsilon_r = \frac{du}{dr}, \quad \varepsilon_\theta = \frac{u}{r}, \quad \varepsilon_r = 0 \tag{8.4}$$

where u is the radial displacement.

The stress boundary condition is expressed

$$\sigma_r|_{r=R} = 0, \ \sigma_r|_{r=0} = finite \tag{8.5}$$

The mechanical equilibrium is established much faster than the diffusion equilibrium [17], due to the atomic diffusion much slower than the elastic deformation in solids. Regardless of the body force, the mechanical equilibrium is written as

$$\frac{d\sigma_r}{dr} + \frac{\sigma_r - \sigma_\theta}{r} = 0 \tag{8.6}$$

Based on Eqs. (8.1)–(8.6), in the plane strain condition the stresses are obtained by

$$\sigma_r = \frac{E\Omega}{3(1-\nu)}\left[\frac{1}{r^2}\int_0^R rC(r,t)dr - \frac{1}{r^2}\int_0^r rC(r,t)dr\right] \tag{8.7}$$

$$\sigma_\theta = \frac{E\Omega}{3(1-\nu)}\left[\frac{1}{r^2}\int_0^R rC(r,t)dr + \frac{1}{r^2}\int_0^r rC(r,t)dr - C(r,t)\right] \tag{8.8}$$

$$\sigma_z = \frac{E\Omega}{3(1-\nu)}\left[\frac{2\nu}{r^2}\int_0^R rC(r,t)dr - C(r,t)\right] \tag{8.9}$$

Figure 8.1c shows the relationship of Li-ion concentration and dislocation density in a radial infinitesimal cylindrical surface layer [18]. Here, at the lower surface of r in the diffused layer, $C(r, t)$ is the Li-ion concentration, and the dislocation density induced by diffusion is $\rho(r, t)$; at the upper surface of $r + \Delta r$, the concentration becomes $C(r, t) + (\partial C/\partial r)dr$. Thus, the dislocation density induced by the Li-ion concentration gradient is expressed [15,16]

$$\rho(r,t) = \frac{\beta}{b}\frac{\partial C(r,t)}{\partial r} \tag{8.10}$$

where b is the Burgers vector, and β is the coefficient dependent upon the partial molar volume with $\Omega/3$.

During the beginning of insertion, the maximum stress occurs at the surface, which triggers the positive edge dislocations nucleated at the surface [18]. For the charging, a high density of dislocations in a nanowire electrode is observed by a transmission electron microscope (TEM) [19]. For the sake of simplicity, DIS from the surface is always higher than the critical stress required for dislocation nucleation. As the diffusion proceeds, the dislocations move into the interior attribute to DIS. The dislocation interacting with diffusion solute is investigated [19,20]. Here, it is assumed that the long-range interaction of the dislocation and diffusing solute is the dominant contribution. The flow stress is proportional to the square root of dislocation density in the following form [21]:

$$\sigma_\tau = M\alpha\mu b\sqrt{\rho} \tag{8.11}$$

where M is the Taylor orientation factor, α is the empirical constant, and μ is the shear modulus.

During the Li-ion oxidation and reduction, the stress produced by dislocation causes a resistance to the diffusion process. Considering the Li-ion diffusion-induced dislocation, the modified DIS is expressed as

$$\sigma_r = \frac{E\Omega}{3(1-\nu)}\left[\frac{1}{r^2}\int_0^R rC(r,t)dr - \frac{1}{r^2}\int_0^r rC(r,t)dr\right] - \sigma_\tau \tag{8.12}$$

Hence, DIS at any location of electrode and time with the dislocation resistance is obtained if the distribution of composition is known.

8.2.2 Model of Spherical Electrolyte

A spherical negative electrode with the diameter $2R$ is considered (Figure 8.2a), as a simple structure of operating principle (Figure 8.2b). To enhance the cycling performance, the new electrode materials are developed [22]. A spherical electrode structure is designed based on surface modification method (Figure 8.3), where a predefined dislocation density is formed by heat treatment to improve the cycling ability. Figure 8.3 presents the arbitrary radial infinitesimal spherical element [18]. The Li-ion concentration of bulk takes place on the lower surface of the diffusion layer, and the diffusion-induced dislocation density is expressed as Eq. (8.10).

To study the structural stability of the electrode, the stress is determined during the charging. Due to the electrode is at nanoscale, the surface effect on the mechanical response is considered [25–27]. Based on the previous work [28–30], the surface relation for a nanoscale electrode is expressed as

$$\sigma_\theta^s = \tau^s + 2(\mu^s + \lambda^s)\varepsilon_\theta^s = \tau^s + k^s\varepsilon_\theta^s \tag{8.13}$$

where σ_θ^s and ε_θ^s are the surface stress and strain components, λ^s and μ^s are the surface Lamé constants, and τ^s is the surface tension.

The Laplace pressure of the nanoscale electrode during cycling is given as

$$\Delta P = 2\gamma/R \tag{8.14}$$

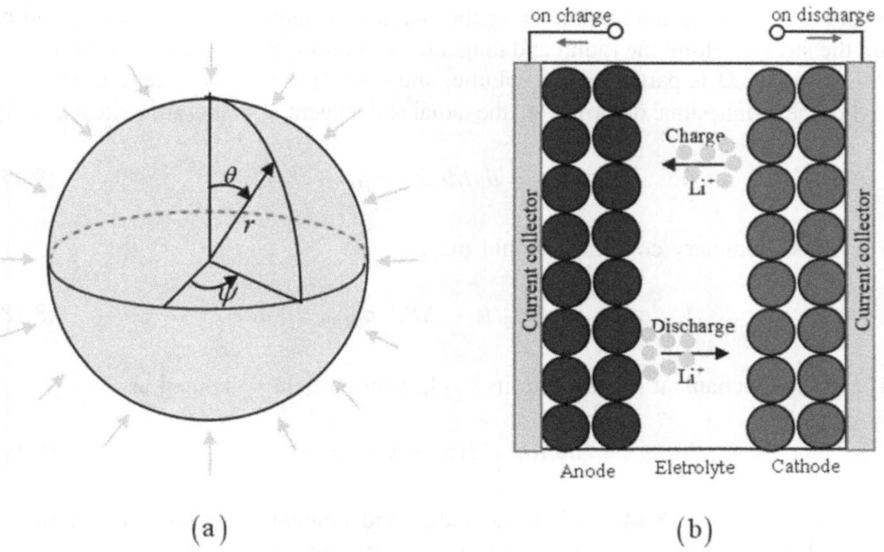

FIGURE 8.2 (a) Schematic illustration for Li-ion diffusion in a spherical electrode, where the diameter is 2 R. (b) Schematic composition of LIB [23,24].

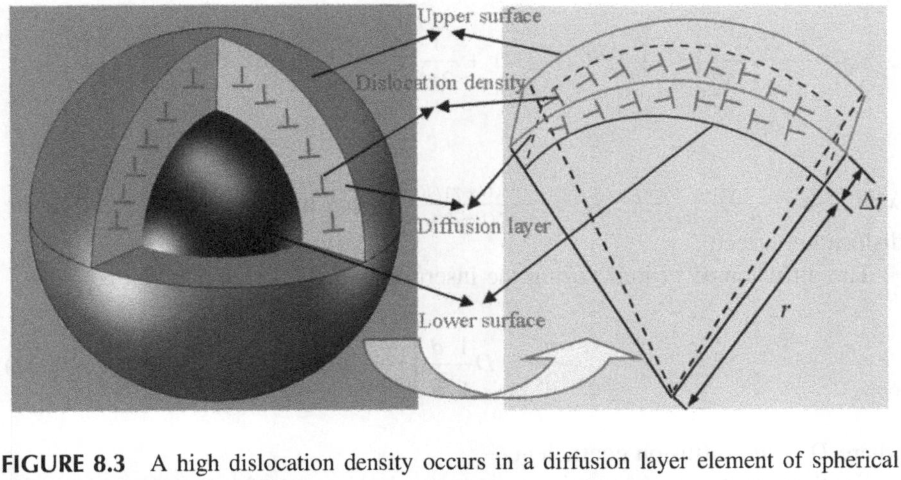

FIGURE 8.3 A high dislocation density occurs in a diffusion layer element of spherical electrode during cycling [23].

where ΔP is Laplace pressure relying on the electrode radius, and γ is surface energy.

The relationship of the stress and strain for the spherical coordinate system is expressed as [11,31–33]:

$$\varepsilon_r = (\sigma_r - 2\nu\sigma_\theta)/E + \Omega C(r, t)/3 \tag{8.15}$$

$$\varepsilon_\theta = [(1 - \nu)\sigma_\theta - \nu\sigma_r]/E + \Omega C(r, t)/3 \tag{8.16}$$

where ε_r and ε_θ are the strains along the radial and tangential direction, σ_r and σ_θ are the stresses along the radial and tangential direction, E is Young's modulus, ν is Poisson ratio, Ω is partial molar volume, and $C(r, t)$ is molar concentration.

For the infinitesimal deformation, the radial and tangential strains are written as [38]:

$$\varepsilon_r = du/dr,\ \varepsilon_\theta = u/r \tag{8.17}$$

The stress boundary condition should meet

$$\sigma_r|_{r=R} = -2\sigma_\theta^s/R - \Delta P, \quad \sigma_r|_{r=0} = finite \tag{8.18}$$

The static mechanical equilibrium in a spherical particle is defined as

$$d\sigma_r/dr + 2(\sigma_r - \sigma_\theta)/r = 0 \tag{8.19}$$

According to Eqs. (8.14)–(8.20), the radial and tangential stresses considering a coupled effect on the spherical electrode is described as:

$$\sigma_r = \frac{2E\Omega}{3(1-\nu)}\left[\frac{S_1}{r^3}\int_0^R r^2C(r,t)dr - \frac{1}{r^3}\int_0^r r^2C(r,t)dr\right] - S_2 - \sigma_\tau \tag{8.20}$$

$$\sigma_\theta = \frac{2E\Omega}{3(1-\nu)}\left[\frac{S_1}{r^3}\int_0^R r^2C(r,t)dr + \frac{1}{2r^3}\int_0^r r^2C(r,t)dr - \frac{1}{2}C(r,t)\right] - S_2 - \sigma_\tau \tag{8.21}$$

with $S_1 = \frac{1-k^s(1+\nu)/(ER)}{1+2k^s(1-2\nu)/(ER)}$, $S_2 = \frac{2(\tau^s+\gamma)/R}{1+2k^s(1-2\nu)/(ER)}$, and the stress σ_τ is related to the dislocation effect.

The diffusion of Li-ions during the insertion and extraction is given by:

$$\frac{\partial C}{\partial t} = D\frac{1}{r}\frac{\partial}{\partial r}\left(r\frac{\partial C}{\partial r}\right) \tag{8.22}$$

where D is the diffusion coefficient.

For the potentiostatic operation (constant voltage and surface concentration), the boundary conditions should meet:

$$C(R,t) = C_R,\ C(0,t) = finite \tag{8.23}$$

During insertion for potentiostatic, the Li-ion concentration is given as [34]:

$$\frac{C(r,t) - C_0}{C_R - C_0} = 1 + 2\sum_{k=1}^{\infty}\frac{(-1)^k\sin(k\pi\phi)}{k\pi\phi}\exp(-\varphi k^2\pi^2) \tag{8.24}$$

where C_0 is the initial Li-ion concentration, $\phi = r/R$ and $\varphi = Dt/R^2$.

Based on Eqs. 8.20, 8.21, and 8.24, the normalization stresses along the radial and tangential direction are presented as:

$$\begin{aligned}\zeta_r &= \frac{\sigma_r}{E\Omega\,(C_R - C_0)\,/\,[3(1-\nu)]} = \frac{2}{3}\frac{C_0}{C_R - C_0}(S_1 - 1) + \frac{S_2}{E\Omega\,(C_R - C_0)\,/\,[3(1-\nu)]}\\ &\quad - 4\sum_{k=1}^{\infty}\left[\frac{S_1}{k^2\pi^2} + (-1)^k\frac{\sin(k\pi\phi) - k\pi\phi\cos(k\pi\phi)}{(k\pi\phi)^3}\right]\exp(-\varphi k^2\pi^2)\end{aligned} \tag{8.25}$$

$$\begin{aligned}\zeta_\theta &= \frac{\sigma_\theta}{E\Omega\,(C_R - C_0)\,/\,[3(1-\nu)]} = \frac{2}{3}\frac{C_0}{C_R - C_0}(S_1 - 1) + \frac{S_2}{E\Omega\,(C_R - C_0)\,/\,[3(1-\nu)]}\\ &\quad - 2\sum_{k=1}^{\infty}\left[\frac{2S_1}{k^2\pi^2} - (-1)^k\frac{\sin(k\pi\phi) - k\pi\phi\cos(k\pi\phi)}{(k\pi\phi)^3} + (-1)^k\frac{\sin(k\pi\phi)}{k\pi\phi}\right]\exp(-\varphi k^2\pi^2)\end{aligned} \tag{8.26}$$

Based on Eq. (8.24) and Eq. (8.11), the normalized stress induced by dislocation is obtained by:

$$\begin{aligned}\zeta_\tau &= \frac{\sigma_\tau}{M\alpha\mu\sqrt{2b\beta\,(C_R - C_0)/R}}\\ &= \left[\sum_{k=1}^{\infty}\ (-1)^k\frac{k\pi\phi\cos(k\pi\phi) - \sin(k\pi\phi)}{k\pi\phi^2}\exp(-\varphi k^2\pi^2)\right]^{1/2}\end{aligned} \tag{8.27}$$

Here, the stress is determined for the arbitrary position and time.

The total strain energy is given as [13]:

$$W = 4\pi\int_0^R wr^2dr \tag{8.28}$$

where w is the strain energy density, which is equal to $w = [\sigma_r^2 + 2\sigma_\theta^2 - 2\nu\sigma_\theta(2\sigma_r + \sigma_\theta)]/2E$. In addition, the total strain energy come from three parts [13]:

$$E_{total} = E_{bulk} + E_{surface} + E_{dislocation} \tag{8.29}$$

where E_{total} is the total elastic energy, E_{bulk} is the strain energy from the bulk deformation, $E_{surface}$ is the strain energy from the surface deformation, and $E_{dislocation}$ is the strain energy from the dislocation.

8.2.3 Model of Hollow Spherical Electrolyte

Based on the Von Mises yield criterion, the plastic deformation occurs [35] when it is satisfied by the following equation:

$$|\sigma_\theta - \sigma_r| = \sigma_Y(r) \tag{8.30}$$

It should be noted that the yield stress is considered as a constant value during charging [36], when the plastic deformation occurs. Here, the yield stress is a linear function of the Li-ion concentration, which is given as:

$$\sigma_Y(r) = \sigma_{Y0} - \varepsilon_1 C(r, t) \tag{8.31}$$

where ε_1 is the slope of the linear function for the Li-ion concentration, and δ_{Y0} is the yield stress. According to the previous work [37,38], the plastic deformation takes place for the outer and inner surfaces (Figure 8.4). When the outer concentration increases, the plastic region spreads outward in the outer surface.

The von Mises yield criterion and the equilibrium equation are given by [36,40]:

$$\frac{d\sigma}{dr} = \frac{2}{r}\sigma_Y(r) \tag{8.32}$$

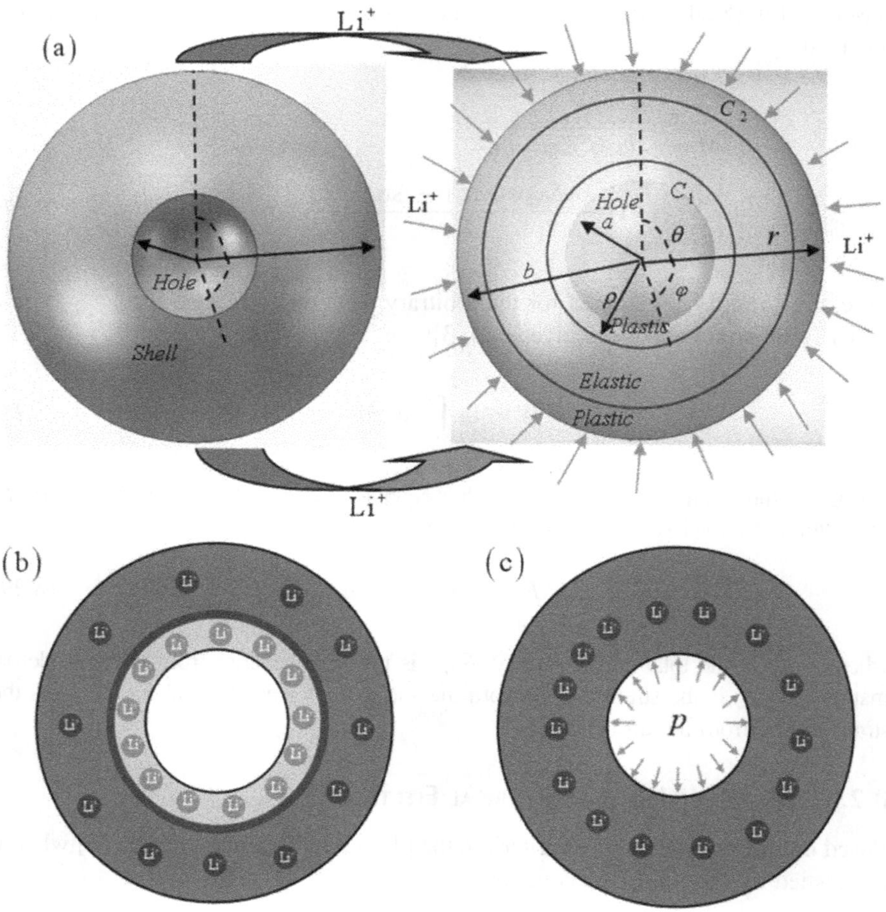

FIGURE 8.4 (a) Schematic illustration for Li-ion diffusion and heat loading at the surface of electrode. a is an inner radius, and b is an outer radius. (b) The inner surface introduced the biological membrane. (c) The inner surface applied compressive stress of p [39].

According to the boundary condition of $\sigma_r(a) = 0$ in the inner surface and $\sigma_r(b) = 0$ in the outer surface, the radial and hoop stresses are obtained in the plastic region:

$$\sigma_r = \int_a^r \frac{2}{r}\delta_Y(r)dr, \quad a \le r \le \rho_1 \tag{8.33}$$

$$\sigma_\theta = \int_a^r \frac{2}{r}\sigma_Y(r)dr + \sigma_Y(r), \quad a \le r \le \rho_1 \tag{8.34}$$

$$\sigma_r = \int_{\rho_2}^r \frac{2}{r}\sigma_Y(r)dr - \int_{\rho_2}^b \frac{2}{r}\sigma_Y(r)dr, \quad \rho_2 \le r \le b \tag{8.35}$$

$$\sigma_\theta = \int_{\rho_2}^r \frac{2}{r}\sigma_Y(r)dr - \int_{\rho_2}^b \frac{2}{r}\sigma_Y(r)dr - \sigma_Y(r), \quad \rho_2 \le r \le b \tag{8.36}$$

Combining Eqs. (8.15) and (8.16), the stresses in the elastic region are written as:

$$\sigma_r = -\frac{2E}{1-\nu}\frac{1}{r^3}\int_{\rho_1}^r G(r)r^2dr + \frac{EB_1}{1-2\nu} - \frac{2EB_2}{1+\nu}\frac{1}{r^3}, \quad \rho_1 \le r \le \rho_2 \tag{8.37}$$

$$\sigma_\theta = \frac{E}{1-\nu}\frac{1}{r^3}\int_{\rho_1}^r G(r)r^2dr + \frac{EB_1}{1-2\nu} + \frac{EB_2}{1+\nu}\frac{1}{r^3} - \frac{EG(r)}{1-\nu}, \quad \rho_1 \le r \le \rho_2 \tag{8.38}$$

Here B_1 and B_2 are the constant, which is determined from the continuity condition at the elastic-plastic interface.

Because the radial stress is continuous for the interface between the elastic and plastic region, the constants B_1 and B_2 are determined from Eqs. (8.33)–(8.36):

$$B_1 = \frac{1-2\nu}{E}\frac{1}{\rho_2^3-\rho_1^3}\left[\frac{2E}{1-\nu}\int_{\rho_1}^{\rho_2} G(r)r^2dr - \rho_2^3\int_{\rho_2}^b \frac{2}{r}\sigma_Y(r)dr - \rho_1^3\int_a^{\rho_1}\frac{2}{r}\sigma_Y(r)dr\right] \tag{8.39}$$

$$B_2 = \frac{1+\nu}{2E}\frac{\rho_2^3\rho_1^3}{\rho_2^3-\rho_1^3}\left[\frac{2E}{1-\nu}\frac{1}{\rho_2^3}\int_{\rho_1}^{\rho_2} G(r)r^2dr - \int_{\rho_2}^b \frac{2}{r}\sigma_Y(r)dr - \int_a^{\rho_1}\frac{2}{r}\sigma_Y(r)dr\right] \tag{8.40}$$

The stress meets the von Mises yield criterion in the elastic-plastic interface:

$$|\sigma_\theta(r) - \sigma_r(r)|_{r=\rho_1} = \sigma_Y(\rho_1) \tag{8.41}$$

$$|\sigma_\theta(r) - \sigma_r(r)|_{r=\rho_2} = \sigma_Y(\rho_2) \tag{8.42}$$

where the plastic radii ρ_1 and ρ_2 are determine by taking Eqs. (8.34)–(8.37) into Eqs. (8.41) and (8.42).

8.3 DIFFUSION-INDUCED DAMAGE

8.3.1 Cylindrical Electrolyte

With and without the dislocation effect as well as the size effect, the distributions of Li-ion concentration and corresponding DIS are calculated under galvanostatic and potentiostatic operations.

8.3.1.1 Galvanostatic (Constant Current and Surface Flux) Operation

The electrode material is regarded as an isotropic linear elastic solid, and the quasi-static deformation is assumed for an insertion and extraction of Li-ion. The diffusion equation is written as Eq (8.22) [34].

The galvanostatic boundary condition is expressed as (Figure 8.1b):

$$-D\frac{\partial C}{\partial r}\bigg|_{r=R} = -\frac{I}{F},\ -D\frac{\partial C}{\partial r}\bigg|_{r=0} = 0 \tag{8.43}$$

where F=96486.7 C mol^{-1} is Faraday's constant. The initial Li-ion concentration is assumed to be zero.

Without the dislocation effect, the Li-ion concentration during the insertion is expressed by [34]:

$$\frac{C(r,t) - C_0}{IR/FD} = -2\varphi - \frac{\phi^2}{4} + \frac{1}{4} + \frac{4}{\phi}\sum_{k=1}^{\infty}\frac{J_1(\phi\lambda_k)\exp(-\varphi\lambda_k^2)}{\lambda_k^2 J_0(\lambda_k)} \tag{8.44}$$

where C_0 is the initial Li-ion concentration, $\phi = r/R$, $\varphi = Dt/R^2$ and J_0 are the Bessel function of the first kind, and λ_k is the root of $J_1(\lambda_k)$.

Without the dislocation influence, the radial stress is:

$$\frac{\sigma_r}{[E\Omega/3(1-\nu)](IR/FD)} = \frac{\phi^2 - 1}{8} - \frac{2}{\phi}\sum_{k=1}^{\infty}\frac{J_1(\phi\lambda_k)\exp(-\varphi\lambda_k^2)}{\lambda_k^3 J_0(\lambda_k)} \tag{8.45}$$

With the time increasing, DIS reaches a steady state for the Li-ion concentration. The DIS is reversible for the intercalation and deintercalation process. During the

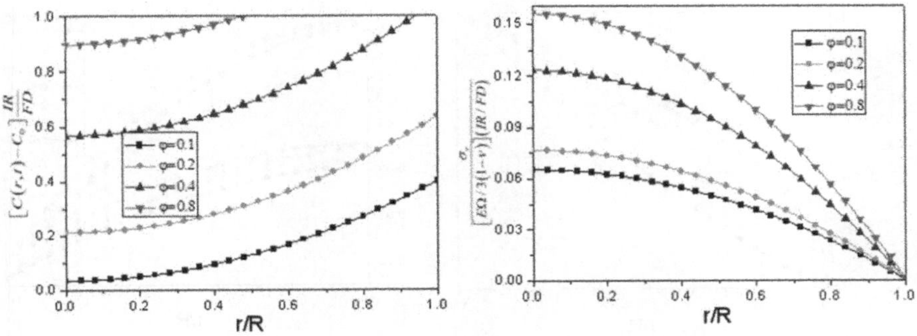

FIGURE 8.5 Distribution of Li-ion concentration (a) and the corresponding radial stress (b) without the dislocation influence [16].

inserting process, the dislocation is nucleated at the surface until the stress reaches a steady state [15,16,41]. When the dislocation effect is absent, DIS shows the same magnitude with opposites signs.

Figure 8.5 shows the normalized concentration and the radial stress with the increase of position and time, where the dislocation effect is ignored. As the time and radial position increase, the Li-ion concentration of a cylindrical electrode goes up correspondingly, and the stress induced by Li-ion migration serves as a barrier to the diffusion. There is a sharp rise for asymptotic behavior at high radial position, and a constant at low position. The slow-to-fast transition of concentration occurs when the radial position exceeds 0.2 [13–15]. The tensile radial stress reduces at the larger radial position. The radial stress keeps zero with the increasing time at the surface [42], and the highest at the center (Figure 8.5b). In addition, there is a sharp rise of asymptotic behavior near to surface. For the insertion and extraction, the value of radial stress is the same but in the opposite direction.

Based on Eqs. (8.22) and (8.11), DIS is obtained by:

$$\frac{\sigma_\tau}{M\alpha\mu\sqrt{\beta bI/FD}} = \left[\frac{4}{\phi}\sum_{k=1}^{\infty}\frac{\partial J_1(\phi\lambda_k)}{\partial\phi}\frac{\exp(-\varphi\lambda_k^2)}{\lambda_k^2 J_0(\lambda_k)} - \frac{\phi}{2} - \frac{4}{\phi^2}\sum_{k=1}^{\infty}\frac{J_1(\phi\lambda_k)\exp(-\varphi\lambda_k^2)}{\lambda_k^2 J_0(\lambda_k)}\right]^{1/2} \tag{8.46}$$

To reveal the coupled DIS during charging, Figure 8.6a shows the DIS with the increased time. The tensile stress induced by dislocation occurs for insertion, which reduces continuously to zero from the surface to the center. The asymptotic behavior also depends upon the time, especially at a high radial position. After a sufficient long time, the tensile stress tends to a stability, and expressed as

$$\sigma_\tau^0\big|_{t\to\infty} = M\alpha\mu\sqrt{\frac{b\beta I\phi}{FD}} \tag{8.47}$$

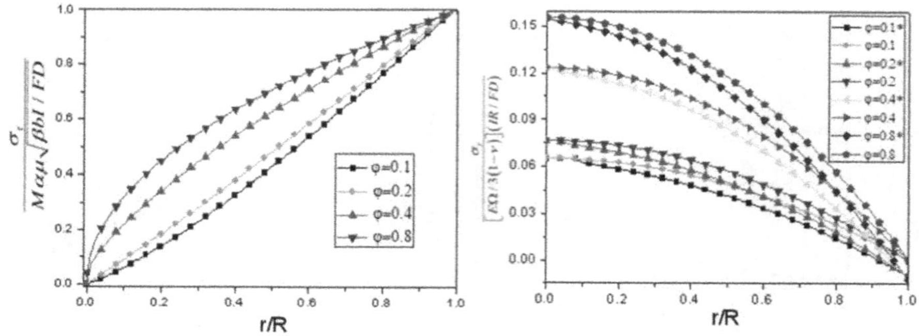

FIGURE 8.6 The distribution of dislocation-induced stress (a) and the corresponding radial stress (b). The dislocation-induced stress is normalized during insertion [16].

TABLE 8.1
Material Properties and Constants

Symbol	Name	Value
E	Young's modulus of Li-Si	30–80 Gpa
b	Burgers vector	2.532 Å
ν	Poisson's ratio	0.22
D	Diffusion coefficient	1 2×10^{-18} m^2 s^{-1}
$C_{s\ max}$	Max stoichiometric Li concentration	2.0152×10^{-4} mol m^{-3}
Ω	Partial molar volume of solute	2 $\times10^{-5}$ m^3 mol^{-1}
M	Taylor orientation factor	1.732–3.06
α	Empirical constant	0.33
a	Lattice constant of Si	5.431 Å

The peak stress from dislocation occurs at the surface, and is computed by

$$\sigma_{\tau\ max} = M\alpha\mu\sqrt{\frac{b\beta I}{FD}} \tag{8.48}$$

To study the role of dislocation on the stress under galvanostatic operation, the parameters are given in Table 8.1 under the constant current of 0.011 Am^{-2} [15,16,42–45]. Figure 8.6b presents the evolution of the radial stress, which is compressive at the surface with the time. However, the tensile stress occurs without the dislocation effect in the classical elasticity. The dislocation effect results in the compressive stress, and even the tensile-to-compressive stress at the nanoscale interface.

8.3.1.2 Potentiostatic (Constant Voltage and Surface Concentration) Operation

The boundary condition for a potentiostatic operation is given as

$$C(R, t) = C_R, \quad C(0, t) = finite \tag{8.49}$$

Here, the Li-ion concentration of the cylindrical electrode is found as [34]

$$\frac{C(r, t) - C_0}{C_R - C_0} = 1 - \frac{4}{\phi}\sum_{k=1}^{\infty}\frac{J_1(\phi\lambda_k)\exp(-\varphi\lambda_k^2)}{\lambda_k^2 J_1(\lambda_k)} \tag{8.50}$$

where C_0 is the initial Li-ion concentration, J_0 and J_1 are the Bessel functions of the first kind of order 0 and 1, respectively. λ_k is the kth zero of the 1st order Bessel function of the first kind.

When the dislocation effect is absent, the associated radial stress is obtained by

$$\frac{\sigma_r}{E\Omega(C_R - C_0)/3(1-\nu)} = \frac{2}{\phi}\sum_{k=1}^{\infty}\frac{J_1(\phi\lambda_k)\exp(-\varphi\lambda_k^2)}{\lambda_k^2 J_1(\lambda_k)} - 2\sum_{k=1}^{\infty}\frac{\exp(-\varphi\lambda_k^2)}{\lambda_k^2} \tag{8.51}$$

Without the dislocation effect, Figure 8.7 shows the distribution of concentration and radial stress during charging. The Li-ion concentration increases with the time, but decreases continuously from the surface to the center (Figure 8.7a). At the time of $\varphi = 0.1$, there is a sharp rise of the asymptotic behavior of the Li-ion concentration at the high radial position, and a constant at the low radial position. During the insertion, the tensile radial stress is the highest at the center, and then reduces continuously until the surface [46–48]. The peak radial stress occurs at the center and the time of $\varphi = 0.8$, once the Li-ion spreads to the center (Figure 8.7b).

Based on Eq. 8.50 and 8.11, the DIS is obtained

$$\frac{\sigma_\tau}{M\alpha\mu\sqrt{\beta b(C_R - C_0)}} = \left[\frac{4}{\phi^2}\sum_{k=1}^{\infty}\frac{J_1(\phi\lambda_k)\exp(-\varphi\lambda_k^2)}{\lambda_k^2 J_1(\lambda_k)} - \frac{4}{\phi}\sum_{k=1}^{\infty}\frac{\partial J_1(\phi\lambda_k)}{\partial\phi}\frac{\exp(-\varphi\lambda_k^2)}{\lambda_k^2 J_1(\lambda_k)}\right]^{1/2} \tag{8.52}$$

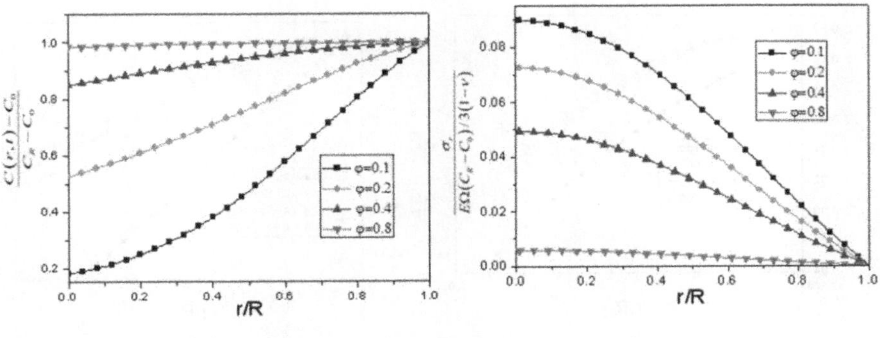

FIGURE 8.7 Distribution of Li-ion concentration (a) and the corresponding radial stress (b) without dislocation effect [16].

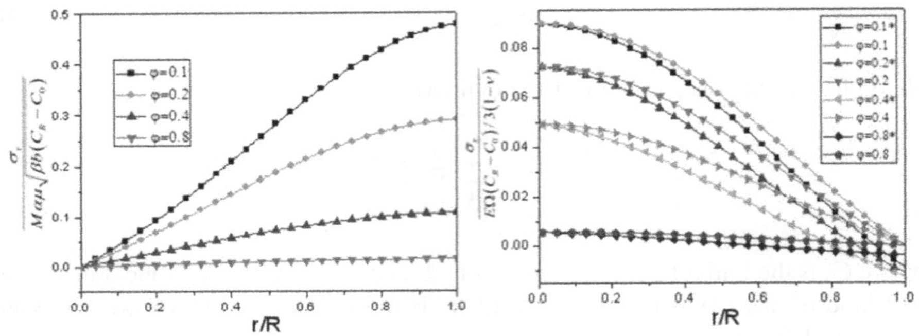

FIGURE 8.8 Distribution of dislocation-induced stress (a) and radial stress (b) for potentiostatic control. The dislocation-induced stress is normalized during insertion. A radius of cylindrical particle is 5 nm. * represents a coupled radial stress (b) [16].

Figure 8.8a shows the normalized DIS for the different radial position and time during discharging. For the insertion, the tensile stress is induced by dislocation in the cylindrical electrode [49]. This stress continuously increases from the surface to the inner near the surface, and the highest DIS goes towards the cylindrical electrode with the increase time [50–52]. The coupled effect of radial stress and DIS on the mechanical behavior is studied using Table 8.1. Figure 8.8b indicates an obvious decrease of the stress with the dislocation effect at the nanoscale level, and the tensile-to-compressive stress happens for different time [53,54].

8.3.1.3 Particle Size Effect

Considering the dislocation effect, Figure 8.9 shows the radial stress under the galvanostatic and potentiostatic cases, for the dimensionless time $\varphi = 0.4$. The dislocation effect plays an important role for the decrease radius. The radial tensile stress is significantly reduced as the radial position increases (Figure 8.9). Similarly, the tensile-to-compressive stress occurs in a nanoscale cylindrical electrode [55,56]. In addition, the elastic strain energy is stored in the cylindrical electrode during deformation process due to the diffusion [57,58], which would cause the crack propagation. While the mobile dislocation nucleated reduces the expansion, owing

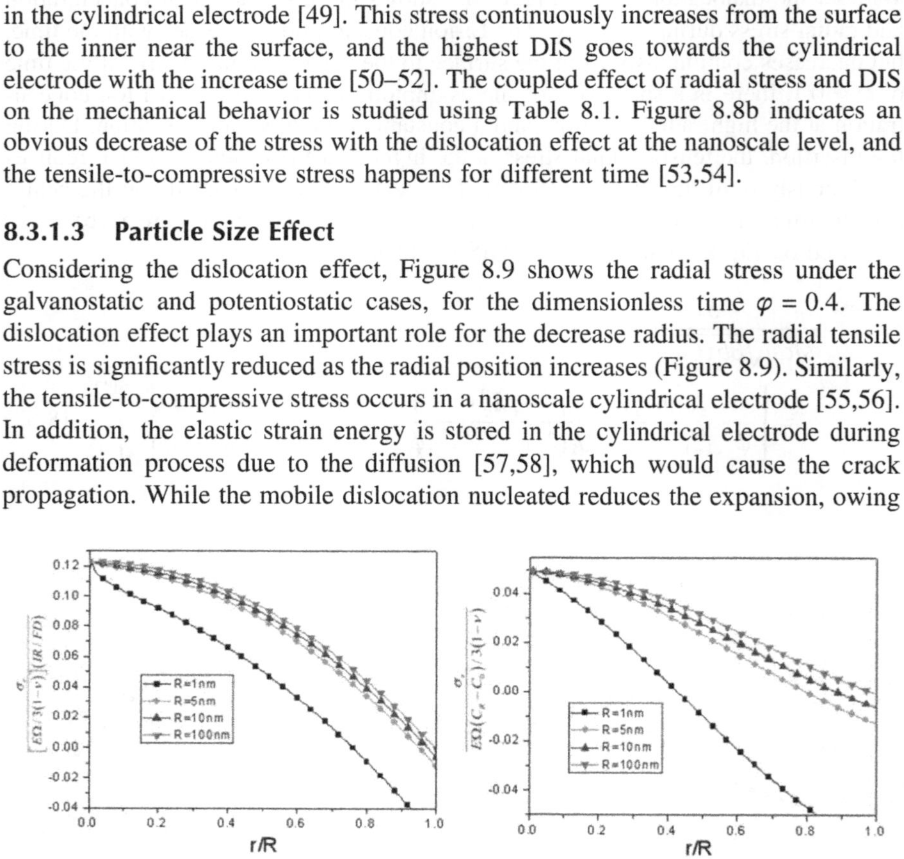

FIGURE 8.9 Radial stress for the dislocation effect and different radius. The dimensionless time is $\varphi = 0.4$ for galvanostatic control (a), and for potentiostatic control (b) [16].

to the release of strain energy [59,60], especially, evolved into the dislocation network requiring more energy. This trend would reduce the probability of crack formation, and then improve the stability of the electrode.

8.3.2 Spherical Electrolyte

In the subsequent calculation, the dimensionless time of $\varphi = Dt/R^2$, the radial stress of σ_r and tangential stress of σ_θ normalized by $E\Omega(C_R - C_0)/[3(1 - \nu)]$, and the initial Li-ion concentration of 0 mol/m^3 are used. Further, a critical radius of the particle for the spherical electrode is estimated based on DIS and strain energy [61,62]. Table 8.2 summarizes the computational parameters to investigate the coupled effect and size effect [15,16,42–45]. The relationship between normalization and real parameter is presented in Table 8.3.

TABLE 8.2
Material Properties Used in Spherical Electrode

Name	Value
Yong's modulus of Li-Si, E	30–80 Gpa
Burgers vector, b	2.532 Å
Diffusion coefficient, D	$1\ 2\times10^{-18}$ m^2 s^{-1}
Max stoichiometric Li concentration, $C_{s\ max}$	2.0152×10^{-4} mol m^{-3}
Partial molar volume of solute, Ω	2×10^{-5} m^3 mol^{-1}
Taylor orientation factor, M	1.732–3.06
Empirical constant, α	0.33
Lattice constant of Si, a	5.431 Å
Surface tension, $E\Omega(C_R - C_0)/[3(1 - \nu)]$	1 J m^{-2}
Surface modulus, $E\Omega(C_R - C_0)/[3(1 - \nu)]$	−5, +5 N m^{-1}
Initial Li-ion concentration, C_0	0 mol m^{-3}
surface energy, γ	1 J m^{-2}

TABLE 8.3
Relationship of Normalization-Real Parameters: Real Parameters= Dimensionless Parameters×Scale

Real Parameters	Scale	Dimensionless Parameters
Time, t	$E\Omega(C_R - C_0)/[3(1 - \nu)]$	φ
Radius, r	R	$E\Omega(C_R - C_0)/[3(1 - \nu)]$
Radial stress,σ_r	$E\Omega(C_R - C_0)/[3(1 - \nu)]$	ξ_r
Tangential stress, σ_θ	$E\Omega(C_R - C_0)/[3(1 - \nu)]$	ξ_θ
Dislocation induced stress, σ_τ	$M\alpha\mu\sqrt{2b\beta(C_R - C_0)/R}$	ξ_τ
Strain energy, W	$2\pi R^3\Omega^2 E(C_R - C_0)/[3(1 - \nu)]^2$	W*

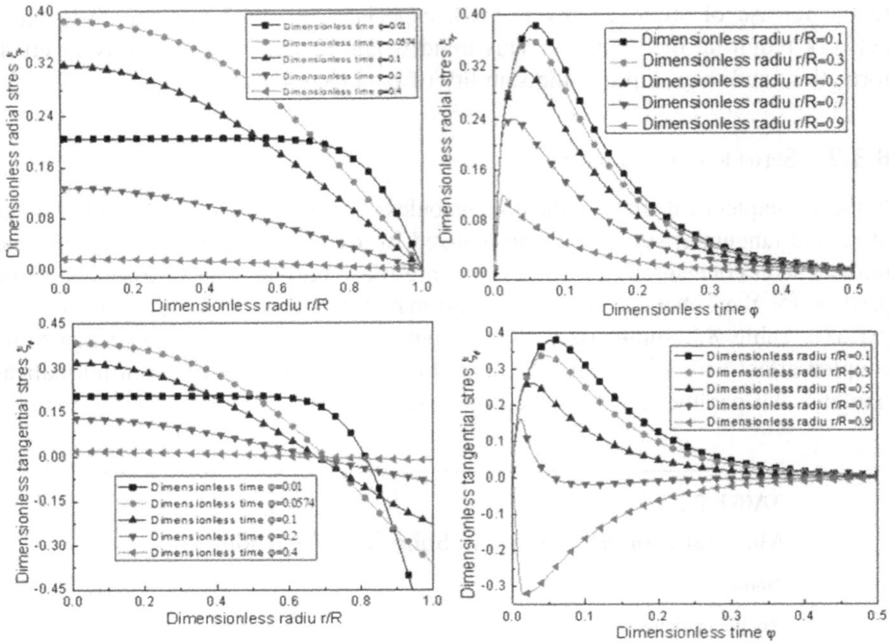

FIGURE 8.10 Without the coupled effect for insertion, the stress versus the radial position (a, c). The stress versus the time (b, d) [23].

8.3.2.1 DIS in a Classic LIB

Without the coupled effect, Figure 8.10 shows the stress variation during charging. Laplace pressure decreases rapidly for the macroscale electrode, but it is very important for the nanoscale electrode [63–65]. The stresses along the radial and tangential direction decrease as the time increases (Figure 8.10). The radial stress for the macroscale electrode is always tensile during cycling. With the increasing time, the tensile radial stress decreases continuously from the center to the surface (Figure 8.10a). This finding agrees with the previous work [16,21]. Compared to different times, the maximum radial stress occurs for the Li-ion in the electrode center, which is deponent upon the radial position [66,67], for the normalized time less than 0.4 (Figure 8.10b,d). Moreover, this influence gradually subsides when the time exceeds 0.4.

The compressive tangential stress takes place at the surface, and then becomes tensile at the center (Figure 8.10c). There is the tensile tangential stress once the Li-ion spreads to the center. In addition, the hydrostatic stress exits at the center of spherical electrode [67–69]. Especially, the maximum compressive tangential stress takes place near the surface. For Li-ion insertion, the stresses show the same magnitude but the opposite direction.

8.3.2.2 Surface Effect

To study the surface effect on stress distribution, Figure 8.11 shows the stress variation with the increase radial position, where Table 8.2 shows the selected electrode material parameters. An additional tensile/compressive stress is generated

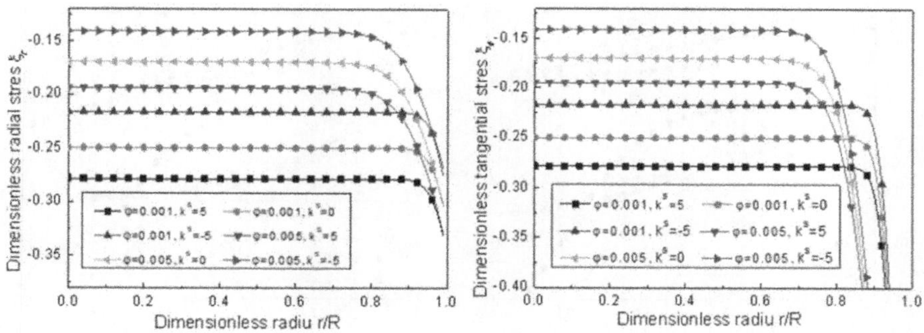

FIGURE 8.11 The radial stress and tangential stress versus the radial position, where the spherical electrode is R = 5 nm, surface tension is τ^s=1 J m^{-2}, and surface energy is γ=1 J m^{-2} [23].

due to the strong surface effect (Figure 8.11), in good agreement with the previous work [11,30]. The positive surface modulus causes the compressive stress of the spherical electrode [70,71]. In addition, for the radial position less than 0.8, the surface modulus plays a key role in stress distribution and direction. In other words, the surface effect reduces the stress amplitude, and then increases stability of the electrode by weakening the volume expansion during cycling.

8.3.2.3 Dislocation Effect

To study the dislocation effect on stress distribution, Figure 8.12a shows the DIS. During cycling, the tensile stress induced by dislocation goes up with the increase position, and then reduces continuously to zero [15,16]. The radial position effect on the stress distribution is presented in Figure 8.12b. As the time increases, dislocation-induced stress goes up firstly, and then reduces, finally tends to be stabilized. This stress strongly relies on the position at the time less than 0.4, however, it has a weak relationship with the radial position at the time exceeds 0.4.

8.3.2.4 Coupled Effect

Figure 8.13 shows the coupled effect on the radial and tangential stresses with the increase of radial position and time. Considering the coupled effect, the stress would

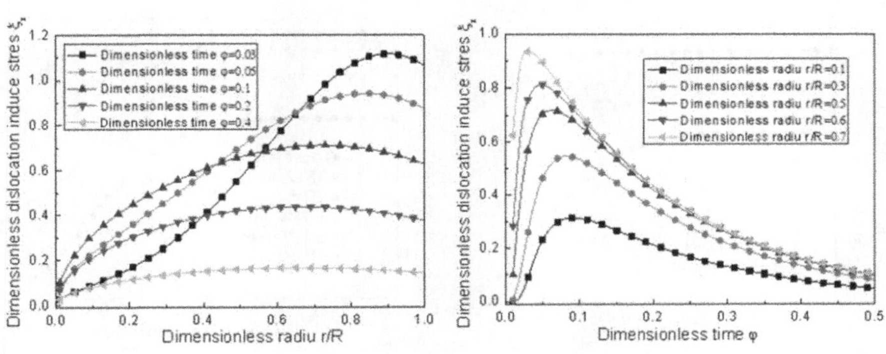

FIGURE 8.12 The dislocation-induced stress versus the radial position [23].

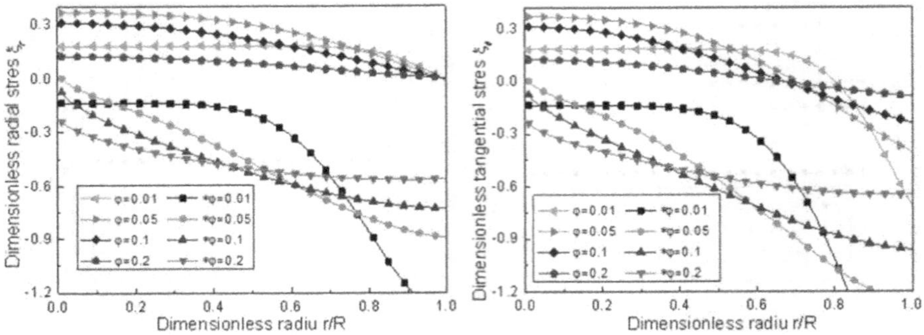

FIGURE 8.13 The tangential and tangential stress versus the radial position [23].

significantly decrease at nanoscale level, and the tensile-to-compressive stress transition may occur (Figure 8.13). This trend could reduce the crack nucleation and propagation owing to the decreasing tensile stress or the tensile-to-compressive stress transformation.

8.3.2.5 Size Effect

Figure 8.14 shows the coupled effect on the radial and tangential stresses under potentiostatic control. With the decrease of the radius, the coupled effect becomes more important. The electrode radius decreases or the radial position increases would obviously reduce the radial tensile stress. In addition, a spherical electrode results in the tensile-to-compressive stress transformation for the case considering the coupled effect. The stored elastic strain energy would be a driving force for the crack propagation in the spherical electrode. The dislocation nucleation/slip releases the strain energy, and thus the stability and reliability of LIB increase [72].

8.3.2.6 Strain Energy

Figure 8.15a shows the normalized total strain energy during potentiostatic process under the spherical nano-electrode and different Poisson's ratios. For the stable material, the value of Poisson's ratio changes from -1 to 0.5. The Poisson's ratio on the total strain energy has a key effect. The strain energy firstly goes up, reaches a peak, and then reduces continuously to zero (Figure 8.10a,b).

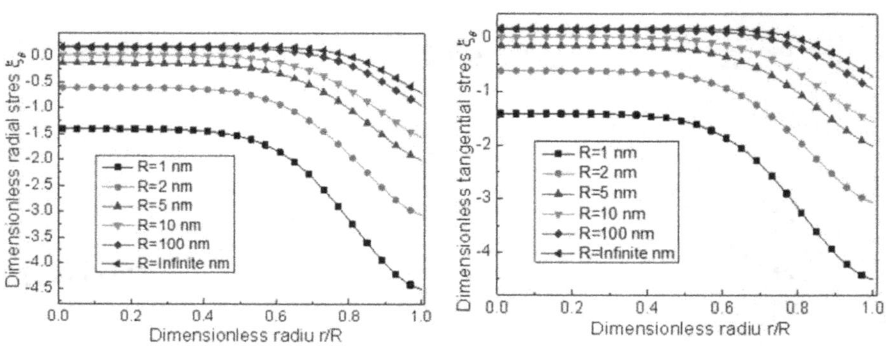

FIGURE 8.14 The tangential and tangential stress versus the radial position [23].

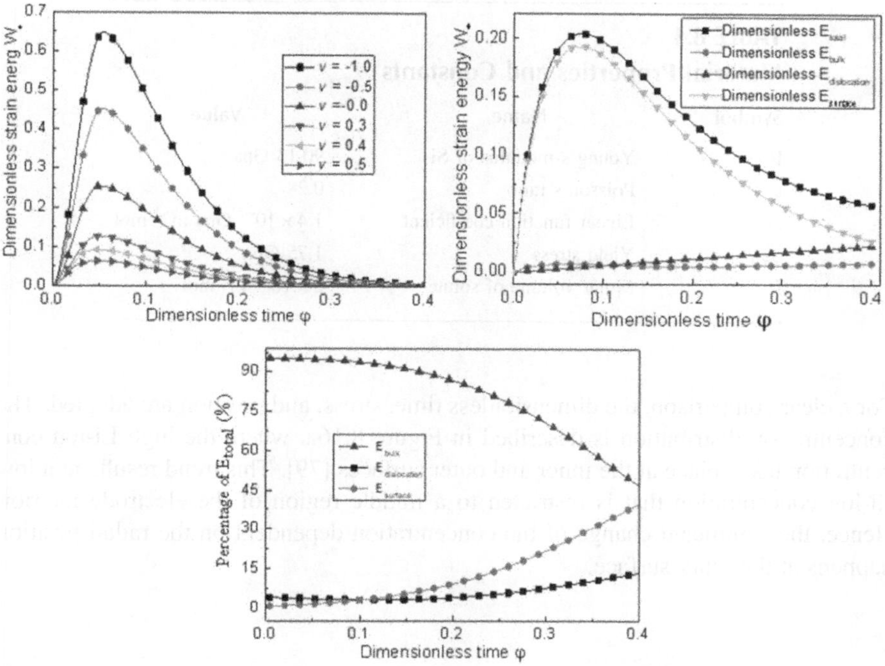

FIGURE 8.15 The total strain energy vs. time (a). Total strain energy with coupled effect with the time (b). The ratio of strain energy to total strain energy with the time (c) [23].

As the time increases, the strain energy from the coupled effect, the bulk, dislocation, and surface versus is presented in Figure 8.15b. As to be expected, the strain energy from the dislocation greater contributes to the total strain energy in comparison with the surface deformation. In addition, the strain energy from surface and dislocation has a comparable magnitude compared with the bulk strain energy. Here, the mechanical model gives the reasonable parameters to enhance stability of electrodes based on the couple effect. Thus, the nanoscale electrode has higher LIB cycle life [16]. The coupled effect reduces the crack nucleation and propagation in nanoscale electrode, comparing with various strain energies [73,74]. Hence, not only bulk strain energy is considered at nanoscale, but also surface strain energy and dislocation strain energy are regarded as the key targets [75–77]. Figure 8.15c shows the ratio of bulk strain energy, dislocation strain energy, surface strain energy to the total strain energy with the increase of time. The dislocation strain energy and surface strain energy always increase at the plane stress condition [78], and bulk strain energy declines continuously (Figure 8.15c). This trend reveals a stronger suppression of the dislocation effect on the crack nucleation compared with the surface effect.

8.3.3 Hollow Spherical Electrolyte

8.3.3.1 Distribution of Concentration

The distribution of Li-ion concentration and DIS of hollow spherical electrolyte are studied. Table 8.4 shows the material constants for the numerical calculation [15,16].

TABLE 8.4
Material Properties and Constants

Symbol	Name	Value
E	Young's modulus of Si	90.13 Gpa
ν	Poisson's ratio	0.28
ε_1	Linear function coefficient	1.4×10^{-7} Gpa m^{-3} mol
σ_{Y0}	Yield stress	1.75 Gpa
Ω	Molar volume of solute	2×10^{-5} m^3 mol^{-1}

For a clear comparison, the dimensionless time, stress, and position are adopted. The concentration distribution is described in Figure 8.16a, where the high Li-ion concentration takes place at the inner and outer surfaces [79]. This trend results in a low Li-ion concentration that is restricted to a middle region of the electrode interior. Hence, the significant change of the concentration dependent on the radial position happens at the inner surface.

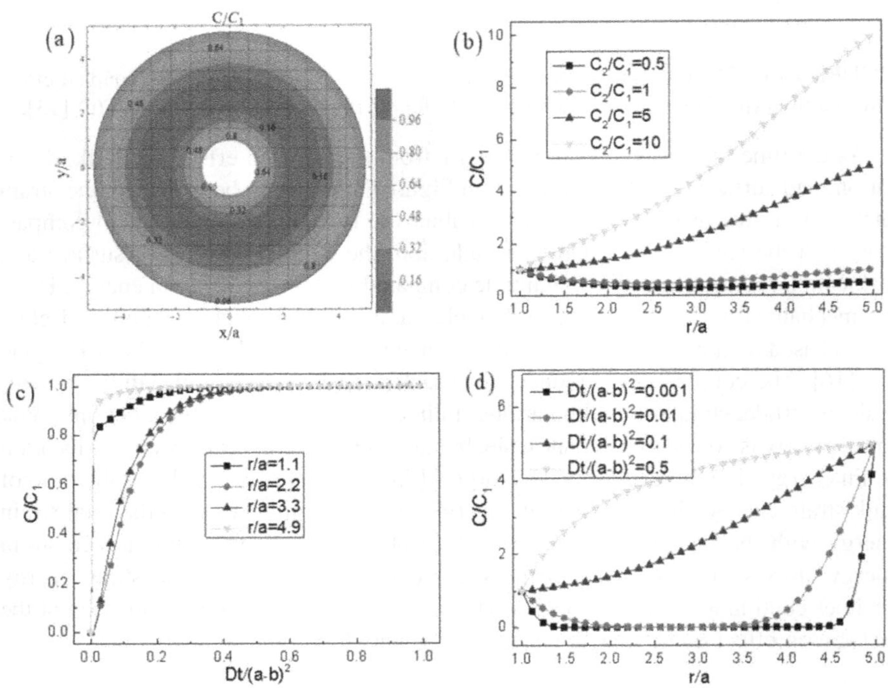

FIGURE 8.16 (a) The concentration distribution, where radius ratio b/a=5, normalized time $\varphi = 0.1$, and concentration ratio C_1/C_2=1. (b) The concentration distribution vs. position for the various concentration ratios and normalized time $\varphi = 0.1$. (c) The concentration vs. diffusion time. (d) The concentration vs. position [39].

Figure 8.16b shows the variation of the concentration distribution at the hollow electrode with the increase position. As the concentration and position increase, the Li-ion concentration goes up. The current concentration distribution causes the crack nucleation at the inner surface owing to the large stress induced by Li-ion diffusion. Similarly, the crack nucleation occurs at outer surface, which is attributed to the high concentration difference [80]. Figure 8.16c shows the concentration with the increase time. The concentration distribution shows the linear relationship to the time at the region of the inner and outer surfaces. Compared with the surface region, the intermediate region exhibits a strong asymptotic behavior dependent on the position in the Li-ion concentration. Furthermore, the interstitial diffusion of Li-ion results in the high concentration. Figure 8.16d shows the high Li-ion concentration distribution, which appears in the inner and outer surfaces. This trend would lead to Li-ion diffusion toward the middle region.

8.3.3.2 Stress Distribution in the Elastic Deformation

To obtain the stress evolution at the elastic deformation, the radial and hoop stress distributions of hollow spherical electrode are presented in Figure 8.17a,b. The tensile radial stress and hoop stress occur at the inner surface, but the compressive hoop stress takes place at the outer surface. The annular middle area presents the high radial stress, and the inner and outer surfaces possess the high hoop stress. In addition, the same trend for the stress distribution happens during the charging [40]. The initial stress can strongly influence the mechanical properties of the electrode [16]. Thus, the high stress from the internal hollow electrode leads to the exhaustion and damage of LIB.

The distributions of the radial and hoop stresses are presented in Figure 8.17c,d. The radial crack is hardly to nucleate at the surface owing to the traction-free condition. The radial stress is tensile when the dimensionless time exceeds 0.1, while the radial stress is compressive when the dimensionless time is below 0.01. The expansion of the inner part happens due to the high Li-ion concentration to diffuse from the inner to the inside surface. A low Li-ion concentration is formed in the middle region (Figure 8.16d). As the diffusion time increases, DIS decreases accordingly due to the uniform Li-ion concentration. The concentration effect on the compressive radial and hoop stress distribution are shown in Figure 8.17e,f. The tensile-to-compressive stress occurs with the increase concentration ratio.

8.3.3.3 Stress Distribution in the Plastic Deformation

The concentration-dependent strength of silicon electrode strongly controls the battery performance [2–5]. Figure 8.18a,b shows the radial and hoop stress singularity and distribution near the inner surface, due to the formation of plastic deformation. Figure 8.18b shows the rapid decrease of stress owing to the concentration effect near the surface. Considered the concentration-dependent strength in the plastic region, the current model could accurately predict stress distribution to reasonably assess damage status. Figure 8.18e,f show the effect of concentration on the stress distributions during stable stage. With the increase of position, the radial stress increases in plastic region up to the highest value in elastic region, and then reduces in plastic region. The high concentration leads to the larger radial stress in

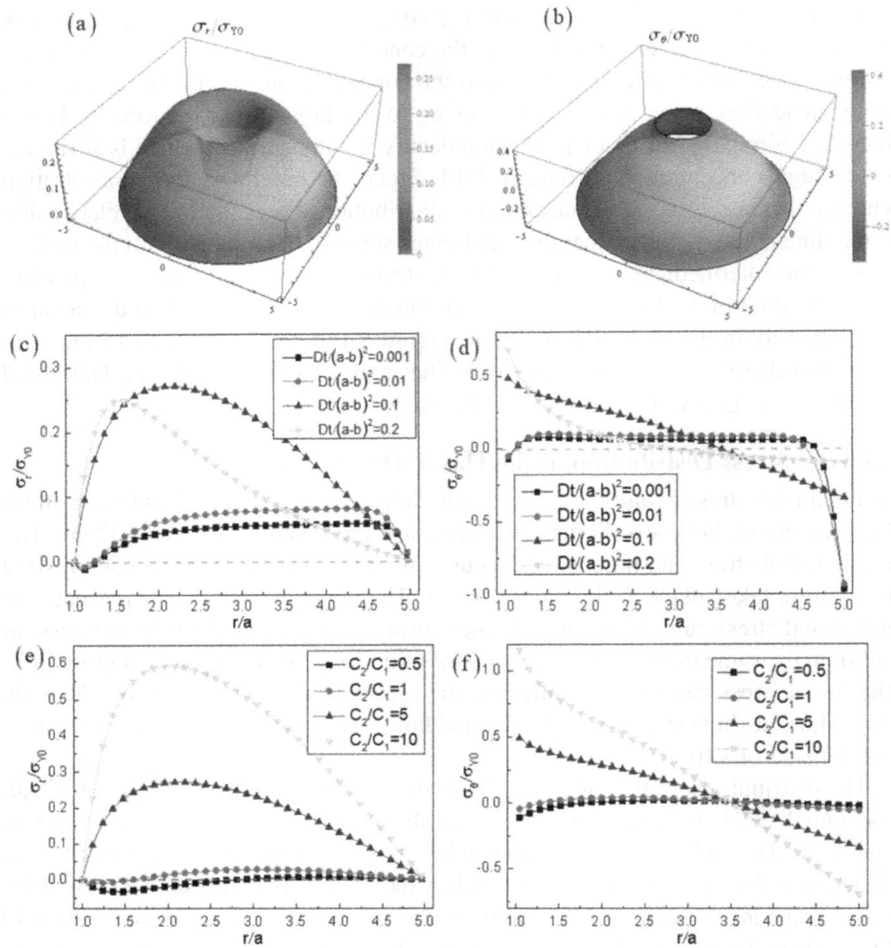

FIGURE 8.17 (a) The radial stress distribution and (b) hoop stress distribution. (c) The radial stress and (d) hoop stress vs. positions with various time. (e) The radial stress and (f) hoop stress vs. positions with various concentrations [39].

hollow spherical electrode, which is benefit for increase the crack nucleation. The concentration-dependence on stress distribution is observed, which is in an agreement with previous work [82]. In fact, the spherical electrodes sometimes serve in a specific of high rate [83–85]. Hence, the uniform stress distribution can prevent and reduce the accelerated damage.

Compared with the classical result without plastic deformation, the decrease radial stress at the plastic deformation is potted in Figure 8.18c,d. Based on this trend, the compressive radial stress is found [81]. With the increase of time, the plastic region increases gradually in inner surface, and decreases slowly at outer surface. After a sufficiently long time, the plastic deformation occurs in the hollow electrode. The decreasing hoop stresses is observed with the increase of position during plastic

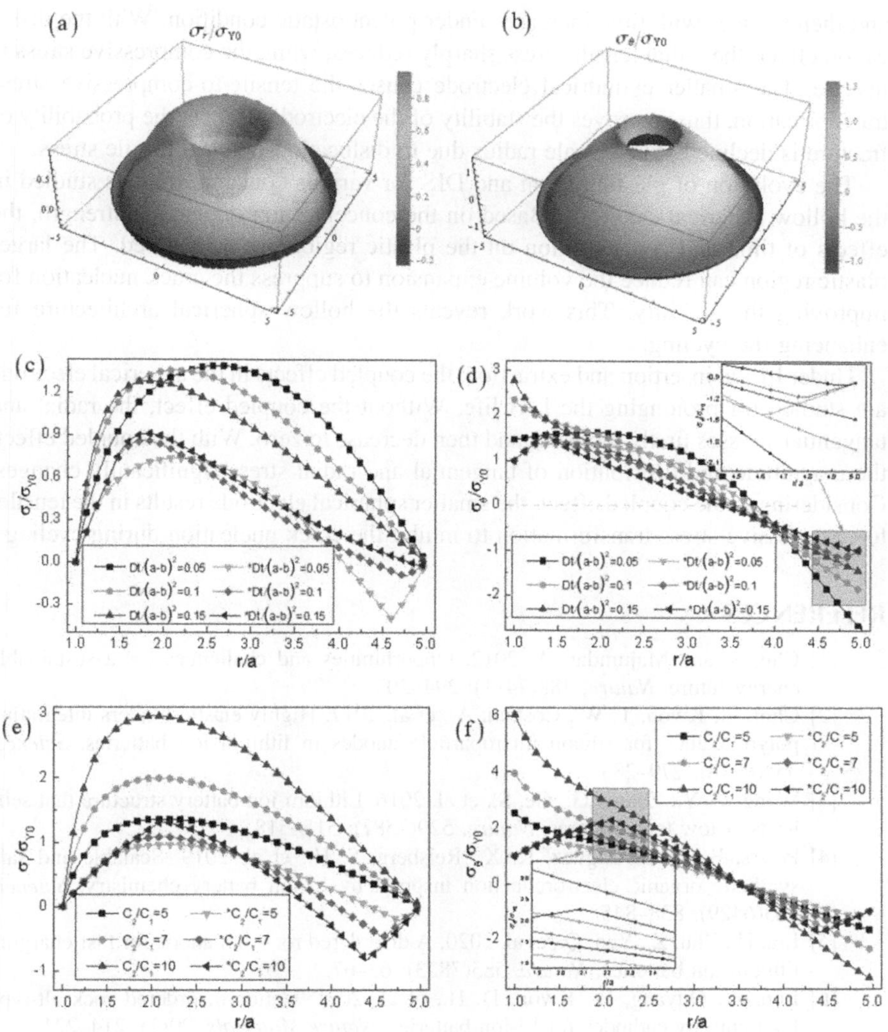

FIGURE 8.18 (a, b) The radial and hoop stress. (c, d) The radial and hoop stress vs. position with various time. (e, f) The radial and hoop stress vs. positions with various concentration ratio. Asterisks represent the plastic deformation, and the rest shows the elastic deformation [39].

deformation (Figure 8.18c). Hence, the plastic deformation behavior can dominate the performance of hollow electrode through inhibiting crack formation.

8.4 CONCLUSIONS

In this chapter, an analytical model is developed to study DIS with the dislocation effect in the cylindrical electrode under galvanostatic and potentiostatic operations. Without the dislocation effect, the radial stress at a given location firstly goes up and then reaches a steady state under galvanostatic condition, while it initially increases and then reduces with time increases under potentiostatic condition. With the dislocation effect, the radial tensile stress sharply reduces, while the compressive stress is inverse. The smaller cylindrical electrode causes the tensile-to-compressive stress transformation, thus improves the stability of the electrode. Hence, the probability of fracture is declined at nanoscale radius due to dislocation-induced tensile stress.

The evolution of plastic region and DIS for various concentrations is studied in the hollow spherical electrode. Based on the concentration-dependent strength, the effects of time and concentration on the plastic region are considered. The larger plastic region can reduce the volume expansion to suppress the crack nucleation for improving the stability. This work reveals the hollow spherical architecture for enhancing the cycling.

Under Li-ion insertion and extraction, the coupled effects in the spherical electrode are studied for prolonging the LIB life. Without the coupled effect, the radial and tangential stresses firstly increase, and then decrease to zero. With the coupled effect, the magnitude and distribution of tangential and radial stress significantly changes. Considering to the coupled effect, the smaller spherical electrode results in the tensile-to-compressive stress transformation to inhibit the crack nucleation during cycling.

REFERENCES

[1] Chu, S. and Majumdar, A. 2012. Opportunities and challenges for a sustainable energy future. *Nature*, 488(7411): 294–303.

[2] Choi, S., Kwon, T. W., Coskun, A., et al. 2017. Highly elastic binders integrating polyrotaxanes for silicon microparticle anodes in lithium ion batteries. *Science*, 357(6348): 279–283.

[3] Wang, C. Y., Zhang, G., Ge, S., et al. 2016. Lithium-ion battery structure that self-heats at low temperatures. *Nature*, 529(7587): 515–518.

[4] Peters, B. K., Rodriguez, K. X., Reisberg, S. H., et al. 2019. Scalable and safe synthetic organic electroreduction inspired by Li-ion battery chemistry. *Science*, 363(6429): 838–845.

[5] Liu, H., Zhu, Z., Yan, Q., et al. 2020. A disordered rock salt anode for fast-charging lithium-ion batteries. *Nature*, 585(7823): 63–67.

[6] Lun, Z., Ouyang, B., Kwon, D. H., et al. 2021. Cation-disordered rocksalt-type high-entropy cathodes for Li-ion batteries. *Nature Materials*, 20(2): 214–221.

[7] Yang, C., Chen, J., Ji, X., et al. 2019. Aqueous Li-ion battery enabled by halogen conversion–intercalation chemistry in graphite. *Nature*, 569(7755): 245–250.

[8] Harper, G., Sommerville, R., Kendrick, E., et al. 2019. Recycling lithium-ion batteries from electric vehicles. *Nature*, 575(7781): 75–86.

[9] Hou, C., Wang, B., Murugadoss, V., et al. 2020. Recent advances in Co3O4 as anode materials for high-performance lithium-ion batteries. *Engineered Science*, 11(5): 19–30.

[10] Zhang, Z., Li, Y., Xu, R., et al. 2022. Capturing the swelling of solid-electrolyte interphase in lithium metal batteries. *Science*, 375(6576): 66–70.

[11] Zhang, X., Shyy, W. and Sastry, A. M. 2007. Numerical simulation of intercalation-induced stress in Li-ion battery electrode particles. *Journal of the Electrochemical Society*, 154(10): A910.

[12] Cheng, Y. T. and Verbrugge, M. W. 2008. The influence of surface mechanics on diffusion induced stresses within spherical nanoparticles. *Journal of Applied Physics*, 104(8): 083521.

[13] Ye, Y., Shi, Y., Cai, N., et al. 2012. Electro-thermal modeling and experimental validation for lithium ion battery. *Journal of Power Sources*, 199: 227–238.
[14] Deshpande, R., Cheng, Y. T. and Verbrugge, M. W. 2010. Modeling diffusion-induced stress in nanowire electrode structures. *Journal of Power Sources*, 195(15): 5081–5088.
[15] Wei, P., Zhou, J., Pang, X., et al. 2014. Effects of dislocation mechanics on diffusion-induced stresses within a spherical insertion particle electrode. *Journal of Materials Chemistry A*, 2(4): 1128–1136.
[16] Li, J., Fang, Q., Liu, F., et al. 2014. Analytical modeling of dislocation effect on diffusion induced stress in a cylindrical lithium ion battery electrode. *Journal of Power Sources*, 272: 121–127.
[17] Cheng, Y. T. and Verbrugge, M. W. 2009. Evolution of stress within a spherical insertion electrode particle under potentiostatic and galvanostatic operation. *Journal of Power Sources*, 190(2): 453–460.
[18] Prussin, S. 1961. The stress effects of the doped elements as discussants in single crystals. *Journal of Applied Physics*, 32(3): 1876–1883.
[19] Huang, J. Y., Zhong, L., Wang, C. M., et al. 2010. In situ observation of the electrochemical lithiation of a single SnO2 nanowire electrode. *Science*, 330(6010): 1515–1520.
[20] Chen, Q., Liu, X. Y. and Biner, S. B. 2008. Solute and dislocation junction interactions. *Acta Materialia*, 56(13): 2937–2947.
[21] Estrin, Y. 1998. Dislocation theory based constitutive modelling: foundations and applications. *Journal of Materials Processing Technology*, 80: 33–39.
[22] Li, C., Zhang, H. P., Fu, L. J., et al. 2006. Cathode materials modified by surface coating for lithium ion batteries. *Electrochimica Acta*, 51(19): 3872–3883.
[23] Li, J., Lu, D., Fang, Q., et al. 2015. Cooperative surface effect and dislocation effect in lithium ion battery electrode. *Solid State Ionics*, 274: 46–54.
[24] Sharma, P., Ganti, S. and Bhate, N. 2003. Effect of surfaces on the size-dependent elastic state of nano-inhomogeneities. *Applied Physics Letters*, 82(4): 535–537.
[25] He, J. and Lilley, C. M. 2008. Surface effect on the elastic behavior of static bending nanowires. *Nano Letters*, 8(7): 1798–1802.
[26] Li, J., Fang, Q. and Liu, Y. 2013. Crack interaction with a second phase nanoscale circular inclusion in an elastic matrix. *International Journal of Engineering Science*, 72: 89–97.
[27] Yang, W., Wang, S., Kang, W., et al. 2023. A unified high-order model for size-dependent vibration of nanobeam based on nonlocal strain/stress gradient elasticity with surface effect. *International Journal of Engineering Science*, 182: 103785.
[28] Fang, Q. H. and Liu, Y. W. 2006. Size-dependent interaction between an edge dislocation and a nanoscale inhomogeneity with interface effects. *Acta Materialia*, 54(16): 4213–4220.
[29] Shodja, H. M., Ahmadzadeh-Bakhshayesh, H. and Gutkin, M. Y. 2012. Size-dependent interaction of an edge dislocation with an elliptical nano-inhomogeneity incorporating interface effects. *International Journal of Solids and Structures*, 49(5): 759–770.
[30] Verbrugge, M. and Cheng, Y. T. 2008. Stress distribution within spherical particles undergoing electrochemical insertion and extraction. *ECS Transactions*, 16(13): 127.
[31] Zhao, Y., Stein, P., Bai, Y., et al. 2019. A review on modeling of electro-chemo-mechanics in lithium-ion batteries. *Journal of Power Sources*, 413: 259–283.
[32] Bhowmick, A. and Chakraborty, J. 2021. Improving predictions of amorphous-crystalline silicon interface velocity through alloying-dealloying reactions in lithium-ion battery anode particles. *International Journal of Solids and Structures*, 224: 111046.

[33] Lu, Y., Soh, A. K., Ni, Y., et al. 2019. Understanding size-dependent migration of a two-phase lithiation front coupled to stress. *Acta Mechanica*, 230: 303–317.
[34] Crank, J. 1979. *The mathematics of diffusion*. Oxford university press.
[35] Timoshenko, S. and Goodier, J. N. 1951. *Theory of elasticity*. McGraw-Hill.
[36] Zhang, X., Wang, Q. J., Harrison, K. L., et al. 2020. Pressure-driven interface evolution in solid-state lithium metal batteries. Cell Reports Physical *Science*, 1(2): 100012.
[37] Kasemchainan, J., Zekoll, S., Spencer Jolly, D., et al. 2019. Critical stripping current leads to dendrite formation on plating in lithium anode solid electrolyte cells. *Nature Materials*, 18(10): 1105–1111.
[38] Wang, M. J., Kazyak, E., Dasgupta, N. P., et al. 2021. Transitioning solid-state batteries from lab to market: Linking electro-chemo-mechanics with practical considerations. *Joule*, 5(6): 1371–1390.
[39] Li, J., Fang, Q., Wu, H., et al. 2015. Investigation into diffusion induced plastic deformation behavior in hollow lithium ion battery electrode revealed by analytical model and atomistic simulation. *Electrochimica Acta*, 178: 597–607.
[40] Yao, Y., McDowell, M. T., Ryu, I., et al. 2011. Interconnected silicon hollow nanospheres for lithium-ion battery anodes with long cycle life. *Nano Letters*, 11(7): 2949–2954.
[41] Zhang, K., Li, B., Zuo, Y., et al. 2019. Voltage decay in layered Li-rich Mn-based cathode materials. *Electrochemical Energy Reviews*, 2: 606–623.
[42] Wang, B. and Aifantis, K. E. 2022. Probing the effect of surface parameters and particle size in the diffusion-induced stress of electrodes during lithium insertion. *International Journal of Mechanical Sciences*, 215: 106917.
[43] Bhandakkar, T. K. and Gao, H. 2010. Cohesive modeling of crack nucleation under diffusion induced stresses in a thin strip: Implications on the critical size for flaw tolerant battery electrodes. *International Journal of Solids and Structures*, 47(10): 1424–1434.
[44] Zhu, L., Ruan, H., Li, X., et al. 2011. Modeling grain size dependent optimal twin spacing for achieving ultimate high strength and related high ductility in nanotwinned metals. *Acta Materialia*, 59(14): 5544–5557.
[45] Kang, K. and Cai, W. 2010. Size and temperature effects on the fracture mechanisms of silicon nanowires: Molecular dynamics simulations. *International Journal of Plasticity*, 26(9): 1387–1401.
[46] Peng, Y., Zhang, K., Zheng, B., et al. 2019. Semi-analytical solution of lithiation-induced stress in a finite cylindrical electrode. *Journal of Energy Storage*, 25: 100834.
[47] Wang, X., Liu, X. and Yang, Q. 2021. Transient analysis of diffusion-induced stress for hollow cylindrical electrode considering the end bending effect. *Acta Mechanica*, 232: 3591–3609.
[48] Li, Y. and Yang, F. 2022. Numerical calculation of lithiation-induced stress in a spherical electrode: Effect of lithiation-induced structural damage. *Materials Science and Engineering: A*, 839: 142873.
[49] Liu, D., Shadike, Z., Lin, R., et al. 2019. Review of recent development of in situ/operando characterization techniques for lithium battery research. *Advanced Materials*, 31(28): 1806620.
[50] Ma, Z., Wu, H., Wang, Y., et al. 2017. An electrochemical-irradiated plasticity model for metallic electrodes in lithium-ion batteries. *International Journal of Plasticity*, 88: 188–203.
[51] Li, Y., Zhang, J., Zhang, K., et al. 2019. A defect-based viscoplastic model for large-deformed thin film electrode of lithium-ion battery. *International Journal of Plasticity*, 115: 293–306.

[52] Reimuth, C., Lin, B., Yang, Y., et al. 2021. Chemo-mechanical study of dislocation mediated ion diffusion in lithium-ion battery materials. *Journal of Applied Physics*, 130(3): 035103.

[53] Lu, Y., Zhang, P., Wang, F., et al. 2018. Reaction-diffusion-stress coupling model for Li-ion batteries: The role of surface effects on electrochemical performance. *Electrochimica Acta*, 274: 359–369.

[54] Wang, Y., Wu, H., Sun, L., et al. 2021. Coupled electrochemical-mechanical modeling with strain gradient plasticity for lithium-ion battery electrodes. *European Journal of Mechanics-A/Solids*, 87: 104230.

[55] Zhao, Y., Stein, P., Bai, Y., et al. 2019. A review on modeling of electro-chemo-mechanics in lithium-ion batteries. *Journal of Power Sources*, 413: 259–283.

[56] Byeon, Y. W., Ahn, J. P. and Lee, J. C. 2020. Diffusion Along Dislocations Mitigates Self-Limiting Na Diffusion in Crystalline Sn. *Small*, 16(52): 2004868.

[57] Li, Y., Zhang, J., Zhang, K., et al. 2019. A defect-based viscoplastic model for large-deformed thin film electrode of lithium-ion battery. *International Journal of Plasticity*, 115: 293–306.

[58] Boyce, A. M., Martínez-Pañeda, E., Wade, A., et al. 2022. Cracking predictions of lithium-ion battery electrodes by X-ray computed tomography and modelling. *Journal of Power Sources*, 526: 231119.

[59] Singh, A. and Pal, S. 2022. Chemo-mechanical modeling of inter-and intra-granular fracture in heterogeneous cathode with polycrystalline particles for lithium-ion battery. *Journal of the Mechanics and Physics of Solids*, 163: 104839.

[60] Shin, J., Chen, L. Y., Sanli, U. T., et al. 2019. Controlling dislocation nucleation-mediated plasticity in nanostructures via surface modification. *Acta Materialia*, 166: 572–586.

[61] Lu, Y., Zhao, C. Z., Yuan, H., et al. 2021. Critical current density in solid-state lithium metal batteries: Mechanism, influences, and strategies. *Advanced Functional Materials*, 31(18): 2009925.

[62] Yuan, H., Ding, X., Liu, T., et al. 2022. A review of concepts and contributions in lithium metal anode development. *Materials Today*, 53: 173–196.

[63] Zhao, L., Wang, Q. J., Zhang, X., et al. 2022. Laplace-Fourier transform solution to the electrochemical kinetics of a symmetric lithium cell affected by interface conformity. *Journal of Power Sources*, 531: 231305.

[64] Zheng, Z. J., Ye, H. and Guo, Z. P. 2020. Recent progress in designing stable composite lithium anodes with improved wettability. *Advanced Science*, 7(22): 2002212.

[65] Gao, Y., Hu, H., Chang, J., et al. 2021. Realizing high-energy and stable wire-type batteries with flexible lithium-metal composite yarns. Advanced Energy *Materials*, 11(40): 2101809.

[66] Wen, J., Wei, Y. and Cheng, Y. T. 2018. Stress evolution in elastic-plastic electrodes during electrochemical processes: A numerical method and its applications. *Journal of the Mechanics and Physics of Solids*, 116: 403–415.

[67] Hu, P., Wang, B., Xiao, D., et al. 2019. Capturing the differences between lithiation and sodiation of nanostructured TiS2 electrodes. *Nano Energy*, 63: 103820.

[68] Bagheri, A., Arghavani, J. and Naghdabadi, R. 2019. On the effects of hydrostatic stress on Li diffusion kinetics and stresses in spherical active particles of Li-ion battery electrodes. *Mechanics of Materials*, 137: 103134.

[69] Hou, J., Qu, S., Yang, M., et al. 2020. Materials and electrode engineering of high capacity anodes in lithium ion batteries. *Journal of Power Sources*, 450, 227697.

[70] Wang, B. and Aifantis, K. E. 2022. Probing the effect of surface parameters and particle size in the diffusion-induced stress of electrodes during lithium insertion. *International Journal of Mechanical Sciences*, 215: 106917.

[71] Jin, C., Li, H., Song, Y., et al. 2019. On stress-induced voltage hysteresis in lithium ion batteries: Impacts of surface effects and interparticle compression. *Science China Technological Sciences*, 62: 1357–1364.

[72] Dong, Y. and Li, J. 2023. Oxide cathodes: functions, instabilities, self healing, and degradation mitigations. *Chemical Reviews*, 123: 811–833.

[73] Yuan, C., Gao, X., Jia, Y., et al. 2021. Coupled crack propagation and dendrite growth in solid electrolyte of all-solid-state battery. *Nano Energy*, 86: 106057.

[74] Ge, X., Cao, S., Lv, Z., et al. 2022. Mechano-Graded Electrodes Mitigate the Mismatch between Mechanical Reliability and Energy Density for Foldable Lithium-Ion Batteries. *Advanced Materials*, 34(45): 2206797.

[75] Chen, Y., Sang, M., Jiang, W., et al. 2021. Fracture predictions based on a coupled chemo-mechanical model with strain gradient plasticity theory for film electrodes of Li-ion batteries. *Engineering Fracture Mechanics*, 253: 107866.

[76] Li, Y., Mao, W., Zhang, K., et al. 2019. Analysis of large-deformed electrode of lithium-ion battery: Effects of defect evolution and solid reaction. *International Journal of Solids and Structures*, 170: 1–10.

[77] Singh, D. K., Fuchs, T., Krempaszky, C., et al. 2023 . Origin of the lithium metal anode instability in solid-state batteries during discharge. *Matter*, 6(5): 1463–1483.

[78] Bi, Y., Tao, J., Wu, Y., et al. 2020. Reversible planar gliding and microcracking in a single-crystalline Ni-rich cathode. *Science*, 370(6522): 1313–1317.

[79] Wang, X., Li, S., and Tong, Q. 2022. Size-and-thickness-dependent fracture patterns of hollow core–shell electrodes during lithiation. *Extreme Mechanics Letters*, 52: 101647.

[80] Zhu, X., Xie, Y., Chen, H., et al. 2021. Numerical analysis of the cyclic mechanical damage of Li-ion battery electrode and experimental validation. *International Journal of Fatigue*, 142: 105915.

[81] Zhao, K., Pharr, M., Cai, S., et al. 2011. Large plastic deformation in high-capacity lithium-ion batteries caused by charge and discharge. *Journal of the American Ceramic Society*, 94: s226–s235.

[82] Ryu, I., Choi, J. W., Cui, Y., et al. 2011. Size-dependent fracture of Si nanowire battery anodes. *Journal of the Mechanics and Physics of Solids*, 59(9): 1717–1730.

[83] Liu, T., Qu, Y., Liu, J., et al. 2021. Core–Shell Structured C@ SiO2 Hollow Spheres Decorated with Nickel Nanoparticles as Anode Materials for Lithium-Ion Batteries. *Small*, 17(49): 2103673.

[84] Geng, L., Yang, D., Gao, S., et al. 2020. Facile Fabrication of Porous Si Microspheres from Low-Cost Precursors for High-Capacity Electrode. *Advanced Materials Interfaces*, 7(3): 1901726.

[85] Wang, H. F., Chen, L., Wang, M., et al. 2021. Hollow spherical superstructure of carbon nanosheets for bifunctional oxygen reduction and evolution electrocatalysis. *Nano Letters*, 21(8): 3640–3648.

9 Nanostructure-Reinforced Composites

9.1 INTRODUCTION

Dislocations display a crucial role in the field of material science, as their behavior dictates the strength and hardening mechanisms of metals and alloys. One important area is studying the interaction between dislocations and nanostructures, such as nanoinclusions, nanopores, and nanowires. Particularly important, the internal forces induced by nanostructures have a significant impact on dislocation mobility [1,2]. Given its importance, researchers have investigated the interaction between dislocations and nanostructures for several decades, with a focus on developing a systematic understanding of this phenomenon [3–6].

Due to the very small size of nanostructures, the influence of surface/interface becomes significant for their interaction with dislocations. The atoms of the surface/interface undergo a distinct physical environment compared with the atoms of bulk, the surface/interface exhibits special geometrical and physical characteristics [13–16]. Especially, the distinct lattice spacing of the surface/interface induces the surface/interface stresses. For the nanostructure with a large characteristic size, the surface/interface stress is negligible since the volume fraction of surface/interface is very small. However, the surface/interface stress would be enhanced when the characteristic size of nanostructure reduces, such as the nanocomposite that has numerous internal interfaces [17,18].

To date, models considering the surface/interface stress are widely accepted to describe the elastic interaction between dislocations and nanostructures [7–12]. In these models, the solid's surface/interface is a unique region characterized by an extremely thin layer [19–23], which adheres to the bulk solid without any slip [24,25]. Meanwhile, due to the physical property of the surface/interface being different from the bulk, non-classical boundary conditions and unique elastic properties are required for them in modelling. These assumptions are the basis for describing the special elastic response in the dislocation-nanostructure interaction [26–35].

In this section, we will introduce the elastic models for the interaction between the edge/screw dislocation and the nanoinclusion, nanopore and nanowire. The core equations of the dislocation-nanostructure interaction model are provided. We analyze the image force acting on the dislocation, as well as the critical shear stress (CSS) and interaction energies for the dislocation movement. The influence of geometrical and physical properties of the dislocation and nanostructure are

DOI: 10.1201/9781003225706-9

systematically studied. in particularl, we uncover the key role of surface/interface stress and its effect on the motion of dislocations.

9.2 INCLUSIONS

For the interaction involved nanoinclusion, the inclusion is generally viewed as a nanoscale inhomogeneity. The inhomogeneity is surrounded by a surface layer, and both the inhomogeneity's inner region and the surface layer are characterized by different elastic properties compared with the matrix. In the following part, we first investigate the elastic interaction between the nano-inclusion and the edge dislocation, and then the case involving the screw dislocation is presented.

9.2.1 Edge Dislocation Interacting with Nanoscale Inhomogeneity

9.2.1.1 Basic Formula

In the physical problem, an infinitely elastic medium containing a circular nanoscale inhomogeneity with radius R is considered. The elastic properties of the nanoscale inhomogeneity and elastic medium are κ_1 and κ_2, respectively. Here, for the plane strain state, $\kappa_j = 3 - 4v_j$ $(j = 1, 2)$, and v_j is the Poisson's ratio. The shear moduli of the nanoscale inhomogeneity and elastic medium are μ_1 and μ_2, respectively. Assume that edge dislocation with Burgers vector (b_x, b_y) is placed at an arbitrary point in an infinite matrix. S^+ and S^- refer to the domains filled by the inhomogeneity and the matrix, respectively. According to prior work [24], the classical elastic differential equation is employed to describe the elastic field within a bulk solid, while the interface is distinguished by its unique elastic constants and constitutive law. Equilibrium and constitutive equations that are suitable for an isotropic solid are established [34], based on the assumption that no sliding occurs between the interface area and the bulk solid and there are no body forces. For a plane elasticity problem, the interface conditions are given by:

$$\begin{cases} u_{x1}^{+}(t) & = & u_{x2}^{-}(t) \\ u_{y1}^{+}(t) & = & u_{y2}^{-}(t) \\ \sigma_{rr1}^{+}(t) - \sigma_{rr2}^{-}(t) & = & -\frac{\sigma_{\theta\theta}^{0}}{R} \\ \sigma_{r\theta1}^{+}(t) - \sigma_{r\theta2}^{-}(t) & = & \frac{1}{R}\frac{\partial \sigma_{\theta\theta}^{0}}{\partial \theta} \end{cases} \tag{9.1}$$

where u_{xi} and u_{yi} $(i = 1, 2)$ denote the displacement in the Cartesian coordinates, σ_{rri} and $\sigma_{r\theta i}$ $(i = 1, 2)$ denote the stresses in polar coordinates system. As z approaches the interface from S^+ and S^-, the boundary values of a physical quantity are denoted by the superscripts + and −, respectively. The regions S^+ and S^- are represented by subscripts 1 and 2. The interface region is denoted by superscript "0". $|t| = R$ represents the points located at the circular arc interface L. Another constitutive equation in the area of the interface is written as follows:

$$\sigma_{\theta\theta}^{0} = (2\mu^{0} + \lambda^{0})\varepsilon_{\theta\theta}^{0} \tag{9.2}$$

where interface stress and interface strain are denoted by $\sigma^0_{\theta\theta}$ and $\varepsilon^0_{\theta\theta}$, respectively. The Lame constants for the interface are represented by μ^0 and λ^0. In our model, the interface is assumed to be coherent. That is because that coherent interfaces are prevalent in materials, and semi-coherent or incoherent interfaces would produce complex stress field [36–38]. Benefiting from the coherent interface, we could exclude the impact of residual stress fields. Meanwhile, the interfacial strain $\varepsilon^0_{\theta\theta}$ is equivalent to the corresponding tangential strain in the adjacent bulk region.

Through the utilization of the additional constitutive equation in the area of the interfacial boundary provided in Eq. 9.2, in conjunction with the subsequent equation pertaining to the bulk solid (inhomogeneity), we can get the following equation:

$$\varepsilon_{\theta\theta 1} = \frac{\lambda_1 + 2\mu_1}{4\mu_1(\lambda_1 + \mu_1)}\sigma_{\theta\theta 1} - \frac{\lambda_1}{4\mu_1(\lambda_1 + \mu_1)}\sigma_{rr1} \tag{9.3}$$

where λ_1 and μ_1 are the Lame constants of the inhomogeneity.

In Eq. 9.1, the stress discontinuity condition at the interface is modified as

$$\begin{cases} \sigma^+_{rr1}(t) - \sigma^-_{rr2}(t) &= -\frac{2\mu^0 + \lambda^0}{4R\mu_1(\lambda_1 + \mu_1)}[(\lambda_1 + 2\mu_1)\sigma_{\theta\theta 1} - \lambda_1\sigma_{rr1}] \\ \sigma^+_{r\theta 1}(t) - \sigma^-_{r\theta 2}(t) &= \frac{2\mu^0 + \lambda^0}{4R\mu_1(\lambda_1 + \mu_1)}\left[(\lambda_1 + 2\mu_1)\frac{\partial\sigma_{\theta\theta 1}}{\partial\theta} - \lambda_1\frac{\partial\sigma_{rr1}}{\partial\theta}\right] \end{cases} \tag{9.4}$$

Equation 9.4 is the non-classical stress boundary condition.

Two Muskhelishvili complex potentials $\Phi(z)$ and $\Psi(z)$ are employed to represent the stress and displacement components in the bulk solids [39]:

$$\begin{cases} \sigma_{rr} + \sigma_{\theta\theta} &= 2[\Phi(z) + \overline{\Phi(z)}] \\ \sigma_{rr} + i\sigma_{r\theta} &= \Phi(z) + \overline{\Phi(z)} - \bar{z}\overline{\Phi'(z)} - (\bar{z}/z)\overline{\Psi(z)} \\ 2\mu(u_x' + iu_y') &= iz[\kappa\Phi(z) - \overline{\Phi(z)} + \bar{z}\overline{\Phi'(z)} + (\bar{z}/z)\overline{\Psi(z)}] \end{cases} \tag{9.5}$$

where $u_x' = \partial u_x/\partial\theta$, $u_y' = \partial u_y/\partial\theta$, $\Phi'(z) = d[\Phi(z)]/dz$, the over-bar represents the complex conjugate.

Considering the interaction between edge dislocations and nanoscale inhomogeneity with surface effects, the complex potentials $\Phi_2(z)$ and $\Psi_2(z)$ in the infinitely matrix domain are expressed as [11]

$$\begin{cases} \Phi_2(z) = \frac{\gamma_2}{z - z_0} + \Phi_2^*(z) & |z| > R \\ \Psi_2(z) = \frac{\overline{\gamma_2}}{z - z_0} + \frac{\gamma_2\bar{z}_0}{(z - z_0)^2} + \Psi_2^*(z) & |z| > R \end{cases} \tag{9.6}$$

where $\gamma_2 = \frac{\mu_2}{\pi(1 + \kappa_2)}(b_y - ib_x)$, The terms caused by the interplay between the edge dislocation and the interface are denoted by $\Phi_2^*(z)$ and $\Psi_2^*(z)$. Similarly, the complex potentials $\Phi_1(z)$ and $\Psi_1(z)$ in the inhomogeneity domain S^+ are holomorphic.

To handle boundary conditions on the interface, the following analytic functions are introduced:

$$\begin{cases} \Omega_1(z) = -\overline{\Phi_1}\left(\frac{R^2}{z}\right) + \frac{R^2}{z}\overline{\Phi_1}'\left(\frac{R^2}{z}\right) + \frac{R^2}{z^2}\overline{\Psi_1}\left(\frac{R^2}{z}\right) & |z| > R \\ \Omega_2(z) = -\overline{\Phi_2}\left(\frac{R^2}{z}\right) + \frac{R^2}{z}\overline{\Phi_2}'\left(\frac{R^2}{z}\right) + \frac{R^2}{z^2}\overline{\Psi_2}\left(\frac{R^2}{z}\right) & |z| < R \end{cases} \tag{9.7}$$

Taking the complex conjugate of Eq. 9.7, it is obtained that

$$\begin{cases} \Psi_1(z) = \frac{R^2}{z^2}\left[\Phi_1(z) + \overline{\Omega_1}\left(\frac{R^2}{z}\right) - z\Phi_1'(z)\right] & |z| < R \\ \Psi_2(z) = \frac{R^2}{z^2}\left[\Phi_2(z) + \overline{\Omega_2}\left(\frac{R^2}{z}\right) - z\Phi_2'(z)\right] & |z| > R \end{cases} \tag{9.8}$$

According to Eq. 9.7, the displacement continuity condition for the interface of circle Eq. 9.1 is expressed as:

$$\left[\frac{\kappa_1}{\mu_1}\Phi_1(t) - \frac{1}{\mu_2}\Omega_2(t)\right]^+ = \left[\frac{\kappa_2}{\mu_2}\Phi_2(t) - \frac{1}{\mu_1}\Omega_1(t)\right]^- \quad t \in L(|t| = R) \tag{9.9}$$

Applying the generalized Liouville's theorem, the following equations are obtained from Eq. 9.9:

$$\begin{cases} \frac{\kappa_1}{\mu_1}\Phi_1(z) - \frac{1}{\mu_2}\Omega_2(z) = h(z) & z \in S^+ \ (|z| < R) \\ \frac{\kappa_2}{\mu_2}\Phi_2(z) - \frac{1}{\mu_1}\Omega_1(z) = h(z) & z \in S^- \ (|z| > R) \end{cases} \tag{9.10}$$

where $h(z) = \frac{\kappa_2}{\mu_2}\frac{\gamma_2}{z - z_0} + \frac{1}{\mu_2}\left[\frac{\gamma_2}{z - z^*} - \frac{\gamma_2}{z} - \frac{\overline{\gamma_2}z^*(z_0 - z^*)}{\overline{z_0}(z - z^*)^2}\right] + D_1$ and $z^* = R^2/\overline{z_0}$.

Solving Eqs. 9.7 and 9.10 for the unknown constant D_1 yields:

$$D_1 = \overline{\Phi_1(0)}/\mu_1 \tag{9.11}$$

The non-classical stress boundary condition Eq. 9.4 is rewritten as follows

$$\begin{aligned} &[\sigma_{rr2}(t) + i\sigma_{r\theta 2}(t)]^- - [\sigma_{r\theta 1}(t) + i\sigma_{rr1}(t)]^+ \\ &= \frac{2\mu^0 + \lambda^0}{4R\mu_1}\left[(\sigma_{\theta\theta 1} - \sigma_{rr1}) + \frac{\mu_1}{\lambda_1 + \mu_1}(\sigma_{\theta\theta 1} + \sigma_{rr1}) \right. \\ &\left. - i\frac{\partial(\sigma_{\theta\theta 1} - \sigma_{rr1})}{\partial\theta} - i\frac{\mu_1}{\lambda_1 + \mu_1}\frac{\partial(\sigma_{\theta\theta 1} + \sigma_{rr1})}{\partial\theta}\right] \end{aligned} \tag{9.12}$$

It can be found from Eq. 9.5 that

$$\begin{cases} \sigma_{rr1} + \sigma_{\theta\theta 1} &= 2[\Phi_1(z) + \overline{\Phi_1(z)}] \\ \sigma_{\theta\theta 1} - \sigma_{rr1} &= z\Phi_1{}'(z) + (z/\bar{z})\Psi_1{}'(z) + \bar{z}\overline{\Phi_1{}'(z)} + (\bar{z}/z)\overline{\Psi_1{}'(z)} \end{cases} \tag{9.13}$$

Substituting Eqs. 9.5 and 9.13 into Eq. 9.12, and taking Eq. 9.8 into account, the following equation is derived:

$$\begin{aligned} &\left[\Phi_1(t) + \Omega_2(t) + (a+b)\Phi_1(t) + (a+b)t\Phi_1{}'(t) + a\overline{\Omega_1}\left(\frac{R^2}{t}\right) - a\frac{R^2}{t}\overline{\Omega_1}{}'\left(\frac{R^2}{t}\right)\right]^+ \\ &= \left[\Phi_2(t) + \Omega_1(t) - a\Omega_1(t) - at\Omega_1{}'(t) - (a+b)\overline{\Phi_1}\left(\frac{R^2}{t}\right) + (a+b)\frac{R^2}{t}\overline{\Phi_1}{}'\left(\frac{R^2}{t}\right)\right]^- \\ &t \in L \ (|t| = R) \end{aligned} \tag{9.14}$$

where $a = \frac{2\mu^0 + \lambda^0}{4R\mu_1}$ and $b = \frac{2\mu_1(2\mu^0 + \lambda^0)}{4R\mu_1(\lambda_1 + \mu_1)}$.

The entire plane can be described by following equation:

$$g(z) = \begin{cases} \Phi_1(z) + \Omega_2(z) + (a+b)\Phi_1(z) + (a+b)z\Phi_1{}'(z) \\ \quad + a\overline{\Omega_1}\left(\frac{R^2}{z}\right) - a\frac{R^2}{z}\overline{\Omega_1}{}'\left(\frac{R^2}{z}\right) & |z| < R \\ \Phi_2(z) + \Omega_1(z) - a\Omega_1(z) - az\Omega_1{}'(z) \\ \quad - (a+b)\overline{\Phi_1}\left(\frac{R^2}{z}\right) + (a+b)\frac{R^2}{z}\overline{\Phi_1}{}'\left(\frac{R^2}{z}\right) & |z| > R \end{cases} \tag{9.15}$$

Applying the generalized Liouville's theorem, we obtain:

$$g(z) = \frac{\gamma_2}{z - z_0} + \frac{\gamma_2}{z} - \frac{\gamma_2}{z - z^*} + \frac{\overline{\gamma_2}z^*(z_0 - z^*)}{\bar{z}_0(z - z^*)^2} + D_2 \tag{9.16}$$

The unknown parameter D_2 in Eq. 9.16 can be derived from Eq. 9.7 and the latter equation in Eq. 9.15 as $z \to \infty$.

$$D_2 = -(1+b)\overline{\Phi_1(0)} \tag{9.17}$$

From Eqs. 9.10, and 9.15, the following two expressions for complex potential $\Phi_1(z)$ is derived as:

$$\begin{cases} \left(\frac{\mu_2\kappa_1}{\mu_1} + 1 + a + b\right)\Phi_1(z) + (a+b)z\Phi_1{}'(z) \\ \quad + a\left[\overline{\Omega_1}\left(\frac{R^2}{z}\right) - \frac{R^2}{z}\overline{\Omega_1}{}'\left(\frac{R^2}{z}\right)\right] \\ \quad = (1+\kappa_2)\frac{\gamma_2}{z - z_0} + \mu_2 D_1 + D_2 & |z| < R \\ \left(\frac{\mu_2\kappa_1}{\mu_1} + 1 - a\right)\Omega_1(z) - az\Omega_1{}'(z) \\ \quad - (a+b)\left[\overline{\Phi_1}\left(\frac{R^2}{z}\right) - \frac{R^2}{z}\overline{\Phi_1}{}'\left(\frac{R^2}{z}\right)\right] \\ \quad = \left(1 + \frac{1}{\kappa_2}\right)\left[\frac{\gamma_2}{z} - \frac{\gamma_2}{z - z^*} + \frac{\overline{\gamma_2}z^*(z_0 - z^*)}{\bar{z}_0(z - z^*)^2}\right] - \frac{\mu_2}{\kappa_2}D_1 + D_2 & |z| > R \end{cases} \tag{9.18}$$

The power series method is utilized to solve the first-order differential equations with changing parameters mentioned above. It's worth to mention that the complex potentials $\Phi_1(z)$ and $\Omega_1(z)$ can be expressed as Eq. 9.19:

$$\begin{cases} \Phi_1(z) = A_0 + \sum\limits_{k=1}^{+\infty} A_k z^k & |z| < R \\ \Omega_1(z) = B_0 + \sum\limits_{k=1}^{+\infty} B_{-k} z^{-k} & |z| > R \end{cases} \tag{9.19}$$

After substituting the Eq. 9.19 into Eq. 9.18 and the coefficients of the unknown terms on the right-hand side of Eq. 9.19 can be determined by comparing them with their corresponding power terms.

$$\begin{cases} A_0 = \frac{c_1}{(c_2)^2 - 1}\left(\frac{c_2\overline{\gamma_2}}{\overline{z_0}} + \frac{\gamma_2}{z_0}\right) \\ B_0 = -\overline{A_0} = \frac{c_1}{1-(c_2)^2}\left(\frac{\overline{\gamma_2}}{\overline{z_0}} + \frac{c_2\gamma_2}{z_0}\right) \\ A_k = -\frac{(1+\kappa_2)}{c_3}\frac{\gamma_2}{(z_0)^{k+1}} \\ \qquad + \frac{c_4}{c_3}\Bigg[\overline{\gamma_2}(\overline{z^*})^{k-1}R^{-2k} - \overline{\gamma_2}R^{-2k}\delta_{1k} \\ \qquad - \frac{\gamma_2\overline{z^*}(\overline{z_0} - \overline{z^*})R^{-2k}(k-1)(\overline{z^*})^{k-2}}{z_0}\Bigg] & k \geq 1 \\ B_{-k} = -\frac{c_6}{c_5}\frac{\overline{\gamma_2}R^{2k}}{(\overline{z_0})^{k+1}} \\ \qquad - \frac{1+\kappa_2}{c_5\kappa_2}\Bigg[\gamma_2(z^*)^{k-1} - \gamma_2\delta_{1k} \\ \qquad - \frac{\overline{\gamma_2}z^*(z_0 - z^*)(k-1)(z^*)^{k-2}}{\overline{z}_0}\Bigg] & k \geq 1 \end{cases} \tag{9.20}$$

where $c_1 = \frac{\mu_1(1+\kappa_2)}{\mu_2\kappa_1 + \mu_1(1+b)}$, $c_2 = \frac{\mu_2 - \mu_1(1+b)}{\mu_2\kappa_1 + \mu_1(1+b)}$, $c_3 = 1 + (a+b)(k+1) + \frac{\mu_2\kappa_1}{\mu_1} + \frac{a(a+b)(1+k)(1-k)}{1 - a + ka + \mu_2/(\kappa_2\mu_1)}$, $c_4 = \frac{(1+\kappa_2)a(1+k)}{\kappa_2[1 - a + ka + \mu_2/(\kappa_2\mu_1)]}$, $c_5 = 1 + a(k-1) + \frac{\mu_2}{\mu_1\kappa_2} + \frac{a(a+b)(1+k)(1-k)}{1 + (a+b)(k+1) + \mu_2\kappa_1/\mu_1}$, $c_6 = \frac{(1+\kappa_2)(a+b)(1-k)}{1 + (a+b)(k+1) + \mu_2\kappa_1/\mu_1}$
and δ_{ij} is the Kronecker delta.

From Eqs. 9.10 and 9.19, the complex potentials $\Omega_2(z)$ and $\Phi_2(z)$ are written by:

$$\begin{cases} \Omega_2(z) = \frac{\kappa_1\mu_2}{\mu_1}\sum\limits_{k=0}^{+\infty} A_k z^k - \frac{\kappa_2\gamma_2}{z - z_0} + \frac{\gamma_2}{z} \\ \qquad - \frac{\gamma_2}{z - z^*} + \frac{\overline{\gamma_2}z^*(z_0 - z^*)}{\overline{z}_0(z - z^*)^2} - \frac{\mu_2}{\mu_1}\overline{A_0} & |z| < R \\ \Phi_2(z) = \frac{\mu_2}{\mu_1\kappa_2}\sum\limits_{k=1}^{+\infty} B_{-k} z^{-k} + \frac{\gamma_2}{z - z_0} \\ \qquad + \frac{1}{\kappa_2}\left[\frac{\gamma_2}{z - z^*} - \frac{\gamma_2}{z} - \frac{\overline{\gamma_2}z^*(z_0 - z^*)}{\overline{z}_0(z - z^*)^2}\right] & |z| > R \end{cases} \tag{9.21}$$

The complex potentials $\Psi_1(z)$ and $\Psi_2(z)$ can be obtained by Eq. 9.8.

9.2.1.2 Image Force

Due to the dependence of the elastic strain energy of the system on the dislocation position, a force (f_x, f_y) acts upon the normalized length of the dislocation line, where f_x, and f_y denote the x and y components of the image force, respectively, which can be computed by the Peach-Koehler formula [40]:

$$f_x - if_y = [\sigma^*_{xy2}(z_0)b_x + \sigma^*_{yy2}(z_0)b_y] + i[\sigma^*_{xx2}(z_0)b_x + \sigma^*_{xy2}(z_0)b_y] \quad (9.22)$$

where the perturbation stress fields at the dislocation point z_0 (located at the matrix) are denoted by $\sigma^*_{xx2}(z_0)$, $\sigma^*_{yy2}(z_0)$ and $\sigma^*_{xy2}(z_0)$. The perturbation complex potentials $\Phi^*_2(z_0)$ and $\Psi^*_2(z_0)$ within the matrix are utilized to compute the perturbation stresses.

The Peach-Koehler formula is expressed in an alternative form as follows [30]:

$$f_x - if_y = \frac{\mu_2(b_y^2 + b_x^2)}{\pi(1 + \kappa_2)}\left[\frac{\Phi^*_2(z_0) + \overline{\Phi^*_2(z_0)}}{\gamma_2} + \frac{\overline{z_0}\Phi^{*\prime}_2(z_0) + \Psi^*_2(z_0)}{\overline{\gamma_2}}\right] \quad (9.23)$$

Based on the work [12] and considering Eq. 9.6, the perturbation complex potentials $\Phi^*_2(z_0)$, $\Phi^{*\prime}_2(z_0)$ and $\Psi^*_2(z_0)$ are represented by the following expression:

$$\begin{cases} \Phi^*_2(z_0) &= \lim\limits_{z\to z_0}[\Phi_2(z) - \Phi_0(z)] \\ \Phi^{*\prime}_2(z_0) &= \lim\limits_{z\to z_0}\dfrac{d[\Phi_2(z) - \Phi_0(z)]}{dz} \\ \Psi^*_2(z_0) &= \lim\limits_{z\to z_0}[\Psi_2(z) - \Psi_0(z)] \end{cases} \quad (9.24)$$

where $\Phi_0(z) = \frac{\gamma_2}{z - z_0}$ and $\Psi_0(z) = \frac{\overline{\gamma_2}}{z - z_0} + \frac{\gamma_2\overline{z_0}}{(z - z_0)^2}$.

To derive the elaborate expression of the complex potential $\Psi_2(z)$, substitute Eq. 9.21 into Eq. 9.8, the following equations are obtained:

$$\begin{aligned} \Psi_2(z) = {} & \frac{R^2}{z^2}[\Phi_2(z) - z\Phi_2{}'(z)] \\ & + \frac{R^2}{z^2}\left\{\frac{\mu_2\kappa_1}{\mu_1}\left[\sum_{k=1}^{+\infty}\overline{A_k}\left(\frac{R^2}{z}\right)^k - \frac{\mu_1(1 + \kappa_2)}{\mu_2\kappa_1 + \mu_1(1 + b)}\frac{\overline{\gamma_2}}{\overline{z_0}}\right]\right. \\ & - \frac{\kappa_2\overline{\gamma_2}z}{R^2 - z\overline{z_0}} + \frac{\overline{\gamma_2}z_0z}{R^2(z - z_0)} + \frac{\overline{\gamma_2}z}{R^2} + \frac{\gamma_2\overline{z^*}(\overline{z_0} - \overline{z^*})z_0z^2}{R^4(z - z_0)^2} \\ & \left. - \frac{\mu_2(1 + \kappa_1)(1 + b)A_0}{\mu_2\kappa_1 + \mu_1(1 + b)}\right\} \end{aligned} \quad (9.25)$$

Eq. 9.23 provides a comprehensive representation of the image force. The image force can be decomposed into two components: the glide force along the direction of the Burges vector and the climb force perpendicular to the Burges vector. These components are expressed as:

$$\begin{cases} f_g = f_x \cos\theta + f_y \sin\theta \\ f_c = f_x \sin\theta - f_y \cos\theta \end{cases} \tag{9.26}$$

where $\theta = \arctan(b_y/b_x)$.

9.2.1.3 Critical Shear Stress

The CSS is the minimum value of shear stress required for the dislocation movement (slip or climb). Generally, the CSS is influenced by the chemical situation of matrix and various microstructures including the inhomogeneities. Here, we only concern the contribution from the nanostructure.

According to previous work [41], we examine composites that are characterized by N inclusions per unit area, with each inclusion having a radius of R. The average distance between the dislocation and the neighboring dislocation is equivalent to half the inclusion spacing $r_0 = (N^{-1/2})/2$. Hence, when $C = N\pi R^2$ denotes the volume fraction of inclusions, the mean force f_{rm} that acts upon the edge dislocation is calculated as:

$$f_{rm} = \frac{\mu_2 b_r^2 (\Pi_1 + \Pi_2)}{\pi(1 + \kappa_2)} \tag{9.27}$$

The special symbols in the above formula are given as:

$$\begin{aligned} \Pi_1 = {} & \frac{4C}{\pi}(\Pi_3 - \Pi_2) - \frac{4C}{\pi}\frac{\mu_2\kappa_1}{\mu_1}\sum_{k=1}^{+\infty}\left\{\frac{1+\kappa_2}{c_3}\left(\frac{4C}{\pi}\right)^k \frac{2\sqrt{C}}{R\sqrt{\pi}}\right. \\ & \left. + \frac{c_4}{c_3}\left[\left(\frac{4C}{\pi}\right)^{k-1} - \delta_{1k} + (k+1)\left(\frac{4C}{\pi}\right)^{k-1}\left(1 - \frac{4C}{\pi}\right)\right]\frac{2\sqrt{C}}{R\sqrt{\pi}}\right\} \\ & - \frac{4C}{\pi}\frac{\mu_2\kappa_1(1+\kappa_2)}{\mu_2\kappa_1 + \mu_1(1+b)}\frac{2\sqrt{C}}{R\sqrt{\pi}} - \frac{4C}{\pi}\left[\frac{2\kappa_2\sqrt{C\pi}}{(4C-\pi)R}\right. \\ & \left. + \frac{\mu_2(1+\kappa_1)(1+b)}{\mu_2\kappa_1 + \mu_1(1+b)}\frac{c_1}{1+c_2}\frac{2\sqrt{C}}{R\sqrt{\pi}} + \left(\frac{\pi}{4C} - 2\right)\frac{2\sqrt{C}}{R\sqrt{\pi}} - \frac{2\sqrt{\pi}}{R\sqrt{C}}\right] \end{aligned} \tag{9.28}$$

$$\begin{aligned} \Pi_2 = {} & \frac{1}{\kappa_2}\left[\frac{2\pi^{3/2}C^{1/2} + 8\pi^{1/2}C^{3/2}}{R(\pi - 4C)^2} - \frac{2\sqrt{C}}{R\sqrt{\pi}}\right] \\ & - \frac{\mu_2}{\mu_1\kappa_2}\sum_{k=1}^{+\infty}\left\{\frac{(1+\kappa_2)k}{c_5\kappa_2}\left[\left(\frac{4C}{\pi}\right)^{k-1} - \delta_{1k} + \left(\frac{4C}{\pi}\right)^{k-1}\left(1 - \frac{4C}{\pi}\right)\right]\frac{2\sqrt{C}}{R\sqrt{\pi}}\right. \\ & \left. - \frac{c_6}{c_5}\left(\frac{4C}{\pi}\right)^k \frac{2k\sqrt{C}}{R\sqrt{\pi}}\right\} \end{aligned} \tag{9.29}$$

According to the research [42], The CSS resulting from this elastic interaction (or its corresponding effect on the material's critical shear stress) can be computed using the mean force described in Eq. 9.20.

$$\Delta \tau_c = \frac{\mu_2 b_r [\Pi_1(R, C) + \Pi_2(R, C)]}{\pi(1 + \kappa_2)} \tag{9.30}$$

9.2.1.4 Numerical Example and Discussion

Utilizing Eq. 9.26, one can analyze the glide/climb force acting on the dislocation, under the influences of various factors. Hence, the influence of the characteristic of the interface is first evaluated. To simplify our analysis, it is assumed that the edge dislocation is situated at the point x_0 on the positive x-axis ($z_0 = x_0 > R$). The glide/climb force are simplified as $f_{go} = \pi R(1 + \kappa_2) f_g / [\mu_2 (b_x^2 + b_y^2)]$ and $f_{co} = \pi R(1 + \kappa_2) f_c / [\mu_2 (b_x^2 + b_y^2)]$ [30]. The parameters are defined as $\alpha = \mu^0/\mu_1$ and $\beta = \lambda^0/\mu_1$. The proportion of the shear modulus of two bulk solids $\eta = \mu_2/\mu_1$. The relative position between the dislocation and the inhomogeneity is $\rho = x_0/R$. Because the interface and surface share similar geometrical and physical properties (a free surface can be regarded as a distinct state of interface), we may estimate the interface constants μ^0 and μ^0/μ using the surface properties [31]. In this case, we get $\mu^0/\mu \approx \lambda^0/\mu \approx 1.0\text{Å}$.

Figure 9.1a shows the change of standardized glide force f_{go} with radius R, under different interface properties. Here, it is worth to note that, under the current material parameters, the force for dislocation climbing is zero. When the interface stress is introduced, extra repulsive or attractive force would be exerted on the dislocation, resulting in an increase or decrease in the glide force. Therefore, we may conclude that the presence of interface stress contributes to the local strengthening and softening of the interface. As the radius of the inhomogeneity decreases, the absolute value of the glide force keeps increasing. Meanwhile, the size dependence becoming

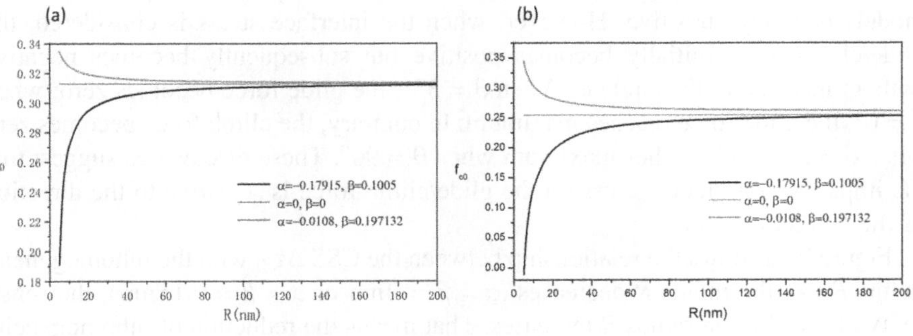

FIGURE 9.1 (a) Variation of the glide force f_{go} as a function of the radius R, for $\eta = 0.9$, $\nu_1 = \nu_2 = 0.3$ and $\rho = 1.15$ ($b_y = 0$); (b) Variation of the climb force f_{co} as a function of the radius R, for $\eta = 0.9$, $\nu_1 = \nu_2 = 0.3$ and $\rho = 1.15$ ($b_x = 0$) [43]. Intrinsic lengths are $\alpha = -0.17915\,\text{nm}$ and $\beta = 0.1005\,\text{nm}$ (Al [100]), and intrinsic lengths $\alpha = -0.01082\,\text{nm}$ and $\beta = 0.197132\,\text{nm}$ (Al [111]) [31,34].

more noticeable when the radius is less than 20 nm. These phenomena cannot be captured by classical models that do not take the surface/interface effect into account. The relationship between the standardized climb force f_{co} with the radius R is shown in Figure 9.1b. Note that the glide force is equal to zero in this case. Obviously, the results involved climb force (Figure 9.1b) are similar to those involved glide force (Figure 9.1a). A different finding is that, when the inhomogeneity radius decreases to 6 nm, the direction of climb force changes.

Figure 9.2a shows the relationship between the force f_{go} and the relative position ρ. Here, the climb force is equal to zero. When the interface stress is introduced, the soft inhomogeneity repels the dislocation that approaches the inhomogeneity ($\eta = \mu_2/\mu_1 = 1.1$). Meanwhile, we found the equilibrium point at the x-axis when the gliding force for dislocation is equal to zero. If the we do not consider the interface effect ($\alpha = \beta = 0$), the soft inhomogeneity always attracts the dislocation. This trend would induce a local strengthening at the interface, because the presence of equilibrium point inhibits the dislocation glide. If there are multiple equilibrium points, the local strengthening would be more pronounced. The relationship between the force f_{co} and relative location ρ is shown in Figure 9.2b. The results are similar to them from Figure 9.2a, suggesting that the interface stress also makes the dislocation climb more difficult.

Figure 9.2c displays the relationship between the glide force f_{go} with the relative location ρ. Interestingly, we found the unstable equilibrium points after introducing the interface stress. The presence of unstable equilibrium points would induce the local softening of interface. Similar results are also found in Figure 9.2d.

Figure 9.3a ($b_y = 0$) and Figure 9.3b ($b_x = 0$) present the glide force f_{go} and climb force f_{co} under different intrinsic lengths α and β. For the case with large intrinsic length, the influence of interface stress on the glide and climb force is more significant. Meanwhile, when the inhomogeneity radius is small, the influence of interface stress is also enhanced.

In Figure 9.3c,d, the relationships between the glide force f_{go} and climb force f_{co} with respect to the Burgers vector direction θ (the angle between the Burgers vector and the x-axis). are presented. The glide/climb force predicted by the classical models is always positive. However, when the interface stress is considered, the glide/climb force initially becomes positive but subsequently becomes negative with an increase in the angle θ. When $\theta = 35°$, the glide force becomes zero; when $\theta = 0°$, the glide force reaches maximum. In contrary, the climb force becomes zero when $\theta = 50°$, and reaches maximum when $\theta = 90°$. These discoveries suggest that the impact of interfacial stress on the glide/climb force is sensitive to the direction of the Burgers vector.

Figure 9.4a shows the relationship between the CSS $\Delta\tau_{c0}$ with the inhomogeneity radius R. As the radius R decreases ($\alpha = \beta = 0\,\text{nm}$ or $\alpha = \beta = 0.15\,\text{nm}$), the sensitivity of the CSS on radius R increases. That means the reduction of inhomogeneity size enhances the influences of interface effect. When $\alpha = \beta = -0.15\,\text{nm}$, the CSS initially rises and subsequently falls as R decreases. This trend suggests a critical inhomogeneity radius that determines the maximal impact on the CSS. Meanwhile, it indicates that there is an optimal inhomogeneity radius to realize the maximum strengthening effect.

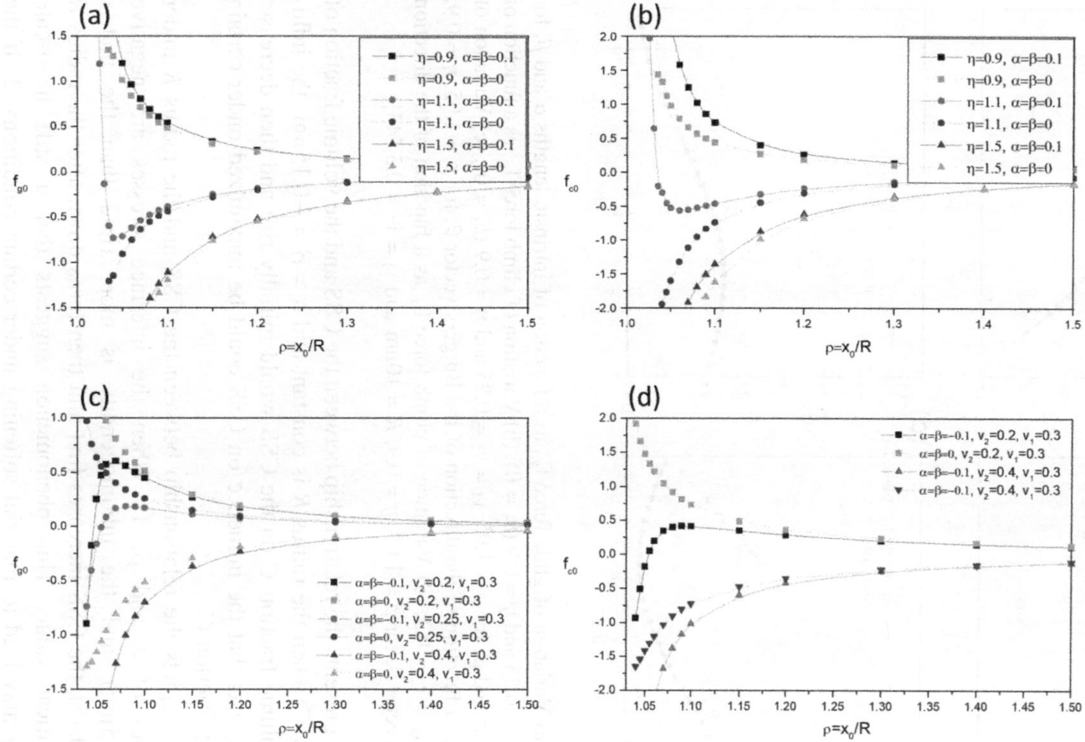

FIGURE 9.2 (a) Variation of glide force f_{go} as a function of the relative position ρ, for $R = 10\text{nm}$, $v_1 = v_2 = 0.25$, $b_y = 0$, $\alpha = 1.0\mathring{A}$ and $\beta = 1.0\mathring{A}$; (b) Variation of climb force f_{co} as a function of the relative position ρ, for $R = 10\text{nm}$ and $v_1 = v_2 = 0.25$ ($b_x = 0$); (c) Variation of glide force f_{go} as a function of the relative position ρ, for $R = 10\text{nm}$, $\eta = 1$, $b_y = 0$, $\alpha = -1.0\text{A}$ and $\beta = -1.0\text{A}$; (d) Variation of climb force f_{co} as a function of the relative position ρ, for $R = 10\text{nm}$ and $\eta = 1$ ($b_x = 0$) [43].

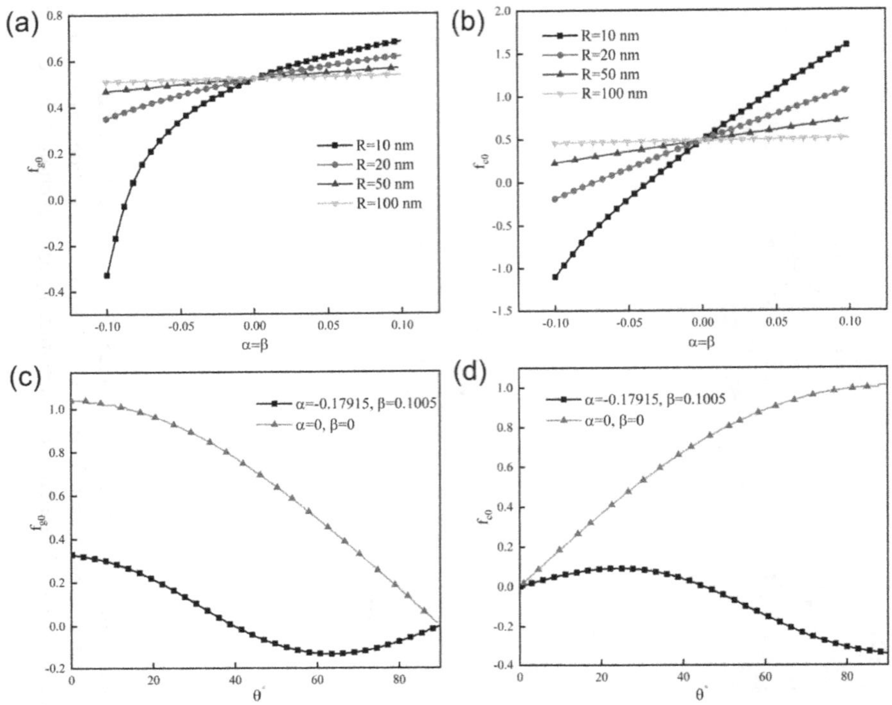

FIGURE 9.3 (a) Variation of glide force f_{go} as a function of intrinsic lengths α and β, for $\rho = 1.05$, $\nu_1 = \nu_2 = 0.25$ and $\eta = 0.9$ ($b_y = 0$); (b) Variation of climb force f_{co} as a function of intrinsic lengths α and β, for $\rho = 1.05$, $\nu_1 = \nu_2 = 0.25$ and $\eta = 0.9$ ($b_x = 0$); (c) Variation of glide force f_{go} as a function of the direction of the Burgers vector θ, for $\rho = 1.05$, $\eta = 0.9$, $R = 10$nm and $\nu_1 = \nu_2 = 0.3$; (d) Variation of climb force f_{co} as a function of the direction of the Burgers vector θ, for $\rho = 1.05$, $\eta = 0.9$, $R = 10$nm and $\nu_1 = \nu_2 = 0.3$ [43].

Figure 9.4b presents the relationship between the CSS and the volume fraction of inhomogeneity C, when the radius R is constant and $\alpha = \beta = -0.15$nm, the influence of the volume fraction C on the CSS would initially rise and then decrease. This trend suggests that the influence on CSS could be maximized under certain critical volume fraction C.

Figure 9.4c depicts the relationship between the CSS and the radius R under different values of α and β. 1) When the interface stresses are negative ($\alpha = \beta = -0.15$nm), and the inhomogeneity is more rigid than the matrix ($\eta = \mu_2/\mu_1 = 0.85$), the CSS increases first and then turns down after the radius R exceeds a critical value. This phenomenon suggests that a stiff nanoscale inhomogeneity may lead to material softening under certain conditions. 2) If the interface stresses are positive ($\alpha = \beta = 0.15$nm), and the stiffness of inhomogeneity is lower than that of the matrix, the CSS is negative initially and then becomes positive as the radius R decreases. This trend implies that a soft nanoscale inhomogeneity may strengthen the material under certain conditions.

The above results are available for the coherent interface. For the semi-coherent and in-coherent interfaces, the elastic strain discontinuity needs to be taken into

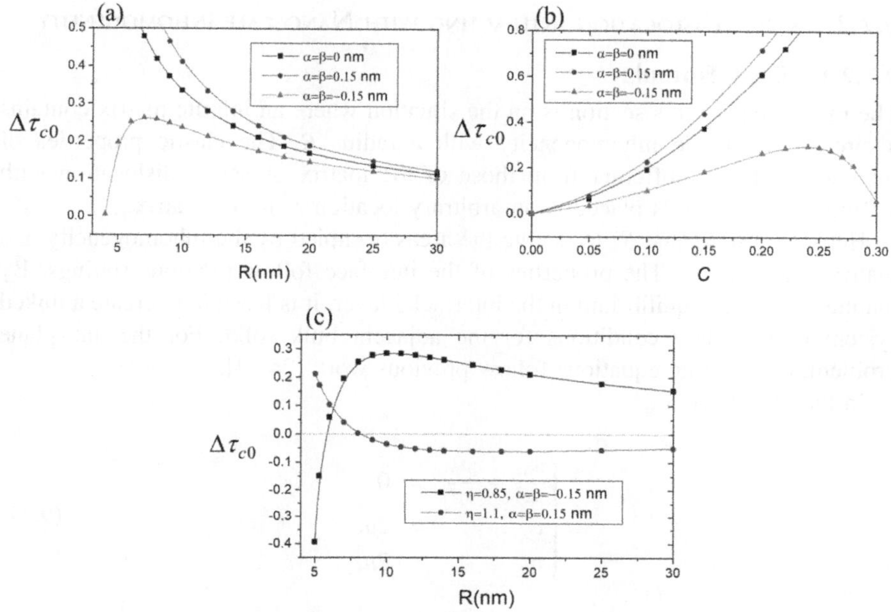

FIGURE 9.4 (a) Variation of the critical shear stress $\Delta\tau_{c0}$ as a function of the radius R with varying values of α and β, for $C = 0.1$ and $\eta = 0.85$; (b) Variation of the critical shear stress $\Delta\tau_{c0}$ as a function of the proportion of volume C with varying values of α and β, for $R = 10$nm and $\eta = 0.85$; (c) Variation of the critical shear stress $\Delta\tau_{c0}$ as a function of the radius R with varying values of the α and β, and the relative shear modulus $\eta = \mu_2/\mu_1$, for $C = 0.1$ [44].

account [45]. The problem about the interaction between dislocation and semi-coherent/incoherent interfaces could be addressed, by revising the related boundary conditions mentioned in Eq. 9.4.

9.2.1.5 Remarks

This section investigates the elastic interaction between edge dislocation and nanoscale inhomogeneity, based on the elastic model considering the interface stress. We obtain the explicit solution under new boundary conditions, and systematically analyze the image force acting on the edge dislocation. The results demonstrate that the interface stress causes a rise (or fall) in the dislocation glide/climb force, thereby attracting (or repelling) the dislocation, and eventually inducing the local hardening (or softening) effect.

The influence of interface stress will be enhanced when the size of inhomogeneity reduces to nanoscale. The presence of interface stress induces the additional equilibrium positions when the dislocation approaching the inhomogeneity. This phenomenon makes the motion of dislocation more difficult, leading to the locally strengthening. In addition, owing to the interface stress, the CSS could be maximized under critical radius of inhomogeneity or critical volume fraction of inhomogeneity.

9.2.2 Screw Dislocation Interacting with Nanoscale Inhomogeneity

9.2.2.1 Basic Formula

The main focus of this section is on the situation where an infinite matrix contains a circular nanoscale inhomogeneity with a radius R. The elastic properties of inhomogeneity are different from those of the matrix. A screw dislocation with a Burger's vector b_z is placed at an arbitrary location z_0 in the matrix.

Here, we use S^+ and S^- to denote the areas occupied by the inhomogeneity and matrix, respectively. The properties of the interface follow previous settings. By maintaining force equilibrium in the interfacial layer, it is feasible to create a linked system of boundary conditions for the adjacent bulk solid. For the anti-plane problem, the ensuing equations follow previous work [24,34].

In the bulk [34]:

$$\begin{cases} \frac{\partial^2 w}{\partial x^2} + \frac{\partial^2 w}{\partial y^2} & = 0 \\ \tau_{rz} & = 2\mu\varepsilon_{rz} \\ \tau_{\theta z} & = 2\mu\varepsilon_{\theta z} \end{cases} \tag{9.31}$$

For the surface [34]:

$$\begin{cases} \tau_{\theta z}^0 & = 2(\mu^0 - \tau^0)\varepsilon_{\theta z}^0 \\ [\tau_{rz}(t)] & = \frac{1}{R}\frac{\partial \tau_{\theta z}^0(t)}{\partial \theta} \end{cases} \tag{9.32}$$

where w denotes anti-plane displacement. In polar coordinates r and θ, τ_{rz} and $\tau_{\theta z}$denote stress components, ε_{rz} and $\varepsilon_{\theta z}$ denote strain components. The superscript "0" refers to interface region, τ^0 refers the residual interface tension. Additionally, the condition $[\tau_{rz}(t)] = \tau_{rz1}(t) - \tau_{rz2}(t)$ signifies the stress discontinuity across the interface, with the subscripts 1 and 2 denoting the inhomogeneity and matrix regions, respectively. For simplification, the interface between nanowire and matrix is assumed to be coherent. Therefore, the interface strain $\varepsilon_{\theta z}^0$ equals to the tangential strain in adjacent block material ($\varepsilon_{\theta z}^0(t) = \varepsilon_{\theta z}(t)$). In this section, we shall not consider the impact of residual stress fields, i.e., $\tau_0 = 0$. In this case, we get $\mu_0 - \tau_0 = \mu_0$ for Eq. 9.32.

Assuming $\varepsilon_{\theta z}^0(t) = \varepsilon_{\theta z}(t)$, we get:

$$[\tau_{rz}(t)] = \frac{(\mu^0 - \tau^0)}{R\mu}\frac{\partial \tau_{\theta z}(t)}{\partial \theta} \tag{9.33}$$

According to the research [39], an analytical function $f(z)$ is utilized to represent the anti-plane displacement w, as well as shear stresses τ_{rz} and $\tau_{\theta z}$ in the bulk solid.

$$\begin{cases} w & = [f(z) + \overline{f(z)}]/2 \\ \tau_{rz} - i\tau_{\theta z} & = \mu e^{i\theta} F(z) \end{cases} \tag{9.34}$$

where the overbar signifies the complex conjugate, while the prime and $F(z) = f'(z)$ denote the derivative corresponding to the parameter z.

The displacement continuity condition and the equilibrium condition at the interface are expressed as [24]:

$$\begin{cases} w_1^+(t) & = w_2^-(t) & t \in L \\ \tau_{rz1}^+(t) - \tau_{rz2}^-(t) & = \frac{\mu_0}{R\mu_2}\frac{\partial \tau_{\theta z2}^-(t)}{\partial \theta} & t \in L \end{cases} \tag{9.35}$$

where μ_2 denotes shear modulus of the matrix, while the superscripts "+" and "–" signify the physical quantity as the variable "z" approaches the interface from regions S^+ and S^-, respectively.

With the utilization of the Riemann-Schwarz symmetry principle, we can introduce two novel complex potentials, denoted by $\Omega_1(z)$ and $\Omega_2(z)$

$$\begin{cases} \Omega_1(z) & = -\frac{R^2}{z^2}\overline{F_1}\left(\frac{R^2}{z}\right) & z \in S^- \\ \Omega_2(z) & = -\frac{R^2}{z^2}\overline{F_2}\left(\frac{R^2}{z}\right) & z \in S^+ \end{cases} \tag{9.36}$$

To address the current problem, we can express the complex potential $F_2(z)$ within the matrix using the following format:

$$F_2(z) = \frac{b_z}{2\pi i}\frac{1}{z - z_0} + \Gamma + F_{20}(z) \quad z \in S^- \tag{9.37}$$

where $F_{20}(z)$ is holomorphic in infinite matrix. The longitudinal stresses at infinity are associated with $\Gamma = (\tau_{xz}^\infty - i\tau_{yz}^\infty)/\mu_2$. Substituting Eq. 9.37 into Eq. 9.36, the complex potential $\Omega_2(z)$ is expressed as:

$$\Omega_2(z) = \frac{b_z}{2\pi i}\left(\frac{1}{z} - \frac{1}{z - z^*}\right) - \frac{R^2}{z^2}\overline{\Gamma} + \Omega_{20}(z) \quad z \in S^+ \tag{9.38}$$

where $z^* = R^2/\overline{z_0}$, $\Omega_{20}(z)$ is conformal inside circular inhomogeneity region.

By differentiating Eq. 9.35 over the variable θ, the continuity condition for displacement is expressed as the following equation:

$$[F_1(t) - \Omega_2(t)]^+ = [F_2(t) - \Omega_1(t)]^- \quad t \in L \tag{9.39}$$

Upon considering Eqs. 9.36–9.38, applying the generalized Liouville's theorem [39], and following the method in Section 9.2.1, we get the following result:

$$\begin{aligned} F_2(z) &= \frac{b_z}{2\pi i}\left(\frac{1}{z-z_0}+\frac{1}{z-z^*}-\frac{1}{z}\right)+\Gamma+\frac{\mu_1-\mu_2+(\mu_0/R)}{\mu_1+\mu_2+(\mu_0/R)}\frac{R^2}{z^2}\overline{\Gamma} \\ &- \frac{b_z}{2\pi i}\sum_{k=0}^{\infty}\frac{2\mu_2}{\mu_1+\mu_2+(1+k)(\mu_0/R)}\left(\frac{R^2}{\overline{z_0}}\right)^{k+1} z^{-k-2} \qquad |z|>R \end{aligned} \tag{9.40}$$

When the interface stress is equal to zero ($\mu_0 - \tau_0 = 0$), Eq. 9.40 is simplified to the following expression:

$$F_2(z) = \frac{b_z}{2\pi i}\frac{1}{z-z_0}+\frac{\mu_1-\mu_2}{\mu_1+\mu_2}\frac{b_z}{2\pi i}\left(\frac{1}{z-z^*}-\frac{1}{z}\right)+\Gamma+\frac{\mu_1-\mu_2}{\mu_1+\mu_2}\frac{R^2}{z^2}\overline{\Gamma} \tag{9.41}$$

This result is consistent with the result in Ref. [46].

The stress and displacement components in both the matrix and circular inhomogeneity zones can be derived using Eq. 9.34.

9.2.2.2 Image Force

The Peach-Koehler formula is employed to calculate the image stress applied to the screw dislocation from the circular inhomogeneity.

$$\mathrm{f}_x - i\mathrm{f}_y = ib_z\left[\widehat{\tau_{xz2}}(z_0) - i\widehat{\tau_{yz2}}(z_0)\right] \tag{9.42}$$

where the force vector in the x-axis and y-axis are f_x and f_y, respectively. The perturbation of stress vectors at the dislocation position are $\widehat{\tau_{xz2}}(z_0)$ and $\widehat{\tau_{yz2}}(z_0)$. According to previous work [10], the explicit equation for the image force is calculated by:

$$\begin{aligned} \mathrm{f}_x - i\mathrm{f}_y &= \frac{\mu_2 b_z^2}{2\pi}\left(\frac{1}{z_0-z^*}-\frac{1}{z_0}\right)+ib_z\mu_2\Gamma+\frac{\mu_2[\mu_1-\mu_2+(\mu_0/R)]}{\mu_1+\mu_2+(\mu_0/R)}\frac{R^2}{z_0^2}(ib_z\overline{\Gamma}) \\ &- \frac{b_z^2}{2\pi}\sum_{k=0}^{\infty}\frac{2\mu_2^2}{\mu_1+\mu_2+(1+k)(\mu_0/R)}\left(\frac{R^2}{\overline{z_0}z_0}\right)^{k+1}\frac{1}{z_0} \end{aligned} \tag{9.43}$$

By evaluating the negative gradient of the interaction energy with respect to the dislocation position, the image force F_r, which acts along the direction of the Burgers vector is determined by $F_r = -\partial W\frac{\partial W}{\partial r_0}$.

The following expression is utilized to calculate the interaction energy [47]:

$$W = \frac{\mu_2 b}{2}\mathrm{Im}\left[\int \Lambda\Phi(z)dz\right] \tag{9.44}$$

where "Im" denotes the imaginary part of the complex variable, and "Λ" corresponds to the remaining complex potential after the dislocation core is eliminated. This operation leads to the following equation:

$$W = \frac{\mu_2 b^2}{4\pi}\left[\ln\frac{r_0^2}{r_0^2 - R^2} - \sum_{k=0}^{\infty}\frac{1}{1+k}\frac{2\mu_2}{\mu_1 + \mu_2 + (1+k)(\mu^s - \tau^s)/R}\left(\frac{R}{r_0}\right)^{2k+2}\right] \tag{9.45}$$

Thus,

$$F_r = -\frac{\partial W}{\partial r_0} = \frac{\mu_2 b^2}{2\pi}\left[\frac{R^2}{r_0(r_0^2 - R^2)} - \sum_{k=0}^{\infty}\frac{2\mu_2}{\mu_1 + \mu_2 + (1+k)(\mu^s - \tau^s)/R}\left(\frac{R^{2k+2}}{r_0^{2k+3}}\right)\right] \tag{9.46}$$

9.2.2.3 Critical Shear Stress

The geometric parameter settings for inclusions are consistent with Section 9.2.1.3. Given a volume fraction of inclusions $f = N\pi R^2$, the mean force F_{rm} that acts on the screw dislocation is expressed as:

$$F_{rm} = \frac{\mu_2 b^2}{R\pi^{3/2}}\left[\frac{4f^{3/2}}{\pi - 4f} - \sum_{k=0}^{\infty}\frac{2\mu_2}{\mu_1 + \mu_2 + (1+k)(\mu^s - \tau^s)/R} f^{1/2}\left(\frac{4f}{\pi}\right)^{k+1}\right] \tag{9.47}$$

Prior study demonstrates that the CSS resulting from elastic interaction is calculated from the mean force [42].

$$\Delta\tau_c = \frac{\mu_2 b}{R\pi^{3/2}}\left[\frac{4f^{3/2}}{\pi - 4f} - \sum_{k=0}^{\infty}\frac{2\mu_2}{\mu_1 + \mu_2 + (1+k)(\mu^s - \tau^s)/R} f^{1/2}\left(\frac{4f}{\pi}\right)^{k+1}\right] \tag{9.48}$$

Utilizing Eq. 9.48, it is feasible to investigate the effect of different factors on the CSS, such as the inclusion radius R, the volume fraction of inclusions, and the interface stress.

9.2.2.4 Numerical Examples and Discussion

The Eq. 9.43 provides the basis for analyzing the image force applied to the screw dislocation. In the following analysis, it is assumed that the remote force is zero. The screw dislocation is located at position x_0 on the positive x-axis ($z_0 = x_0 > R$). In this case, the force component along the Y axis is $f_y = 0$, and the X axis component of the standardized image force is defined as $f_{xo} = 2\pi R f_x/\mu_2 b_z^2$. Meanwhile, we define relative shear modulus $\alpha = \mu_1/\mu_2$, the intrinsic length $\beta = \mu_0/\mu_2$, the relative location $\delta = x_0/R$, the remaining interfacial tension $\tau^0 = 0$, and the Burgers vector $b = 0.25\,\text{nm}$.

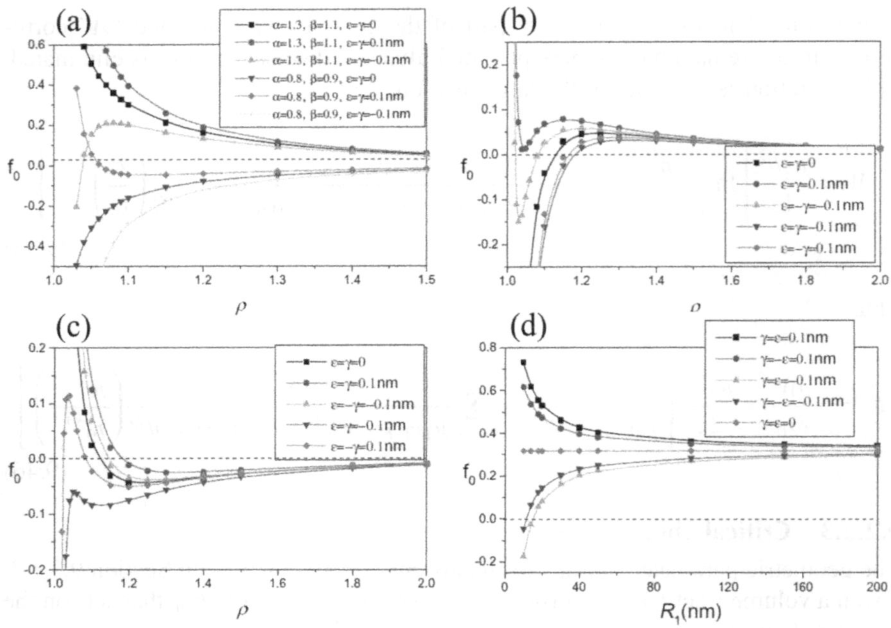

FIGURE 9.5 (a) Variation of f_{xo} as a function of the relative location δ, for $\beta = 1.0\,\mathring{A}$ and $R = 15\,\text{nm}$; (b) Variation of f_{xo} as a function of the relative location δ, for $\beta = -1.0\,\mathring{A}$ and $R = 15\,\text{nm}$; (c) Variation of f_{xo} as a function of the radius R, for $\delta = 1.05$; (d) Variation of f_{xo} as a function of intrinsic length β, for $\delta = 1.05$ and $\alpha = 2$ [43].

Figure 9.5a,b depicts the relationship between the image force f_{xo} and relative position δ, under different value of α and β. Figure 9.5a demonstrates that the soft nanoscale inhomogeneity with interfacial stress can lead to the repulsion of the screw dislocation when $\alpha = 0.7$. This differs from classical elasticity, where soft inclusions without interface stress normally attract the screw dislocation. Furthermore, a stable equilibrium point exists at the x-axis, indicating that the motion of dislocation becomes harder than that case without interface stress. From Figure 9.5b, a hard nanoscale inhomogeneity with interfacial stress would attract the screw dislocation, and an unstable equilibrium point exists in the vicinity of the inhomogeneity.

Figure 9.5c displays the relationship between the force f_{xo} and the radius R, under various values of α and β. Due to the presence of the interface stress, the positive β produces a repulsive force on the screw dislocation, while the negative β leads to an attractive force. The influence of the interface stress becomes significant when the inhomogeneity radius reduces to 50 nm, which is different with the results from classical model. Figure 9.5d displays the relationship between the force f_{xo} and the intrinsic length β. Similar with the results in Figure 9.5c, the reduction of the intrinsic length β enhances the influence of the interface stress on the image force.

The relationship between the standardized CSS $\Delta\tau_{c0} = (\Delta\tau_c * 10^4)/\mu_2$ and the radius R is displayed in Figure 9.6a. The reduction of intrinsic length enhances the influence of the interface stress on the CSS, especially when $\alpha = \mu^s/\mu_2 = 0.1\,\text{nm}$

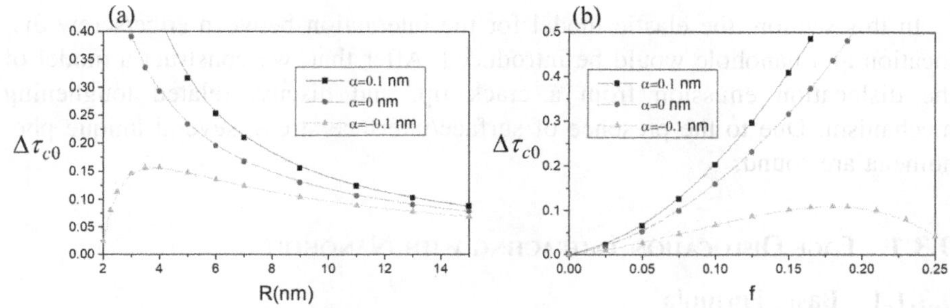

FIGURE 9.6 (a) Variation of the critical shear stress $\Delta\tau_{c0}$ as a function of the radius R with various values of intrinsic length $\alpha = \mu^s/\mu_2$, for $\mu_1/\mu_2 = 1.2$, $f = 0.1$ and Burgers vector $b = 0.25\,\text{nm}$; (b) Variation of the critical shear stress $\Delta\tau_{c0}$ as a function of the volume proportion f various values of intrinsic length $\alpha = \mu^s/\mu_2$, for $\mu_1/\mu_2 = 1.2$, $R = 5\,\text{nm}$ and Burgers vector $b = 0.25\,\text{nm}$ [48].

and $f = 0.1$. For certain conditions, such as $\alpha = -0.1\,\text{nm}$ and $f = 0.1$, the CSS initially increases as the radius R decreases, and then decreases after reaching a critical radius. Similar results are also observed in Figure 9.6b, which shows the relationship between CSS and the volume fraction of inhomogeneity.

9.2.2.5 Remarks

In this section, the elastic interaction between a circular nanoscale inhomogeneity and a screw dislocation is investigated, based on the complex potential method. Most results are similar with the results involving edge dislocation in Section 9.2.1. Specifically, the presence of interface stress changes the image force acting on the dislocation. The dislocation may be attracted (or repelled) by the nanoscale inhomogeneity due to the interface stress, contributing to the local softening (or strengthening) of materials. The interface stress also induces the equilibrium point when the dislocation approaches the inhomogeneity, thus make the dislocation motion more difficult. In addition, critical values of the radius and volume fraction of inhomogeneity are found. Under the critical radius, the influence of interface stress on the CSS would be enhanced.

9.3 NANOPORES

During the forming process, materials inevitably produce micro defects such as voids and microcracks, which seriously weaken the strength and toughness of the material [49–52]. Among them, the presence of nanopores in nanocomposites generates a strong local stress field, which affects the movement of nearby dislocations and is an important factor in the local strengthening of the material. Microcracks, as a special form of pores, propagate through the emission of dislocations and are key microstructures that control material toughness. Exploring the interaction between nanopores and dislocations is an important issue for understanding the strength and toughness of materials.

In this section, the elastic model for the interaction between edge/screw dislocation and nanohole would be introduced. After that, we construct a model of the dislocation emission from a crack tip, and discuss related toughening mechanism. Due to the presence of surface/interface stress, several unique phenomena are found.

9.3.1 Edge Dislocation Interacting with Nanohole

9.3.1.1 Basic Formula

In this part, an elastic medium with infinite dimensions and a circular hole with radius R is considered. The elastic medium has elastic properties $\kappa = 3 - 4\upsilon$ and shear modulus μ, where υ is the Poisson's ratio. Multiple parallel edge dislocations are present in the matrix and extend infinitely in a straight line at arbitrary points. The boundary conditions at the surface of circular hole are simplified by incorporating the surface/interface stress model as follows [24,34,53]:

$$\sigma_{rr}(t) = \frac{\sigma_{\theta\theta}^{0}(t)}{R} \quad \sigma_{r\theta}(t) = -\frac{1}{R}\frac{\partial\sigma_{\theta\theta}^{0}(t)}{\partial\theta} \quad |t| = R \tag{9.49}$$

where σ_{rr} and $\sigma_{r\theta}$ represent the stress components within a polar coordinate system. The superscript symbol "0" signifies the surface, and $|t| = R$ stands for the circular arc of the hole. An extra constitutive equation related to the surface region is introduced [53].

$$\sigma_{\theta\theta}^{0} = \tau^{0} + (2\mu^{0} + \lambda^{0} - \tau^{0})\varepsilon_{\theta\theta}^{0} \tag{9.50}$$

where $\sigma_{\theta\theta}^{0}$ and $\varepsilon_{\theta\theta}^{0}$ represent the stress and strain on the surface, respectively, μ^{0} and λ^{0} are the Lamé constants on the surface. τ^{0} is the residual stress tension of the surface. $\varepsilon_{\theta\theta}^{0}$ is the surface strain, which can be viewed as the tangential strain of the adjacent bulk region.

The impact of the supplementary constitutive equation for the surface region, as well as the subsequent equation for the bulk solid, needs to be taken into account:

$$\varepsilon_{\theta\theta} = \frac{\lambda + 2\mu}{4\mu(\lambda + \mu)}\sigma_{\theta\theta} - \frac{\lambda}{4\mu(\lambda + \mu)}\sigma_{rr} \tag{9.51}$$

The stress boundary conditions specified in Eq. 9.49 at the surface are expressed in a different form as follows

$$\begin{cases} \sigma_{rr}(t) = \dfrac{\tau^{0}}{R} + \dfrac{(2\mu^{0} + \lambda^{0} - \tau^{s0})}{4R\mu(\lambda + \mu)}[(\lambda + 2\mu)\sigma_{\theta\theta}(t) - \lambda\sigma_{rr}(t)] & |t| = R \\ \sigma_{r\theta}(t) = \dfrac{(2\mu^{0} + \lambda^{0} - \tau^{0})}{4R\mu(\lambda + \mu)}\left[(\lambda + 2\mu)\dfrac{\partial\sigma_{\theta\theta}(t)}{\partial\theta} - \lambda\dfrac{\partial\sigma_{rr}(t)}{\partial\theta}\right] & |t| = R \end{cases} \tag{9.52}$$

where μ and λ represent the Lamé constants of the bulk.

The $\Phi(z)$ within the matrix region is given by:

$$\Phi(z) = \frac{\gamma_1}{z - z_1} + \sum_{k=0}^{+\infty} A_{-k} z^{-k} \quad |z| > R \tag{9.53}$$

Besides, we also get:

$$\begin{aligned}\Psi(z) &= \frac{R^2}{z^2}[\Phi(z) - z\Phi'(z)] + \frac{R^2}{z^2}\left[\sum_{k=0}^{+\infty} \overline{B_k}\left(\frac{R^2}{z}\right)^k + \frac{\overline{\gamma_1} z_1 z}{R^2(z - z_1)}\right.\\ &\quad \left. + \frac{\overline{\gamma_1} z}{R^2} + \frac{\gamma_1 \overline{z_1^*}(\overline{z_1} - \overline{z_1^*}) z_1 z^2}{R^4 (z - z_1)^2}\right] \quad |z| > R\end{aligned} \tag{9.54}$$

The parameters in the above equations are given by:

$$\begin{cases} A_0 = \dfrac{-2a - b}{1 - b}\dfrac{\overline{\gamma_1}}{\overline{z_0}} + \dfrac{a}{1 - b}\dfrac{\tau^s}{R} \\ B_0 = \dfrac{2a + 2b - 1}{1 - b}\dfrac{\gamma_1}{z_0} + \dfrac{1 - a - b}{1 - b}\dfrac{\tau^s}{R} \\ A_{-k} = \dfrac{(a + b)(k - 1) + c_2}{c_1}\dfrac{\overline{\gamma_1} R^{2k}}{(\overline{z_1})^{k+1}} \\ \quad + \dfrac{(ka - a - 1) + c_3}{c_1} \\ \quad \times \left[\gamma_1 (z_1^*)^{k-1} - \gamma_1 \delta_{1k} - \dfrac{\overline{\gamma_1} z_1^*(z_1 - z_1^*)(k - 1)(z_1^*)^{k-2}}{\overline{z_1}}\right] \quad k \geq 1 \\ B_k = \dfrac{c_5}{c_4}\dfrac{\gamma_1}{(z_1)^{k+1}} + \dfrac{a + ak + c_6}{c_4}\left[\dfrac{\overline{\gamma_1}(\overline{z_1^*})^{k-1}}{R^{2k}} - \dfrac{\overline{\gamma_1}\delta_{1k}}{R^{2k}}\right. \\ \quad \left. - \dfrac{\gamma_1 \overline{z_1^*}(\overline{z_1} - \overline{z_1^*})(k - 1)(\overline{z_1^*})^{k-2}}{z_1 R^{2k}}\right] \quad k \geq 1 \end{cases} \tag{9.55}$$

where $c_1 = 1 + (a + b)(k - 1) + \frac{a(a+b)(1+k)(1-k)}{1+a+ka}$, $c_2 = \frac{a(1-k)[(a+b)(k+1)-1]}{1+a+ka}$, $c_3 = \frac{a^2(1-k)(1+k)}{1+a+ka}$, $c_4 = 1 + a(k + 1) + \frac{a(a+b)(1+k)(1-k)}{1+(a+b)(k-1)}$, $c_5 = (a + b)(k + 1) - 1 + \frac{(a+b)^2(1+k)(k-1)}{(a+b)(1-k)-1}$, $c_6 = \frac{(a+b)(1+k)(1+a-ak)}{1+(a+b)(k-1)}$ and δ_{ij} is the Kronecker delta.

When the surface stress is ignored ($\mu^0 = \lambda^0 = 0$), Eq. 9.53 is simplified to the classical elasticity solution [15],

$$\Phi(z) = \frac{\gamma_1}{z - z_1} - \frac{\gamma_1}{z - z_1^*} + \frac{\gamma_1}{z} + \frac{\overline{\gamma_1} z_1^*(z_1 - z_1^*)}{\overline{z_1}(z - z_1^*)^2} \quad |z| > R \tag{9.56}$$

Using the stress components in Eq. 9.5 and the derived $\Phi(z)$ and $\Psi(z)$, the stress fields within the infinite matrix is readily computed. Eq. 9.5 represents the explicit forms of Green's functions for the present model under the influence of a single edge dislocation positioned within the matrix. For the case with multiple dislocations, the solutions can be easily obtained via superposing the Green's functions.

9.3.1.2 Image Force

For a single edge dislocation with the Burgers vector $(b_{1x},\ b_{1y})$ and located at z_1, the image force is computed by the Peach-Koehler formula [40]:

$$\mathrm{f}_x - i\mathrm{f}_y = [\sigma^*_{xy}(z_1)b_{1x} + \sigma^*_{yy}(z_1)b_{1y}] + i[\sigma^*_{xx}(z_1)b_{1x} + \sigma^*_{xy}(z_1)b_{1y}] \tag{9.57}$$

Here, f_x and f_y denote the forces in the x^- and y^- orientation, respectively. $\sigma^*_{yy}(z_1)$ and $\sigma^*_{xy}(z_1)$ represent the components of the perturbation stress fields at point z_1. The stresses are determined by the $\Phi^*(z_1)$ and $\Psi^*(z_1)$ within the matrix. The Peach-Koehler formula is expressed in the following form:

$$\mathrm{f}_x - i\mathrm{f}_y = (b_{1y} + ib_{1x})[\Phi^*(z_1) + \overline{\Phi^*(z_1)}] + (b_{1y} - ib_{1x})[\overline{z_1}\Phi^{*'}(z_1) + \Psi^*(z_1)] \tag{9.58}$$

According to previous work [12], the $\Phi^*(z_1)$, $\Phi^{*'}(z_1)$ and $\Psi^*(z_1)$ are given by:

$$\begin{cases} \Phi^*(z_1) = \sum\limits_{k=0}^{+\infty} A_{-k}(z_1)^{-k} \\ \Phi^{*'}(z_1) = -\sum\limits_{k=1}^{+\infty} kA_{-k}(z_1)^{-k-1} \\ \Psi^*(z_1) = \dfrac{R^2}{z_1^2}\left[\sum\limits_{k=0}^{+\infty} A_{-k}(z_1)^{-k} + \sum\limits_{k=1}^{+\infty} \overline{B_k}R^{2k}(z_1)^{-k} \right. \\ \qquad \left. + \sum\limits_{k=0}^{+\infty} A_{-k}k(z_1)^{-k} + \dfrac{\gamma_1(z_1\overline{z_1} - R^2)}{R^2 z_1} + \dfrac{2\overline{\gamma_1}z_1}{R^2}\right] - \dfrac{2\overline{\gamma_1}}{z_1} - \dfrac{\gamma_1\overline{z_1}}{z_1^2} \end{cases} \tag{9.59}$$

We can obtain the details of the image force based on Eq. 9.83 in conjunction with Eq. 9.59.

In the absence of surface stress ($\mu^0 = \lambda^0 = 0$), for the dislocation with Burgers vector $(b_x, 0)$ and situates in point x_0, the image force is reduced to the classical result [53]:

$$\begin{aligned} \mathrm{f}_x = \frac{\mu b_x^2}{\pi(1+\kappa)}\Bigg[\frac{1}{x_0} - \left(x_0^3 + 2R^2x_0 + \frac{R^6 - 4R^4x_0^2}{x_0^3}\right)\frac{1}{(x_0^2 - R^2)^2} \\ - \frac{2R^2}{x_0(R^2 - x_0^2)} - \frac{2R^2}{x_0^2} + \frac{R^4}{x_0^3(R^2 - x_0^2)}\Bigg] \end{aligned} \tag{9.60}$$

By stacking the above equation, we can further deduce the image forces when multiple parallel edge dislocations are taken into consideration. Here, the case with two parallel edge dislocations is displaced as an example. The two dislocations are characterized by different Burgers vectors (b_{1x}, b_{1y}) and (b_{2x}, b_{2y}), and different position z_1 and z_2. Therefore, the overall image force impacting on z_1 dislocation can be computed by summing the Eq. 9.57 and image force exerted on z_1 dislocation by the z_2 dislocation. Following the above method, the $\Phi^*(z_1)$, $\Phi^{*'}(z_1)$ and $\Psi^*(z_1)$ in Eq. 9.58 are expressed as follows:

$$\Phi^*(z_1) = \sum_{k=0}^{+\infty} A_{-k}(z_1)^{-k} + \sum_{k=0}^{+\infty} C_{-k}(z_1)^{-k} + \frac{\gamma_2}{z_1 - z_2} \tag{9.61}$$

$$\Phi^{*'}(z_1) = -\sum_{k=1}^{+\infty} kA_{-k}(z_1)^{-k-1} - \sum_{k=0}^{+\infty} kC_{-k}(z_1)^{-k-1} - \frac{\gamma_2}{(z_1 - z_2)^2} \tag{9.62}$$

$$\begin{aligned}\Psi^*(z_1) = {} & \frac{R^2}{z_1^2}\left[\sum_{k=0}^{+\infty} A_{-k}(z_1)^{-k} + \sum_{k=1}^{+\infty} \overline{B_k} R^{2k}(z_1)^{-k} + \sum_{k=0}^{+\infty} A_{-k}k(z_1)^{-k}\right. \\ & \left. + \frac{\gamma_1(z_1\overline{z_1} - R^2)}{R^2 z_1} + \frac{\overline{2\gamma_1 z_1}}{R^2}\right] - \frac{2\overline{\gamma_1}}{z_1} - \frac{\gamma_1\overline{z_1}}{z_1^2} \\ & + \frac{R^2}{z_1^2}\left[\sum_{k=0}^{+\infty} C_{-k}(z_1)^{-k} + \frac{\gamma_2}{z_1 - z_2} + \sum_{k=0}^{+\infty} kC_{-k}(z_1)^{-k}\right. \\ & + \frac{\gamma_2 z_1}{(z_1 - z_2)^2} + \sum_{k=0}^{+\infty} \overline{D_k}\left(\frac{R^2}{z_1}\right)^k + \frac{\overline{\gamma_2} z_2 z_1}{R^2(z_1 - z_2)} + \frac{\overline{\gamma_2} z_1}{R^2} \\ & \left. + \frac{\gamma_2 \overline{z_2^*}(\overline{z_2} - \overline{z_2^*}) z_2 z_1{}^2}{R^4(z_1 - z_2)^2}\right]\end{aligned} \tag{9.63}$$

where

$$C_0 = \frac{-2a - b}{1 - b}\frac{\overline{\gamma_2}}{\overline{z_2}} + \frac{a}{1 - b}\frac{\tau^s}{R} \tag{9.64}$$

$$D_0 = \frac{2a + 2b - 1}{1 - b}\frac{\gamma_2}{z_2} + \frac{1 - a - b}{1 - b}\frac{\tau^s}{R} \tag{9.65}$$

$$\begin{aligned}C_{-k} = {} & \frac{(a + b)(k - 1) + c_2}{c_1}\frac{\overline{\gamma_2} R^{2k}}{(\overline{z_2})^{k+1}} \\ & + \frac{(ka - a - 1) + c_3}{c_1}\left[\gamma_2(z_2^*)^{k-1} - \gamma_2\delta_{1k}\right. \\ & \left. - \frac{\overline{\gamma_2} z_2^*(z_2 - z_2^*)(k - 1)(z_2^*)^{k-2}}{\overline{z_2}}\right] \quad k \geq 1\end{aligned} \tag{9.66}$$

$$D_k = \frac{c_5}{c_4}\frac{\gamma_1}{(z_2)^{k+1}} + \frac{a+ak+c_6}{c_4}\left[\frac{\overline{\gamma_2}(\overline{z_2^*})^{k-1}}{R^{2k}} - \frac{\overline{\gamma_2}\delta_{1k}}{R^{2k}} - \frac{\gamma_2\overline{z_2^*}(\overline{z_2} - \overline{z_2^*})(k-1)(\overline{z_2^*})^{k-2}}{z_2R^{2k}}\right] \quad k \geq 1 \tag{9.67}$$

with $\gamma_2 = \frac{\mu}{\pi(1+\kappa)}(b_{2y} - ib_{2x})$ and $z_2^* = R^2/\overline{z_2}$

By substituting Eqs. 9.62–9.63 into Eq. 9.58, we can determine the total image force impacting on z_1 dislocation. The force is further divided into two mutually perpendicular components [30], 1) the glide force along the vector, and 2) the climb force perpendicular to the vector:

$$\begin{cases} \mathrm{f}_g = \mathrm{f}_x\cos\theta + \mathrm{f}_y\sin\theta \\ \mathrm{f}_c = \mathrm{f}_x\sin\theta - \mathrm{f}_y\cos\theta \end{cases} \tag{9.68}$$

where $\theta = \arctan(b_y/b_x)$.

9.3.1.3 Numerical Examples

In the section, we analyze the image forces that act on edge dislocations. Initially, we examine the effect of surface properties and circular hole size on these forces. The glide and climb forces for the dislocation with a Burgers vector (b_{1x}, b_{1y}) are separately discussed based on Eq. 9.68. As representative case, we assume an edge dislocation situated in point x_1 on the positive x-axis ($z_1 = x_1 > R$). For the sake of convenience in subsequent discussions, we define the standardized glide force and climb force as $\mathrm{f}_{go} = \pi R(1+\kappa)\mathrm{f}_g/[\mu(b_{1x}^2 + b_{1y}^2)]$ and $\mathrm{f}_{co} = \pi R(1+\kappa)\mathrm{f}_c/[\mu(b_{1x}^2 + b_{1y}^2)]$.

Two sets of surface properties are used for numerical calculations: intrinsic lengths $\alpha = -0.01082$ nm and $\beta = 0.197132$ nm (for Al [111]) [34], and intrinsic lengths $\alpha = 0.035714$ nm and $\beta = -0.114286$ nm (for freshly cleaved iron) [52]. The relationship between the f_{go} and the radius R is illustrated in Figure 9.7a. It is worth noting that the climb force is zero in the considered case. When the size of hole is very small, the contribution of the size effect becomes pronounced. Additionally, the surface effect becomes more significant for the case with freshly cleaved iron surface. The presence of the surface effect leads to an extra attractive/repulsive force that acts on the edge dislocation in the glide direction (the x^+-axis), resulting in an improvement/reduction of the glide force.

Figure 9.7b displays the variations of the f_{co} under different hole radius R and identical $b_x = 0$. The glide force is zero in the current case. Obviously, the climb force exhibits similar trend compared with the case of glide force. The surface effect is more notable for Al [111] surface, different with the case of glide force. The standardized glide (climb) force increases (decreases) as the hole radius decreases, and the size effect is enhanced when the hole is small. This result implies that the stress required to facilitate dislocation motion may be affected by the hole size, which is in agreement with the previous work [54].

The relationship between f_{go} (or f_{co}) and the direction angle θ is illustrated in Figure 9.7c,d. The surface stress produces an additional image force, which repels

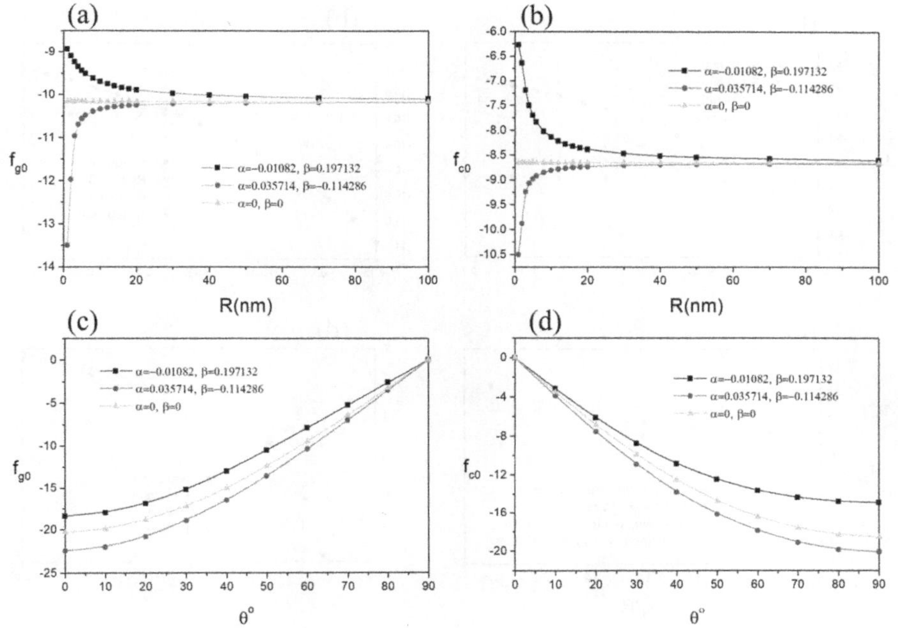

FIGURE 9.7 (a) Variation of glide force f_{go} as a function of the radius R, for $\rho = 1.1$ ($b_y = 0$); (b) Variation of climb force f_{co} as a function of the radius R, for $\rho = 1.1(b_x = 0)$; (c) Variation of glide force f_{go} as a function of the orientation of the Burgers vector θ, for $\rho = 1.05$ and $R = 10\text{nm}$; (d) Variation of climb force f_{co} as a function of the orientation of the Burgers vector θ, for $\rho = 1.05$ and $R = 10\text{nm}$ [55].

the edge dislocation for the freshly cleaved iron surface, but attracts the edge dislocation for the Al [111] surface. The repulsive/attractive force reduces as the angle θ increases for the case of glide force. However, the repulsive/attractive force reduces as the angle θ decreases for the case of climb force. The effect of surface stress on the glide force (climb force) reaches a peak when $\theta = 0°$ ($\theta = 90°$). These results demonstrate that the effect of surface stress on the glide or climb force relies on the dislocation direction.

The standardized f_{go} as the functions of α and β are presented in Figures 9.8a,b, under different radius R of hole. The results demonstrate that the surface effect becomes significant as the hole size R decrease. But the surface effect is negligible when $R > 100\text{nm}$.

Next, we demonstrate how the image forces between two dislocations are influenced by the surface properties of a nanohole. The considered dislocations are parallel to each other, and characterized by different Burgers vectors (b_{1x}, b_{1y}) and (b_{2x}, b_{2y}), and different position z_1 ($z_1 = x_1 > 0$) and z_2 ($z_2 = x_2 > R$). The relative position of the z_2 dislocation to the hole is $\varepsilon = x_2/R$. Figure 9.8c presents the standardized glide force f_{go} (dislocation z_1) as a function of $\rho = x_1/R$. When the nanohole disappears, the glide force impacting the z_1 dislocation produced by the z_2 dislocation is always positive. That means the z_2 dislocation intrinsically attracts the z_1 dislocation. However, the presence of nanohole induces the

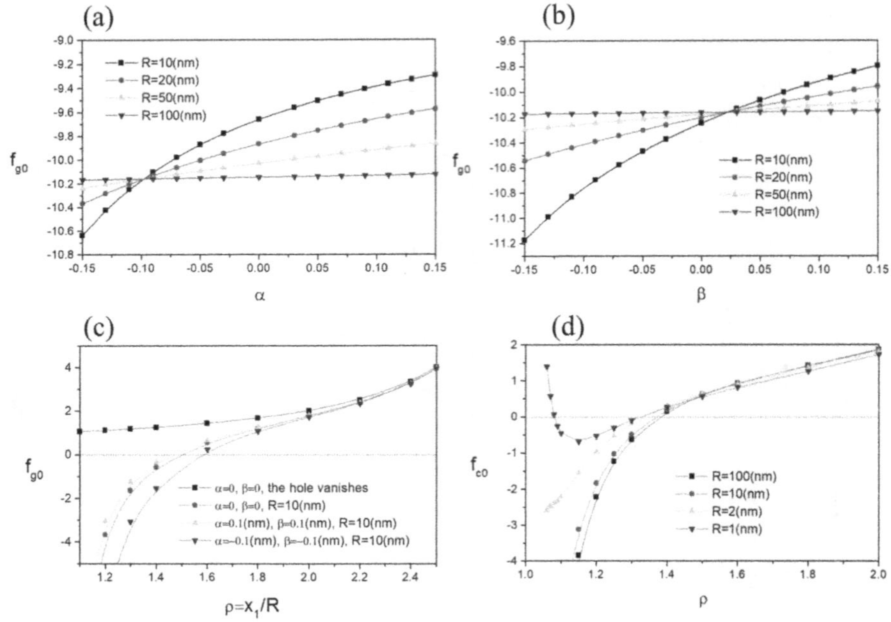

FIGURE 9.8 (a) Variation of glide force f_{go} as a function of the inherent length α, for $\rho = 1.1$ ($\beta = 0.197132$, $b_y = 0$); (b) Variation of glide force f_{go} as a function of the inherent length β, for $\rho = 1.1$ ($\alpha = -0.01082$, $b_y = 0$) (c) Variation of glide force f_{go} as a function of ρ, for $\varepsilon = 3$ ($b_{2x} = -b_{1x}$, $b_{1y} = b_{2y} = 0$); (d) Variation of climb force f_{co} as a function of ρ, for $\varepsilon = 3$ and $\alpha = \beta = 0.1\,\text{nm}$ ($b_{2y} = -b_{1y}$, $b_{1x} = b_{2x} = 0$ and $\upsilon = 0.25$) [55].

negative glide force when the z_1 dislocation approaching to the surface of the hole. An equilibrium position for the z_1 dislocation exists on the x-axis. If the surface effect is taken into account, the value of the glide force would change, and the influence of the negative surface properties would be enhanced. Nevertheless, the surface effect does not change the interaction mechanism between dislocations and the hole.

Figure 9.8d shows the f_{co} (dislocation z_1) versus $\rho = x_1/R$ under different radius R. It is demonstrated that the effect of surface properties on the climb force becomes significant with the decrease of the size of the nanohole. If the nanohole is very small and the surface properties are positive (such as $R = 1\,\text{nm}$ and $\alpha = \beta = 0.1\,\text{nm}$), the movement of z_1 dislocation to the surface of the hole would be inhibited by the repelling force. This behavior is attributed to the increased climb force in the positive x-axis. Note that the total climb force in the positive x-axis derives from the surface stress and the z_2 dislocation, and the climb force in the negative x-axis is produced by the hole. When the total climb force in the positive x-axis is greater than the climb force in the negative x-axis, the direction of the climb force will reverse. When the climb forces in the positive x-axis is equal to that in the negative x-axis, the dislocation reaches an equilibrium point. For the present case, two equilibrium points are found, including a stable point near the hole and an unstable point away from the hole.

9.3.1.4 Remarks

This section proposes a new elastic model for the interaction between edge dislocations and a single circular hole, by taking into account the surface stress. The theoretical results demonstrate that the standardized glide/climb force is affected by the hole size, which is different from the classical results. The effect of the surface stress will be enhanced as the magnitude of the hole decreases. The change of the surface property would produce an extra repulsive (or attractive) force for the dislocation movement, resulting in the local hardening (or softening) at the hole surface. The nanohole with the surface stress change the elastic interaction between nearby dislocations when the dislocations are close to the hole surface.

9.3.2 Screw Dislocation Interacting with Nanohole

9.3.2.1 Basic Formula

For the interaction model for screw dislocation and nanohole, the surface region of the solid follows the identical settlement of the edge dislocation-nanohole interaction in Section 9.3.1. Following the derivation process of Section 9.2.2.1, for a screw dislocation located at position z_0 with Burgers vector b_z, we may get the complex potential $F(z)$:

$$\begin{aligned} F(z) &= \frac{b_z}{2\pi i}\left(\frac{1}{z-z_0}+\frac{1}{z-z^*}-\frac{1}{z}\right)+\Gamma-\frac{\mu-(\mu^0/R)}{\mu+(\mu^0/R)}\frac{R^2}{z^2}\overline{\Gamma} \\ &\quad -\frac{b_z}{2\pi i}\sum_{k=0}^{\infty}\frac{2\mu}{\mu+(1+k)(\mu^0/R)}\left(\frac{R^2}{\overline{z_0}}\right)^{k+1} z^{-(k+2)} \end{aligned} \tag{9.69}$$

The interaction between a screw dislocation and a nanohole under the influence of surface stress can be solved using similar approach. As example, the complex potential for the infinite matrix is expressed as:

$$\begin{aligned} F_2(z) &= \frac{b_z}{2\pi i}\left(\frac{1}{z-z_0}+\frac{1}{z-z^*}-\frac{1}{z}\right)+\Gamma+\frac{\mu_1-\mu_2+(\mu^0/R)}{\mu_1+\mu_2+(\mu^0/R)}\frac{R^2}{z^2}\overline{\Gamma} \\ &\quad -\frac{b_z}{2\pi i}\sum_{k=0}^{\infty}\frac{2\mu_2}{\mu_1+\mu_2+(1+k)(\mu^0/R)}\left(\frac{R^2}{\overline{z_0}}\right)^{k+1} z^{-(k+2)} \end{aligned} \tag{9.70}$$

where μ^0 is the shear modulus of surface. μ_1 and μ_2 are the shear constant of the hole and matrix regions, respectively.

9.3.2.2 Image Force

The image force on the screw dislocation produced by the hole is computed based on the Peach-Koehler formula:

$$f_x - i f_y = i b_z\left[\widehat{\tau_{xz}}(z_0) - i\widehat{\tau_{yz}}(z_0)\right] \tag{9.71}$$

where f_x (f_y) is the force component in the X axis (Y axis) direction. $\widehat{\tau_{xz}}(z_0)$ and $\widehat{\tau_{yz}}(z_0)$ are the perturbation stress components for the dislocation.

Following general method [10], the explicit equation for the image force impacting on the screw dislocation is computed using Eq. 9.102

$$\begin{aligned} f_x - if_y &= \frac{\mu b_z^2}{2\pi}\left(\frac{1}{z_0 - z^*} - \frac{1}{z_0}\right) + ib_z\mu\Gamma - \frac{\mu[\mu - (\mu^s/R)]}{\mu + (\mu^s/R)}\frac{R^2}{z_0^2}(ib_z\overline{\Gamma}) \\ &- \frac{\mu b_z^2}{2\pi}\sum_{k=0}^{\infty}\frac{2\mu^2}{\mu + (1+k)(\mu^s/R)}\left(\frac{R^2}{\overline{z_0}z_0}\right)^{k+1}\frac{1}{z_0} \end{aligned} \tag{9.72}$$

Finally, the image force for the screw dislocation from the nanohole can be obtained using Eqs. 9.103 and 9.104.

9.3.2.3 Numerical Examples and Discussion

Based on the Eq. 9.105, the influence of nanohole on the image force for a screw dislocation is investigated. Figure 9.9a plots the standardized image force f_{xo} as a function of the distance x_0/R for different hole radius in freshly cleaved iron (μ^s = 2.5 N/m) [52]. Due to the presence of surface stress, the image force reduces as the size of the hole decreases and the dislocation approaches the surface of hole. Figure 9.9b illustrates the influence of hole radius R on the standardized image force f_{x0}, under different surface properties [45,48]. When $\mu^s > 0$, the attraction image force decreases as hole radius decreases, while the opposite trend is observed if the surface elastic is negative. This result is different with the classical case that do not consider the surface stress, in which the image force is not sensitive to the hole size. These results emphasize the significant influence of surface stress on the elastic interaction between screw dislocations and nanoscale holes. That is consistence with the case with edge dislocations.

9.3.2.4 Remarks

The section investigates the elastic interaction between screw dislocation and a circular nanohole by considering the influence of surface stress. The analysis expression for the image force impacting the screw dislocation is obtained, based on

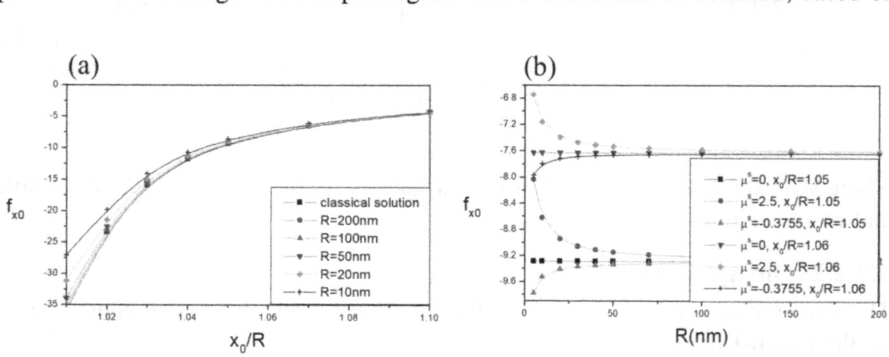

FIGURE 9.9 (a) f_{xo} as a function of x_0/R with different radius; (b) f_{xo} as a function of the radius R [43].

the complex variable method. Particularly, the solution for the problem of screw dislocation-circular hole interaction is obtained. Theoretical results show that as the size of hole decreases, the influence of surface stress on the dislocation motion is enhanced. The image force is sensitive to the size of hole, as revealed by the size-dependent classical elastic solution. These results are consistence with the case with edge dislocations, indicating that both edge dislocations and screw dislocations are strongly affected by the surface effect of nanohole.

9.3.3 Dislocation Emission from a Crack Tip

9.3.3.1 Basic Formula

The dislocation emitted from the crack tip controls the evolution of crack and the fracture of materials. The model for crack-dislocation interaction involves an edge dislocation (Burgers vector is $b_r e^{i\theta}$ and located at position $re^{i\theta}$), and a nearby elliptically blunted crack tip. The crack tip is characterized by a radius $\rho = b^2/a$. The origin of the polar coordinates (r, θ) used in the present model is settled behind the crack root. For the present problem, a two-dimensional plane strain is applied for the model.

The stress tensor of surface $\sigma^s_{\alpha\beta}$ varies with the surface energy $\Gamma(\varepsilon_{\alpha\beta})$ that sensitive to the material deformation, i.e., $\sigma^s_{\alpha\beta} = \tau_0\delta_{\alpha\beta} + \partial\Gamma/\partial\varepsilon_{\alpha\beta}$. Here, Γ and $\varepsilon_{\alpha\beta}$ are the deformation-independent surface tension and surface strain tensor, respectively. For simplification, it is assumed that the surface energy is independent with the elastic strain. The equation $\sigma^s_{\alpha\beta} = \tau_0\delta_{\alpha\beta}$ will be used in our analysis [56].

First, the stress field around the crack tip induced by the surface stress is obtained. Then, to transit $z = x + iy$ plane into the $\zeta = \xi + i\eta$ plane, the following transformation function is applied:

$$z = w(\zeta) = R(\zeta + m/\zeta) \tag{9.73}$$

where $m = (a - b)/(a + b)$ and $R = (a + b)/2$.

The complex potential functions of $\zeta = \xi + i\eta$ plane can be obtained based on general method [39], then the functions on the $z = x + iy$ plane are subsequently derived using the transformation function Eq. 9.73. In this section, we present the stress fields in the z-plane.

$$\sigma_{xx} - i\sigma_{xy} = \frac{\tau_0}{2R}\left[\frac{(1 - m^2)G(z)}{\sqrt{1 - mG^2(z)}\sqrt[3]{(G^2(z) - m)^2}} - \frac{G(z)}{\sqrt{z^2 - l^2}}\right] \tag{9.74}$$

where $G(z) = z + \sqrt{z^2 - l^2}$ and $l^2 = a^2 - b^2$.

Using $z_1 = z - a + \rho/2$, the stress field of the x-y coordinate system is transformed to the coordinate system of x_1-y_1 (Figure 9.10). In polar coordinate (r, θ), the stress field is rewritten as follows

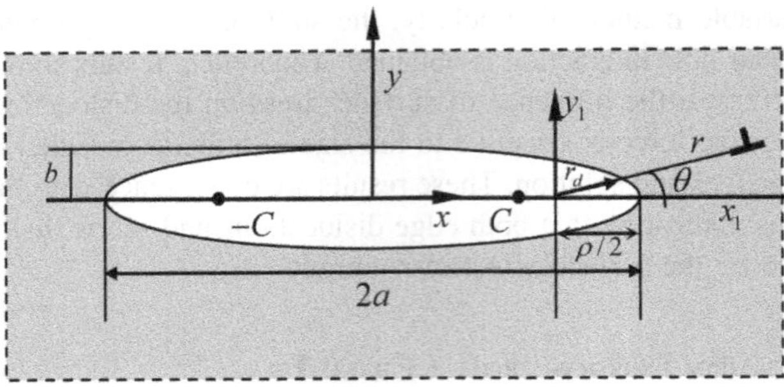

FIGURE 9.10 Schematic representation of an edge dislocation around a finite blunted crack [57].

$$\sigma_{rr} - i\sigma_{r\theta} = \frac{\tau_0}{2R}\left[\frac{(1-m^2)G(re^{i\theta})}{\sqrt{1-mG^2(re^{i\theta})}\sqrt[3]{(G^2(re^{i\theta})-m)^2}}\frac{G(re^{i\theta})}{\sqrt{(re^{i\theta}+a-\rho/2)^2-l^2}}e^{i2\theta}\right] \tag{9.75}$$

where $G(re^{i\theta}) = re^{i\theta} + a - \rho/2 + \sqrt{(re^{i\theta} + a - \rho/2)^2 - l^2}$.

9.3.3.2 Forces Impacting the Edge Dislocation and the Critical Applied SIFs

The glide force that results entirely from the surface stress is obtained from Eq. 9.75. Specifically, it is given by:

$$\begin{aligned} f_s &= b_r\sigma_{r\theta} \\ &= \frac{b_r\tau_0}{2R}\mathrm{Im}\left[\frac{(1-m^2)G(re^{i\theta})}{\sqrt{1-mG^2(re^{i\theta})}\sqrt[3]{(G^2(re^{i\theta})-m)^2}}\frac{G(re^{i\theta})}{\sqrt{(z_1+a_1)^2-l^2}}e^{i2\theta}\right] \end{aligned} \tag{9.76}$$

According to the work [58], the elliptically blunted crack can produce the image force f_i for edge dislocation:

$$f_i = \frac{\mu(b_r)^2}{4\pi(1-\nu)(r-r_d)} \tag{9.77}$$

Here, μ represents the shear modulus, while ν denotes its Poisson's ratio. The distance between the dislocation and the crack surface is $(r - r_d)$, where r_d refers the distance from the original point in the x_1-y_1 coordinate system to the blunted crack surface. This distance is expressed as

$$r_d = \frac{\sqrt{a^2 - 0.25a\rho(\sin\theta)^2} - (a-\rho/2)\cos\theta}{\rho(\cos\theta)^2 + a(\sin\theta)^2}\rho \tag{9.78}$$

Assuming that the plane exhibits linear elastic property, the equations developed by Creager and Paris [59] are employed to assess the stress fields associated with an elliptically blunted tip. Based on their solutions, one can obtain the glide force f_e for the edge dislocation under the loading modes I and II, i.e.,

$$\begin{aligned} f_e = b_r\sigma_{r\theta} &= \frac{b_r K_I}{2\sqrt{2\pi r}}\sin\frac{\theta}{2}\left(1 + \cos\theta + \frac{\rho}{r}\right) \\ &+ \frac{b_r K_{II}}{\sqrt{2\pi r}}\left(\cos\frac{3\theta}{2} + \frac{1}{2}\sin\theta\sin\frac{\theta}{2} - \frac{\rho}{2r}\cos\frac{\theta}{2}\right) \end{aligned} \tag{9.79}$$

where $\sigma_{r\theta}$ denotes the in-plane shear stress.

In this study, we examine the impact of the friction force $f_r = b_r\sigma_f$ (which represents the lattice resistance to dislocation motion) in plane strain conditions. Consequently, the glide force is composed by four parts, originating from the applied stress, surface stress, the dislocation image force, and the friction force σ_f.

The equilibrium distance r_d for a mode I crack, that corresponds to the stable point (total force = 0) can be computed using the following equation:

$$\begin{aligned} &\frac{b_r K_I}{2\sqrt{2\pi r_0}}\sin\frac{\theta}{2}\left(1 + \cos\theta + \frac{\rho}{r_0}\right) \\ &- \frac{\mu(b_r)^2}{4\pi(1-\nu)(r_0 - r_d)} - \frac{b_r\tau_0}{2R}\mathrm{Im}[F(r_0)] - b_r\sigma_f = 0 \end{aligned} \tag{9.80}$$

where $F(r_0) = \frac{(1-m^2)G(r_0e^{i\theta})}{\sqrt[3]{[G^2(r_0e^{i\theta}) - m]^2}\sqrt{1 - mG^2(r_0e^{i\theta})}}\frac{G(r_0e^{i\theta})}{\sqrt{(r_0e^{i\theta} + a - \rho/2)^2 - l^2}}e^{i2\theta}$.

When the distance between the crack tip and the stable point is smaller than the radius of dislocation core, the crack tip would emit a new dislocation [60]. Here, it is considered that the dislocation core radius is equal to the Burgers vector.

Substituting the relation $r_0 = r_d + b_r$ into Eq. 9.113, the critical applied SIF K_I^C under mode I loading is given by:

$$K_I^C = \left\{\frac{\mu}{2\pi(1-\nu)} + \frac{\tau_0}{R}\mathrm{Im}[F(r_d + b_r)] + \frac{\sigma_f}{2}\right\}\frac{\sqrt{2\pi(r_d + b_r)}}{\sin\frac{\theta}{2}\left(1 + \cos\theta + \frac{\rho}{r_d + b_r}\right)} \tag{9.81}$$

New dislocation would nucleate from the crack tip if the applied SIF $K_I > K_I^C$. Rearranging Eq. 9.81, it gives standardized critical applied SIF for mode I loading:

$$\begin{aligned} \frac{K_I^C}{\mu\sqrt{b_r}} &= \left\{\frac{1}{\sqrt{2\pi}(1-\nu)} + \frac{\sqrt{2\pi}\tau_0}{\mu R}\mathrm{Im}[F(r_d + b_r)] + \frac{\sqrt{2\pi}\sigma_f}{2\mu}\right\} \\ &\times \frac{\sqrt{(1 + r_d/b_r)}}{\sin\frac{\theta}{2}\left(1 + \cos\theta + \frac{\rho}{r_d + b_r}\right)} \end{aligned} \tag{9.82}$$

When ignoring friction force and the surface stress ($\sigma_f = 0$ and $\tau_0 = 0$), the critical applied mode I SIF would agree with the classical result [58].

Similar to the mode I crack, the standardized critical applied mode II SIF is given by:

$$\frac{K_{II}^C}{\mu\sqrt{b_r}} = \left\{\frac{1}{2\sqrt{2\pi}(1-\nu)} + \frac{\sqrt{2\pi}\tau_0}{2\mu R}\mathrm{Im}[F(r_d+b_r)] + \frac{\sqrt{2\pi}\sigma_f}{4\mu}\right\} \times \frac{\sqrt{(1+r_d/b_r)}}{\left(\cos\frac{3\theta}{2} + \frac{1}{2}\sin\theta\sin\frac{\theta}{2} - \frac{\rho}{2(r_d+b_r)}\cos\frac{\theta}{2}\right)} \tag{9.83}$$

9.3.3.3 Numerical Examples and Discussion

Based on Eqs. 9.82 and 9.83, we can assess the impact of surface stress on the critical applied SIFs. To describe the surface stress, we require the constant τ_0. According to previous atomic simulation [31], the inherent length $\beta = \tau_0/\mu$ is about 0.01nm. In the subsequent calculations, the friction force is ignored since the magnitude of σ_f/μ in Eqs. 9.82 and 9.83 is negligible. The elliptically blunted crack is characterized by the parameter a and the ratio ρ/a.

Figure 9.11a,b shows the relationship between the critical standardized SIF and the angle θ of dislocation emission, under different surface stresses $\beta = \tau_0/\mu$ and crack length a. With the increase of emission angle θ, the critical standardized SIF first decreases and then increases after a critical emission angle is reached. When the surface stress is positive (Figure 9.11a), the increment of emission angle θ results in the increase of minimal critical SIF K_I^C and corresponding emission angle θ; When the surface stress is positive (Figure 9.11b), the increase of critical emission angle θ results in the decrease of minimal critical SIF K_I^C and corresponding critical angle θ. When the crack length is very small, the surface stress may induce complicated relationship between the critical SIF and dislocation emission angle θ. From Figure 9.11b, when the surface stress becomes negative and the crack becomes very short ($a = 80b_r$), the SIF is negative when the emission angle is within the range of $7° < \theta < 18°$. That means dislocation may spontaneously emit from the crack at the angle $7° < \theta < 18°$, driven solely by the negative surface stress.

The relationship between the critical SIF and the curvature radius of crack tip for mode I applied stress with varying surface stress $\beta = \tau_0/\mu$ is presented in Figure 9.11c. When the curvature radius is small, the effect of surface stress on the K_I^C-θ relationship is complicate. When the surface stress is negative, the SIF decreases firstly and then increases once the curvature radius exceeds a critical value; if the surface stress becomes positive, the SIF increases firstly and then decreases once the curvature radius exceeds a critical value. For both cases, the critical curvature radius ρ/a is within the range from 0.0025 to 0.005. In addition, when the curvature radius is relatively large, the increase in the curvature radius always raises the critical SIF of dislocation emission.

9.3.3.4 Remarks

This section investigates the effects of surface stress on the dislocation emitted from an elliptically blunted crack subjected to mode I and mode II loading conditions.

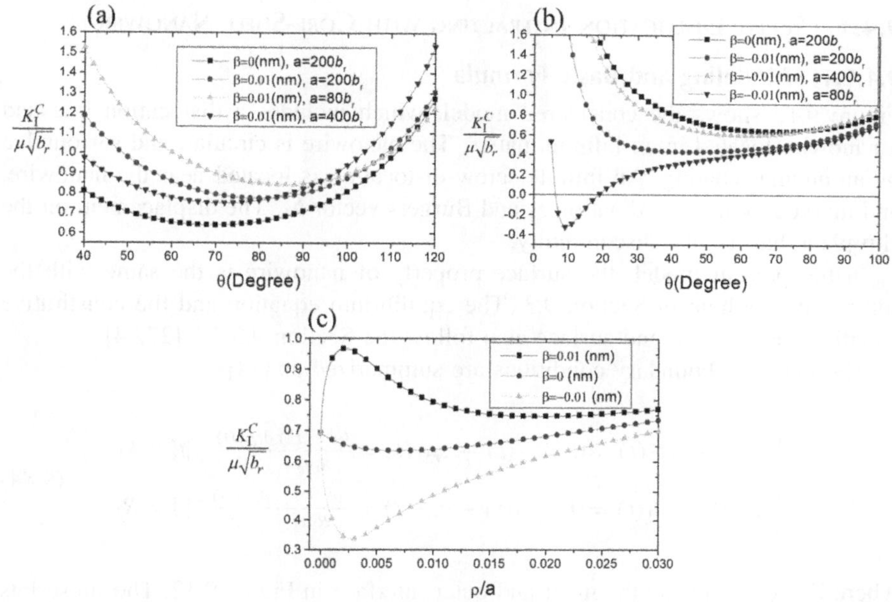

FIGURE 9.11 (a) Variation of the critical standardized SIF as a function of emission angle θ, for varying crack length a under positive surface stress ($\beta = \tau_0/\mu = 0.01\,\text{nm}$, $\rho/a = 0.01$ and $b_r = 0.25\,\text{nm}$); (b) Variation of the critical standardized SIF as a function of emission angle θ, for varying crack length a under negative surface stress ($\beta = \tau_0/\mu = -0.01\,\text{nm}$, $\rho/a = 0.01$ and $b_r = 0.25\,\text{nm}$); (c) Variation of the critical standardized SIF as a function of the radius of crack tip under different $\beta = \tau_0/\mu$, for emission angle $\theta = 65°$ ($a = 200b_r$ and $b_r = 0.25\,\text{nm}$) [57].

The critical SIF for dislocation emission is explicitly derived. Our analysis reveals that the effect of surface stress on the critical SIF becomes prominent with the decrease of crack length, typically on the nanometers. When the surface stress is negative, a minimal SIF would occur at a critical curvature radius of crack tips. These results are different from previous work that does not consider the surface stress, thereby emphasizing the key role of surface effect on the crack evolution.

9.4 CORE-SHELL NANOWIRE

Nanocomposites containing high-density nanowire structures, such as carbon nanotube-reinforced composites, exhibit various unique physical and chemical properties, and are widely used in advanced engineering fields such as intelligent electronic device and electromagnetic shielding devices [61–65]. The interaction between nanowire and dislocation is critical for the long-time structure stability of materials [66–78]. In this section, the elastic interaction between screw dislocation and two kinds of nanowires is studied, including the coated (core-shell) nanowire and embedded nanowire. Considering the surface/interface stress, we explore the CSS of the dislocations near the nanowire, and reveal the influence of surface stress on the CSS.

9.4.1 Screw Dislocation Interacting with Core-Shell Nanowire

9.4.1.1 Modeling and Basic Formula

Figure 9.12 shows the considered model, which includes a dislocation line and a nanowire located in an infinite matrix. The nanowire is circular, and surrounded by an annular coating. An infinite screw dislocation is located near the nanowire, and marked by the coordination z_1 and Burgers vector $\boldsymbol{b}_1$. The displacement on the slip plane has a finite discontinuity.

In the present model, the surface property of nanowire is the same with the setting of nanohole in Section 9.3. The equilibrium equation and the constitutive equation for the bulk and surface also follow the Section 9.3.2.1 [27,34].

The interface boundary conditions are summarized as [34]

$$\begin{cases} w_1(t) - w_2(t) = 0 \quad \tau_{rz1}(t) - \tau_{rz2}(t) = \dfrac{(\mu^{\Gamma} - \tau^{\Gamma})}{R_1\mu_2}\dfrac{\partial \tau_{\theta z2}(t)}{\partial \theta} & |t| = R_1 \\ w_2(t) - w_3(t) = 0 \quad \tau_{rz2}(t) - \tau_{rz3}(t) = \dfrac{(\mu^{\Omega} - \tau^{\Omega})}{R_2\mu_2}\dfrac{\partial \tau_{\theta z2}(t)}{\partial \theta} & |t| = R_2 \end{cases} \tag{9.84}$$

where Γ and Ω indicate the inner and outer interface in Figure 9.12. The subscripts 1, 2 and 3 indicate the nanowire, nano-coating layer and the matrix areas, respectively.

In bulk solids, the anti-plane displacement w as well as the shear stresses $\tau_{\theta z}$ and τ_{rz} are given in Eq. 9.34.

Using boundary conditions, the analytic functions $f_1(z)$, $f_2(z)$ and $f_3(z)$ for the nanowires, nano-coatings, and matrix regions are obtained. For the screw

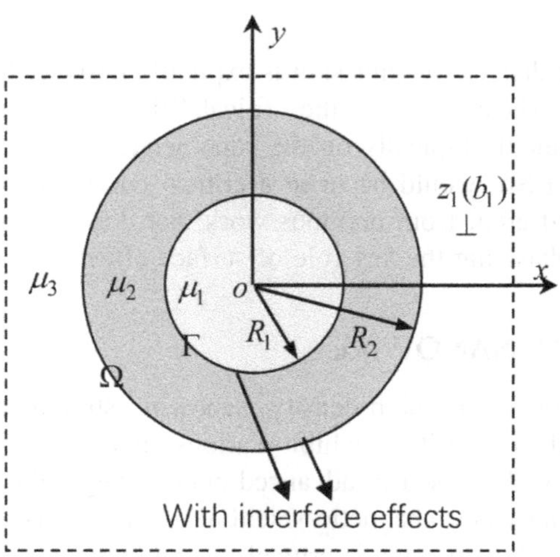

FIGURE 9.12 Schematic representation of the interaction of dislocations with coated nanowires. The nanowire which extends indefinitely is straight and perpendicular to the xy-plane. Consider the interface stress of the coated outer interface (Ω) and inner interface (Γ). The inner and outer radius of the coating annulus are R_1 and R_2 respectively [79].

dislocation has the Burgers vector b_1 and located at $z_1 (z_1 = x_1 + iy_1)$ in the matrix, the analysis function of the matrix region can be expressed as follows:

$$f_3(z) = \frac{b_1}{2\pi i} \ln(z - z_0) + f_{30}(z) \quad |z| > R_2 \tag{9.85}$$

where the function vector $f_{30}(z)$ is holomorphic in the area $|z| > R_2$.

The analysis function $f_2(z)$ can be extended to the Laurent series in the ring region, as long as the constant term is ignored, i.e.,

$$f_2(z) = \sum_{k=0}^{\infty} c_k z^{-(k+1)} + \sum_{k=0}^{\infty} d_k z^{k+1} \quad R_1 < |z| < R_2 \tag{9.86}$$

On the basis of the Schwarz symmetry principle, we adopt several analytical functions for all regions, i.e.,

$$\begin{cases} F_3(z) = zf'_3(z) = \frac{b_1}{2\pi i}\frac{z}{z - z_0} + F_{30}(z) & |z| > R_2 \\ F_{3*}(z) = \overline{F_3}(R_2^2/z) = \frac{b_1}{2\pi i}\left(\frac{z}{z - z_1^*} - 1\right) + F_{3*0}(z) & |z| > R_2 \\ F_2(z) = zf'_2(z) = G_N(z) + G_P(z) & R_1 < |z| < R_2 \\ F_{2*}(z) = \overline{F_2}(R_1^2/z) = \overline{G}_N(R_1^2/z) + \overline{G}_P(R_1^2/z) & R_1^2/R_2 < |z| < R_1 \\ F_{2**}(z) = \overline{F_2}(R_2^2/z) = \overline{G}_N(R_2^2/z) + \overline{G}_P(R_2^2/z) & R_2 < |z| < R_2^2/R_1 \end{cases} \tag{9.87}$$

where $z_1^* = R_2^2/\overline{z_1}$, $G_N(z) = -\sum_{k=0}^{\infty} (k+1)c_k z^{-(k+1)}$ and $G_P(z) = \sum_{k=0}^{\infty} (k+1)d_k z^{(k+1)}$

Because the complex potential $F_1(z) = zf'_1(z)$ is holomorphic when $|z| < R_1$, it is expressed as:

$$F_{1*}(z) = \overline{F_1}(R_1^2/z) \quad |z| > R_1 \tag{9.88}$$

where $F_{1*}(z)$ is holomorphic when $|z| > R_1$.

With the help of Eqs. 9.87–9.88 and using Eq. 9.84, the displacement boundary conditions are given as:

$$\begin{cases} [F_1(t) + F_{2*}(t) + F_{3*}(t)]^N = [F_2(t) + F_{1*}(t) + F_{3*}(t)]^C & |t| = R_1 \\ [F_2(t) + F_{3*}(t) + F_{1*}(t)]^C = [F_3(t) + F_{2**}(t) + F_{1*}(t)]^M & |t| = R_2 \end{cases} \tag{9.89}$$

where the superscript N, M and C denote the boundary physical quantity when z approaches the interface from the regions of nanowire, matrix and the coating layer, respectively. Eq. 9.89 is expressed as a continuously resolved complex potential $F_w(z)$, i.e.,

$$F_w(z) = \begin{cases} F_1(z) + F_{3*}(z) & |z| < R_1^2/R_2 \\ F_1(z) + F_{2*}(z) + F_{3*}(z) & R_1^2/R_2 < |z| < R_1 \\ F_2(z) + F_{1*}(z) + F_{3*}(z) & R_1 < |z| < R_2 \\ F_3(z) + F_{2**}(z) + F_{1*}(z) & R_2 < |z| < R_2^2/R_1 \\ F_3(z) + F_{1*}(z) & |z| > R_2^2/R_1 \end{cases} \tag{9.90}$$

By utilizing Eqs. 9.87–9.88 and analyzing the singularity of $F_w(z)$ [81], it is derived that

$$F_w^+(t) - F_w^-(t) = \begin{cases} \overline{G_P}(R_1^2/t) + \overline{G_N}(R_1^2/t) + \frac{b_1}{2\pi i}\left(\frac{t}{t - z_1^*} - 1\right) & |t| = R_1^2/R_2 \\ \overline{G_P}(R_1^2/t) + \overline{G_N}(R_2^2/t) + \frac{b_1}{2\pi i}\frac{t}{t - z_1} & |t| = R_2^2/R_1 \end{cases} \tag{9.91}$$

where $F_w(z)$ is a function that is sectional holomorphic on the z-plane, $\overline{G_P}(R_1^2/t) = \sum_{k=0}^{\infty}(k+1)\overline{d_k}(R_1^2/t)^{(k+1)}$ and $\overline{G_N}(R_1^2/t) = -\sum_{k=0}^{\infty}(k+1)\overline{c_k}(t/R_1^2)^{(k+1)}$.

As a result of the aforementioned work [80], explicit solutions of the Eq. 9.91 can be obtained

$$F_1(z) + F_{3*0}(z) = -\overline{G_N}(R_1^2/z) + \overline{G_N}(R_2^2/z) + \frac{b_1}{2\pi i}\frac{z}{z - z_1} \tag{9.92}$$

$$F_s(z) = \begin{cases} \mu_1 F_1(z) - \mu_3 F_{3*}(z) & |z| < R_1^2/R_2 \\ \mu_1 F_1(z) - \mu_2 F_{2*}(z) & \\ \quad - \mu_3 F_{3*}(z) - F'_{2*}(t) z(\mu^{\Gamma} - \tau^{\Gamma})/R_1 & R_1^2/R_2 < |z| < R_1 \\ \mu_2 F_2(z) - \mu_1 F_{1*}(z) & \\ \quad - \mu_3 F_{3*}(z) - F'_2(z) z(\mu^{\Gamma} - \tau^{\Gamma})/R_1 & R_1 < |z| < R_2 \\ \mu_3 F_3(z) - \mu_2 F_{2**}(z) & \\ \quad - \mu_1 F_{1*}(z) + F'_{2**}(z) z(\mu^{\Omega} - \tau^{\Omega})/R_2 & R_2 < |z| < R_2^2/R_1 \\ \mu_3 F_3(z) - \mu_1 F_{1*}(z) & |z| > R_2^2/R_1 \end{cases} \tag{9.93}$$

$$\begin{aligned} F_{30}(z) + F_{1*}(z) &= \overline{G_P}(R_1^2/z) \\ &\quad - \overline{G_P}(R_2^2/z) + \frac{b_1}{2\pi i}\left(\frac{z}{z - z_1^*} - 1\right) \quad |z| > R_2^2/R_1 \end{aligned} \tag{9.94}$$

For the second boundary condition, we follow similar method, and get the following equations:

$$\begin{aligned} F_1(z) = & \left[\frac{\mu_2-\mu_3}{\mu_1+\mu_3} + \frac{(1+k)(\mu^\Gamma-\tau^\Gamma)/R_1}{\mu_1+\mu_3}\right]\overline{G_N}(R_1^2/z) \\ & + \left[\frac{\mu_3-\mu_2}{\mu_1+\mu_3} + \frac{(1+k)(\mu^\Omega-\tau^\Omega)/R_2}{\mu_1+\mu_3}\right]\overline{G_N}(R_2^2/z) \\ & + \frac{\mu_3 b_1}{(\mu_1+\mu_3)\pi i}\frac{z}{z-z_1} \\ & - \left[\frac{(\mu^\Gamma-\tau^\Gamma)/R_1+(\mu^\Omega-\tau^\Omega)/R_2}{\mu_1+\mu_3}\right](1+k)G_P(z) \end{aligned} \tag{9.95}$$

It can be obtained from Eq. 9.94

$$\begin{aligned} F_3(z) = & \left[\frac{\mu_1-\mu_2}{\mu_1+\mu_3} + \frac{(1+k)(\mu^\Gamma-\tau^\Gamma)/R_1}{\mu_1+\mu_3}\right]\overline{G_P}(R_1^2/z) \\ & + \left[\frac{\mu_2-\mu_1}{\mu_1+\mu_3} + \frac{(1+k)(\mu^\Omega-\tau^\Omega)/R_2}{\mu_1+\mu_3}\right]\overline{G_P}(R_2^2/z) \\ & + \frac{b_1}{2\pi i}\left(\frac{z}{z-z_1} + \frac{\mu_1-\mu_3}{\mu_1+\mu_3}\frac{z_1^*}{z-z_1^*}\right) \\ & - \left[\frac{(\mu^\Gamma-\tau^\Gamma)/R_1+(\mu^\Omega-\tau^\Omega)/R_2}{\mu_1+\mu_3}\right](1+k)G_N(z) \end{aligned} \tag{9.96}$$

According to Eqs. 9.95 and 9.96, the complex potentials $F_1(z)$ and $F_3(z)$ are related to the function $F_2(z) = G_N(z) + G_P(z)$. The next step is to ascertain $F_2(z)$. Based on Eq. 9.93, the following equation is obtained:

$$\begin{aligned} & (\mu_1+\mu_2)F_2(z) + (\mu_1-\mu_3)F_{3*}(z) - F'_2(z)z(\mu^\Gamma-\tau^\Gamma)/R_1 \\ & = \frac{(\mu_1+\mu_3)b_1}{2\pi i}\frac{z}{z-z_1} + [(1+k)(\mu^\Gamma-\tau^\Gamma)/R_1 + \mu_1 - \mu_2]\overline{G_P}(R_1^2/z) \\ & + \frac{(\mu_1-\mu_3)b_1}{2\pi i}\frac{z_1^*}{z-z_1^*} + [(1+k)(\mu^\Omega-\tau^\Omega)/R_2 + \mu_1 - \mu_2]\overline{G_N}(R_2^2/z) \\ & - [(\mu^\Gamma-\tau^\Gamma)/R_1 + (\mu^\Omega-\tau^\Omega)/R_2](1+k)G_P(z) \end{aligned} \tag{9.97}$$

By ignoring the constant term that represents the rigid displacement, and substituting Eqs. 9.87 and 9.96 into the Eq. 9.97, the coefficients with identical power terms are compared and it is derived

$$d_k = -\frac{b_1}{2\pi i}\frac{1}{1+k}\frac{1}{z_1^{(1+k)}}\frac{4\mu_1\mu_3}{\Pi_1+\Pi_2-\Pi_3} \tag{9.98}$$

$$c_k = \frac{\mu_2-\mu_1-(1+k)(\mu^\Gamma-\tau^\Gamma)/R_1}{\mu_2+\mu_1+(1+k)(\mu^\Gamma-\tau^\Gamma)/R_1}R_1^{2(1+k)}\overline{d_k} \tag{9.99}$$

where $\Pi_1 = 2\mu_1[\mu_2+\mu_3+(1+k)(\mu^\Omega-\tau^\Omega)/R_2]$, $\Pi_2 = (\mu_1-\mu_3)[\mu_1-\mu_2+(1+k)(\mu^\Gamma-\tau^\Gamma)/R_1](R_1/R_2)^{2(k+1)}$

$$\Pi_3 = \frac{[(\mu_1+\mu_3)(\mu_2-\mu_1)-(\mu_1-\mu_3)(1+k)(\mu^\Gamma-\tau^\Gamma)/R_1-2\mu_1(1+k)(\mu^\Omega-\tau^\Omega)/R_2]}{[\mu_2+\mu_1+(1+k)(\mu^\Gamma-\tau^\Gamma)/R_1](R_2/R_1)^{2(k+1)}}$$
$$\times[\mu_2-\mu_1-(1+k)(\mu^\Gamma-\tau^\Gamma)/R_1]$$

The analytical functions $F_1(z)$, $F_2(z)$ and $F_3(z)$ are explicitly expressed as derived from Eqs. 9.87, 9.95 and 9.96 together with Eqs. 9.98 and 9.99. Finally, considering the relations of $F_j(z)$ and $f_j(z)$ ($j = 1, 2, 3$) in Eq. 9.87, we can deduce the solutions of the stress and displacement fields for the interaction between a screw dislocation and a nanowire:

$$\begin{cases} w_j = \mathrm{Re}[f_j(z)] & j = 1, 2, 3 \\ \tau_{xzj} - i\tau_{yzj} = \mu_j f_j'(z) & j = 1, 2, 3 \end{cases} \tag{9.100}$$

where the subscripts $j = 1$, 2 and 3 denote the regions of nanowire, the coating and the matrix, respectively. As an example, the stress fields in the matrix region are given as:

$$\begin{aligned}\tau_{xz3} - i\tau_{yz3} = {} & \frac{\mu_3 b_1}{2\pi i}\left[\frac{1}{z-z_1} + \frac{\mu_1-\mu_3}{\mu_1+\mu_3}\frac{z_1^*}{(z-z_1^*)z}\right] \\ & + \frac{\mu_3}{\mu_1+\mu_3}[(\mu^\Gamma-\tau^\Gamma)/R_1+(\mu^\Omega-\tau^\Omega)/R_2]\sum_{k=0}^{\infty}(k+1)^2 c_k z^{-(k+2)} \\ & + \frac{\mu_3}{\mu_1+\mu_3}\sum_{k=0}^{\infty}(k+1)[\mu_1-\mu_2+(1+k)(\mu^\Gamma-\tau^\Gamma)/R_1]\overline{d_k}R_1^{2(1+k)}z^{-(2+k)} \\ & + \frac{\mu_3}{\mu_1+\mu_3}\sum_{k=0}^{\infty}(k+1)[\mu_2-\mu_1+(1+k)(\mu^\Omega-\tau^\Omega)/R_2]\overline{d_k}R_2^{2(1+k)}z^{-(2+k)}\end{aligned} \tag{9.101}$$

Eq. 9.100 provides explicit expressions of Green's functions for the case with a single screw dislocation. For the case with multiple parallel dislocations, and solution is easily obtained by superimposing the Green's functions [82]. When the interface stress is ignored ($\mu^\Gamma = \tau^\Gamma = \mu^\Omega = \tau^\Omega = 0$), the complex potentials solutions $f_j(z)$ can be reduced to the results in previous model [72].

9.4.1.2 Interaction Energy, Interaction Force and the Critical Shear Stress

The strain energy induced by the dislocation is equal to the work done for inserting the dislocation in a clean matrix. The displacement along the dislocation is equal to its Burgers vector. The total force exerted on the dislocation is obtained from the stress fields. The interaction energy is the difference between the total strain energy and the self-energy of the dislocation. For a screw dislocation with Burgers vector b_1, the work done for inserting a screw dislocation is computed by [47]:

$$W = \frac{\mu_3 b_1}{2}\mathrm{Im}[\Delta f_3(z)] \tag{9.102}$$

where "Im" denotes the imaginary component of the concerned complex value. "Δ" signifies the complex potential after removing the dislocation singularity. The force f_r exerted on the screw dislocation is given by:

$$\begin{aligned}
f_r = & -\frac{\partial W}{\partial r_1} = \frac{\mu_3 b_1^2}{2\pi}\frac{\mu_1-\mu_2}{\mu_1+\mu_3}\frac{R_2^2}{r_1(r_1^2-R_2^2)} \\
& + \frac{2b_1^2\mu_3}{(\mu_1+\mu_3)\pi}\left\{\sum_{k=0}^{\infty}[\mu_1-\mu_2+(1+k)(\mu^\Gamma-\tau^\Gamma)/R_1]\frac{R_1^{2(1+k)}}{r_1^{2k+3}}\right. \\
& \times\frac{\mu_1\mu_3}{\Pi_1+\Pi_2-\Pi_3} \\
& + \sum_{k=0}^{\infty}[\mu_2-\mu_1+(1+k)(\mu^\Omega-\tau^\Omega)/R_2]\frac{R_2^{2(1+k)}}{r_1^{2k+3}}\frac{\mu_1\mu_3}{\Pi_1+\Pi_2-\Pi_3} \\
& + \sum_{k=0}^{\infty}(1+k)[(\mu^\Gamma-\tau^\Gamma)/R_1+(\mu^\Omega-\tau^\Omega)/R_2] \\
& \left.\times\frac{\mu_2-\mu_1-(1+k)(\mu^\Gamma-\tau^\Gamma)/R_1}{\mu_2+\mu_1+(1+k)(\mu^\Gamma-\tau^\Gamma)/R_1}\frac{\mu_1\mu_3}{\Pi_1+\Pi_2-\Pi_3}\frac{R_1^{2(1+k)}}{r_1^{2k+3}}\right\}
\end{aligned} \tag{9.103}$$

Here, if $\mu_1 = 0$, a new solution can be derived for the interaction among the nanohole with surface coating and the screw dislocation in nanocomposites.

$$\begin{aligned}
f_r = & -\frac{\mu_3 b_1^2}{2\pi}\frac{\mu_2}{\mu_3}\frac{R_2^2}{r_1(r_1^2-R_2^2)} + \frac{b_1^2\mu_3}{(\mu_1+\mu_3)\pi} \\
& \times\left\{\sum_{k=0}^{\infty}[-\mu_2+(1+k)(\mu^\Gamma-\tau^\Gamma)/R_1]\Pi_4\frac{R_1^{2(1+k)}}{r_1^{2k+3}}\right. \\
& + \sum_{k=0}^{\infty}[\mu_2+(1+k)(\mu^\Omega-\tau^\Omega)/R_2]\Pi_4\frac{R_2^{2(1+k)}}{r_1^{2k+3}} \\
& + \sum_{k=0}^{\infty}(1+k)[(\mu^\Gamma-\tau^\Gamma)/R_1+(\mu^\Omega-\tau^\Omega)/R_2] \\
& \left.\times\Pi_4\frac{\mu_2-(1+k)(\mu^\Gamma-\tau^\Gamma)/R_1}{\mu_2+(1+k)(\mu^\Gamma-\tau^\Gamma)/R_1}\frac{R_1^{2(1+k)}}{r_1^{2k+3}}\right\}
\end{aligned} \tag{9.104}$$

where

$$\begin{aligned}
\Pi_4 = & \mu_3[\mu_2+(1+k)(\mu^\Gamma-\tau^\Gamma)/R_1] \\
& /\{[\mu_2+\mu_3+(1+k)(\mu^\Omega-\tau^\Omega)/R_2][\mu_2+(1+k)(\mu^\Gamma-\tau^\Gamma)/R_1] \\
& +[\mu_2-\mu_3-(1+k)(\mu^\Omega-\tau^\Omega)/R_2][-\mu_2+(1+k)(\mu^\Gamma-\tau^\Gamma)/R_1]\}
\end{aligned}$$

For the composite that contains N core-shell nanowires (radius is R_2) per unit area [48], the average distance between a dislocation and the nearest nanowire is equal to $r_0 = (N^{-1/2})/2$. Assuming that the volume fraction of nanowires is $g = N\pi R_2^2$, then the force f_{rm} acting on the screw dislocation is given by:

$$
\begin{aligned}
\mathrm{f}_{rm} = {} & \frac{4b_z^2\mu_3(\mu_1-\mu_2)}{\mu_1+\mu_3}\frac{g^{3/2}}{\pi^{3/2}(\pi-4f)R_2} \\
& + \frac{4b_z^2\mu_3}{(\mu_1+\mu_3)}\left\{\sum_{k=0}^{\infty}[\mu_1-\mu_2+(1+k)(\mu^{\Gamma}-\tau^{\Gamma})/R_1]\frac{R_1^{2(1+k)}}{R_2^{2(k+1)}}\frac{\mu_1\mu_3}{P_1+P_2-P_3}\right. \\
& \times \frac{g^{1/2}}{\pi^{3/2}R_2}\left(\frac{4g}{\pi}\right)^{k+1} + \sum_{k=0}^{\infty}[\mu_2-\mu_1+(1+k)(\mu^{\Omega}-\tau^{\Omega})/R_2] \\
& \times \frac{\mu_1\mu_3}{P_1+P_2-P_3}\frac{g^{1/2}}{\pi^{3/2}R_2}\left(\frac{4f}{\pi}\right)^{k+1} \\
& + \sum_{k=0}^{\infty}(1+k)[(\mu^{\Gamma}-\tau^{\Gamma})/R_1+(\mu^{\Omega}-\tau^{\Omega})/R_2] \\
& \left.\times \frac{\mu_2-\mu_1-(1+k)(\mu^{\Gamma}-\tau^{\Gamma})/R_1}{\mu_2+\mu_1+(1+k)(\mu^{\Gamma}-\tau^{\Gamma})/R_1}\frac{\mu_1\mu_3}{P_1+P_2-P_3}\frac{g^{1/2}}{\pi^{3/2}R_2}\left(\frac{4g}{\pi}\right)^{k+1}\right\}
\end{aligned}
\tag{9.105}
$$

Through analyzing the force f_{rm}, the contribution of the dislocation-nanowire elastic interaction to the CSS is obtained [42]:

$$
\begin{aligned}
\Delta\tau_c = {} & \frac{4b_z\mu_3(\mu_1-\mu_2)}{\mu_1+\mu_3}\frac{g^{3/2}}{\pi^{3/2}(\pi-4f)R_2} \\
& + \frac{4b_z\mu_3}{(\mu_1+\mu_3)}\left\{\sum_{k=0}^{\infty}[\mu_1-\mu_2+(1+k)(\mu^{\Gamma}-\tau^{\Gamma})/R_1]\right. \\
& \times \frac{R_1^{2(1+k)}}{R_2^{2(k+1)}}\frac{\mu_1\mu_3}{P_1+P_2-P_3} \\
& \times \frac{g^{1/2}}{\pi^{3/2}R_2}\left(\frac{4g}{\pi}\right)^{k+1} + \sum_{k=0}^{\infty}[\mu_2-\mu_1+(1+k)(\mu^{\Omega}-\tau^{\Omega})/R_2] \\
& \times \frac{\mu_1\mu_3}{P_1+P_2-P_3}\frac{g^{1/2}}{\pi^{3/2}R_2}\left(\frac{4g}{\pi}\right)^{k+1} \\
& + \sum_{k=0}^{\infty}(1+k)[(\mu^{\Gamma}-\tau^{\Gamma})/R_1+(\mu^{\Omega}-\tau^{\Omega})/R_2] \\
& \left.\times \frac{\mu_2-\mu_1-(1+k)(\mu^{\Gamma}-\tau^{\Gamma})/R_1}{\mu_2+\mu_1+(1+k)(\mu^{\Gamma}-\tau^{\Gamma})/R_1}\frac{\mu_1\mu_3}{P_1+P_2-P_3}\frac{g^{1/2}}{\pi^{3/2}R_2}\left(\frac{4g}{\pi}\right)^{k+1}\right\}
\end{aligned}
\tag{9.106}
$$

Based on Eq. 9.106, one can easily explore the CSS affected by the size, interface stress, and the volume fraction of nanowire.

9.4.1.3 Numerical Examples and Discussion

This part examines the force exerted on the screw dislocation and the equilibrium position of the dislocation near the nanowire, using the Eq. 9.103. For subsequent numerical calculations, the residual interface tension is ignored, i.e., $\tau^{\Gamma} = \tau^{\Omega} = 0$. Meanwhile, the relative shear modulus $\alpha = \mu_1/\mu_3$ and $\beta = \mu_2/\mu_3$, relative coating thickness $\lambda = R_1/R_2$ are defined. The intrinsic lengths $\varepsilon = \mu^{\Gamma}/\mu_3$ and $\gamma = \mu^{\Omega}/\mu_3$ are introduced, and their values are approximately 0.1 nm [31]. No loss of generality, it is assumed that the screw dislocation is located at the point x_1 on the x-axis ($z_1 = x_1 = r_1 > R_2$). $f_0 = 2\pi R_2 f_r/(\mu_3 b_1^2)$ is defined as the normalized force, and $\rho = r_1/R_2 = x_1/R_2$ is the relative position of the dislocation.

Figure 9.13a–c displays how the value of f_{x0} changes as the parameter ρ varies. From Figure 9.13a, several cases for the screw dislocation may occur. 1) When the interface stress becomes zero, the nanowire may attract the dislocation if the coating layer and the nanowire are softer than the matrix. 2) When interface stress is positive, the screw dislocation is initially drawn towards the nanowire and then pushed away, eventually reaching a stable equilibrium position; conversely, when the interface stress is negative, the attractive force exerted on the dislocation increases, and exceeds the attractive force in the case without interface stress. 3) If the coating layer and nanowire are stronger than the matrix, and the interface stress is negative, the nearby screw dislocation would be pushed away and then drawn

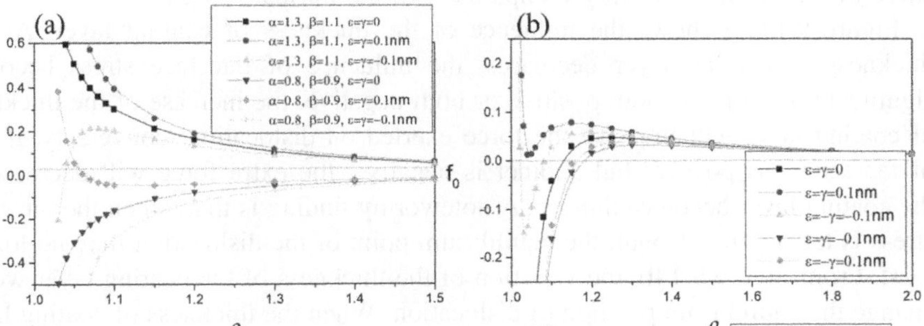

FIGURE 9.13 (a) Variation of normalized force f_0 as a function of ρ, for $\lambda = 0.9$ and $R_1 = 15$ nm; (b) Variation of normalized force f_0 as a function of ρ, for $\alpha = 1.3$, $\beta = 0.8$, $\lambda = 0.9$ and $R_1 = 15$ nm; (c) Variation of normalized force f_0 vas a function of ρ, for $\alpha = 0.8$, $\beta = 1.2$, $\lambda = 0.9$ and $R_1 = 15$ nm; (d) Variation of normalized force f_0 as a function of R_1, for $\alpha = 0.8$, $\beta = 1.2$, $\lambda = 0.9$ and $\rho = 1.05$ [79].

back to the nanowire, and consequently reaches an unstable equilibrium point in the vicinity of the nanowire.

From Figure 9.13b, if the matrix is softer than the nanowire but more rigid than the coating layer, the dislocation would reach an unstable equilibrium point when the interface stress becomes zero ($\varepsilon = 0$ and $\gamma = 0$). When $\varepsilon = -0.1\,\text{nm}$ and $\gamma = 0.1\text{nm}$, there are two equilibrium points for the dislocation: the point near the nanowire is stable while the farter point is unstable. From Figure 9.13c, if the matrix is more rigid than the nanowire but less rigid than the coating layer, there would be a stable equilibrium point when the interface stress becomes zero ($\varepsilon = 0$ and $\gamma = 0$). When $\varepsilon = 0.1\,\text{nm}$ and $\gamma = -0.1\,\text{nm}$, there are two equilibrium points for the dislocation: the point near the nanowire is unstable while the farter point is stable. Comparing with the classical solution (without interface stress), the presence of interface effects leads to more equilibrium positions for the dislocation.

Figure 9.13d shows the normalized image force f_{x0} changes with the radius R_1. It is observed that positive ε or γ causes the interface repelling the screw dislocation, but negative ε or γ causes the interface attracting the screw dislocation. In addition, the decrease of the radius of nanowire would improve the normalized image force, while λ and ρ remain constant. Interestingly, when the radius of the nanowire decreases to approximately 15 nm and $\gamma = -0.1\text{nm}$, the direction of the image force reverses. At this time, the dislocation exactly arrives the direction of the image force will reverse. Additionally, it is also found that the image force is more sensitive to the intrinsic length γ compared with the intrinsic length ε.

Figure 9.14a,b shows the influence of the thickness of coating layer. As the thickness of coating layer decreases, the influence of interface stress becomes significant. If ε or γ is both positive or both negative, the increase of the thickness of coating layer will improve the force exerted on dislocation; conversely, if one of the ε or γ is positive but another is negative, the extra force will decrease as the coating layer becomes thinner. A noteworthy finding is that, when the interface stress is taken into account, the equilibrium point of the dislocation may no longer exist. From Figure 9.14b, the variation of the thickness of the coating layer would change the equilibrium position of dislocation. When the thickness of coating layer is too large, the equilibrium position of dislocation is no longer existing.

The case with two parallel dislocations is further studied. Here, it is assumed that the screw dislocations are characterized by the Burgers vectors b_1 and b_2. The dislocations locate at the points $z_1\,(z_1 = x_1)$ and $z_2\,(z_2 = x_2 > R_2)$ on the positive x-axis. Then, we can explore the interaction between two dislocations, under the influence of nanowire and the surface property. The total force exerting on the z_1 dislocation is equal to the sum of the force resulting from Eq. 9.103 and the force derived from z_2 dislocation. The force acting on z_1 dislocation is split into the force component along the X axis (f_x), and the force component along the Y axis (f_y). The distance between the z_2 dislocation and the inclusion is defined as $\delta = x_2/R_2$.

Figure 9.14c illustrates the normalized force f_0 acting on dislocation z_1 as a function of $\rho = x_1/R_2$, under different interface properties. When the coated nanowire is absent, the part of the force acting on z_1 dislocation from the z_2 dislocation is always negative, indicating that the z_1 dislocation is repelled by the z_2 dislocation. However, when the nanowire is inserted in the matrix, the force acting on the z_1

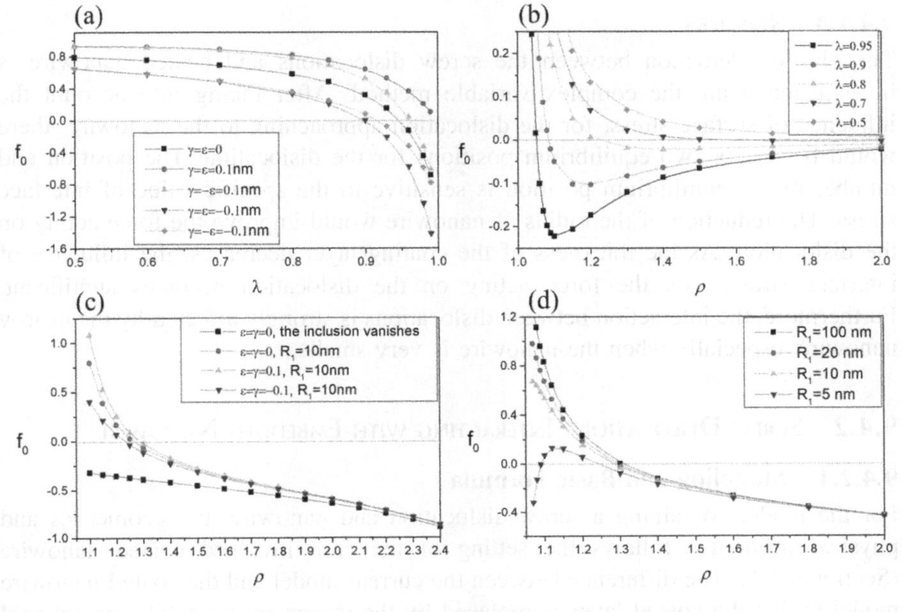

FIGURE 9.14 (a) Variation of normalized force f_0 as a function of λ, for $\alpha = 0.8$, $\beta = 1.2$, $R_1 = 15\text{nm}$ and $\rho = 1.05$; (b) Variation of normalized force f_0 as a function of ρ, for $\alpha = 0.8$, $\beta = 1.2$, $R_1 = 15\text{nm}$ and $\gamma = -\varepsilon = 0.1\text{nm}$; (c) Variation of normalized force f_0 as a function of ρ, for $\delta = 3$, $\alpha = 2$, $\beta = 1.5$, $\lambda = 0.9$ and $b_2 = b_1$; (d) Variation of normalized force f_0 as a function of ρ, for $\delta = 3$, $\alpha = 2$, $\beta = 1.5$, $\lambda = 0.9$ $\varepsilon = \gamma = -0.1\text{nm}$ and $b_2 = b_1$ [79].

dislocation would become positive if the z_1 dislocation is close to the nanowire. This leads to a stable equilibrium position at the x-axis for the z_1 dislocation. Taking into account the stress of interface can alter the value of the force. However, in this case, the mechanism by which the two dislocations interact with the nanowire remains unchanged.

Figure 9.14d depicts the normalized force f_0 versus $\rho = x_1/R_2$ using various values of the radius R_1. The impact of interface properties on the force becomes more pronounced as the radius of the nanowire decreases. A fascinating finding is that, in situations where the interface properties are negative and the nanowire is very small ($R_1 = 5\text{nm}$ and $\varepsilon = \gamma = -0.1\text{nm}$), the stiff coated nanowire can attract the z_1 dislocation as it approaches the nanowire. The attractive force comes from the interface stress, whose force component along the x-axis is negative, and increases with the decrease of nanowire radius. The competition of the force along the negative x-axis (derived from the interface stress and z_2 dislocation) and the force along the positive x-axis (produced by the stiff coated nanowire) determines the total force acting on the z_1 dislocation. When the force along the negative z_1-axis is higher than the force along the positive x-axis, the direction of the total force acting on the dislocation would reverse. In addition, we found that there are two equilibrium points in the vicinity of the nanowire: the one closer to the nanowire is stable, while the other far away from the nanowire is unstable.

9.4.1.4 Remarks

The elastic interaction between the screw dislocations and coated nanowire is investigated using the complex variable method. After taking into account the influence of surface stress, for the dislocation approaching to the nanowire, there would be one or two equilibrium positions for the dislocation. The position and number of the equilibrium position is sensitive to the specific value of interface stress. The reduction of the radius of nanowire would improve the force acting on the dislocation. As the thickness of the coating layer decreases, the influence of interface stresses on the force acting on the dislocation becomes significant. Furthermore, the interaction between dislocations is strongly affected by the nearby nanowire, especially when the nanowire is very small.

9.4.2 Screw Dislocations Interacting with Embedded Nanowire

9.4.2.1 Modeling and Basic Formula

For the model containing a screw dislocation and nanowire, the geometries and physical properties follows the setting of the case involved coated nanowire (Section 9.4.1). The difference between the current model and the coated-nanowire model is that the coated layer is replaced by the matrix phase, while the original matrix region in coated-nanowire model is replaced by the composite. The boundary conditions for the interfaces are [34,53]:

$$\begin{cases} w_1(t) - w_2(t) = 0 \quad \tau_{rz1}(t) - \tau_{rz2}(t) = \frac{1}{R_1}\frac{\partial \tau_{\theta z}^0(t)}{\partial \theta} & |t| = R_1 \\ w_2(t) - w_3(t) = 0 \quad \tau_{rz2}(t) - \tau_{rz3}(t) = 0 & |t| = R_2 \end{cases} \tag{9.107}$$

Following general method, the interfaces are assumed to be coherent. Then the additional constitutive equation of the interface region $|t| = R_1$ is [53]:

$$\tau_{\theta z}^0(t) = 2(\mu^0 - \tau^0)\varepsilon_{\theta z}^0(t) \quad |t| = R_1 \tag{9.108}$$

An analytical function $f(z)$ is used to describe the relationship between the anti-plane displacement w, shear stresses τ_{rz} and $\tau_{\theta z}$ for the bulk solid [80]. The non-classical stress boundary condition on the interface using Eq. 9.108 and Eq. 9.107. Using boundary conditions, one can determine the complex potentials $f_1(z)$, $f_2(z)$and $f_3(z)$ for different regions.

For the case of a single screw dislocation with a Burgers vector b_1 and located at the point z_1 ($z_1 = x_1 + iy_1$) in the matrix, the complex potential in the matrix region can be found from [79]. By comparing the coefficients of the same power terms, there are

$$\begin{aligned} c_k &= \frac{b_1}{2\pi i}\frac{\mu_2 - \mu_1 - (1+k)(\mu^0 - \tau^0)/R_1}{1+k} \\ &\times \frac{(\mu_2 + \mu_3)(R_1^2 R_2^2/\bar{z}_1)^{(k+1)} + (\mu_3 - \mu_2)R_1^{2(1+k)} z_1^{(k+1)}}{M} \end{aligned} \tag{9.109}$$

$$d_k = \frac{b_1}{2\pi i}\frac{\mu_2 - \mu_3}{1+k}\frac{[\mu_1 + \mu_2 + (1+k)(\mu^0 - \tau^0)/R_1]\bar{z}_1^{(k+1)}}{M} + \frac{[\mu_1 - \mu_2 + (1+k)(\mu^0 - \tau^0)/R_1](R_1^2/z_1)^{(k+1)}}{M} \quad (9.110)$$

where
$$\begin{aligned} M = {} & (\mu_3 - \mu_2)[\mu_2 - \mu_1 - (1+k)(\mu^0 - \tau^0)/R_1]R_1^{2(k+1)} \\ & + (\mu_2 + \mu_3)[\mu_2 + \mu_1 + (1+k)(\mu^0 - \tau^0)/R_1]R_2^{2(k+1)} \end{aligned}.$$

Using Eqs. 9.109 and 9.110, the solution for the analytical function $F_2(z)$ is determined by:

$$F_2(z) = \frac{b_1}{2\pi i}\frac{z}{z - z_1} - \sum_{k=0}^{\infty}(k+1)c_k z^{-(k+1)} + \sum_{k=0}^{\infty}(k+1)d_k z^{(k+1)} \quad (9.111)$$

Finally, the solutions for analytic functions $F_1(z)$ and $F_3(z)$ are given by:

$$\begin{aligned} F_1(z) = {} & \frac{b_1}{2\pi i}\frac{2\mu_2}{\mu_1 + \mu_2}\frac{z}{z - z_1} + \frac{2\mu_2}{\mu_1 + \mu_2}G_P(z) \\ & - \frac{b_1}{2\pi i}\frac{(\mu^0 - \tau^0)}{R_1(\mu_1 + \mu_2)}\left[\frac{z}{z - z_1} - \frac{z^2}{(z - z_1)^2}\right] \\ & - \frac{(\mu^0 - \tau^0)}{R_1(\mu_1 + \mu_2)}[zG'_P(z) - z\overline{G'}_N(R_1^2/z)] \end{aligned} \quad (9.112)$$

$$F_3(z) = \frac{b_1}{2\pi i}\frac{2\mu_2}{\mu_3 + \mu_2}\frac{z}{z - z_1} + \frac{b_1}{2\pi i}\frac{\mu_3 - \mu_2}{\mu_3 + \mu_2} + \frac{2\mu_2}{\mu_3 + \mu_2}G_N(z) \quad (9.113)$$

In conclusion, taking into account the relations between $F_j(z)$ and $f_j(z)(j = 1, 2, 3)$, the complete solutions of displacement and stress fields are acquired.

$$\begin{cases} w_j = \mathrm{Re}[f_j(z)] & (j = 1, 2, 3) \\ \tau_{xzj} - i\tau_{yzj} = \mu_j f_j'(z) & (j = 1, 2, 3) \end{cases} \quad (9.114)$$

where the subscripts $j = 1$, 2 and 3 correspond to the nanoinclusion, matrix and the effective medium regions, respectively.

The above equations present the Green's function solutions for the case with only one screw dislocation. For the case involved multiple parallel dislocations, the solution can be easily obtained by superimposing Green's functions.

Notably, from Eqs. 9.111–9.113, if we do not take into account the interface stress ($\mu^0 = \tau^0 = 0$), the complex potentials solutions $f_j(z)$ would be reduced to the classical result [84]. Additionally, when $\mu_3 = 0$, we may get the new solution for the two-phase cylinder model composed by a finite thickness matrix and cylindrical nanowire.

9.4.2.2 Image Force on Screw Dislocations

Following general method, the image force acting on a screw dislocation at the point z_1 is given by [85].

$$f_x - if_y = ib_1[\widehat{\tau_{xz2}}(z_1) - i\widehat{\tau_{yz2}}(z_1)] \tag{9.115}$$

These dislocation points are obtained by subtracting the corresponding contributions of dislocations in the infinite uniform medium obtained in Eq. 9.114, and then taking the limit as z approaches z_1. According to reference [86], the analytical expression for the image force is given by

$$\begin{aligned} f_x - if_y = & \frac{\mu_2 b_1^2(\mu_2-\mu_3)}{2\pi}\sum_{k=0}^{\infty}\left\{\frac{[\mu_1+\mu_2+(1+k)(\mu^0-\tau^0)/R_1]\bar{z}_1^{(k+1)}}{z_1^{-k}M}\right. \\ & \times\left.\frac{+[\mu_1-\mu_2+(1+k)(\mu^0-\tau^0)/R_1](R_1^2/z_1)^{(k+1)}}{z_1^{-k}M}\right\} \\ & -\frac{\mu_2 b_1^2}{2\pi}\left\{\sum_{k=0}^{\infty}\frac{[\mu_2-\mu_1-(1+k)(\mu^0-\tau^0)/R_1]}{z_1^{(k+2)}M}\right. \\ & \times\left.\frac{[(\mu_2+\mu_3)(R_1^2R_2^2/\bar{z}_1)^{(k+1)}+(\mu_3-\mu_2)R_1^{2(1+k)}z_1^{(k+1)}]}{z_1^{(k+2)}M}\right\} \end{aligned} \tag{9.116}$$

For the case with two parallel dislocations, marked by Burgers vectors b_1 and b_2 and position z_1 and z_2, the image force acting on z_1-dislocation produced from the z_2-dislocation is given by:

$$\begin{aligned} f_x - if_y = & \frac{\mu_2 b_1^2(\mu_2-\mu_3)}{2\pi}\sum_{k=0}^{\infty}\left\{\frac{[\mu_1+\mu_2+(1+k)(\mu^0-\tau^0)/R_1]\bar{z}_1^{(k+1)}}{z_1^{-k}M}\right. \\ & +\left.\frac{[\mu_1-\mu_2+(1+k)(\mu^0-\tau^0)/R_1](R_1^2/z_1)^{(k+1)}}{z_1^{-k}M}\right\} \\ & -\frac{\mu_2 b_1^2}{2\pi}\sum_{k=0}^{\infty}\left\{\frac{[\mu_2-\mu_1-(1+k)(\mu^0-\tau^0)/R_1]}{z_1^{(k+2)}M}\right. \\ & +\left.\frac{[(\mu_2+\mu_3)(R_1^2R_2^2/\bar{z}_1)^{(k+1)}+(\mu_3-\mu_2)R_1^{2(1+k)}z_1^{(k+1)}]}{z_1^{(k+2)}M}\right\} + \frac{\mu_2 b_1 b_2}{2\pi(z_1-z_2)} \\ & +\frac{\mu_2 b_1 b_2(\mu_2-\mu_3)}{2\pi}\sum_{k=0}^{\infty}\left\{\frac{[\mu_1+\mu_2+(1+k)(\mu^0-\tau^0)/R_1]\bar{z}_2^{(k+1)}}{z_1^{-k}M}\right. \\ & +\left.\frac{[\mu_1-\mu_2+(1+k)(\mu^0-\tau^0)/R_1](R_1^2/z_2)^{(k+1)}}{z_1^{-k}M}\right\} \\ & -\frac{\mu_2 b_1 b_2}{2\pi}\sum_{k=0}^{\infty}\left\{\frac{[\mu_2-\mu_1-(1+k)(\mu^0-\tau^0)/R_1]}{z_1^{(k+2)}M}\right. \\ & \times\left.\frac{[(\mu_2+\mu_3)(R_1^2R_2^2/\bar{z}_2)^{(k+1)}+(\mu_3-\mu_2)R_1^{2(1+k)}z_2^{(k+1)}]}{z_1^{(k+2)}M}\right\} \end{aligned} \tag{9.117}$$

9.4.2.3 Numerical Examples and Discussion

Using Eqs. 9.116 and 9.117, we analyze the image force acting on a single screw dislocation situated in the matrix. Without loss of generality, it is assumed that the screw dislocation is located at the point x_1 ($R_1 < x_1 < R_2$) on the x-axis.

Figure 9.15a,b displays the relationship between the force f_{x0} and the intrinsic length γ and parameter δ. If the dislocation is far away from the nanowire and the effective medium is softer than the matrix, the dislocation would be attracted by the effective medium. When the dislocation is close to the medium-matrix interfaces, the attraction force would be improved. Conversely, when the effective medium becomes harder than the matrix, it repels the screw dislocation. The presence of interface stress would enhance the dislocation mobility compared with the classical case without interface stress.

From Figure 9.15a, for the case when the matrix is stiffer than the effective medium but softer than the nanowire, the disappearance of the interface stress ($\gamma = 0$) would cause the vanish of the equilibrium point; if $\gamma = 1.0\text{Å}$, there will be a single unstable equilibrium point. In addition, for the case when the matrix is softer than both the effective medium and the nanowire, the disappearance of the interface stress ($\gamma = 0$) induce an unstable equilibrium point; if $\gamma = 1.0\text{Å}$, one stable equilibrium point exists near the nanowire and one unstable equilibrium point exists at a farer position.

From Figure 9.15b, for the case when the matrix is stiffer than the effective medium but softer than the nanowire, no equilibrium point exists if $\gamma = 0$ and one single stable equilibrium point if $\gamma = 1.0\text{Å}$. In addition, for the case when the matrix is softer than both the effective medium and the nanowire, one equilibrium point exists if $\gamma = 0$ and one unstable equilibrium point exists near the nanowire and one stable equilibrium point exists at a farer position if $\gamma = 1.0\text{Å}$.

Figure 9.16a depicts the change of the image force f_{x0} with the radius R_1. The interface repels the dislocation if γ is positive, but attract the dislocation if γ is negative. The repelling and attraction of dislocation are corresponding to the local hardening and softening of the interface. As the radius of the nanowire decreases,

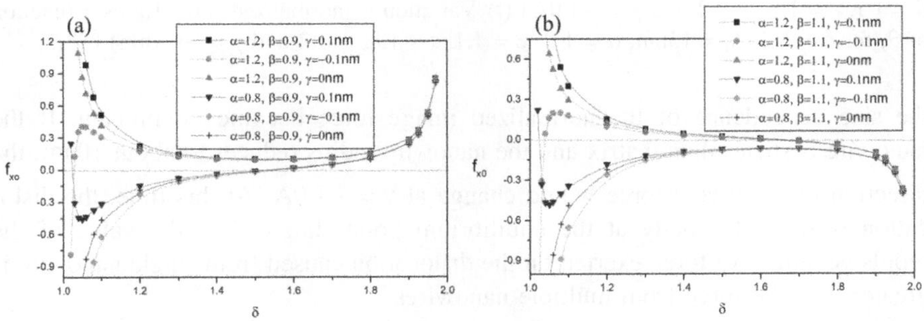

FIGURE 9.15 (a) Variation of normalized force f_{x0} as a function of δ, for $\beta = 0.9$, $\lambda = 2$ and $R_1 = 15\text{nm}$; (b) Variation of normalized force f_{x0} as a function of δ, for $\beta = 1.1$, $\lambda = 2$ and $R_1 = 15\text{nm}$ [83].

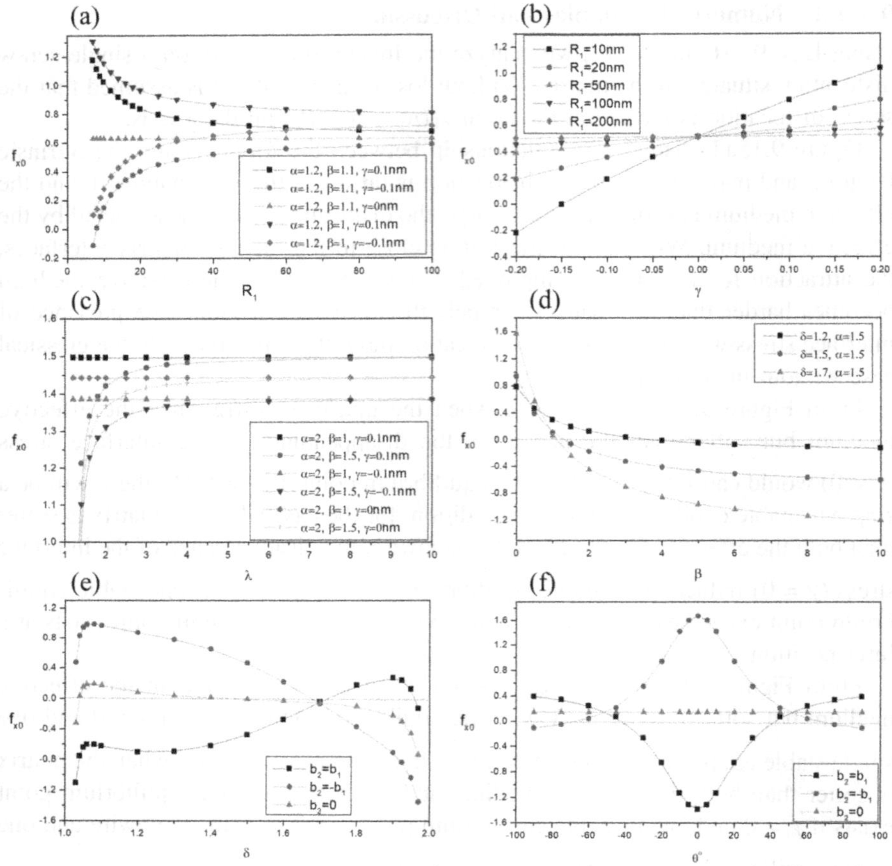

FIGURE 9.16 (a) Variation of normalized force f_{x0} as a function of R_1, for $\lambda = 2$ and $\delta = 1.05$; (b) Variation of normalized force f_{x0} as a function of γ, for $\lambda = 2$, $\delta = 1.06$, $\alpha = 1.2$ and $\beta = 1.1$; (c) Variation of normalized force f_{x0} as a function of λ, for $f_{x0}$and $\lambda = R_2/R_1$; (d) Variation of normalized force f_{x0} as a function of $\beta = \mu_3/\mu_2$, for $R_1 = 15$nm, $\lambda = 2$, $\gamma = 1.0$A and $\alpha = 1.5$. (e) Variation of normalized force f_{x0} as a function of δ, for $R_1 = 15$nm ($\alpha = 1.2$, $\beta = 1.1$, $\varepsilon = 1.8$, $\lambda = 2$ and $\gamma = -1.0$A) (f) Variation of normalized force f_{x0} as a function of θ, for $\delta = 1.2$ ($R_1 = 15$nm, $\alpha = 1.2$, $\beta = 1.1$, $\varepsilon = 1.8$, $\lambda = 2$ and $\gamma = -0.1$nm) [83].

the size dependence of the normalized image force becomes significant. If the nanowire is stiffer than matrix and the nanowire radius reduces to about 10nm, the direction of the image force would change at $\gamma = -1.0\mathring{A}$. At this time, the dislocation is located exactly at the equilibrium point. Interesting, the value of the repulsive/attractive force exerted on the dislocation caused from single nanowire is greater than the force from multiple nanowires.

Figure 9.16b shows the normalized image force f_{x0} under different R_1. As the absolute magnitude of intrinsic length λ gets smaller, the influence of the interface stress on image force is weaken. A special finding is that when the nanowire radius is close to 10nm, stiff nanowire first repels the dislocation and then attracts it as the negative γ increases. At this time, the dislocation is exactly located at the equilibrium point.

Figure 9.16c shows the normalized image force f_{x0} under different $\lambda = R_2/R_1$. As the value of $\lambda = R_2/R_1$ rises, the model gives solutions that are progressively closer to the solution given by the two-phase model. The increase of the number of nanowires weakens the repulsive force exerted on the dislocation. Figure 9.16d depicts the normalized image force f_{x0} under different values of δ. When the effective medium is less rigid, the dislocation would be more easily repelled by the nanowire; when the effective medium gets stiffer, the dislocations are repelled by the effective medium. Meanwhile, we observe a threshold value of β that changes the direction of image force. The value of β would increase with the decrease of the distance between dislocation and nanowire.

Next, we discuss the effect of the closer parallel screw dislocations. The distance between the z_2 dislocation and the nanowire is prescribed as $\varepsilon = r/R_1$. Figure 9.16e depicts the normalized image force f_{x0} under different values of δ. When δ reaches a critical value, the direction of the image force exerted by the z_2 dislocation on the z_1 dislocation would change when $b_2 = b_1$ and $b_2 = -b_1$. The unstable equilibrium point of z_1 dislocation near the nanowire would disappear due to the influence of z_2 dislocation. When $b_2 = b_1$, the dislocation z_1 may produce one stable equilibrium point close to the effective medium, or two equilibrium points in the region between the effective medium and matrix. When $b_2 = -b_1$, one equilibrium point in the region of effective medium, but no equilibrium point in the region between the effective medium and matrix.

Figure 9.16f presents the normalized image force f_{x0} varies with θ under different values of b_2. When the absolute value of the angle θ reaches a critical value, the direction of the total image force acting on the dislocation z_1 changes. If both z_1 dislocation and z_2 dislocation lie on the x-axis, the image force on z_1 dislocation from the z_2 dislocation would increase.

9.4.2.4 Remarks

This section investigates the elastic interaction between screw dislocations and the nanowires subjected to the interface stresses. The interface stress induces multiple equilibrium points for the dislocation when the dislocation approaches the nanowire. When the radius of nanowire decreases, the influence of the interface stress on the image force is enhanced. In addition, the magnitude of the repulsive/attractive force on the dislocation generated by multiple nanowires is smaller than that caused by one single nanowire. When two dislocations are taken into account, the equilibrium position of the dislocation would be influenced by the other dislocation. Especially, if the original equilibrium point disappears, a new equilibrium point may appear induced by the closer dislocations.

REFERENCES

[1] Bulatov V., Abraham F.F., Kubin L., et al. 1998. Connecting atomistic and meso-scale simulations of crystal plasticity. *Nature*, 391(6668): 669–672.

[2] Bulatov V.V., Hsiung L.L., Tang M., et al. 2006. Dislocation multi-junctions and strain hardening. *Nature*, 440(7088): 1174–1178.

[3] Chen L.Y., He M.-R., Shin J., et al. 2015. Measuring surface dislocation nucleation in defect-scarce nanostructures. *Nature Materials*, 14(7): 707–713.

[4] Yamakov V., Wolf D., Phillpot S.R., et al. 2002. Dislocation processes in the deformation of nanocrystalline aluminium by molecular-dynamics simulation. *Nature Materials*, 1(1): 45–49.

[5] Huang X., Hansen N. and Tsuji N. 2006. Hardening by annealing and softening by deformation in nanostructured metals. *Science*, 312(5771): 249–251.

[6] Liu G.-R., Zhang G.J., Jiang F.-j., et al. 2013. Nanostructured high-strength molybdenum alloys with unprecedented tensile ductility. *Nature Materials*, 12(4): 344–350.

[7] Dundurs J. 1967. On the interaction of a screw dislocation with inhomogeneities. *Recent Advances in Engineering Science*, 2: 223–233.

[8] Dundurs J., Mura T.J.J.O.T.M. and Solids P.O. 1964. Interaction between an edge dislocation and a circular inclusion. 12(3): 177–189.

[9] Fang Q.H., Liu Y.W. and Jiang C.P. 2003. Edge dislocation interacting with an interfacial crack along a circular inhomogeneity. *International Journal of Solids and Structures*, 40(21): 5781–5797.

[10] Fang Q.H., Liu Y.W. and Jiang C.P. 2005. Electroelastic interaction between a piezoelectric screw dislocation and an elliptical inclusion with interfacial cracks. *Physica Status Solidi (b)*, 242(14): 2775–2794.

[11] Luo H.A. and Chen Y. 1991. An edge dislocation in a three-phase composite cylinder model.

[12] Qaissaunee M.T. and Santare M.H. 1995. Edge dislocation interacting with an elliptical inclusion surrounded by an interfacial zone. *The Quarterly Journal of Mechanics and Applied Mathematics*, 48(3): 465–482.

[13] Santare M.H. and Keer L.M. 1986. Interaction between an edge dislocation and a rigid elliptical inclusion.

[14] Stagni L. 1993. Edge dislocation near an elliptic inhomogeneity with either an adhering or a slipping interface a comparative study. *Philosophical Magazine A*, 68(1): 49–57.

[15] Wang X. and Shen Y.-P.J.J.A.M. 2002. An edge dislocation in a three-phase composite cylinder model with a sliding interface. 69(4): 527–538.

[16] Xiao Z.M. and Chen B.J. 2001. On the interaction between an edge dislocation and a coated inclusion. *International Journal of Solids and Structures*, 38(15): 2533–2548.

[17] Wong E.W., Sheehan P.E. and Lieber C.M. 1997. Nanobeam mechanics: elasticity, strength, and toughness of nanorods and nanotubes. *Science*, 277(5334): 1971–1975.

[18] Zhou L.G. and Huang H. 2004. Are surfaces elastically softer or stiffer? *Applied Physics Letters*, 84(11): 1940–1942.

[19] Gibbs J.W. 1948. *The collected works of J. Willard Gibbs*. Yale Univ. Press.

[20] Cammarata R.C. 1994. Surface and interface stress effects in thin films. *Progress in Surface Science*, 46(1): 1–38.

[21] Ibach H. 1997. The role of surface stress in reconstruction, epitaxial growth and stabilization of mesoscopic structures. *Surface Science Reports*, 29(5–6): 195–263.

[22] Orowan E. 1970. Surface energy and surface tension in solids and liquids. *Proceedings of the Royal Society of London. A. Mathematical and Physical Sciences*, 316(1527): 473–491.

[23] Shuttleworth R. 1950. The surface tension of solids. *Proceedings of the Physical Society, London, Section A*, 63: 444.

[24] Gurtin M.E. and Ian Murdoch A. 1975. A continuum theory of elastic material surfaces. *Archive for Rational Mechanics and Analysis*, 57: 291–323.
[25] Gurtin M.E., Weissmüller J. and Larche F. 1998. A general theory of curved deformable interfaces in solids at equilibrium. *Philosophical Magazine A*, 78(5): 1093–1109.
[26] Cammarata R.C., Sieradzki K. and Spaepen F. 2000. Simple model for interface stresses with application to misfit dislocation generation in epitaxial thin films. *Journal of Applied Physics*, 87(3): 1227–1234.
[27] Duan H.L., Wang J.-X., Huang Z.P., et al. 2005. Size-dependent effective elastic constants of solids containing nano-inhomogeneities with interface stress. *Journal of the Mechanics and Physics of Solids*, 53(7): 1574–1596.
[28] He L.H., Lim C.W. and Wu B.S. 2004. A continuum model for size-dependent deformation of elastic films of nano-scale thickness. *International Journal of Solids and Structures*, 41(3–4): 847–857.
[29] Kukta R.V., Kouris D. and Sieradzki K. 2003. Adatoms and their relation to surface stress. *Journal of the Mechanics and Physics of Solids*, 51(7): 1243–1266.
[30] Lim C.W., Li Z.R. and He L.H. 2006. Size dependent, non-uniform elastic field inside a nano-scale spherical inclusion due to interface stress. *International Journal of Solids and Structures*, 43(17): 5055–5065.
[31] Miller R.E. and Shenoy V.B. 2000. Size-dependent elastic properties of nanosized structural elements. *Nanotechnology*, 11(3): 139.
[32] Sharma P. and Ganti S. 2002. Interfacial elasticity corrections to size-dependent strain-state of embedded quantum dots. *Physica Status Solidi (b)*, 234(3): R10–R12.
[33] Sharma P. and Ganti S. 2004. Size-dependent Eshelby's tensor for embedded nano-inclusions incorporating surface/interface energies. *J. Appl. Mech.*, 71(5): 663–671.
[34] Sharma P., Ganti S. and Bhate N. 2003. Effect of surfaces on the size-dependent elastic state of nano-inhomogeneities. *Applied Physics Letters*, 82(4): 535–537.
[35] Zhang X. and Sharma P. 2005. Size dependency of strain in arbitrary shaped anisotropic embedded quantum dots due to nonlocal dispersive effects. *Physical Review B*, 72(19): 195345.
[36] Romanov A.E., Wagner T. and Rühle M. 1998. Coherent to incoherent transition in mismatched interfaces. *Scripta Materialia*, 38(6): 869–875.
[37] Sutton A.P. and Balluff R.W. *Interfaces in crystalline materials*. Oxford: Clarendon Press, 1996.
[38] Rottman C. 1988. Landau theory of coherent interphase interfaces. *Physical Review B*, 38(16): 12031.
[39] Schumann W., *Some basic problems of mathematical theory of elasticity*, BIRKHAUSER VERLAG AG VIADUKSTRASSE 40-44, PO BOX 133, CH-4010 BASEL, SWITZERLAND, 1965.
[40] Hirth J.P. and Lothe J. 1982. *Theory of Dislocations* 2nd edn. p 76. New York: Wiley, Interscience.
[41] Mott N.F. and Nabarro F.R.N. 1940. An attempt to estimate the degree of precipitation hardening, with a simple model. *Proceedings of the Physical Society*, 52(1): 86.
[42] Nembach E. and Nembach E. 1997. *Particle strengthening of metals and alloys* Vol. 1. New York: Wiley.
[43] Fang Q.H. and Liu Y.W. 2006. Size-dependent interaction between an edge dislocation and a nanoscale inhomogeneity with interface effects. *Acta Materialia*, 54(16): 4213–4220.
[44] Fang Q.H., Chen J.M., Liu Y.W., et al. 2010. Critical shear stress produced by interaction of edge dislocation with nanoscale inhomogeneity. *Bulletin of Materials Science*, 33: 123–127.

[45] Romanov A.E. and Wagner T. 2001. On the universal misfit parameter at mismatched interfaces. *Scripta Materialia*, 45(3): 325–331.

[46] Smith E. 1968. The interaction between dislocations and inhomogeneities—I. *International Journal of Engineering Science*, 6(3): 129–143.

[47] Gong S.X. and Meguid S.A. 1994. A screw dislocation interacting with an elastic elliptical inhomogeneity. *International Journal of Engineering Science*, 32(8): 1221–1228.

[48] Fang Q.H., Liu Y., Huang B.Y., et al. 2008. Contribution to critical shear stress of nanocomposites produced by interaction of screw dislocation with nanoscale inclusion. *Materials Letters*, 62(20): 3521–3523.

[49] Ovid'ko I. and Sheinerman A.J.J.O.P.C.M. 2006. Nanoparticles as dislocation sources in nanocomposites. 18(19): L225.

[50] Stagni L. and Lizzio R. 1983. Shape effects in the interaction between an edge dislocation and an elliptical inhomogeneity. *Applied Physics A*, 30: 217–221.

[51] Warren W.E. 1983. The edge dislocation inside an elliptical inclusion. *Mechanics of Materials*, 2(4): 319–330.

[52] Gurtin M.E. and Murdoch A.I. 1978. Surface stress in solids. *International Journal of Solids and Structures*, 14(6): 431–440.

[53] Povstenko Y.Z. 1993. Theoretical investigation of phenomena caused by heterogeneous surface tension in solids. *Journal of the Mechanics and Physics of Solids*, 41(9): 1499–1514.

[54] Osetsky Y.N. and Bacon D.J. 2003. Void and precipitate strengthening in α-iron: what can we learn from atomic-level modelling? *Journal of Nuclear Materials*, 323(2–3): 268–280.

[55] Fang Q.H., Li B. and Liu Y.W. 2007. Interaction between edge dislocations and a circular hole with surface stress. *Physica Status Solidi (b)*, 244(7): 2576–2588.

[56] Wang G.F. and Wang T.J. 2006. Deformation around a nanosized elliptical hole with surface effect. *Applied Physics Letters*, 89(16): 161901.

[57] Fang Q.H., Liu Y., Liu Y.W., et al. 2009. Dislocation emission from an elliptically blunted crack tip with surface effects. *Physica B: Condensed Matter*, 404(20): 3421–3424.

[58] Huang M. and Li Z. 2004. Dislocation emission criterion from a blunt crack tip. *Journal of the Mechanics and Physics of Solids*, 52(9): 1991–2003.

[59] Creager M. and Paris P.C. 1967. Elastic field equations for blunt cracks with reference to stress corrosion cracking. *International Journal of Fracture Mechanics*, 3: 247–252.

[60] Rice J.R. and Thomson R. 1974. Ductile versus brittle behaviour of crystals. *The Philosophical Magazine: A Journal of Theoretical Experimental and Applied Physics*, 29(1): 73–97.

[61] Brands M., Carl A., Posth O., et al. 2005. Electron-electron interaction in carbon-coated ferromagnetic nanowires. *Physical Review B*, 72(8): 085457.

[62] Link S. and El-Sayed M.A.J.A.R.O.P.C. 2003. Optical properties and ultrafast dynamics of metallic nanocrystals. *Annual Review of Physical Chemistry*, 54(1): 331–366.

[63] Vollath D. and Szabó D.V.J.A.E.M. 2004. Synthesis and properties of nanocomposites. *Synthesis and Properties of Nanocomposites*, 6(3): 117–127.

[64] Martin C.R. and Siwy Z.J.N.M. 2004. Pores within pores. *Nature Materials*, 3(5): 284–285.

[65] Duan H., Wang J., Karihaloo B.L., et al. 2006. Nanoporous materials can be made stiffer than non-porous counterparts by surface modification. *Acta Materialia*, 54(11): 2983–2990.

[66] Hu T., Grosberg A.Y. and Shklovskii B.J.P.R.B. 2006. Conductivity of a suspension of nanowires in a weakly conducting medium. *Physical Review B*, 73(15): 155434.

[67] Meden V. and Schollwöck U.J.P.R.B. 2003. Conductance of interacting nanowires. *Physical Review B*, 67(19): 193303.
[68] Petrov V., Srinivasan G., Bichurin M., et al. 2007. Theory of magnetoelectric effects in ferrite piezoelectric nanocomposites. *Physical Review B*, 75(22): 224407.
[69] Fang Q., Liu Y. and Chen J.J.A.P.L. 2008. Misfit dislocation dipoles and critical parameters of buried strained nanoscale inhomogeneity. *Applied Physics Letters*, 92(12): 121923.
[70] Takahashi A., Ghoniem N.M.J.J.O.T.M. and Solids P.O. 2008. A computational method for dislocation–precipitate interaction. *Journal of the Mechanics and Physics of Solids*, 56(4): 1534–1553.
[71] Tsuchida E., Ohno M. and Kouris D.J.A.P.A. 1991. Effects of an inhomogeneous elliptic insert on the elastic field of an edge dislocation. *Applied Physics A*, 53: 285–291.
[72] Xiao Z. and Chen B.J.M.O.M. 2000. A screw dislocation interacting with a coated fiber. *Mechanics of Materials*, 32(8): 485–494.
[73] Honein E., Rai H., Najjar M.J.I.J.O.S., et al. 2006. The material force acting on a screw dislocation in the presence of a multi-layered circular inclusion. *International Journal of Solids and Structures*, 43(7–8): 2422–2440.
[74] Shen M.-H.J.E.J.O.M.-A.S. 2008. A magnetoelectric screw dislocation interacting with a circular layered inclusion. *European Journal of Mechanics-A/Solids*, 27(3): 429–442.
[75] Wang X. and Sudak L. 2006. Interaction of a screw dislocation with an arbitrary shaped elastic inhomogeneity. *Journal of Applied Mechanics*, 73(2): 206-211.
[76] Gutkin M.Y., Ovid'ko I. and Sheinerman A.J.J.O.P.C.M. 2003. Misfit dislocations in composites with nanowires. *Journal of Physics: Condensed Matter*, 15(21): 3539.
[77] Kolesnikova A. and Romanov* A.J.P.M.L. 2004. Misfit dislocation loops and critical parameters of quantum dots and wires. *Philosophical Magazine Letters*, 84(8): 501–506.
[78] Chen T., Dvorak G.J., Yu C.J.I.J.O.S., et al. 2007. Solids containing spherical nano-inclusions with interface stresses: effective properties and thermal–mechanical connections. *International Journal of Solids and Structures*, 44(3–4): 941–955.
[79] Fang Q., Liu Y., Jin B., et al. 2009. Interaction between a dislocation and a core–shell nanowire with interface effects. *International Journal of Solids and Structures*, 46(6): 1539–1546.
[80] Mushkelishvili N.I. 1953. Some basic problems of mathematical theory of elasticity. (Vol. 15). Groningen: Noordhoff.
[81] Liu Y., Fang Q., Jiang C.J.I.J.O.S., et al. 2004. A piezoelectric screw dislocation interacting with an interphase layer between a circular inclusion and the matrix. *International Journal of Solids and Structures*, 41(11–12): 3255–3274.
[82] Ma C.-C. and Lu H.-T.J.P.R.B. 2006. Theoretical analysis of screw dislocations and image forces in anisotropic multilayered media. *Physical Review B*, 73(14): 144102.
[83] Fang Q., Liu Y. and Wen P.J.J.O.A.M. 2008. Screw dislocations in a three-phase composite cylinder model with interface stress. *Journal of Applied Mechanics*, 75(4): 041019.
[84] Xiao Z. and Chen B.J.A.M. 2002. A screw dislocation interacting with inclusions in fiber-reinforced composites. *Acta Mechanica*, 155(3–4): 203–214.
[85] Shi J. and Li Z.J.M.R.C. 2006. An approximate solution of the interaction between an edge dislocation and an inclusion of arbitrary shape. *Mechanics Research Communications*, 33(6): 804–810.
[86] Lee S.J.E.F.M. 1987. The image force on the screw dislocation around a crack of finite size. *Engineering Fracture Mechanics*, 27(5): 539–545.

10 Challenges and Opportunities

10.1 INTRODUCTION

With significant progress in recent years, modeling and simulation have become essential tools for explaining and/or predicting the mechanical behaviors of advanced materials. Nevertheless, the utilization of these modeling and simulation methods in the future is accompanied by numerous challenges. For example, material performance usually depends on the synergy effects of deformation mechanisms at multiple space and time scales. However, the current multiscale modeling and simulation methods are still constrained to a specific scale. The cross-scale calculation methods for investigating the mechanical response of materials have not been systematically established. Furthermore, there is an urgent need to break the trade-off relationship between efficiency and accuracy in multiscale simulation methods. Meanwhile, the emergence of artificial intelligence tools, such as machine learning methods, has brought tremendous opportunities not only for solving these challenges, but also for promoting modeling/simulation-driven low-cost and high-efficiency material design. In this chapter, these challenges and opportunities are presented from two aspects: modeling and simulation methods, and model/simulation-driven material design.

10.2 MODELING AND SIMULATION

Modeling in advanced materials typically involves multiple scales, such as microscale (atoms, electron), mesoscale (nanocrystalline structure) and macroscale (mechanical properties and deformation behaviors). However, establishing a precise relationship between microscale mechanisms and macroscopic mechanical properties by effectively coupling models at different scales remains a challenging problem [1–3], such as the cross-scale modeling of mechanical behavior in the high-entropy alloys (HEAs) under irradiation condition (Figure 10.1). Different scales are governed by unique physical laws and mechanical models. To ensure precise scale transformation and prevent loss of information caused by model distortion, it is crucial to establish effective connections between models at different spatio-temporal scales [4–6]. In the past years, several cross-scale modeling techniques have been developed to bridge the microscale, mesoscale, and macroscale phenomena of materials [7–10]. For example, the combination of macroscopic finite element method and crystal plastic finite element method is used to investigate the texture evolution of aluminum alloys, revealing the influence of different strain states on both texture evolution

DOI: 10.1201/9781003225706-10

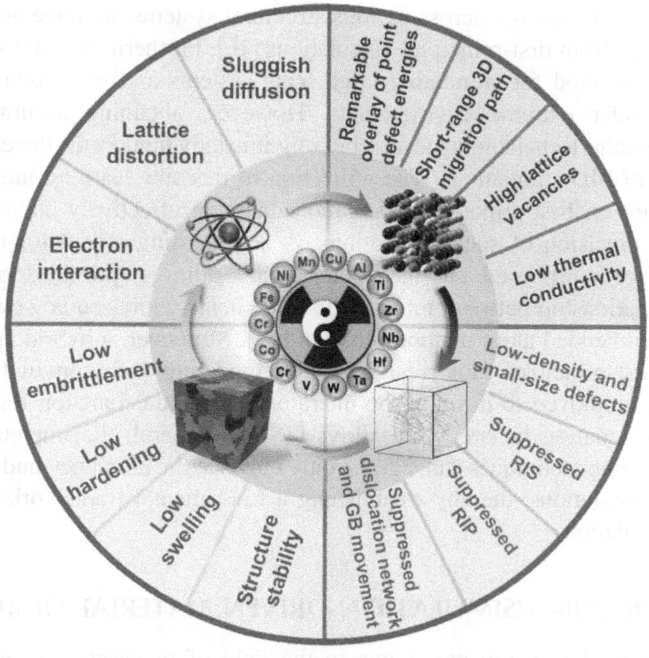

FIGURE 10.1 A cross-scale modeling and simulations of HEAs mechanical behavior under irradiation condition [3].

and slip mode [11]. Nevertheless, cross-scale modeling is still an emerging field, and suggestions for future work are summarized into two directions: (i) using artificial intelligence, data mining and other technologies to better understand the interaction and coupling mechanism between different scales, thus improving the accuracy and reliability of the models, and (ii) enhancing the computational efficiency of models through the utilization of high-performance computing resources.

The intersection of materials science and computer science has promoted the advancement of multiscale simulation. Nonetheless, the simulation method at each scale has its own limitations. For instance, first-principles calculation is confined to the atomic scale and exhibits poor accuracy under high-temperature, high-pressure, or strong magnetic field conditions. MD simulations significantly rely on precise potential functions. Finite element simulations need accurate boundary conditions and constitutive model. To address these limitations, the integrated approach of combining machine learning methods with multi-scale simulations has emerged as a promising avenue for advancing material multiscale simulations in recent years [12–15]. As the most accurate material calculation method, first-principles calculation data is commonly utilized for supervised learning training to predict material properties. For example, by integrating first-principles calculations with artificial neural networks, a surrogate model is established to correlate the local atomic environment configuration and migration energy. This model is applicable for the computation of vacancy migration energy across various structural systems, in agreement with outcomes derived from first-principles calculations [16]. Furthermore, MD simulation is an effective method for simulating large scale systems and predicting mechanical responses under extreme environments. However, obtaining accurate potential functions remains a challenging task, which significantly hinders the development and application of MD simulations. The utilization of machine learning interatomic potentials based on first-principles calculation data can effectively enhance the efficiency and precision of such simulations. A self-learning model for the potential energy surface is proposed, and the clustering algorithm is optimized to capture the nonlinear relationship between interatomic force fields with accuracy comparable to that of the embedded atom method function [17]. Moreover, a hybrid approach that combines first-principles molecular dynamics modeling with convolutional neural networks is employed to develop the interatomic potential function for martensitic phase transformation in zirconium alloys [18]. In general, the implementation of machine learning techniques can significantly enhance the efficiency and precision of multiscale simulations, thereby establishing a fundamental framework for efficient material simulation.

10.3 MODELING/SIMULATION-DRIVEN MATERIAL DESIGN

The evolution of research paradigms in the field of materials science is closely linked to advancements in human cognition and progress in science and technology. Figure 10.2 summarizes the development history of material science research paradigm. From the Stone Age to the 17th Century, the discovery of new materials mainly depended on trial and error methods. Meanwhile, such strategy also provided humans with the accumulation of primitive metallurgical knowledge and

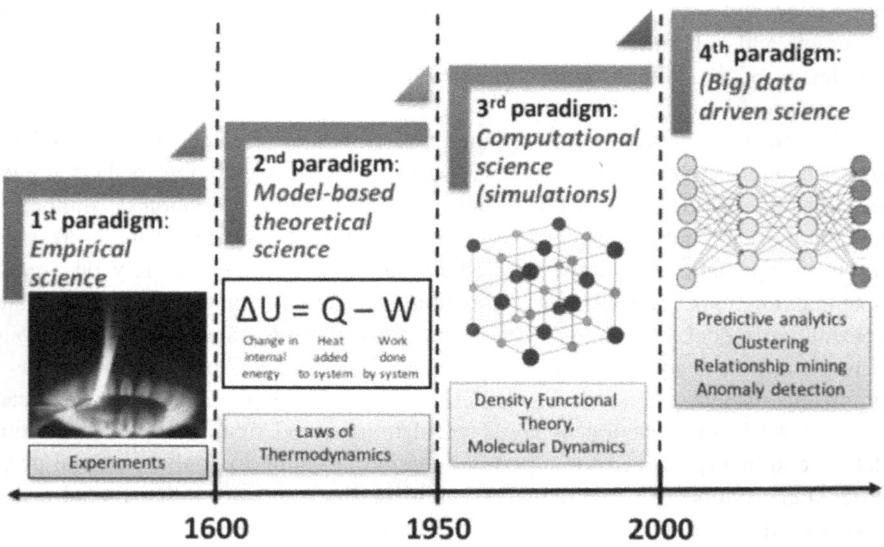

FIGURE 10.2 Development paradigms of materials science [19].

experience. The maturity of mathematical and physical methods enabled these empirical knowledge to be rapidly promoted in the form of a theoretical framework, such as thermodynamic law. However, as the complexity of the material system increased, the analytical solutions of some theoretical models were no longer reasonable. Along with the development of computer science and computational materials science theory, computer simulation methods represented by first-principles calculations and MD have enabled humans to uncover the evolution of material microstructure and predict their properties. In addition, the tremendous data produced by experiments, modeling and simulations has promoted the precise material design [19–21].

Due to the complex composition space and microstructure of advanced materials, conventional material design methods that rely on the experience of domain experts encounter significant challenges. In recent years, material design strategies driven by modeling and simulation have demonstrated greater efficiency and lower costs compared to traditional methods [22–25]. For example, high-throughput MD simulation combined with machine learning is proposed to investigate the influence of element concentration on mechanical properties in multi-principal element alloys. The optimal element ratio of alloys with high strength and low cost/low density is determined. This approach not only facilitates rapid acquisition of the data required, but also establishes an accurate quantitative relationship between composition and mechanical properties [26]. Importantly, the proposed method provides the initial references for experimental preparation (Figure 10.3a). Moreover, the chemical-element-distribution induced uncertainty of microstructure greatly affects the deformation mechanism and mechanical properties. Based on the defined element anisotropy factor, MD simulations are integrated with machine learning to construct an accurate mapping between the chemical element distribution and the macroscopic mechanical properties (Figure 10.3b). The findings indicate that HEAs with high strength and good plasticity can be achieved by adjusting the chemical element distribution. With the new insights gained from this section, it is possible to create high-performance materials by customizing the chemical element distribution through heat treatment [27].

However, conducting multiscale simulations is both computationally expensive and time-consuming, which poses challenges in obtaining the necessary data for machine learning. Recently, physical modeling has been combined with machine learning to overcome these challenges and create efficient and robust predictive models for material design. For example, in order to reveal the strengthening mechanism of heterogeneous HEAs, a material structure optimization method has been proposed, which integrates physical modeling, atomistic simulations, and machine learning [28]. Using a physics-based strength model to extend the results from atomistic simulations, data covering various microstructure is obtained for constructing a comprehensive structure-property dataset. Machine learning is then used to determine a critical grain size of the transition from the Hall-Petch relationship to the inverse Hall-Petch relationship, agreeing with the prediction from atomistic simulations (Figure 10.4a). In addition, a multi-stage material design method mixing machine learning, physical laws, and mechanical modeling has been developed (Figure 10.4b). Compared to existing physical models or machine

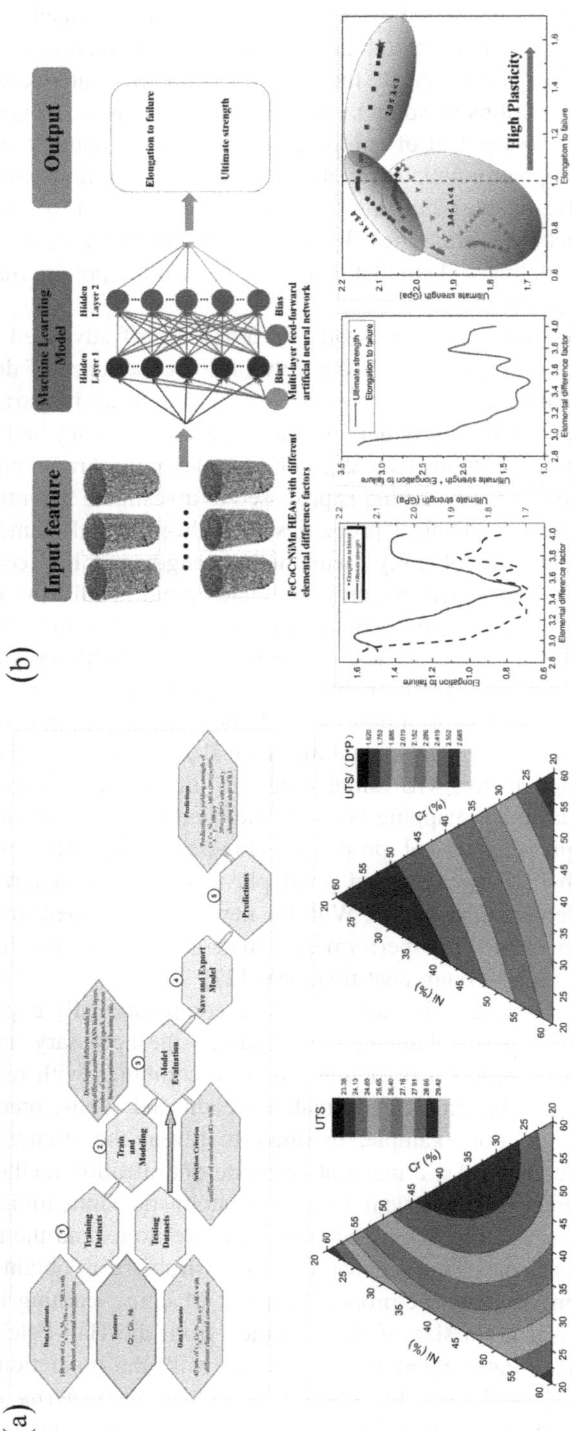

FIGURE 10.3 (a) High-throughput MD simulation combining with machine learning method to search high-strength, low-cost/low-density alloys. (b) An accurate mapping between the element anisotropy factor and the macroscopic mechanical properties via MD simulations and machine learning [26,27].

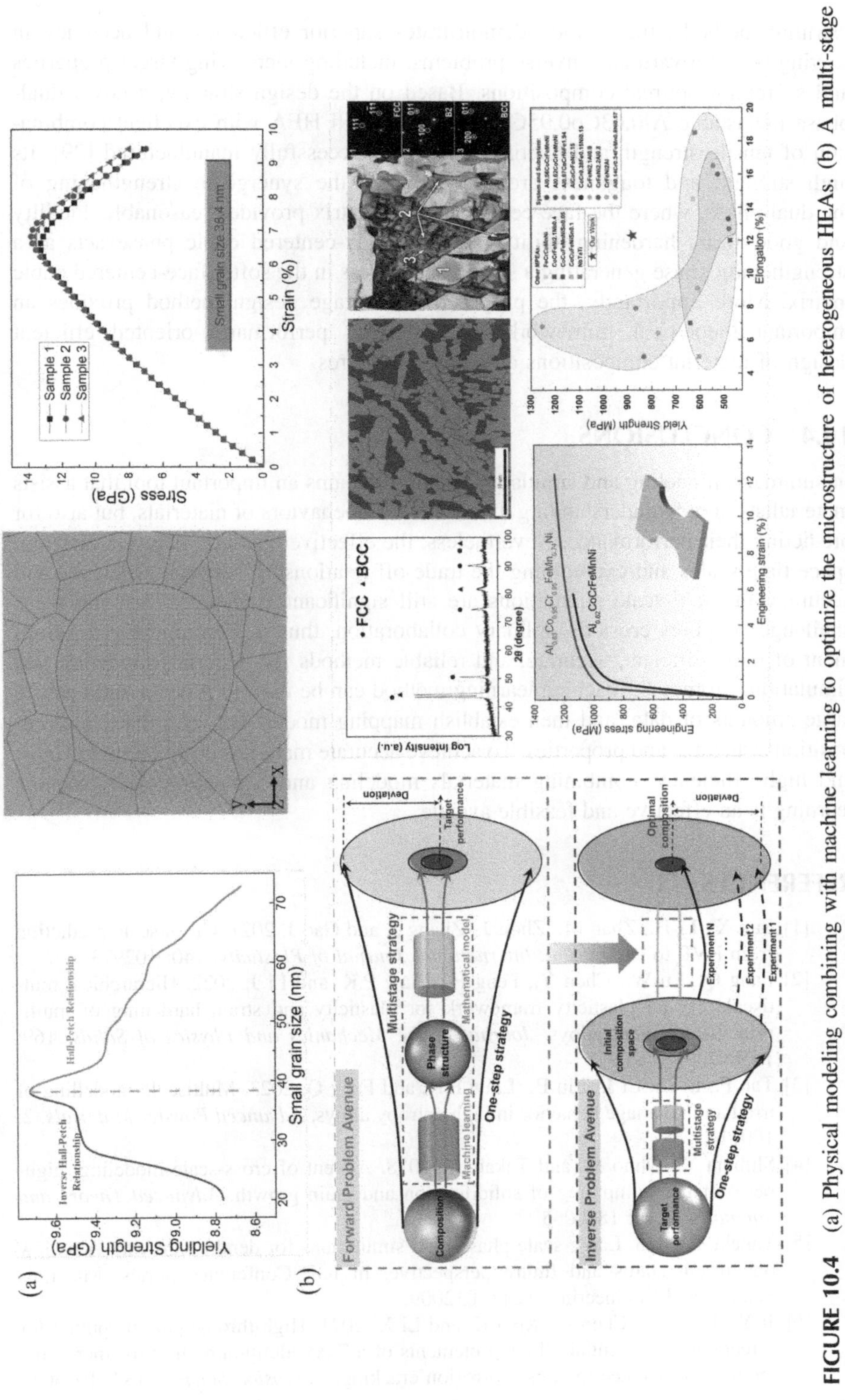

FIGURE 10.4 (a) Physical modeling combining with machine learning to optimize the microstructure of heterogeneous HEAs. (b) A multi-stage material design method combining machine learning, physical laws, and mechanical modeling [28,29].

learning methods, the method demonstrates superior efficiency and accuracy in solving both forward and inverse problems, including identifying target properties and searching optimal compositions. Based on the design strategy, a novel dual-phase metastable Al0.63Co0.95Cr0.95FeMn0.74Ni HEA with excellent combination of tensile strength and ductility has been successfully manufactured [29]. Its high strength and toughness are derived from the synergistic strengthening of the dual-phase, where the face-centered cubic matrix provides reasonable ductility and good strain hardening ability, and the body-centered cubic phase acts as a strengthening phase generating a strong back stress in the softer face-centered cubic matrix. More importantly, the proposed multi-stage design method provides an important theoretical framework for achieving performance-oriented efficient design of material compositions and microstructures.

10.4 CONCLUSIONS

In summary, modeling and simulation method remains an important tool that assists materialists in only understanding the mechanical behaviors of materials, but also for predicting their performance. Nevertheless, the effective coupling between different space-time scales and overcoming the trade-off relationship between efficiency and accuracy in multi-scale simulations are still significant challenges. Solving these challenges requires cross-disciplinary collaboration, thus accelerating the development of more efficient, accurate, and reliable methods for materials modeling and simulation. Moreover, machine learning method can be used to process and analyze large amounts of data, and then establish mapping models between material composition/structures and properties. To achieve accurate materials design with low cost and high efficiency, combining materials modeling and simulation with machine learning is an effective and feasible avenue.

REFERENCES

[1] Sun X., Li H., Zhan M., Zhou J., Zhang J. and Gao J. 2021. Cross-scale prediction from RVE to component. *International Journal of Plasticity*, 140: 102973.

[2] Fang Q., Lu W., Chen Y., Feng H., Liaw P.K. and Li J. 2022. Hierarchical multiscale crystal plasticity framework for plasticity and strain hardening of multi-principal element alloys. *Journal of the Mechanics and Physics of Solids*, 169: 105067.

[3] Tan F., Li L., Li J., Liu B., Liaw P.K. and Fang Q. 2023. Multiscale modelling of irradiation damage behavior in high entropy alloys. *Advanced Powder Materials*, 2: 100114.

[4] Shibuta Y., Ohno M. and Takaki T. 2018. Advent of cross-scale modeling: high-performance computing of solidification and grain growth. *Advanced Theory and Simulations*, 19: 1800065.

[5] Takaki T. 2023. Large-scale phase-field simulations for dendrite growth: A review on current status and future perspective. In IOP Conference Series: Materials Science and Engineering, 1247: 012009.

[6] Ji Y., Dong C., Chen L., Xiao K. and Li X. 2021. High-throughput computing for screening the potential alloying elements of a 7xxx aluminum alloy for increasing the alloy resistance to stress corrosion cracking. *Corrosion Science*, 183: 109304.

[7] Gao X., Meng X., Cui L. and Zhu M. 2019. Cross-scale simulation for MnS precipitation of Fe-C alloy with cooling rate variation. *Materials Research Express*, 69: 096583.
[8] Meng X., Gao X., Huang S. and Zhu M. 2018. Cross-scale modeling of mns precipitation for steel solidification. *Metals*, 87: 529–533.
[9] Zeng X., Han T., Guo Y. and Wang F. 2018. Molecular dynamics modeling of crack propagation in titanium alloys by using an experiment-based Monte Carlo model. *Engineering Fracture Mechanics*, 190: 120–133.
[10] Dong Q., Li H. and Yin Z. 2022. A cross-scale analysis method for predicting the compressive mechanical properties of tin-based bearing alloy. *Journal of Materials Science*, 1–12.
[11] Pei Y., Hao Y., Zhao J., Yang J. and Teng B. 2023. Texture evolution prediction of 2219 aluminum alloy sheet under hydro-bulging using cross-scale numerical modeling. *Journal of Materials Science and Technology*, 149: 190–204.
[12] Lunghi A. and Sanvito S. 2022. Computational design of magnetic molecules and their environment using quantum chemistry, machine learning and multiscale simulations. *Nature Reviews Chemistry*, 611: 761–781.
[13] Qiao L., Liu Y. and Zhu J. 2021. A focused review on machine learning aided high-throughput methods in high entropy alloy. *Journal of Alloys and Compounds*, 877: 160295.
[14] Samaei A. and Chaudhuri S. 2021. Mechanical performance of zirconia-silica bilayer coating on aluminum alloys with varying porosities: Deep learning and microstructure-based FEM. *Materials and Design*, 207: 109860.
[15] Fish J., Wagner G.J. and Keten S. 2021. Mesoscopic and multiscale modelling in materials. *Nature Materials*, 206: 774–786.
[16] Terentyev D., Bonny G., Castin N., Domain C., Malerba L., Olsson P., Moloddtsov V. and Pasianot R.C. 2011. Further development of large-scale atomistic modelling techniques for Fe-Cr alloys. *Journal of Nuclear Materials*, 4092: 167–175.
[17] Bernstein N., Csányi G. and Deringer V.L. 2019. De novo exploration and self-guided learning of potential-energy surfaces. *npj Computational Materials*, 51: 99–108.
[18] Zong H., Pilania G., Ding X., Ackland G.J. and Lookman T. 2018. Developing an interatomic potential for martensitic phase transformations in zirconium by machine learning. *npj Computational Materials*, 41: 48–56.
[19] Agrawal A. and Choudhary A. 2016. Perspective: Materials informatics and big data: Realization of the "fourth paradigm" of science in materials science. *APL Materials*, 45: 053208.
[20] Ward C.H., Warren J.A. and Hanisch R.J. 2014. Making materials science and engineering data more valuable research products. *Integrating Materials And Manufacturing Innovation*, 3: 1–17.
[21] Kalidindi S.R. and Graef M.D. 2015. Materials data science: Current status and future outlook. *Annual Review of Materials Research*, 45: 171–193.
[22] Guo K., Yang Z., Yu C.H. and Buehler M.J. 2021. Artificial intelligence and machine learning in design of mechanical materials. *Materials Horizons*, 84: 1153–1172.
[23] Tao Q., Xu P., Li M. and Lu W. 2021. Machine learning for perovskite materials design and discovery. *npj Computational Materials*, 71: 23–41.
[24] Wei J., Chu X., Sun X.Y., Xu K., Deng H.X., Chen J., Wei Z.M. and Lei M. 2019. Machine learning in materials science. *InfoMat*, 13: 338–358.
[25] Butler K.T., Davies D.W., Cartwright H., Isayev O. and Walsh A. 2018. Machine learning for molecular and materials science. *Nature*, 5597715: 547–555.

[26] Li J., Xie B., Fang Q., Liu B., Liu Y. and Liaw P.K. 2021. High-throughput simulation combined machine learning search for optimum elemental composition in medium entropy alloy. *Journal of Materials Science and Technology*, 68: 70–75.

[27] Li J., Xie B., He Q., Liu B., Zeng X., Liaw P.K., Fang Q.H., Yang Y. and Liu Y. 2022. Chemical-element-distribution-mediated deformation partitioning and its control mechanical behavior in high-entropy alloys. *Journal of Materials Science and Technology*, 120: 99–107.

[28] Li L., Xie B., Fang Q. and Li J. 2021. Machine learning approach to design high entropy alloys with heterogeneous grain structures. *Metallurgical and Materials Transactions A*, 52: 439–448.

[29] Li J., Xie B., Li L., Liu B., Liu Y., Shaysultanov D., Fang Q.H., Stepanov N. and Liaw P.K. 2022. Performance-oriented multistage design for multi-principal element alloys with low cost yet high efficiency. *Materials Horizons*, 95: 1518–1525.

Index

Accumulated strain, 215, 223
Al0.63Co0.95Cr0.95FeMn0.74Ni HEA, 318
AlCoCrCuFeNi HEA, 132, 134–136, 138
Amorphous high-entropy alloys, 118
Amorphous structure, 125, 126, 128–132, 134–136, 138, 139
ANN. *see* Artificial neural network (ANN)
Annihilation, 2, 3, 17, 53, 65, 83, 87, 88
Annihilation parameter, 88
Anti-phase boundary energy, 20, 146, 152, 159
Arrhenius-type law, 93
Artificial intelligence, 7, 312, 313
Artificial neural network (ANN), 58
Asymptotic behavior, 241, 243, 251
Asymptotic hardening rate, 216
Atomic binding energy, 187, 194
Atomic displacement, 124
Atomic size mismatch, 16, 77
Atomistic simulations, 315

Body-centered cubic, 318
Boltzmann constant, 17, 34, 54, 96, 191
Boundary layer, 119, 120

Capillary force, 201–203
Cartesian coordinates, 260
Cauchy-Green elastic strain tensor, 94
Challenges and opportunities, 7, 312
Chemomechanical modeling, 231
Climbing mechanism, 145, 166–168, 170, 171, 173
CNA. *see* Common-neighbor analysis (CNA)
Coherent, 45, 52, 62, 64, 65, 146, 160, 261, 270–272, 302
Common-neighbor analysis (CNA), 47
Complex variable method, 287, 302
Computational materials science, 315
Computational solid mechanics, 1
Computer science, 313, 315
Configurational force, 203
Conjugate twin, 194, 201
Constitutive model, 1, 6, 13, 31, 32, 34, 96, 313
Continuum mechanics model, 13
Continuum model, 1
Contraction twinning (CTW), 213, 220
Cooling rate, 43, 131–136, 138, 139, 145, 151, 152, 156–158, 172
Core-shell nanowires, 298
Couple effect, 249
CP. *see* Crystal plasticity (CP)
CPFE. *see* Crystal plasticity finite element (CPFE)
Crack nucleation, 186, 189, 231, 248, 249, 251, 252, 254
Crack propagation, 6, 28, 45, 244,248
Creep rate, 7, 166, 168, 169, 170, 173
Critical resolved shear stress (CRSS), 33, 51, 81, 147, 201, 216
Critical shear stress, 2, 3, 17, 35, 58, 211, 212, 225, 259, 266, 267, 271, 275, 277, 296
Critical size, 3, 18, 19
Cross-scale calculation, 312
Cross-scale coupling, 7
Cross-slip, 35, 51, 93, 94
CRSS. *see* Critical resolved shear stress (CRSS)
Crystal plasticity (CP), 95
Crystal plasticity finite element (CPFE), 1, 24, 30, 91
Crystallographic theory, 195
CTW. *see* Contraction twinning (CTW)
Cu29Zr32Ti15Al5Ni19 amorphous HEA, 118, 119, 120, 121, 122, 123, 124
Cubic-octahedron void, 83, 85
Cubic void, 83, 84, 85
Cumulative distribution function, 161
Cutting mechanism, 18, 19, 74, 146, 148, 166, 167, 168, 170, 171, 172, 173
Cylindrical electrolyte, 23, 231, 232, 240

DD. *see* Dislocation dynamics (DD)
DDD. *see* Discrete dislocation dynamics (DDD)
Deformation twinning (DT), 194, 200, 225
Density functional theory, 24
Diagram of the double twinning (DTW), 213, 214, 216, 225
Diffusion coefficient, 236, 242, 245
Diffusion-induced stresses, 4
Diffusion layer, 232, 265
Discrete dislocation dynamics (DDD), 91, 93, 159
Dislocation dissociation, 45, 204, 205, 208
Dislocation dynamics (DD), 5
Dislocation effect, 3, 22, 236, 240, 241, 242, 243, 244, 247, 249, 253, 254
Dislocation-glide plane, 163
Dislocation hardening coefficient, 86,103
Dislocation line tensor, 15, 164
Dislocation-loop strength parameter, 88
Dislocation motion path, 73, 75
Dislocation movement, 2, 3, 4, 6, 14, 16, 17, 26, 49, 53, 83, 95, 103, 145, 164, 259, 266, 285

Dislocation-nanostructure interaction model, 4, 23, 259
Dislocation nucleation, 22, 41, 43, 45, 56, 62, 65,95, 234, 248, 308
Dislocation pile-up, 3, 17, 45, 206, 212
Dislocation-segment, 93, 164
Dislocation singularity, 297
Dislocation strengthening, 2, 3, 16, 45, 91, 150
Disperse particle, 18
Dispersion strengthening, 3, 18, 52
DT. *see* Deformation twinning (DT)
DTW. *see* Diagram of the double twinning (DTW)
Dynamic recrystallization (DRX), 213, 214, 215, 216, 217, 218, 219, 220, 221, 222, 223, 224, 225, 226

EAM potential, 27, 134
Edge dislocation, 26, 35, 52, 81, 98, 99, 159, 161, 260, 261, 266, 267, 271, 277, 278, 280, 282, 283, 285, 287, 288, 289
Elastic chemical potential, 198, 200
Elastic deformation, 22, 31, 94, 121, 233, 251
Elastic interaction, 2, 14, 23, 24, 209, 259, 260, 267, 271, 275, 277, 285, 286, 291, 298, 302, 307
Elastic stage, 13, 62, 110, 121
Elastic stiffness coefficients, 197
Elastic strain energy, 29, 197, 200, 244, 248, 265
Electrochemical energy storage, 231
Electrode characteristics, 4
Electromagnetic interference resistance, 179
Electromagnetic shielding, 291
Element segregation, 49, 120
Elemental distribution, 120
Elliptically blunted crack, 287, 288, 290
Embedded nanowire, 291, 302
Escaig stress, 93
ETW. *see* Extension twinning
Extension twinning (ETW), 213, 214, 215, 216, 217, 218, 219, 221, 223, 225

Face-centered cubic (FCC), 16, 73, 132, 179, 318
FDD. *see* Fermi-Dirac distribution (FDD)
FeCoCrNi amorphous HEA, 126
Fermi-Dirac distribution (FDD), 181, 192, 194
First-principles calculation, 5, 6, 24, 25, 26, 43, 186, 190, 313, 314, 315
Fleischer model, 2
Flow stress, 2, 14, 50, 61, 88, 124, 128, 138, 139, 213, 223, 234
Forest strength parameter, 96
Formation enthalpy, 54, 182, 191, 192
Forward and inverse problems, 318
Fourier function, 198, 201, 202
Fourth-rank elastic tensor, 95
Frank dislocation, 55

Galvanostatic operation, 242
Generalized-gradient approximation, 181
Geometric standard deviation, 160
GFA. *see* Glass formation ability (GFA)
Gibbs free energy, 29, 200
Glass formation ability (GFA), 125
Grain boundary strengthening, 2, 3, 17, 18, 49, 52, 74, 75, 150
Ground-state properties, 24, 25, 26

Hall-Petch coefficient, 49
Hall-Petch relationship, 3, 61, 62, 315
Hamiltonian equation, 92
Hardening coefficient, 17, 86, 87, 92, 103, 107, 214
Hardening rate, 6, 107, 108, 123, 216, 220, 221, 223, 225
Hardness, 2, 41, 87, 91, 122, 123, 139, 151, 179
HEA. *see* High-entropy alloys (HEAs)
Hertzian elastic theory, 121
Hexagonal close packing (HCP), 54, 195, 196, 205, 207, 211, 213, 225
Hexagon void, 83
Hexahedral finite element, 105
Hierarchical multiscale approach, 91
High-density nanowire, 291
High-entropy alloys (HEAs), 14, 41, 72
High-temperature diffusion, 191, 194
High-throughput MD simulation, 315
Hirth dislocation, 42
Hollow electrode, 251
Homogenization theory, 198
Hoop stress, 239, 251, 252

Indentation depth, 122, 123, 124
Infinite matrix, 260, 272, 273, 279, 285, 292
Infinitely matrix domain, 261
Interaction and coupling mechanism, 313
Interaction energy, 180, 182, 186, 187, 191, 192, 194, 209, 210, 274, 296
Interatomic interactions, 120, 127, 134
Interatomic potentials, 314
Interface governing equation, 156
Intrinsic stacking fault energy (ISFE), 45, 190, 191, 192, 193
Inverse pole figure (IPF), 224
IPF. *see* Inverse pole figure (IPF)
Irradiation condition, 53, 85, 87, 88, 312
ISFE. *see* Intrinsic stacking fault energy (ISFE)

Kocks Mecking coefficient, 86, 88
Kocks-Mecking model, 2
Kronecker delta, 33, 95, 264, 279

Labusch model, 2
Lamé constant, 234, 278

Laplace pressure, 234, 235, 246
Large-scale Atomic/Molecular Massively Parallel Simulator (LAMMPS), 25, 46, 98, 118, 126, 132, 144, 150
LIB. *see* Lithium-ion battery (LIB)
Li-ion concentration, 231, 232, 233, 234, 236, 237, 238, 240, 241, 243, 245, 249, 250, 352
Lithium-ion battery (LIB), 4, 20, 21, 231, 232, 248, 249, 252, 254
Lithium-ion concentration, 22
Logarithmic normal distribution, 154, 159, 161, 163
Lomer–Cottrell lock, 42
Long-period service, 4
Long-term stress field, 160

Machine learning, 67, 312–315, 318
Material pileup, 124
Materials science, 1
Matrix phase, 302
MD. *see* Molecular dynamics (MD)
Mechanical behaviors, 1, 72, 118, 122, 124, 127, 128, 129, 139, 312, 318
Mechanical equilibrium, 22, 233, 236
Mechanical response, 1, 4, 26, 30, 31, 36, 54, 103, 108, 111, 112, 122, 234, 314
Medium-matrix interface, 305
Metallic glass, 118,119, 127, 138
Microcrack nucleation, 186, 189
Microstructure-based model, 2, 13–14
Microstructure-based strength models, 2
Microstructure evolution, 1, 6, 20, 26, 29, 41
Minimum spanning tree, 145, 164
Mismatch parameter, 16, 73, 74, 77, 78
Modeling and simulation, 1, 7, 13, 36, 145, 312–313, 315, 318
Modulus mismatch, 16, 72, 73, 74, 77, 89
Molecular dynamics (MD), 1, 24, 91, 118
Mott-Nabarro model, 2
MPEAs. *see* Multi-principal-element alloys (MPEAs)
Multilayer nanotwin, 179, 180, 194
Multi-modal, 145
Multi-principal-element alloys (MPEAs), 49
Multi-scale, 36, 67, 145, 166, 172, 313, 318
Multiscale simulation, 1, 3, 5, 7, 13, 24, 72, 312, 313, 314, 315
Muskhelishvili complex potential, 261

Nanocrystalline HEA, 133, 136, 138
Nanoindentation model, 118, 120
Nanoscale electrode, 231, 232, 234, 246, 249
Nanoscratching, 45
Nanosize effect, 123
Nanostructure, 1, 4, 23, 24, 259, 266, 269
Nanostructured-reinforced composites, 4
Newton-Raphson method, 97
Newton's second law, 120
Nickel-based superalloys, 145, 146, 150, 155, 158, 159, 166, 170, 172, 173
9R phase, 190, 191, 193, 194
Normalized-activation energy, 86, 88
Normalized force, 299, 300, 301

Orientation-dependent interface free energy (ODIE), 198, 200
Orowan loop mechanism, 145, 170
Orowan mechanism, 18, 148, 149, 160, 166, 167, 168, 170, 172, 173
Orowan model, 3
Oxide particle strengthening, 3, 4

ParaDiS, 100, 159
Partial molar volume, 21, 232, 233, 236
Partial radial distribution function (RDF), 125, 136, 139
Particle-stimulated nucleation (PSN), 2, 17, 53, 213
Prismatic/basal (PB), 194, 198
Peach-Koehler force, 93
Peach-Koehler formula, 265, 274, 280, 285
Peierls-Nabarro model, 2
Perdew–Burke–Ernzerhof, 181
Perfect dislocation, 55, 207
Perfect supercell, 179, 180, 184, 186
Phase field model, 28, 29, 194, 197, 203
Phase field simulation, 7, 28, 30
Phase transformation, 26, 31, 34, 35, 36, 46, 48, 48, 61, 314
Phenomenological constitutive model, 31, 32, 34
Physics-based constitutive model, 34
Plastic stage, 13, 62, 121, 124
Poisson's ratio, 21, 85, 121–122, 209, 211, 248, 260, 278, 288
Post-yield softening, 83
Potential energy, 25–27, 43, 53, 92, 134, 314
Potential energy surface, 314
Potentiostatic operation, 236, 240, 242, 253
Precipitate radius, 146, 148, 157–161
Precipitation strengthening, 3, 8, 18, 52, 74–75
Probability-related model, 152, 154
Proliferation, 2, 17, 53, 103, 223
Proliferation and annihilation, 2, 17, 53
PSN. *see* Particle-stimulated nucleation (PSN)

Quasi-static loading, 111
Quenching, 41, 118, 120, 125, 132

Radial position, 241, 243–244, 246–248, 250
Radial stress, 239–247, 251–254

Redistribution, 139, 156
Reference strain rate, 17, 111
Renewable energy, 231
Representative volume element (RVE), 104, 198
Resolved shear stress, 33, 35, 51, 81, 95, 103, 147, 198, 201, 214–216
Rhombic void, 81–83
Riemann-Schwarz symmetry principle, 273
Resolved shear stress (RSS), 33, 35, 81, 95, 103, 147, 198, 201, 214, 215, 216

Saturation strain, 215, 217
Scale transformation, 312
Schmid factor, 102
Schmid tensor, 33, 95, 216
Schwarz symmetry principle, 273, 293
SD. *see* Subgrain development (SD)
Second phase particle, 14, 18, 19, 20
Second phase strengthening, 2, 3, 18, 52
Second-rank tensor, 196
Self-learning model, 314
Semi-linear manner, 314
Shear bands, 129, 130, 138
Shear strain, 32, 33, 94, 101, 102, 123, 124, 126, 129, 130, 138, 139, 201, 216
Shear zones, 124
Shearing mechanism, 145
Shockley dislocation, 56
Shockley partial dislocation, 55, 56, 129, 194, 205, 207
Short-range order, 49, 50, 67
SIF, 288, 289, 290, 291
Silicon electrode, 251
Solid solution enthalpy, 180
Solid solution strengthening, 2, 14, 15, 34, 51, 72, 90, 96, 150
Solidification, 29, 132, 133, 134, 139
Solute atoms, 2, 14, 96, 180, 181, 182, 186, 189, 190, 191, 193, 194
Solute concentration, 14, 156, 180, 181, 182, 186, 189, 191, 192, 193, 194
Space and time scales, 159, 312
Spatial distribution, 159, 162, 164, 166, 167, 168
Spatial uncertainty, 145, 172, 173
Spherical electrolyte, 23, 234, 237, 245, 249
Spherical indenter, 45, 118, 120
Stacking-fault energy, 5
Stair-rod dislocations, 42, 205
Stair-rod partial dislocation, 56
Stiffness, 122, 123, 179, 197, 270
Strain energy, 29, 80–81, 197, 200, 213, 216, 231, 237, 244, 245, 248, 249, 265, 296
Strain-rate sensitivity, 14, 86, 88
Strengthening mechanism, 6–7, 14–15, 18, 28, 41, 45, 49, 91, 145, 158, 159, 315
Stress concentrations, 130, 212, 213, 223
Stress distribution, 109, 135, 138, 246, 247, 251, 252
Strong-pair coupling mechanism, 20, 146, 147, 152, 154, 167
Structural stability, 138, 234
Subgrain development (SD), 213
Surface energy, 5, 184, 186–190, 194, 195, 198, 199, 201, 208, 225, 235, 245, 247, 287
Surface stress, 234, 279, 280, 282–291, 302
Sustainable energy, 231
Synergy effects of deformation mechanisms, 312

Tangential strain, 233, 236, 261, 272, 278
Tangential stresses, 236, 247, 248, 254
Taylor coefficient, 150
Taylor factor, 97, 98, 104, 152, 168, 219
Taylor law, 103
Taylor model, 2
Taylor relation, 17
TB. *see* Twin boundary (TB)
TBE. *see* Twin boundary energy (TBE)
TDRX. *see* Twin-induced dynamic recrystallization (TDRX)
Tensile model, 126, 127
Tensile stress, 4, 53, 60, 241, 242, 244, 247, 248, 254
Threshold stress, 54, 167, 168, 170, 171, 216
Threshold-dislocation structure, 175
Total deformation gradient tensor, 94
Trade-off relationship, 312, 318
Triangular void, 81–83
Twin boundary (TB), 45, 49, 62, 64, 65, 135, 179, 197
Twin boundary energy (TBE), 179, 181, 182, 184–186, 194, 204, 211
Twin-induced dynamic recrystallization (TDRX), 213, 215–225
Twin interface, 196, 198, 210

Uniform distribution (UFD), 120, 164, 181

ASP. *see* Vienna ab initio simulation package (VASP)
VCA. *see* Virtual crystal approximation (VCA)
Vegard's law, 15, 74
Velocity gradient tensor, 32, 94
Velocity-Verlet algorithm, 42
VFs. *see* Volume fractions (VFs)
Vibration damping performance, 179
Vienna ab initio simulation package (VASP), 181
Virtual crystal approximation (VCA), 43, 44
Viscoplastic self-consistent (VPSC), 215–217, 219, 221, 224, 225
Voce hardening, 216, 219, 220, 221

Volume fractions (VFs), 34, 49, 153, 154, 215
Von Mises yield criterion, 237, 238, 240
Voronoi polyhedron, 131
VPSC. *see* Viscoplastic self-consistent (VPSC)

Weak-pair coupling mechanism, 19, 146, 147, 152, 154, 167
Weierstrass-Mandelbrot fractal function, 100

Yield stress, 2, 56, 58, 60, 75–91, 98, 102, 104, 111, 161, 214, 220, 238, 250
Young's modulus, 122, 123, 139, 232, 236, 242, 250